Quantum Science and Technology

The book series Quantum Science and Technology is dedicated to one of today's most active and rapidly expanding fields of research and development. In particular, the series will be a showcase for the growing number of experimental implementations and practical applications of quantum systems. These will include, but are not restricted to: quantum information processing, quantum computing, and quantum simulation; quantum communication and quantum cryptography; entanglement and other quantum resources; quantum interfaces and hybrid quantum systems; quantum memories and quantum repeaters; measurement-based quantum control and quantum feedback; quantum nanomechanics, quantum optomechanics and quantum transducers; quantum sensing and quantum metrology; as well as quantum effects in biology. Last but not least, the series will include books on the theoretical and mathematical questions relevant to designing and understanding these systems and devices, as well as foundational issues concerning the quantum phenomena themselves. Written and edited by leading experts, the treatments will be designed for graduate students and other researchers already working in, or intending to enter the field of quantum science and technology.

Muhammad Usman
Editor

Quantum Robustness in Artificial Intelligence

Principles and Applications

Springer

Editor
Muhammad Usman
CSIRO
Melbourne, Australia

The University of Melbourne
Melbourne, VIC, Australia

ISSN 2364-9054 ISSN 2364-9062 (electronic)
Quantum Science and Technology
ISBN 978-3-032-11152-4 ISBN 978-3-032-11153-1 (eBook)
https://doi.org/10.1007/978-3-032-11153-1

This Springer imprint is published by the registered company Springer Nature Switzerland AG
The registered company address is: Gewerbestrasse 11, 6330 Cham, Switzerland

Preface

For Artificial Intelligence Systems, Trust Is Not Optional.

Artificial Intelligence is nowadays increasingly deployed in many real-life decision-making processes which include scenarios where trust or robustness is a key parameter of interest. Self-driving vehicles, autonomous military weapon systems, and medical diagnostics are just some of the key examples from a vast range of applications where any mistake by the underpinning Artificial Intelligence algorithm could lead to serious consequences. It is also well-known that conventional or classical Artificial Intelligence models—despite their efficiency and speed—are vulnerable to data manipulations, noise, and poisoning. Even manipulating a few pixels in an image could lead a well-trained state-of-the-art Artificial Intelligence model to completely mislabel the image, for example, identifying the image of a "STOP" signal as a "YIELD" signal or misinterpret Green light signals as Red light signals. Such mistakes could result in highly damaging consequences including the loss of life. Despite tremendous amount of research on evaluating the robustness of Artificial Intelligence models and developing strategies to mitigate their vulnerabilities, a clear universal solution is still an open challenge.

Quantum Artificial Intelligence has rapidly emerged as a new field of research with the ambition to integrate unique properties of quantum mechanics such as superposition and entanglement into Artificial Intelligence, leading to fundamentally new algorithms and architectures which can address challenges faced by the conventional classical models. Indeed, recent work has investigated the robustness of Quantum Artificial Intelligence algorithms in the context of data spoofing and manipulations. Both empirical and analytical results have demonstrated the potential for achieving reliability in Quantum Artificial Intelligence, leading to the trustworthy autonomous systems in the future. If true, this could be a new avenue for quantum advantage beyond computational speed-up and superior classification accuracies which had been the original inspiration for developing Quantum Artificial Intelligence models.

This book is, to date, the first collection of research with a primary focus on *Quantum Robustness in Artificial Intelligence.* It presents invited work from the leading global research community who have made significant contributions to advance the field. The content list includes a broad range of topics which cover not only fundamental aspects and open questions around Robustness in Quantum Artificial Intelligence but also address some of the key challenges facing its practical implementation such as efficient data encoding and mitigating quantum hardware errors. Although adversarial robustness of Quantum Artificial Intelligence remains the primary focus of the book, a few chapters discuss emerging topics such as Geometric Quantum Machine Learning, Quantum Transfer Learning, Differential Privacy, and Distributed Quantum Machine Learning to give the readers a flavor of the diversity in the field of Quantum Artificial Intelligence.

The book starts with Chap. 1 which provides a high-level introduction and definitions of robustness in classical and quantum artificial intelligence. This chapter is meant to be accessible for a general reader and set the tone for the rest of the book. Chapter 2 introduces classical and quantum adversarial attacks and their impact on artificial intelligence models. Specifically, this chapter covers the important concept of transferability of adversarial attacks under the black-box setting, i.e., attacks generated from one model (classical or quantum) impact the performance of a different model. Chapters 3 and 4 broadly cover adversarial attacks in the context of quantum machine learning and provide a literature survey, as well as related challenges and opportunities. Chapters 6 and 7 aim to provide mathematical formulations underpinning quantum adversarial machine learning and layout analytical bounds for quantum robustness in quantum models. Together, the first seven chapters comprehensively cover foundational aspects pertaining the field of quantum adversarial machine learning and serve as a useful resource for any new researcher or industry engineer who would like to start in the field.

Advanced topics from the latest research are covered in Chaps. 8, 9, and 10, which highlight some new research results from the literature. Chapters 11 and 12 discuss adversarial robustness in special cases of distributed quantum machine learning and transfer learning. In particular, the authors in Chap. 12 show that quantum robustness is retained when learning is transferred from classical to quantum models, leading to the possibility that classical models can be trained on complex datasets and the learning can be transferred to quantum models allowing the best of both worlds. Chapter 13 addresses one of the main challenges hindering practical implementation of quantum machine learning which stems from encoding classical data in quantum states, leading to deep and resource intensive quantum circuits. In this chapter, the authors propose an efficient method based on matrix product states to approximately prepare quantum state which allows significant resource reduction without compromising accuracy and robustness of quantum models. Chapters 14, 15, and 16 broaden the scope of the book beyond adversarial quantum machine learning and cover interesting recent work from the literature. Chapters 17 and 18 discuss the impact of hardware noise on the performance of quantum machine learning models in the NISQ era and the effectiveness of noise mitigation and correction methods. Specifically, Chap. 17 reports an important discovery that quantum machine learning models may not require full quantum error correction and only partial error correction may be sufficient to train quantum machine learning with high fidelities. This will drastically reduce qubit requirements for the fault-tolerant implementation of quantum machine learning models.

As for any other application of quantum computing in the current era, Quantum Artificial Intelligence also faces significant challenges before its real-world applications can be realized. These include limitations imposed by the current generation of small and noisy quantum hardware, intensive resource requirements for data encoding and error correction, and the issues related to trainability of models such as barren plateaus. Therefore, the research and development in the field of Quantum Artificial Intelligence is primarily focused on the theoretical and simulation fronts with only a handful of small-scale experimental demonstrations to date. Nevertheless, with tremendous progress in both hardware and software fronts as well as ambitious quantum processor roadmaps pursued by many developers around the world, there is a strong reason to be optimistic about Quantum Artificial Intelligence transitioning from lab research to practical workflows in the coming years.

Special thanks to all authors who have shared their work and insights in putting together this book, which will be a key reference and a useful learning resource for academic researchers and industry engineers alike. For anyone who is worried about the reliability of Artificial Intelligence models, this book hopefully offers a detailed understanding of how quantum computing could address such challenges and may usher the world into an era of trustable autonomous systems.

Last but not the least, I would like to dedicate this book to my fabulous family Sadia, Maryam, and Muizz.

December 2025

Professor Muhammad Usman
Head of Quantum Systems
CSIRO
Melbourne, Australia

Contents

Contributors

Eromanga Adermann Quantum Systems, CSIRO, Marsfield, NSW, Australia

Julian Berberich Institute for Systems Theory and Automatic Control and Center for Integrated Quantum Science and Technology (IQST), University of Stuttgart, Stuttgart, Germany

Gurcan Comert Computational Data Science and Engineering Department, North Carolina A&T State University, Greensboro, NC, USA

Neil Dowling Institut für Theoretische Physik, Universität zu Köln, Köln, Germany ;
School of Physics and Astronomy, Monash University, Clayton, VIC, Australia

Tobias Fellner Institute for Computational Physics, University of Stuttgart, Stuttgart, Germany

Nicola Franco Fraunhofer Institute for Cognitive Systems IKS, Munich, Germany

Petros Georgiou University of Birmingham, Birmingham, United Kingdom

Archisman Ghosh Pennsylvania State University, University Park, PA, USA

Swaroop Ghosh Pennsylvania State University, University Park, PA, USA

Ji Guan Key Laboratory of System Software (Chinese Academy of Sciences), Institute of Software, Chinese Academy of Sciences, Beijing, China

Stephan Günnemann School of Computation, Information and Technology and Munich Data Science Institute, Technical University of Munich, Garching, Germany

Masahito Hayashi School of Data Science, The Chinese University of Hong Kong, Longgang District, Shenzhen, China;
International Quantum Academy, Futian District, Shenzhen, China;
Graduate School of Mathematics, Nagoya University, Nagoya, Japan

Jamie Heredge School of Physics, The University of Melbourne, Parkville, VIC, Australia;
Global Technology Applied Research, New York, NY, USA

Christian Holm Institute for Computational Physics, University of Stuttgart, Stuttgart, Germany

Hans-Arno Jacobsen Department of Electrical and Computer Engineering, University of Toronto, Toronto, Canada

Alex Jin Centre for Quantum Technologies, National University of Singapore, Singapore, Singapore

Sharu Theresa Jose University of Birmingham, Birmingham, United Kingdom

Mahesh Kamepalli The Beacom College of Computer and Cyber Sciences, Dakota State University, Madison, SD, USA

Pouya Kananian Department of Electrical and Computer Engineering, University of Toronto, Toronto, Canada

Haiyue Kang School of Physics, University of Melbourne, Parkville, Australia

Amena Khatun Quantum Systems, Data61, CSIRO, Clayton, Australia

Satwik Kundu Pennsylvania State University, University Park, PA, USA

Zihao Li State Key Laboratory of Surface Physics and Department of Physics, Fudan University, Shanghai, China

Nana Liu Institute of Natural Sciences, Shanghai Jiao Tong University, Shanghai, China

Jeanette Miriam Lorenz Fraunhofer Institute for Cognitive Systems IKS, Munich, Germany

Kavan Modi School of Physics and Astronomy, Monash University, Clayton, VIC, Australia;
Science, Mathematics and Technology Cluster, Singapore University of Technology and Design, Singapore, Singapore;
Quantum for New South Wales, Sydney, NSW, Australia

Manas Mukherjee Centre for Quantum Technologies, National University of Singapore, Singapore, Singapore

Azar C. Nakhl School of Physics, The University of Melbourne, Parkville, VIC, Australia

Lirandë Pira Centre for Quantum Technologies, National University of Singapore, Singapore, Singapore

Patrick Rebentrost Centre for Quantum Technologies, National University of Singapore, Singapore, Singapore

Anthony Rizi Augusta University, Augusta, CA, USA

Alona Sakhnenko Fraunhofer Institute for Cognitive Systems IKS, Munich, Germany

Aman Saxena School of Computation, Information and Technology and Munich Data Science Institute, Technical University of Munich, Garching, Germany

Martin Sevior School of Physics, The University of Melbourne, Parkville, VIC, Australia

Angus Southwell Centre for Quantum Technology, Transport for New South Wales, Sydney, NSW, Australia

Muhammad Usman Quantum Systems, Data61, CSIRO, Clayton, Australia; School of Physics, The University of Melbourne, Parkville, Victoria, Australia

Varghese Vaidyan The Beacom College of Computer and Cyber Sciences, Dakota State University, Madison, SD, USA

Yong Wang Department of Computer Science, University of Idaho, Moscow, ID, USA

Maxwell West School of Physics, The University of Melbourne, Parkville, VIC, Australia

Tom Wollschläger School of Computation, Information and Technology and Munich Data Science Institute, Technical University of Munich, Garching, Germany

Mingsheng Ying Centre for Quantum Software and Information, University of Technology Sydney, Ultimo, NSW, Australia

Eric Yocam The Beacom College of Computer and Cyber Sciences, Dakota State University, Madison, SD, USA

Huangjun Zhu State Key Laboratory of Surface Physics and Department of Physics, Fudan University, Shanghai, China

Chapter 1
Fundamentals of Quantum Machine Learning and Robustness

Lirandë Pira and Patrick Rebentrost

Abstract Quantum machine learning (QML) sits at the intersection of quantum computing and classical machine learning, offering the prospect of new computational paradigms and advantages for processing complex data. This chapter introduces the fundamentals of QML for readers from both communities, establishing a shared conceptual foundation. We connect the worst-case adversarial perspective from theoretical computer science with the physical principles of quantum systems, highlighting how superposition, entanglement, and measurement collapse influence learning and robustness. Special attention is given to adversarial robustness, understood as the ability of QML models to resist inputs crafted to cause failure. We motivate the study of QML in adversarial settings, outline key distinctions between classical and quantum data and computation, and propose an extended taxonomy incorporating the adversary as a core element. This chapter serves as a conceptual entry point to quantum adversarial machine learning and lays the groundwork for deeper discussions on quantum robustness in subsequent chapters.

1.1 Introduction

Quantum machine learning (QML) is rapidly emerging as one of the most intriguing disciplinary intersections, standing between quantum computing and artificial intelligence [1, 2]. Machine learning has transformed multiple industrial sectors, achieving remarkable success in applications such as image and speech recognition, medical diagnosis, financial forecasting, and autonomous systems [3, 4]. Its ability to extract meaningful patterns from large datasets and improve performance with experience has made it a cornerstone of modern technology and computer science.

L. Pira (✉) · P. Rebentrost
Centre for Quantum Technologies, National University of Singapore, Singapore, Singapore
e-mail: lpira@nus.edu.sg

P. Rebentrost
e-mail: cqtfpr@nus.edu.sg

M. Usman (ed.), *Quantum Robustness in Artificial Intelligence*, Quantum Science and Technology, https://doi.org/10.1007/978-3-032-11153-1_1

Independently, a new paradigm of computation has emerged, based on the principles of quantum mechanics [5]. These principles characterized in superposition, entanglement, and quantum interference are used to process information in fundamentally new ways. As a consequence of this ability, quantum algorithms have demonstrated the potential to solve certain problems exponentially faster than their classical counterparts. A prominent example is Shor's algorithm for integer factorization, which can factor large numbers in polynomial time, thereby posing a theoretical threat to classical cryptographic systems [6]. This breakthrough, among others, has sparked widespread interest in identifying domains where quantum speedups could be achieved [7].

Machine learning, with its heavy reliance on linear algebra calculations and optimization, is a natural candidate for quantum enhancement. This has led to explorations of quantum subroutines within machine learning pipelines, particularly in the context of model training and inference. Notably, QML is motivated by the promise that quantum systems can offer computational advantages, either in speed, representational capacity, or data efficiency, for learning tasks that are classically difficult. Yet, realizing these advantages in practice requires more than raw computational power. Increasing system complexity brings forward questions of reliability, robustness, and interpretability. The same features that empower quantum systems, such as superposition, entanglement, and interference, also make them highly sensitive to noise and perturbations. Robustness therefore becomes a central concern in the development of QML.

A model that performs well under ideal conditions but fails under small perturbations is of limited practical use, especially in sensitive domains. This concern is amplified by the fact that quantum systems are themselves noisy and sensitive to small errors. If we hope to deploy QML systems in real-world settings, they must be able to handle uncertain environments and potential manipulation. This practical concern motivates the growing field of quantum machine learning that explicitly addresses robustness [8].

This chapter aims to establish the foundational context for QML, offering readers a conceptual framework that connects classical machine learning with principles of quantum computation. Equally, an emphasis will be placed on adversarial learning as a paradigm within learning systems, while considering the broader approaches to robustness. By doing so, the goal of this chapter is to serve as an ideal starting point for a deeper exploration of adversarial quantum learning systems and robustness in QML systems more broadly.

1.2 Learning and Computation

1.2.1 Role of Machine Learning in Modern Computation

Machine learning has greatly influenced both the practical and theoretical landscapes of modern computation [9, 10]. Beyond enabling and reshaping large-scale data-driven industrial applications, it has introduced new computational paradigms that challenge and enrich algorithmic thinking.

On a theoretical level, machine learning establishes a connection between statistics and computer science. It formalizes the process of learning as an optimization problem over hypothesis spaces with guarantees on inference and generalization [11]. This connection has expanded the scope of computation beyond deterministic procedures, embracing probabilistic reasoning and approximation methods as intrinsic components. Additionally, machine learning theory explores questions of model complexity and expressivity, investigating how models generalize from finite samples and the computational resources required to learn certain classes of functions. Such insights have led to the development of novel algorithmic frameworks that accommodate uncertainty, and interaction with evolving environments. For example, online and reinforcement learning models embody computational systems that learn and evolve over time, transcending the static input-output nature of algorithms. Along these lines, robustness and stability, broadly defined as how computational models withstand noise, adversarial inputs, and uncertainty, have emerged as essential theoretical concerns. Addressing these challenges has led to refined complexity measures and formal definitions, including average-case and data-dependent analyses.

Historically, progress in machine learning has unfolded through reinforcing conceptual layers, each addressing bottlenecks of the previous era. Historically, machine learning has grown in waves, each clearing a practical hurdle and setting up the next. Early ideas on the perceptron, formal notions of what can be learned, and effective backpropagation showed that adjusting many parameters could still yield reliable generalization [11–14]. The 1990s brought support vector machines, boosting, and kernels, giving practitioners tools that worked well out of the box [15–17]. Around the 2000s, probabilistic graphical models and faster training routines made it easier to encode structure and uncertainty [18–20]. A turning point followed with unsupervised layer-wise pre-training and large convolutional networks [21, 22], which helped push widespread use of GPU (graphical processing unit) computation. Stability techniques, such as normalization, residual links, and adaptive optimizers, made very deep models practical [23–25]. Self-attention and the Transformer [26] simplified handling long-range context, while observed scaling laws [27, 28] encouraged larger models. Most recently, foundation and diffusion models [29, 30] unify these advances through broader pre-training and emergent in-context behavior, further encouraging large-scale training. Each stage thus reinterprets learning: from formal feasibility, to deep representation, to broad flexible pre-trained systems—an arc that underlies today's robustness questions and informs how we think about future (including quantum) learning pipelines.

In summary, machine learning has transformed what computers can do, all the while expanding how we conceive and define computation itself. Algorithms are no longer static input-output mappings but adaptive optimization processes whose reliability depends on capacity control, stable training dynamics, and statistical validation. Crucially for this book, these theoretical foundations provide a natural springboard into quantum settings. Quantum algorithms introduce fundamentally different mechanisms for information processing that promise new approaches to learning tasks. Beyond quantum-classical heuristics, the interplay between quantum mechanics and learning theory is a rapidly developing area within QML research [31]. Below, we outline the broad motivation and current state of quantum computation as the hardware platform on which QML is built.

1.2.2 Present-Day Quantum Computation

Quantum computing departs from classical computation by exploiting quantum mechanical principles such as superposition, entanglement, and unitarity [5, 32]. The fundamental unit of information is the qubit, which unlike a classical bit, can encode a continuum of states between 0 and 1. Multiple qubits form a quantum register whose state is described by a vector in a high-dimensional Hilbert space, with dimension exponential in the number of qubits. Quantum states evolve through the application of unitary transformations, which are reversible operations that preserve the norm of the state vector. In practice, these transformations are implemented as sequences of quantum gates acting on single or pairs of qubits. Computation proceeds by preparing an initial quantum state, applying a meaningful gate sequence, such as a quantum algorithm, and finally measuring the result in the computational basis to extract classical outcomes.

Once a theoretical curiosity of the 1980s [33], quantum computing has progressed to an experimental science with increasingly capable hardware. Platforms range from superconducting qubits to trapped ions and photonics. Current quantum hardware, known as noisy intermediate-scale quantum (NISQ) near-term devices, provides access to tens to hundreds of physical qubits [34]. NISQ systems are limited by decoherence, gate infidelity, and noise accumulation, which restrict the depth of circuits and algorithmic complexity that can be executed. Despite these constraints, these system sizes offer a testbed for exploring certain quantum algorithms, quantum control techniques, and hybrid quantum-classical routines.

It is worth highlighting that the main driver of progress in quantum computing is the pursuit of quantum advantage, the point at which a quantum device performs a computational task demonstrably better than any classical counterpart [35–37]. This could mean faster runtime, lower memory requirements, or the ability to solve problems that are intractable for classical algorithms. While early demonstrations have focused on narrow, contrived tasks, the broader goal is to realize advantage in meaningful applications, including optimization, simulation, and machine learning. This is where fault-tolerant quantum computing comes into the picture. Fault-tolerant quantum computation is the paradigm that refers to large-scale quantum computers with meaningful processing capacities, capable of achieving quantum advantage [38], as opposed to limited NISQ devices. Achieving fault tolerance relies on quantum error correction, which encodes logical qubits into many physical ones to protect against decoherence.

Before full fault tolerance is available, much of the focus in the NISQ era is on the near-term algorithms that can run on noisy hardware. Variational quantum algorithms, such as the variational quantum eigensolver [39] and quantum approximate optimization algorithm [40], are designed to be resilient to some degree of noise, using classical optimizers to tune quantum circuits. Within QML, such hybrid approaches manifest as quantum neural networks, which combine quantum state evolution with classical gradient-based training.

Efforts across academia and industry are focused on scaling quantum systems to larger and more coherent qubit arrays. Approaches vary across platforms, ranging from superconducting circuits, trapped ions, neutral atoms, and photonics. Currently, they share common bottlenecks such as gate fidelity, readout precision, and qubit stability, among others. It is widely accepted that achieving meaningful quantum advantage in practical tasks will require both an increase in the number of qubits and improvements in their quality manifested in improved coherence times, reduction of gate errors, and the development of architectures that support error correction at scale. At the same time, on the theoretical front, it remains equally important to understand what can actually be learned, computed, or inferred with limited quantum resources. This understanding plays a significant role in guiding the development of quantum algorithms today.

1.3 Understanding Robustness in Machine Learning

Computer science fundamentally approaches problems from a worst-case perspective. Unlike many fields that study typical or average scenarios, theoretical computer science asks: *How hard is the problem in the most difficult or adversarial instance?* This worst-case mindset is central to several pillars of the discipline. For instance, algorithmic complexity is defined by the hardest possible inputs an algorithm might face. A problem is deemed "hard" not because it is difficult on average, but because it is intractable for some worst-case instances. Similarly, security and cryptography are built on the assumption that adversaries can be highly powerful and strategic, so systems must remain secure even against the most damaging attacks. In this way, the worst-case analysis suggests that adversarial robustness is a fundamental concern in algorithm design.

Robustness, in the learning setting, asks how performance degrades when reality departs, slightly or substantially, from idealized modeling assumptions. This can occur when inputs are perturbed, distributions shift, labels are corrupted, data are strategically selected, or the model itself is probed for information leakage. Historically, robustness-oriented thinking in learning predates modern deep systems. Robust statistics introduced the idea that estimators should retain stability under a fraction of arbitrary contamination [41, 42]. As machine learning met adversarial behavior in applied settings, early spam filtering and intrusion detection research exposed the strategic nature of evasion and feature manipulation [43, 44]. Domain adaptation theory subsequently clarified how one can transfer performance guarantees across related but shifted distributions [45]. Distributionally robust optimization crystallized these themes in min-max formulations over uncertainty sets, anticipating the empirical saddle-point structure of contemporary adversarial training [46–48].

In understanding learning and computation under uncertainty, it is helpful to view different frameworks along a spectrum defined by three axes: the nature of complexity analysis, the computational mode, and how data is generated or accessed, as captured in Fig. 1.1. Complexity theory traditionally begins with worst-case analysis, asking how algorithms behave under the hardest possible inputs. Average-case complexity and probabilistic analysis relax this analysis by assuming distributions over inputs. Similarly, computational models range from online settings, where decisions are made sequentially without future knowledge, to offline models that assume full data access. Finally, data generation spans from adversarial or deterministic sequences to stochastic models, such as i.i.d. sampling. As an example, multi-armed bandits illustrate this gradient well, with stochastic bandits representing average-case learning under uncertainty, and adversarial bandits capturing worst-case sequential interaction. To exemplify, seen this way, online learning, adversarial machine learning, and robust optimization sit toward the worst-case sequential, adaptive end of the spectrum, while classical supervised learning lives on the average-case, offline, i.i.d. side.

And finally, in real-world scenarios, drawing sharp boundaries between concepts is rarely natural. Theoretical distinctions such as worst-case versus average-case

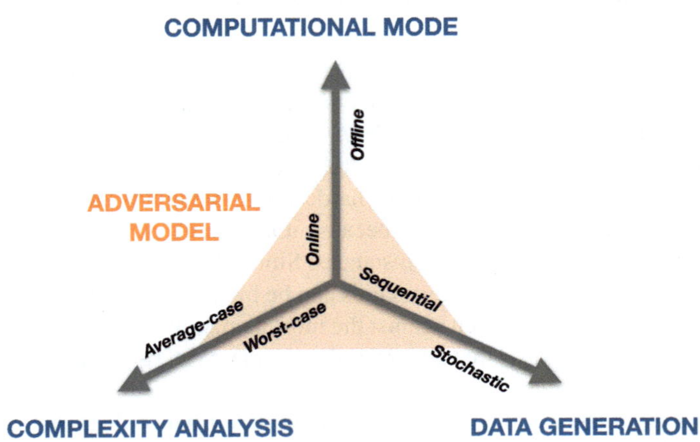

Fig. 1.1 A conceptual map positions learning settings along data generation, computation mode, and complexity analysis dimensions, showing where adversarial models naturally arise (such as the shaded area in orange) and why robustness considerations may vary across this spectrum

analysis, or online versus offline models, are useful abstractions, but they rarely capture the full nuance of practical problems. In reality, there's a spectrum. It is worth noting that worst-case complexity bounds are also problem specific. Thus, the structure of the problem changes the nature of worst-case analysis. As settings move from full information (such as offline) to partial, sequential, or reactive settings (such as online), worst-case analysis becomes both more relevant and more subtle. In adversarial contexts, these distinctions become important when designing robust algorithms, including quantum ones. This perspective helps clarify how different fields relate, and why worst-case thinking becomes essential in adversarial and high-stakes environments, which is often referred to in this book.

In models of online algorithms and adversarial learning, inputs may be chosen sequentially by an adversary aiming to cause the worst possible outcome. A compelling example is when the adversary may select the very first data point in a way that maximally misguides the learning algorithm, potentially causing cascading errors downstream. This perspective reflects the reality of many practical high-stakes settings. These adversarial environments are exemplified in security, finance, and autonomous systems, where worst-case scenarios are not rare edge cases, they are the very situations that must be handled robustly.

1.4 Core Concepts in Quantum Machine Learning

1.4.1 Motivation

Machine learning is fundamentally about using data to make predictions or optimize decisions. The typical approach involves training a model to minimize a given loss function, which measures how far its predictions deviate from expected outcomes. While modern machine learning systems, especially deep learning models, can be powerful, they often come with high computational and data requirements. Real-world practical deployment of machine learning systems often requires large and scalable models, which in turn increases the cost of training and inference as the models scale. This fact, coupled with the rapid emergence and promises of quantum computing as a growing field, has led to growing interest in whether quantum computing could help. The motivation for QML nowadays goes beyond speed, although that is often cited. Equally appealing is the idea that quantum systems can represent and manipulate information in ways that classical systems cannot, thereby capturing structural representations intractable classically. For instance, a quantum system with n qubits has access to a Hilbert space of size 2^n, which allows, at least in principle, compact representations of very complex functions or data distributions. This might enable more efficient algorithms for certain tasks such as kernel evaluation, sampling, or optimization, which are common bottlenecks in classical machine learning. Unsurprisingly, these computational bottlenecks became some of the earliest targets for exploration in QML research. These early explorations naturally gave rise to a broader conceptual mapping of QML, clarifying the different ways quantum and classical components can be combined in learning systems.

1.4.2 Paradigms

In light of these motivations, QML research has converged on several broad paradigms. The first main approach is *quantum-enhanced machine learning* which covers scenarios where quantum computing is used to improve or accelerate standard machine learning tasks. Here, data may be classical, but quantum devices are employed to perform operations such as kernel evaluation, matrix inversion, optimization, or sampling. The second one can be viewed as *machine learning for quantum data*. Here, the learning task itself involves quantum states as input or output. Examples include learning properties of quantum systems, extracting information from quantum experiments, or modeling outcomes of quantum processes. These distinctions are not mutually exclusive, and similar to broader QML, the boundaries are often fluid. Beyond the two broad views discussed above, there are at least two other perspectives worth highlighting. First, there is *machine learning for quantum systems*, where classical learning methods are used to assist in the development and control of quantum devices themselves. This includes tasks such as quantum error correction or pulse shaping. In these settings, the learning aspect supports the underlying challenges on quantum hardware. Secondly, *quantum-inspired classical models* carry ideas or techniques from quantum computing, such as tensor networks or Hilbert space embeddings, for purely classical algorithms [49]. This approach aims to replicate some of the efficiency or structure of quantum systems using classical resources, without requiring access to a quantum device.

More generally, this taxonomy and broad categorization helps understand what parts of the QML landscape are physically realizable today, which are targets for fault-tolerant quantum computation, and which lie outside hardware constraints but still inform algorithmic design. The nature of the input data, whether classical or quantum, dictates what computational resources are needed and what types of learning models are feasible.

This taxonomy, summarized in Fig. 1.2, organizes QML methods along two orthogonal dimensions: the nature of the data and the computational resource or model. The following two subsections explore these axes in more detail, beginning with the data dimension.

1.4.3 Data Representation and Encoding

In conventional machine learning, data points are typically structured arrays over real or discrete fields, for example, vectors, sequences, or images with clear physical interpretation. Quantum data, by contrast, consists of pure or mixed quantum states in

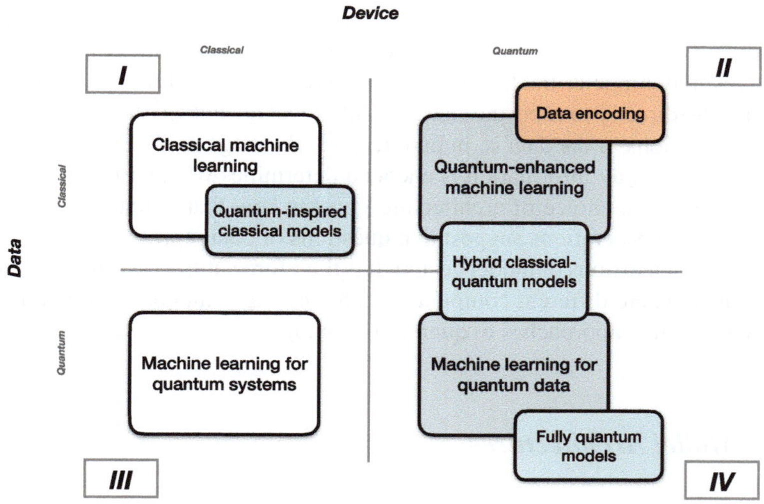

Fig. 1.2 Taxonomy of quantum machine learning approaches. This diagram maps QML methods along two axes: the nature of the data (classical or quantum) and the computational resource (classical or quantum). The top-right quadrant corresponds to quantum-enhanced machine learning, where quantum models are applied to classical data. The bottom-right represents machine learning for quantum data, where quantum systems or outputs are analyzed. Classical machine learning occupies the top-left along with quantum-inspired classical models, while the bottom-left includes classical models used to support quantum computing tasks such as control and error correction

a Hilbert space. Such states may exhibit superposition and entanglement, properties with no straightforward classical analogs, and are subject to constraints such as the no-cloning theorem and the collapse of the wavefunction upon measurement.

This difference in representation is both a challenge and an opportunity. Quantum data can encode correlations, interference patterns, and high-dimensional structures that classical systems struggle to represent efficiently. However, it is also fragile: extracting information risks disturbing or destroying the state, and classical simulation of large quantum systems quickly becomes computationally prohibitive.

When working with classical datasets on a quantum computer, an additional step is required: encoding the classical data into quantum states [1, 50]. Several approaches exist. In *basis encoding*, each classical bitstring is mapped directly to a computational basis state $|x\rangle$. *Amplitude encoding* embeds a normalized real or complex vector into the amplitudes of a quantum state, allowing n qubits to represent 2^n features compactly. *Angle* or *rotation encoding* instead maps classical values to the rotation angles of single-qubit gates, a technique common in variational circuits. More elaborate *hybrid schemes* combine these strategies or exploit problem-specific symmetries to optimize resource usage. A refinement particularly relevant for rotation-based schemes is *data re-uploading* method, which repeatedly inserts the input features across layers to hold qubit count fixed [51].

This raises a natural question: to what extent do speedups depend on the nature of the data itself? Research on classical-quantum separations suggests that certain advantages may hinge critically on whether the input is quantum or classical [52]. This connects directly to robustness: a computational advantage that vanishes under small perturbations in the data is, in practice, no advantage at all.

Yet, while the way information is encoded determines what structure is accessible to a learner, the choice of architecture governs how that structure is exploited. Together, these observations suggest that questions of data representation cannot be separated from questions of model design. It is therefore natural to turn next to model architectures, where different computational assumptions and hardware constraints give rise to distinct approaches to quantum learning.

1.4.4 Model Architectures

Work on model architectures divides broadly into two regimes. A first line of QML research targets the fully fault-tolerant regime, where quantum error correction enables deep, precise circuits and algorithms with formal asymptotic guarantees, albeit often under strong assumptions about data access and state preparation [53]. At a high level, these models frequently draw on quantum linear algebra primitives, such as procedures for solving linear systems, estimating eigenvalues, or transforming singular values, that form the backbone of many classical machine learning methods. Early examples include algorithms such as Harrow-Hassidim-Lloyd (HHL) for linear systems [54] and related routines for tasks such as principal component analysis [55] and kernel methods [56]. More recently, the framework of quantum singular value transformation (QSVT) [57] has emerged as a unifying abstraction for these linear-algebraic building blocks. QSVT makes it possible to implement polynomial transformations of singular values with optimal query complexity, providing a clean theoretical foundation for many large-scale QML algorithms that are actively reshaped [58].

That said, most near-term quantum machine learning research does not aim for full quantum advantage in the theoretical sense. Instead, it focuses on hybrid models that use quantum circuits for parts of the computation, together with classical optimization or preprocessing. These efforts are broadly characterized as variational quantum algorithms [59, 60], which are explicitly designed for noisy NISQ devices.

Within the broader variational algorithms, a commonly studied model architecture is quantum neural network, which attempts to mimic the behavior of classical neural network by introducing a layered architecture. These models implement parameterized quantum circuits in order to tune the learnable parameters. While the model itself may include quantum components such as the parameterized circuits, the parameters are often updated using classical optimization algorithms. Training typically involves a hybrid loop where a quantum processor evaluates the circuit and a classical optimizer updates parameters.

A central technical question is how to obtain gradients of the loss with respect to circuit parameters. Unlike classical neural networks, backpropagation cannot be applied directly due to the nature of quantum measurements. Instead, the most common approach is the parameter-shift rule [61, 62], which exploits the trigonometric structure of many quantum gates to compute exact gradients by evaluating the circuit at two shifted parameter values. This makes it possible to apply gradient-based optimizers such as stochastic gradient descent in a quantum-classical loop. However, while the parameter-shift rule enables proof-of-concept training, it does not fully address optimization and scalability challenges, which remain ongoing open questions.

In parallel, another practical line of work centers on quantum kernel methods [63, 64]. Here, quantum circuits are used not as trainable models but as feature maps, embedding classical data into high-dimensional Hilbert spaces where classical kernel methods can operate.

One trend that currently stands out is the push to identify where quantum machine learning is not only viable but genuinely useful, often while drawing comparisons with classical counterparts [65]. To that end, research focus relies on a growing set of toy problems and benchmarks such as low-dimensional datasets, small-scale classification tasks, or quantum-native problems such as state discrimination, to test model behavior and scaling. There is particular interest in applications to quantum chemistry, where the structure of the problem maps naturally onto quantum systems, and in problems where sampling from high-dimensional distributions is computationally expensive [66]. As these models grow in complexity, questions about their expressivity and capacity become increasingly important. Inevitably, understanding what these models can represent, and how that compares to classical baselines, remain central to evaluating their potential in practice. These benchmark settings often expose both the strengths and current limitations of QML approaches.

In summary, QML exists at the intersection of two distinct paradigms: quantum computation and statistical learning. Each brings its own strengths, and together they open up opportunities that classical approaches alone struggle to match. At the same time, QML inherits limitations from both fields, along with a set of unique obstacles that we are only beginning to understand. Below, we outline the potential benefits of quantum approaches to learning, while also being candid about the barriers ahead.

1.5 Opportunities and Challenges

1.5.1 Algorithmic Speedups and Representational Capacity

One of the most anticipated advantages of QML is the potential for algorithmic speedups. Certain quantum algorithms offer runtime improvements for linear algebra subroutines under constraints, which are central to many machine learning models [67]. As mentioned previously, techniques, such as the HHL algorithm for solving linear systems or the QSVT framework, suggest that quantum models might reduce the computational complexity of training and inference in certain settings. While these speedups are typically proven under idealized assumptions, they motivate the search for practical QML models that can offer meaningful improvements on real-world problems.

Beyond raw speed, quantum representations may allow for richer encodings of data. Quantum states naturally capture high-dimensional structure through entanglement and superposition. This can enable feature spaces that are exponentially large in the number of qubits, allowing quantum models to access complex decision boundaries that would be difficult to learn classically. Quantum feature maps have been proposed in this context, aiming to embed classical data into Hilbert spaces that better expose class structure [63, 64]. Another promising direction is unsupervised learning. Classical unsupervised methods, such as clustering or dimensionality reduction, often struggle in high-dimensional or noisy settings. Quantum models, especially those using variational circuits or quantum annealing, may offer new ways to explore data structure, optimize clustering objectives, or sample from complex distributions. Explorations into applications are emerging across domains. In quantum chemistry and materials science, QML techniques are explored to model molecular properties more efficiently [66]. In finance, quantum algorithms have been applied to examples of portfolio optimization and risk modeling [68]. However, as we will discuss in the following, these opportunities must be viewed alongside serious present-day challenges.

1.5.2 Hardware and Optimization Limitations

Currently, quantum hardware remains noisy and error prone with decoherence, imperfect gate fidelities, and limited qubit connectivity. Many QML proposals assume ideal unitary operations that are far from what current hardware can support. Another limitation arises in data encoding. As we noted, for a quantum algorithm to process classical data, the data must first be embedded into quantum states. This encoding step is often costly and may offset any potential quantum advantage. Furthermore, the amount of data that can be encoded is constrained by the number of available qubits, placing practical limits on input size and resolution. Optimization is also a major bottleneck. Variational quantum algorithms, which are central to many QML architectures, suffer from barren plateaus, or regions of the parameter space where gradients vanish exponentially with system size [69, 70]. This makes training deep or expressive models difficult. Hybrid optimization strategies are being actively explored, however presently, scalable and stable training remains an open problem. There is also a lack of standardized datasets and benchmarks that are native to the quantum domain. Much of the current research relies on synthetic datasets or repurposed classical data to benchmark toy examples. All of these constraints limit the ability to evaluate models rigorously and meaningfully [65]. The same dataset constraints can be further noted in adversarial contexts where fine-grained control over the input distribution is often required.

Finally, the theoretical foundations of QML are still developing with questions around expressivity, generalization, and learnability [71]. While some results provide bounds on what quantum models can represent or learn, these are often limited in scope or dependent on strong assumptions. In summary, QML holds considerable promise, particularly for tasks that involve complex data structures or adversarial interactions. However, realizing that promise will require overcoming substantial technical hurdles, both in hardware and in algorithm design. The remainder of this book addresses one of the most important open questions in this field: how quantum models behave under adversarial conditions, and whether their unique properties offer any advantage in terms of robustness, or more conceptually, what further unique hurdles do QML systems bring the broader adversarial machine learning.

1.6 Robustness and Quantum Adversarial Learning

The motivation for robust learning systems extends into quantum learning [8, 72, 73]. This book is motivated by a central hypothesis: quantum information and computation provide fundamentally new avenues for exploring robustness in machine learning. Furthermore, quantum systems exhibit properties such as state collapse upon measurement, entanglement, and the no-cloning theorem, all features with no classical counterparts. These characteristics open the possibility of designing learning models with novel forms of resilience and security. Quantum cryptographic techniques such as quantum key distribution already demonstrate how quantum principles can secure classical communication, for example, by using measurement collapse to detect eavesdropping.

In a similar spirit, QML models may exhibit forms of robustness inherently tied to the physics of quantum systems [74]. This inherent property of QML systems can potentially offer better defenses against specific classes of adversarial attacks. At the same time, as QML continues to advance and potentially find more relevance in applications, it becomes increasingly important to examine its robustness in adversarial settings [75]. The very quantum mechanical properties that offer the expected advantages may also introduce new kinds of vulnerabilities. These susceptibilities can emerge from sensitivity to noise and decoherence to the possibility of adversaries exploiting quantum interference or circuit design. In this light, QML represents both an opportunity and a challenge. It offers promising new tools to address adversarial threats in machine learning, but it also demands new theoretical and practical frameworks to understand its behavior under worst-case conditions. Studying the robustness of quantum learning systems is, therefore, essential to the responsible development of the field.

Throughout this book, robustness often refers to adversarial robustness. This is defined to the ability of quantum machine learning models to maintain performance in the presence of carefully crafted, maliciously chosen inputs designed to cause system digressions or failure. In the QML context, these adversarial examples may exploit flaws in data encoding, circuit design, or optimization routines [76]. This notion of robustness can be clearly distinguished from other uses of the term, such as in optimization theory which manifests as the sensitivity to perturbations in cost functions, or statistical learning where robustness refers to that toward outliers or distributions.

If we imagine plotting the role of the adversary in QML systems, it can be captured by adding a third dimension to the widely used data-device framework, as illustrated in Fig. 1.3. Traditionally, QML studies focus on whether the model and data are classical or quantum, often exploring the advantages of quantum models acting on classical or quantum data. One can extend the third axes to conceptual analysis of the adversary in each relevant scenarios.

By doing so, we add a new layer of analysis: each quadrant in this cube represents a data-device-adversary pairing, capturing the nature and capabilities of the adversary as the novel extension. For instance, an adversary restricted to classical computation may be unable to efficiently craft perturbations against quantum data encoded in high-dimensional Hilbert spaces, while a quantum adversary might exploit entanglement

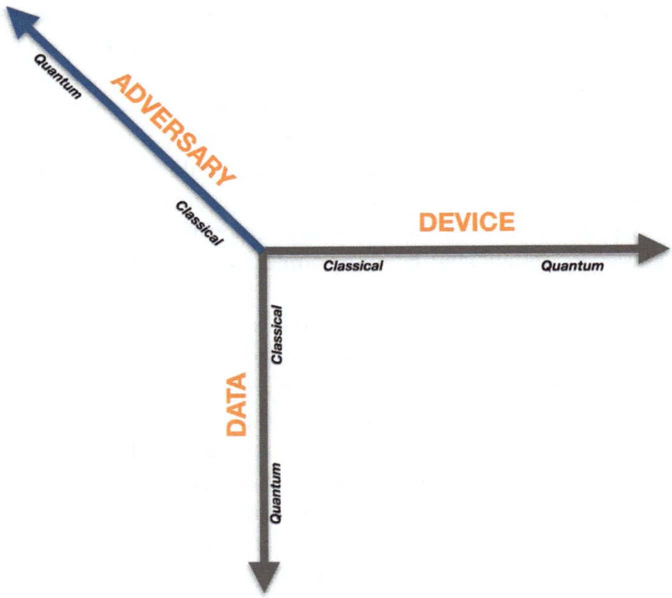

Fig. 1.3 Extension of the quantum-classical data-device paradigm to adversarial settings. The figure illustrates classical versus quantum distinctions along the well-known data (vertical) and device (horizontal) axes, and introduces the adversary as a third component. This extension makes it possible to ask systematic questions such as: *How does the nature of the adversary change when data or computation is quantum? Which attack surfaces exist in each quadrant?* In this way, the framework helps map out where robustness challenges and opportunities arise in hybrid quantum-classical learning systems

or interference to mount entirely new types of attacks. Similarly, a classical device operating on quantum data may face vulnerabilities in the measurement interface, whereas a quantum device processing classical data may be more exposed through its optimization routines.

Viewing QML robustness in this three-dimensional space re-frames familiar questions. Instead of asking simply: *What advantage does a quantum model offer on classical data?*, one can ask: *Does this advantage persist under the capabilities of a quantum adversary, and if not, why*? To further expand, this cubic framework allows us to systematically examine the attack surfaces that exist in each region and analyze how the adversary's computational model interacts with the data and device to shape vulnerability.

In this sense, this conceptual framework underlines that quantum computing may offer new tools to address robustness challenges, either through fundamentally different model behavior or through novel algorithmic defenses. These new computational capabilities as well as new sources of fragility define the next stage of research in the field. The next chapters will cover more in depth the vulnerabilities quantum phenomena introduce into learning, as well as the techniques used to mitigate these vulnerabilities and enhance model robustness.

1.7 Conclusion and Outlook

This chapter has outlined the motivations, foundations, and interdisciplinary nature of quantum machine learning. QML is propelled by the successes and limitations of classical machine learning, and it is grounded on the quantum computational paradigm that is still being actively shaped. As outlined in this chapter, this dual identity brings both promise and uncertainty. The promise comes from the opportunities quantum structures presents. As we elaborated above, one such promise is that of computational speedups, inherited to QML models via the fundamental linear algebra subroutines. Another important opportunity is brought forth by quantum computers' expected ability to capture data representations in higher-dimensional Hilbert spaces, potentially allowing for greater pattern emergence and more expressive models.

However, its practical realization is hindered by today's constrained hardware capacities which suffers from noise, decoherence, and limited qubit connectivity. Many theoretical models assume operations that far exceed today's capabilities requiring the full processing power of fault-tolerant quantum computers. Furthermore, within the quantum-enhanced machine learning paradigm, data encoding as a preprocessing step is a resource-intensive process. Optimization of quantum and hybrid learning models remains a major bottleneck, with variational algorithms commonly plagued by barren plateaus that hinder effective training. Moreover, the lack of standardized quantum-native datasets limits rigorous evaluation and understanding of what quantum phenomena can bring to the learning process.

Nevertheless, research in QML remains active and determined, driven by long-term technological promise, interdisciplinary momentum, and theoretical curiosity. While significant obstacles persist, these very challenges have spurred a rich body of work aimed at both understanding and overcoming them. On an algorithmic level, a growing line of work focuses on understanding and mitigating barren plateaus via architectural design, initialization strategies, and local cost functions. These efforts are part of the broader initiative to develop trainable, scalable, and expressive quantum models, without losing sight of how they measure up to classical techniques. At a more fundamental level, an existing avenue explores theoretical bounds that characterize what quantum models can and cannot learn efficiently, again, often compared to their classical counterparts. These characterizations range from sample complexity to expressive power and quantum generalization bounds. Meanwhile, across the broader landscape of quantum computing there is an active push for error correction, error mitigation, and coherent control strategies at scale. These foundational advancements are expected to benefit a wide range of broader quantum applications, including those enabled or advanced by QML.

Given the practical challenges posed by quantum hardware and algorithmic instability, QML systems demand careful benchmarking and thorough exploration. One important aspect of this evaluation is robustness: the capacity of a QML model to perform reliably in the presence of noise, or intentional perturbations. And because quantum systems are inherently sensitive to environmental factors such as decoherence and noise, this fragility makes the study of robustness especially relevant. Among the various dimensions of robustness, adversarial robustness is of particular interest. We hope this chapter has equipped readers with a foundational understanding of QML as a rapidly evolving interdisciplinary field, with a particular focus on the issue of robustness. In the chapters that follow, we will delve deeper into the specific vulnerabilities introduced by quantum phenomena, explore formal definitions and frameworks for robustness in QML, and examine current approaches to adversarial defense. As we move forward, the next chapters will dissect quantum-specific vulnerabilities, develop formal frameworks for robustness, and explore defenses that harness quantum principles rather than merely tolerate them. By the end of the book, readers will gain a comprehensive understanding of what it entails to design and develop robust QML systems.

Acknowledgements We thank Debbie Lim and Josep Lumbreras for discussions. Portions of this manuscript were drafted or edited with the assistance of ChatGPT to improve clarity and style.

References

1. Schuld M, Petruccione F (2021) Machine learning with quantum computers. In: Quantum science and technology, 2nd edn. Springer, Cham, Switzerland
2. Biamonte J, Wittek P, Pancotti N, Rebentrost P, Wiebe N, Lloyd S (2017) Quantum machine learning. Nature 549(7671):195–202

3. LeCun Y, Bengio Y, Hinton G (2015) Deep learning. Nature 521(7553):436–444
4. Goodfellow I, Bengio Y, Courville A (2016) Deep learning. MIT Press, Cambridge, MA
5. Nielsen MA, Chuang IL (2010) Quantum computation and quantum information. Cambridge University Press, Cambridge, UK. 10th anniversary edition
6. Shor PW (1997) Polynomial-time algorithms for prime factorization and discrete logarithms on a quantum computer. SIAM J Comput 26:1484–1509. Society for Industrial and Applied Mathematics
7. Montanaro A (2016) Quantum algorithms: an overview. npj Quantum Inf 2(1)
8. West MT, Tsang SL, Low JS et al (2023) Towards quantum enhanced adversarial robustness in machine learning. Nat Mach Intell 5:581–589
9. Bishop CM (2006) Pattern recognition and machine learning. Springer
10. Hastie T, Tibshirani R, Friedman J (2009) The elements of statistical learning, 2nd edn. Springer
11. Valiant LG (1984) A theory of the learnable. Commun ACM 27(11):1134–1142
12. Rosenblatt F (1958) The perceptron: a probabilistic model for information storage and organization in the brain. Psychol Rev 65(6):386–408
13. Vapnik V, Chervonenkis A (1971) On the uniform convergence of relative frequencies of events to their probabilities. Theory Probab & Appl 16(2):264–280
14. Rumelhart DE, Hinton GE, Williams RJ (1986) Learning representations by back-propagating errors. Nature 323:533–536
15. Cortes C, Vapnik V (1995) Support-vector networks. Mach Learn 20:273–297
16. Freund Y, Schapire RE (1997) A decision-theoretic generalization of on-line learning and an application to boosting. J Comput Syst Sci 55(1):119–139
17. Schölkopf B, Smola AJ (2002) Learning with Kernels. MIT Press
18. Robbins H, Monro S (1951) A stochastic approximation method. Ann Math Stat 22(3):400–407
19. Nesterov Y (1983) A method of solving a convex programming problem with convergence rate $o(1/k^2)$. Sov Math Dokl 27:372–376
20. Koller D, Friedman N (2009) Probabilistic graphical models: principles and techniques. MIT Press
21. Hinton GE, Osindero S, Teh Y-W (2006) A fast learning algorithm for deep belief nets. Neural Comput 18(7):1527–1554
22. Krizhevsky A, Sutskever I, Hinton GE (2012) Imagenet classification with deep convolutional neural networks. Adv Neural Inf Process Syst
23. Ioffe S, Szegedy C (2015) Batch normalization: accelerating deep network training by reducing internal covariate shift. In: International conference on machine learning
24. He K, Zhang X, Ren S, Sun J (2016) Deep residual learning for image recognition. In: IEEE conference on computer vision and pattern recognition (CVPR), pp 770–778
25. Kingma DP, Ba J (2015) Adam: a method for stochastic optimization. In: International conference on learning representations
26. Vaswani A, Shazeer N, Parmar N, Uszkoreit J, Jones L, Gomez AN, Kaiser Ł, Polosukhin I (2017) Attention is all you need. Adv Neural Inf Process Syst
27. Kaplan J, McCandlish S, Henighan T et al (2020) Scaling laws for neural language models. arXiv:2001.08361
28. Belkin M, Hsu D, Ma S, Mandal S (2019) Reconciling modern machine-learning practice and the classical bias-variance trade-off. Proc Natl Acad Sci 116(32):15849–15854
29. Brown T, Mann B, Ryder N et al (2020) Language models are few-shot learners. Adv Neural Inf Process Syst
30. Ho J, Jain A, Abbeel P (2020) Denoising diffusion probabilistic models. Adv Neural Inf Process Syst
31. Arunachalam S, de Wolf R (2017) A survey of quantum learning theory. arXiv:1701.06806
32. Feynman RP (1982) Simulating physics with computers. Int J Theor Phys 21(6–7):467–488
33. Benioff P (1980) The computer as a physical system: a microscopic quantum mechanical Hamiltonian model of computers as represented by Turing machines. J Stat Phys 22(5):563–591
34. Preskill J (2018) Quantum computing in the NISQ era and beyond. Quantum 2:79

35. Harrow AW, Montanaro A (2017) Quantum computational supremacy. Nature 549(7671):203–209
36. Arute F, Arya K, Babbush R et al (2019) Quantum supremacy using a programmable superconducting processor. Nature 574:505–510
37. Zhong H-S, Wang H, Deng Y-H, Chen M-C, Peng L-C, Luo Y-H, Qin J, Dian W, Ding X, Yi H, Peng H, Yang X-Y, Zhang W-J, Li H, Li Y, Jiang X, Gan L, Yang G, You L, Wang Z, Li L, Liu N-L, Chao-Yang L, Pan J-W (2020) Quantum computational advantage using photons. Science 370(6523):1460–1463
38. Shor PW (1997) Fault-tolerant quantum computation
39. Peruzzo A, McClean J, Shadbolt P, Yung M-H, Zhou X-Q, Love PJ, Aspuru-Guzik A, O'Brien JL (2014) A variational eigenvalue solver on a photonic quantum processor. Nat Commun 5:4213
40. Farhi E, Goldstone J, Gutmann S (2014) A quantum approximate optimization algorithm. arXiv:1411.4028
41. Huber PJ (1964) Robust estimation of a location parameter. Ann Math Stat 35(1):73–101
42. Tukey JW (1960) A survey of sampling from contaminated distributions. In: Contributions to probability and statistics, pp 448–485. Stanford University Press
43. Dalvi N, Domingos P, Sanghai S, Verma D (2004) Adversarial classification. In: ACM SIGKDD international conference on knowledge discovery and data mining, pp 99–108
44. Lowd D, Meek C (2005) Adversarial learning. In: ACM SIGKDD international conference on knowledge discovery and data mining, pp 641–647
45. Ben-David S, Blitzer J, Crammer K, Kulesza A, Pereira F, Vaughan JW (2010) A theory of learning from different domains. Mach Learn 79(1):151–175
46. Ben-Tal A, El Ghaoui L, Nemirovski A (2009) Robust optimization. Princeton University Press
47. Duchi JC, Namkoong H (2022) Statistics of robust optimization: a generalized empirical likelihood approach. Math Oper Res 47(2):882–918. (Original preprint 2016)
48. Namkoong H, Duchi JC (2017) Variance-based regularization with convex objectives. Adv Neural Inf Process Syst
49. Tang E (2019) A quantum-inspired classical algorithm for recommendation systems. In: Proceedings of the 51st annual ACM SIGACT symposium on theory of computing, STOC '19, pp 217–228. ACM
50. LaRose R, Coyle B (2020) Robust data encodings for quantum classifiers. Phys Rev A 102:032420
51. Pérez-Salinas A, Cervera-Lierta A, Gil-Fuster E, Latorre JI (2020) Data re-uploading for a universal quantum classifier. Quantum 4:226
52. Huang H-Y, Broughton M, Mohseni M, Babbush R, Boixo S, Neven H, McClean JR (2021) Power of data in quantum machine learning. Nat Commun 12(1)
53. Aaronson S (2015) Read the fine print. Nat Phys 11(4):291–293
54. Harrow AW, Hassidim A, Lloyd S (2009) Quantum algorithm for linear systems of equations. Phys Rev Lett 103(15)
55. Lloyd S, Mohseni M, Rebentrost P (2014) Quantum principal component analysis. Nat Phys 10(9):631–633
56. Rebentrost P, Mohseni M, Lloyd S (2014) Quantum support vector machine for big data classification. Phys Rev Lett 113(13)
57. Gilyén A, Su Y, Low GH, Wiebe N (2019) Quantum singular value transformation and beyond: exponential improvements for quantum matrix arithmetics. In: Proceedings of the 51st annual ACM SIGACT symposium on theory of computing, STOC '19, pp 193–204. ACM
58. Ivashkov P, Huang P-W, Koor K, Pira L, Rebentrost P (2024) QKAN: quantum kolmogorov-arnold networks
59. Cerezo M, Arrasmith A, Babbush R, Benjamin SC, Endo S, Fujii K, McClean JR, Mitarai K, Yuan X, Cincio Ł, Coles PJ (2021) Variational quantum algorithms. Nat Rev Phys 3:625–644
60. Benedetti M, Lloyd E, Sack S, Fiorentini M (2019) Parameterized quantum circuits as machine learning models. Quantum Sci Technol 4(4):043001

61. Mitarai K, Negoro M, Kitagawa M, Fujii K (2018) Quantum circuit learning. Phys Rev A 98:032309
62. Schuld M, Bergholm V, Gogolin C, Izaac J, Killoran N (2019) Evaluating analytic gradients on quantum hardware. Phys Rev A 99(3)
63. Havlíček V, Córcoles AD, Temme K, Harrow AW, Kandala A, Chow JM, Gambetta JM (2019) Supervised learning with quantum-enhanced feature spaces. Nature 567:209–212
64. Schuld M, Killoran N (2019) Quantum machine learning in feature Hilbert spaces. Phys Rev Lett 122(4)
65. Bowles J, Ahmed S, Schuld M (2024) Better than classical? the subtle art of benchmarking quantum machine learning models
66. Cao Y, Romero J, Olson JP, Degroote M, Johnson PD, Kieferová M, Kivlichan ID, Menke T, Peropadre B, Sawaya NPD, Sim S, Veis L, Aspuru-Guzik A (2019) Quantum chemistry in the age of quantum computing. Chem Rev 119(19):10856–10915
67. Morales MES, Pira L, Schleich P, Koor K, Costa PCS, An D, Aspuru-Guzik A, Lin L, Rebentrost P, Berry DW (2025) Quantum linear system solvers: a survey of algorithms and applications
68. Orús R, Mugel S, Lizaso E (2019) Quantum computing for finance: overview and prospects. Rev Phys 4:100028
69. McClean JR, Boixo S, Smelyanskiy VN, Babbush R, Neven H (2018) Barren plateaus in quantum neural network training landscapes. Nat Commun 9(1):4812
70. Larocca M, Thanasilp S, Wang S, Sharma K, Biamonte J, Coles PJ, Cincio L, McClean JR, Holmes Z, Cerezo M (2025) Barren plateaus in variational quantum computing. Nat Rev Phys 7(4):174–189
71. Caro MC, Huang HY, Cerezo M, Sharma K, Sornborger A, Cincio L, Coles PJ (2022) Generalization in quantum machine learning from few training data. Nat Commun 13:4919
72. Sirui L, Duan L-M, Deng D-L (2020) Quantum adversarial machine learning. Phys Rev Res 2:033212
73. Ren W, Li W, Xu S et al (2022) Experimental quantum adversarial learning with programmable superconducting qubits. Nat Comput Sci 2:711–717
74. Gao X, Anschuetz ER, Wang S-T, Cirac JI, Lukin MD (2022) Enhancing generative models via quantum correlations. Phys Rev X 12(2):021037
75. Liu N, Wittek P (2020) Vulnerability of quantum classification to adversarial perturbations. Phys Rev A 101:062331
76. West MT, Nakhl AC, Heredge J, Creevey FM, Hollenberg LC, Sevior M, Usman M (2024) Drastic circuit depth reductions with preserved adversarial robustness by approximate encoding for quantum machine learning. Intell Comput 3

Chapter 2
Adversarial Attack Transferability of Quantum and Classical Classifiers

Muhammad Usman

Abstract This chapter investigates transferability of classical and quantum adversarial attacks, i.e., the performance of classical classifiers against quantum adversarial attacks and vice versa. The empirical results indicate that quantum adversarial attacks are able to trick classical classifiers significantly reducing their accuracy. However, the converse is not true: the adversarial attacks generated from classical neural networks only slightly impact the accuracy of quantum variational circuits. We argue that the enhanced resilience of quantum classifiers to classical attacks is due to the fundamentally different nature of attacks which is exhibited by visible structures in quantum perturbations whereas classical perturbations tend to show no such structural information. This introduces a potential new avenue for quantum advantage in the form of adversarial robustness against classically attacked datasets.

2.1 Introduction

Machine learning (ML) has rapidly become ubiquitous and has started to profoundly impact every facet of daily life. Despite remarkable accuracy and efficiency of classical ML systems, unlocking their full potential is still plagued by significant challenges, such as tremendous demands for computational power to train large-scale ML models and the vulnerabilities of ML algorithms to data manipulations or cyber-attacks [1]. For example, the computational cost of training state-of-the-art large-scale ML models has already surpassed millions of dollars and requires huge amount of energy with detrimental environmental effects. On the other hand, the vulnerability of ML systems to data spoofing attacks is of grave concern for their adaptation in security-sensitive applications such as autonomous vehicles and robotic military

M. Usman (✉)
Data61, CSIRO, Research Way Clayton 3168 Australia and School of Physics,
The University of Melbourne, Victoria, Australia
e-mail: muhammad.usman@data61.csiro.au

M. Usman (ed.), *Quantum Robustness in Artificial Intelligence*, Quantum Science and Technology, https://doi.org/10.1007/978-3-032-11153-1_2

weapons where any mistake made by an ML algorithm could lead to the loss of lives. A paradigm shift is needed to develop future ML systems which can be efficiently trained and trusted.

Recent progress in quantum computing [2]—both on hardware and software fronts—has accelerated the field of quantum machine learning (QML) in which the ambition is to integrate the power of quantum computing in designing fundamentally new ML models and architectures [3–6] which can offer superior performance, either by exhibiting better training accuracies [7] or by demonstrating stronger resilience to adversarial scenarios [8–10]. Indeed, the field of adversarial QML is being developed as a potential new avenue for quantum advantage. Specifically of interest to this book chapter is the transferability of adversarial attacks from one ML model to the other model [8]. It has been shown in the previous work [8] that adversarial attacks are generally very effective in transferring across networks which is known as block-box setting in the literature. For example, adversarial images generated from a convolutional neural network (ConvNet) were able to trick a separately trained classical ResNet model. The details of the ConvNet and ResNet implementations are provided in Ref. [8]. The strong transferability of adversarial attacks across classical ML models raises serious concerns around their robustness as an adversary can generate powerful attacks without requiring any knowledge of the underpinning models being attacked. A corresponding obvious question is whether similar black-box vulnerabilities exist when attacks are transferred between quantum and classical ML models. This chapter provides an overview of such scenarios and discusses recent results which have investigated transferability of adversarial attacks between classical and quantum networks.

Figure 2.1 illustrates the construction of standard classical and quantum ML models and the generation of adversarial attacks. In part (a), example images from two datasets MNIST [11] and FMNIST [12] are shown which are widely used as a benchmark in the ML literature. Schematics of classical neural network (in this chapter, we train convolutational neural network and ResNet) and quantum neural network (variational quantum circuit) are plotted in parts (b) and (c). Finally, parts (d) and (e) show adversarial perturbations generated from the classical and quantum models. The corresponding perturbed images will be passed to trained classical and quantum models in white-box and black-box settings to evaluate their performance.

2.2 Attack Scenarios

Classical adversarial machine learning literature is very rich and there are several adversarial attack generation scenarios which have been developed over the years. However, the field of quantum adversarial attacks is still in its infancy and no natively quantum adversarial attack methods exist at present. The current state-of-the-art is therefore based on adapting classical adversarial methods for quantum variational circuits and analyzing their performance for the classical and quantum classification tasks. In this context, this chapter focuses on two of the most widely investigated

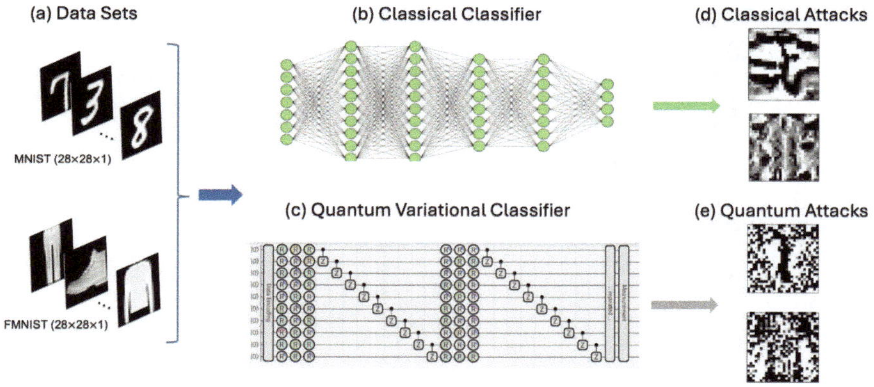

Fig. 2.1 **a** Example images from two standard datasets MNIST and FMNIST which are used to study transferability of adversarial attacks. **b** Schematic diagram illustrating a classical ML model. In this work, we investigate two classical ML models namely convolutional neural network and ResNet. **c** Schematic diagram of a quantum ML model. In this work, we investigate quantum variational circuit (QVC) which consists of a number of repeating layers of single qubit parametrised gates and two-qubit CNOT gates. **d** Examples of adversarial perturbations generated from a classical network. **e** Examples of adversarial perturbations generated from a QVC model

adversarial attacks namely PGD and FGSM. These attacks are described in the classical literature [13, 14] and their implementation in the quantum ML context is outlined in the recent work [8].

2.3 Transferring Classical Adversarial Attacks to Quantum Classifiers

We begin the study of attack transferability by first investigating the attacks generated from the classical ML models and transferring them to QVC. In order to show the relative attack performance, we also transfer them to an independent classical ML model. In this regard, we first train two classical ML systems namely ConvNet (a convolutional neural network) and ResNet. The adversarial attacks are generated by the ConvNet model and first passed to the ResNet. Figure 2.2 top row plots the accuracy of ConvNet (red color) and ResNet (blue color) under a PGD attack. The accuracy of both classical models rapidly drops for the attacked datasets indicating that the perturbations from the PGD attack are clever enough to trick both models under white-box and black-box settings.

The same attacked images are then passed to a pre-trained QVC200 model—a QVC model with 200 variational layers. In both MNIST and FMNIST cases, the accuracy is the QVC200 model drops only slightly. This indicates that the classical adversarial attacks are not effective on quantum ML models under the black-box setting.

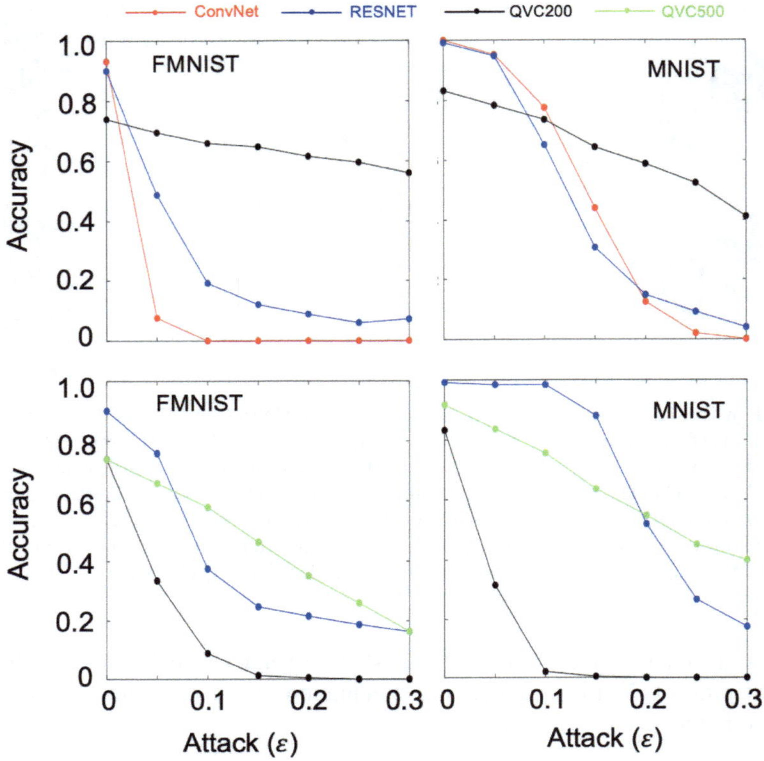

Fig. 2.2 The accuracy achieved by classical and quantum networks on sets of 250 adversarially attacked test images from each of the considered datasets in the cases of white-box PGD attacks on the convolutional network (top row), and 200 layer quantum variational classifier (QVC200, bottom row) as a function of attack strength ε (measured with the l_∞ norm). In both cases we see the accuracies of the networks under attack decrease sharply. The tendency of the accuracy of the independent networks to also decrease is a manifestation of the transferability of adversarial examples—they are typically capable of fooling even networks which they were not explicitly designed to attack. We see an exception to this in the top row, with the quantum classifier largely resisting the attacks generated with respect to the convolutional neural network

2.4 Transferring Quantum Adversarial Attacks to Classical Classifiers

In this section, we investigate the effectiveness of quantum adversarial attacks under the white-box and black-box settings. The PGD attacks are implemented through QVC200 model which is trained on MNIST and FMNIST datasets. The attacked images are then passed to classical ResNet model and a separately trained QVC500 model under black-box setting. Figure 2.2 bottom row plots the training accuracy in both scenarios under the attacks. First we notice that the accuracy of the ResNet model drops sharply (blue color) indicating that the quantum attacks are effective on the classical ML models. Interestingly, the results also show that the attacks generated from the QVC200 model are also able to trick QVC500 model (green color). This implies that the attacked images from a quantum model are not only effective in tricking classical models but also other quantum models. The defence of quantum ML models against quantum adversarial attacks has been explored in recent work which have shown that adversarial training [8] and auto-encoder purification [15] methods could provide useful defence mechanisms.

2.5 Discussion and Outlook

In the last two sections, we have investigated the transferability of both classical and quantum adversarial attacks across classical and quantum models. The results are summarized in Fig. 2.3, indicating that attacked generated by quantum models are effective against both classical and quantum models, but the classical attacks only work against classical models.

In order to understand the difference between quantum and classical adversarial perturbations, Fig. 2.4 plots perturbed images generated from classical and quantum models. Interestingly, the adversarial perturbations from the classical model do not exhibit any structure and indicate that the model is manipulating datasets at the pixel level without considering general features in the images. Contrarily, the perturbations generated from the QVC500 model visibly show feature level modifications. For example, in the FMNIST images the QVC500 model is changing pixel corresponding to clothings to black color in order to trick the classification process. The clothing structures are evident from the perturbation images. Same is true for the handwritten digit images from the MNIST dataset. This stark contrast between classical and quantum adversarial perturbations plays an important role in

	Quantum ML	Classical ML
Quantum Attack	Yes	Yes
Classical Attack	No	Yes

Fig. 2.3 Summary of quantum and classical adversarial attack scenarios and the performance of the corresponding quantum and classical ML models. Yes means the attack is successful in tracking the model and No means the attack is largely unsuccessful

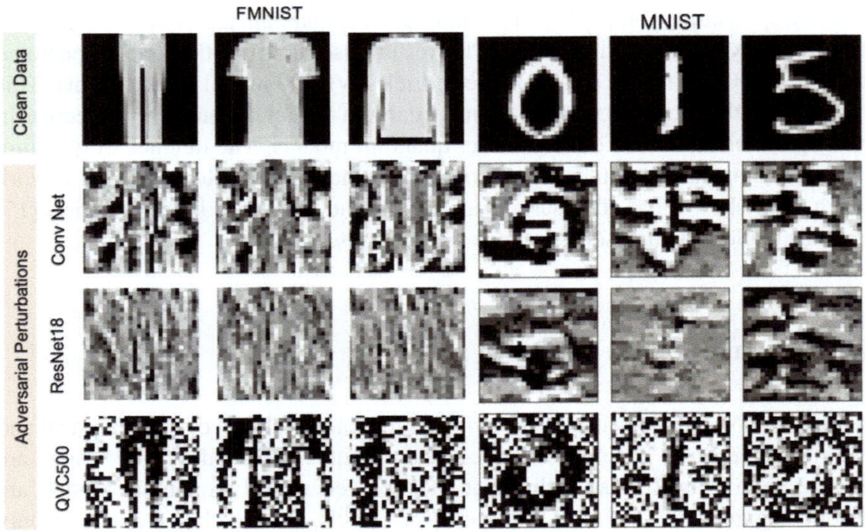

Fig. 2.4 The adversarial perturbations generated by $\varepsilon = 0.1$ PGD attacks on the convolutional neural network, ResNet and the 500 layer quantum variational classifier (QVC500) are shown for several examples from the FMNIST and MNIST databases. In the case of FMNIST the task is to distinguish between classes of images representing various types of clothing, e.g., t-shirts, pants, shoes, while in the case of MNIST the classification problem is to determine a hand written digit. The perturbations generated by the classical networks are incomprehensible to humans, exploiting the highly abstract features discovered by these networks. The attack on the QVC, on the other hand, yields perturbations whose semantic content is clear; they constitute concrete steps toward actually changing the label of the clean image (i.e., filling in the gap between the legs of the pants, adding sleeves to the t-shirt, and removing sleeves from the jumper)

determining the black-box accuracies of the investigated models. Recent analytical work has further established scenarios for quantum adversarial robustness in ML systems [16, 17].

Although Fig. 2.3 has shown a promising potential for quantum advantage which stems from the resilience of QVC models against classical adversarial attacks while at the same time generating quantum attacks which can trick classical ML models, the presented results are based on simulations with ideal or noise-free environment. The current generation of quantum processors however are noisy and hardware errors significantly impact their performance. Therefore, future implementations of QVC models on quantum processors will require error mitigation or correction techniques which could cancel the effect of noise and implement QML models with high fidelities commensurate with the classical models. Recent work has explored such ideas albeit at the proof-of-concept scale in which QML models are trained with quantum error detection [18] and partial quantum error correction [19], exhibiting promising results. Further work is needed to demonstrate their performance on the current generation of hardware devices with scalability.

Another key challenge for variational QML models is the presence of barren plateaus which cause trainability issues [20]. Some recent work has been reported on developing novel QML architectures with provable trainability [7, 21, 22], however their robustness against adversarial attacks is still an open question. A different but interesting line of research is quantum transfer learning [23] in which quantum-classical hybrid models are developed based on the principle of transfer learning—a classical model learns complex datasets and the transfer learning is performed to a quantum model which retains its adversarial robustness properties.

Data encoding is also a key hurdle in practical implementation of QML models. The encoding of classical datasets in quantum state is resource intensive task. For example, qubit-efficient amplitude encoding require deep quantum circuits with thousands of quantum gates which prohibit their implementation on the near term quantum processors due to high levels of noise. On the other hand, angle encoding technique requires many qubits which are not yet available on hardware devices. Recent work has explored approximate encoding methods which significantly reduce resource requirements without compromising accuracies or robustness [24], however further work is needed to enable experimental implementation of QML models with reasonable scale commensurate with the requirements of practical applications. Despite several outstanding challenges, small-scale experimental implementation of adversarial QML has already been reported in the literature [25], indicating that in future larger-scale implementations can be expected. Finally, quantum ML on quantum data may naturally lead to quantum advantage in ML models [26] but no such useful experimental implementation has so far been achieved.

2.6 Conclusion

In summary, this chapter has investigated transferability of adversarial attacks in white-box and black-box settings. The results have shown that adversarial attacks generated from classical ML models hardly impact the accuracy of quantum ML models. However, the converse is not true—quantum adversarial attacks strongly impact the accuracy of classical ML models. This presents an opportunity for potential quantum advantage in ML applications where robustness is the primary parameter of interest such as self-driving vehicles and military robotic weapons.

Acknowledgements This book chapter is adapted from previously published research work in Ref. [8].

References

1. West MT et al (2023) Towards quantum enhanced adversarial robustness in machine learning. Nat Mach Intell 5:581–589
2. Gill SS, Kumar A, Singh H, Singh M, Kaur K, Usman M, Buyya R (2022) Quantum computing: a taxonomy, systematic review and future directions. Softw Pract Exp 52(1):66–114
3. Wang Z, van der Laan T, Usman M (2025) Self-adaptive quantum kernel principal component analysis for compact readout of chemiresistive sensor arrays. Adv Sci 12(15):2411573
4. Tsang SL, West MT, Erfani SM, Usman M (2023) Hybrid quantum–classical generative adversarial network for high-resolution image generation. IEEE Trans Quantum Eng 4:1–19
5. West MT, Sevior M, Usman M (2023) Boosted ensembles of qubit and continuous variable quantum support vector machines for b meson flavor tagging. Adv Quantum Technol 6(10):2300130
6. Wang Z, Wang F, Li L, Wang Z, van der Laan T, Leon RC, Huang JK, Usman M (2025) Quantum kernel learning for small dataset modeling in semiconductor fabrication: application to ohmic contact. Adv Sci e06213
7. West MT et al (2023) Provably trainable rotationally equivariant quantum machine learning. arXiv:2311.05873
8. West MT et al (2023) Benchmarking adversarially robust quantum machine learning at scale. Phys Rev Res 5:023186
9. Sirui L, Duan L-M, Deng D-L (2020) Quantum adversarial machine learning. Phys Rev Res 2(3):033212
10. Wu Y, Adermann E, Thapa C, Camtepe S, Suzuki H, Usman M (2023) Radio signal classification by adversarially robust quantum machine learning. arXiv:2312.07821
11. Deng L (2012) The mnist database of handwritten digit images for machine learning research [best of the web]. IEEE Signal Process Mag 29(6):141–142
12. Xiao H, Rasul K, Vollgraf R (2017) Fashion-mnist: a novel image dataset for benchmarking machine learning algorithms. arXiv:1708.07747
13. Goodfellow IJ, Shlens J, Szegedy C (2014) Explaining and harnessing adversarial examples. arXiv:1412.6572
14. Madry A, Makelov A, Schmidt L, Tsipras D, Vladu A (2017) Towards deep learning models resistant to adversarial attacks. arXiv:1706.06083
15. Khatun A, Usman M (2025) Classical autoencoder distillation of quantum adversarial manipulations. Phys Rev Res 7:L042054
16. Dowling N, West MT, Southwell A, Nakhl AC, Sevior M, Usman M, Modi K (2024) Adversarial robustness guarantees for quantum classifiers. Quant Inform 12:16
17. Guan J, Fang W, Ying M (2021) Robustness verification of quantum classifiers. In: International conference on computer aided verification, pp 151–174. Springer
18. Adermann E, Suzuki H, Usman M (2025) Variational quantum machine learning with quantum error detection. Quantum Machine Intelligence 8(1)
19. Kang H, Kim Y, Adermann E, Sevior M, Usman M (2025) Almost fault–tolerant quantum machine learning with drastic overhead reduction. arXiv:2507.18954
20. Cerezo M et al (2022) Challenges and opportunities in quantum machine learning. Nat Comput Sci 2:567–576
21. West MT et al (2023) Reflection equivariant quantum neural networks for enhanced image classification. Mach Learn Sci Technol 4:035027
22. Heredge J, West M, Hollenberg L, Sevior M (2025) Nonunitary quantum machine learning. Phys Rev Appl 23(4):044046
23. Khatun A, Usman, M (2024) Quantum transfer learning with adversarial robustness for classification of high-resolution image datasets. Advanced Quantum Technologies 8:2400268
24. West MT, Nakhl AC, Heredge J, Creevey FM, Hollenberg LC, Sevior M, Usman M (2024) Drastic circuit depth reductions and preserved adversarial robustness by approximate encoding for quantum machine learning. Intelligent Computing 3:100

25. Ren W, Li W, Xu S, Wang K, Jiang W, Jin F, Zhu X, Chen J, Song Z, Zhang P et al (2022) Experimental quantum adversarial learning with programmable superconducting qubits. arXiv:2204.01738
26. Huang H-Y et al (2022) Quantum advantage in learning from experiments. Science 376:1182–1186

Chapter 3
Adversarial Threats in Quantum Machine Learning: A Survey of Attacks and Defenses

Archisman Ghosh, Satwik Kundu, and Swaroop Ghosh

Abstract Quantum Machine Learning (QML) integrates quantum computing with classical machine learning, primarily to solve classification, regression, and generative tasks. However, its rapid development raises critical security challenges in the Noisy Intermediate-Scale Quantum (NISQ) era. This chapter examines adversarial threats unique to QML systems, focusing on vulnerabilities in cloud-based deployments, hybrid architectures, and quantum generative models. Key attack vectors include model stealing via transpilation or output extraction, data poisoning through quantum-specific perturbations, reverse engineering of proprietary variational quantum circuits, and backdoor attacks. Adversaries exploit noise-prone quantum hardware and insufficiently secured QML-as-a-Service (QMLaaS) workflows to compromise model integrity, ownership, and functionality. Defense mechanisms leverage quantum properties to counter these threats. Noise signatures from training hardware act as non-invasive watermarks, while hardware-aware obfuscation techniques and ensemble strategies disrupt cloning attempts. Emerging solutions also adapt classical adversarial training and differential privacy to quantum settings, addressing vulnerabilities in quantum neural networks and generative architectures. However, securing QML requires addressing open challenges such as balancing noise levels for reliability and security, mitigating cross-platform attacks, and developing quantum-classical trust frameworks. This chapter summarizes recent advances in attacks and defenses, offering a roadmap for researchers and practitioners to build robust, trustworthy QML systems resilient to evolving adversarial landscapes.

A. Ghosh (✉) · S. Kundu · S. Ghosh
Pennsylvania State University, University Park, PA, USA
e-mail: apg6127@psu.edu

S. Kundu
e-mail: satwik@psu.edu

S. Ghosh
e-mail: szg212@psu.edu

M. Usman (ed.), *Quantum Robustness in Artificial Intelligence*, Quantum Science and Technology, https://doi.org/10.1007/978-3-032-11153-1_3

3.1 Introduction

As QML rapidly matures from theoretical promise to experimental and early-stage practical deployment, concerns around its robustness and security have gained significant urgency [1–3]. In the current NISQ era [4], QML systems are inherently constrained by hardware limitations, noisy gate operations, and limited qubit connectivity. These physical limitations are further complicated by the architectural trends that dominate QML today: hybrid classical-quantum designs, cloud-based quantum computing access [5], and variational quantum circuits (VQCs) [6] trained via classical optimizers. While these developments make QML feasible and scalable in the near term, they also introduce a diverse and novel attack surface. Adversarial threats that target classical ML pipelines do not transfer trivially into the quantum domain, but instead manifest in new forms due to the fundamentally different representations, computational models, and physical substrates involved. Crucially, QML systems inherit and amplify certain vulnerabilities, such as model theft, data poisoning, and integrity breaches, particularly when deployed via QML-as-a-Service (QMLaaS) platforms or accessed through remote execution interfaces [7]. This chapter is motivated by the urgent need to understand, categorize, and mitigate adversarial threats specific to QML. As with any emerging computational paradigm, security considerations must evolve in parallel with functionality. QML is not exempt from this imperative.

3.1.1 Vulnerabilities in the Execution Flow of QML

QML enables advanced data-driven inference and decision-making using machine learning paradigms in the quantum Hilbert space. While promising, this integration introduces distinct security challenges stemming from the inherent features of quantum systems, such as superposition and entanglement, and the practical limitations of NISQ devices. These factors expose QML pipelines to a range of adversarial threats, necessitating the development of tailored security mechanisms (Fig. 3.1). The QML workflow begins with preprocessing and encoding classical data into quantum states,

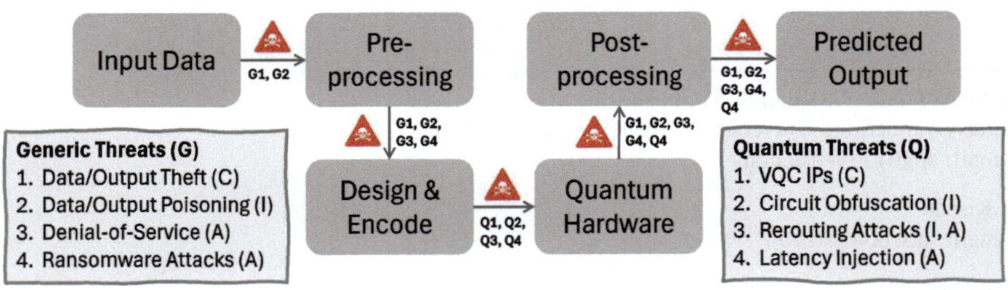

Fig. 3.1 Key threats to confidentiality (C), integrity (I), and availability (A) in the QML pipeline [7]

often via schemes like amplitude or angle encoding [8–10]. This stage is vulnerable to inference attacks, where adversaries with access to circuits or their transpiled forms can deduce the data-encoding methods, compromising input confidentiality. Additionally, data poisoning attacks can be mounted during this phase by injecting malicious perturbations that disrupt training dynamics and degrade model reliability [11]. Following data encoding, the construction and initialization of parameterized quantum circuits (PQCs) form the computational core of QML models. These circuits are susceptible to model extraction, wherein unauthorized access leads to replication or approximation of proprietary designs. Moreover, adversaries may introduce stealthy backdoors, malicious circuit modifications that trigger incorrect outputs under specific conditions, posing risks to the integrity and trustworthiness of the deployed models [12].

QML models often adopt a hybrid architecture, wherein quantum circuits are evaluated by quantum processors while classical optimizers iteratively update circuit parameters. This interplay between quantum and classical components introduces unique security risks. Data exchanged during this loop, such as intermediate outputs, gradients, or loss values, may be intercepted, enabling adversaries to infer sensitive model parameters or internal structures [13]. Furthermore, manipulation of the classical optimization process itself can misguide the training trajectory, resulting in degraded performance or the embedding of malicious behavior within the model [14]. The deployment setup of QML models on cloud-based platforms, while scalable, exposes the system to inference-time threats. Publicly accessible APIs allow adversaries to probe the model and perform functionality reconstruction or model extraction attacks. Additionally, side-channel vulnerabilities may arise from quantum hardware characteristics, such as execution timing, thermal output, or noise signatures, that can inadvertently leak sensitive information about the model's structure or computations [15, 16].

These multifaceted vulnerabilities across the QML execution flow highlight the urgent need for holistic, quantum-aware security strategies. Safeguarding QML requires addressing both classical and quantum threats to ensure robust and trustworthy model deployment.

3.1.2 *Why Is Theft of QML Models a Concern?*

The vulnerabilities outlined across the QML execution pipeline are not merely technical limitations, they represent tangible risks with significant implications. The theft or manipulation of QML models can compromise the confidentiality of sensitive input data, erode trust in quantum-enabled inference, and undermine the proprietary value of carefully engineered quantum circuits. Critically, both quantum data and model training incur substantial costs: preparing high-quality quantum datasets often involves complex preprocessing and domain-specific encoding strategies [17], while training variational quantum circuits demands extensive quantum-classical iteration, subject to hardware constraints and limited quantum coherence times [18]. These investments make QML systems highly valuable assets and, consequently, attractive targets for adversaries. As deployment increasingly shifts to cloud-hosted platforms and QMLaaS offerings, the incentives for unauthorized access, replication, or sabotage are amplified. The inherent opacity of quantum operations further complicates detection and attribution, making attacks stealthier and potentially more damaging.

3.2 Attack Models

Developing and applying well-defined attack models is essential for advancing the study of adversarial robustness in QML. Given the unique architectural, algorithmic, and physical characteristics of QML systems, security research in this space requires a clear understanding of adversarial assumptions, access levels, and objectives. Carefully constructed attack models enable researchers to map existing vulnerabilities, identify overlooked threat vectors, and avoid proposing unrealistic or impractical defenses. As with other domains of security research, such as hardware Trojans or classical adversarial ML, choosing the appropriate attack model is a foundational step. It not only informs the design of effective countermeasures but also guides reproducibility, benchmarking, and comparison across studies. In what follows, we describe a set of representative attack models for QML that serve to categorize current adversarial efforts, surface emerging trends, and offer a framework for evaluating the security posture of QML architectures across deployment contexts.

3.2.1 Taxonomy of the Attack Models

Attack models in QML can be broadly categorized according to the adversary's level of access to the model and execution environment (Table 3.1). These categories—black box, gray box, and white box—provide a foundational framework for assessing the feasibility, impact, and countermeasures for a range of threat vectors.

3.2.1.1 Black-Box Attack Models

In the black-box setting, adversaries do not possess internal knowledge of the QML model architecture, circuit parameters, or training data. Access is limited to querying the model via an external interface, common in QMLaaS or cloud-based deployments. Even with these constraints, attackers can mount effective threats.

- **Model Extraction and Counterfeit Generation**: A commonly encountered threat in QMLaaS deployments is the adversary modeled as an external user who lacks access to the internal architecture, parameterization, training data, or compilation artifacts of the target Quantum Neural Networks (QNNs) [18] (Fig. 3.2). The

Table 3.1 Taxonomy of quantum adversarial attack models

Threat model	Attacker profile	Target artifacts	Capabilities	Related works
Black box	External user/API client; No internal access	QML outputs	Model stealing, functionality replication	[15, 18–20]
Gray box	Cloud provider/semi-privileged adversary; partial access	Transpiled circuits	Manipulate inputs or IRs, estimate parameters or logic	[21, 22]
White box	Transpiler/ infrastructure insider; full access to model internals	PQC, Transpiled gates, pulse schedules	Analyze and RE circuits, inject Trojans	[11, 20, 23–25]

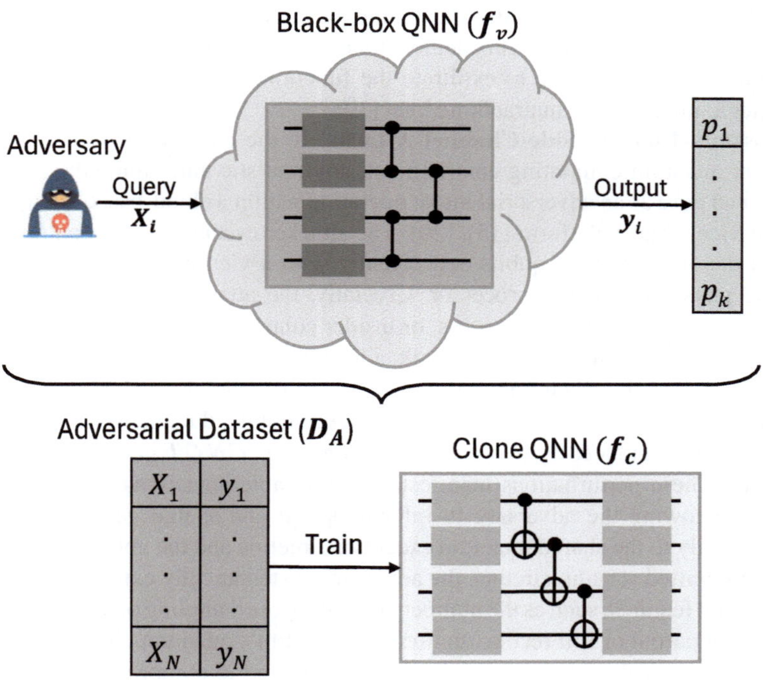

Fig. 3.2 An adversary sends a query vector $X_i = (x_{i1}, x_{i2}, ..., x_{id})$ to a cloud-based victim QNN (f_v), receiving a vector of class probabilities $(f_v(X_i) = y_i = (p_{i1}, p_{i2}, ..., p_{ik}))$ in response. The adversary repeats this to build an attacker dataset D_A and trains a substitute model f_c to clone f_v's functionality [18]

attacker interacts with the model solely through a public inference API, submitting classical inputs and receiving output distributions, typically Top-1 predictions or full class probability vectors (Top-k). Despite the lack of introspective access, the adversary can mount a model extraction attack by systematically querying the QNN with a large corpus of input samples and collecting the corresponding outputs to construct a surrogate dataset that approximates the decision boundary of the original model. To improve fidelity in the presence of NISQ-induced noise, the adversary may issue repeated queries across different time intervals and apply a variance-based consistency check to filter unreliable outputs. This denoised dataset is then used to train a counterfeit QNN locally. Advanced techniques such as quantum-domain contrastive learning are employed to pre-train a feature encoder on auxiliary datasets, while quantum transfer learning is used to fine-tune a classifier based on the cleaned query outputs. The attacker's objective is to construct a functionally equivalent substitute model that replicates the behavior of the original QNN with high fidelity, thereby bypassing service access controls, evading usage metering, and undermining the confidentiality and economic value of the proprietary model. This demonstrates that even under restricted black-box conditions, it is feasible to exfiltrate the functional essence of a QNN through strategic input-output interaction alone [19].
- **Crosstalk-Induced Side-Channel Attacks**: In the current NISQ era of multi-tenant quantum computing environment, any realistic and minimally privileged user can pose as an adversarial threat operating within a cloud-based Quantum-as-a-Service (QaaS) platform [15], [20]. The attacker executes their quantum circuit on a disjoint subset of qubits concurrently with a victim user, both sharing the same physical quantum processor. Crucially, the attacker does not require co-location of qubits, elevated access, or insider control over the quantum hardware. Instead, the attack leverages crosstalk, a form of physical interference inherent in NISQ devices, between qubits to create a passive side channel. By strategically positioning idle "snooping" qubits near victim qubits, the attacker detects subtle perturbations induced by two-qubit operations (e.g., $CNOT$ gates) in the victim's circuit. These perturbations manifest as measurable deviations in the idle qubit states, allowing the adversary to infer temporal and spatial gate activity. With access only to the shared quantum execution timeline and the ability to repeatedly submit crafted sensing circuits, the adversary systematically extracts circuit-level structural features, such as the number and timing of entangling operations, thereby enabling partial or full reconstruction of the victim's quantum circuit, underscoring that physical-layer interactions can be exploited to compromise user privacy without violating system access controls.

3.2.1.2 Gray-Box Attack Models

Gray-box attackers possess partial knowledge of the QML system, such as access to certain components or intermediate data. This level of access enables more sophisticated attacks.

- **Data Poisoning in Quantum Pipelines**: The adversary is assumed to operate within the quantum cloud infrastructure where a victim outsources the training of a hybrid quantum-classical model, having access to the victim's preprocessed and labeled training dataset and partial visibility into the QNN architecture, specifically, the classical-to-quantum data-encoding circuit responsible for mapping classical inputs to quantum states in the Hilbert space (Fig. 3.3). However, the adversary does not possess knowledge of the PQC, the loss function, the optimizer, or other training dynamics. Within this constrained view, the adversary conducts an indiscriminate data poisoning attack by modifying the labels of a subset of the training data based on a quantum-specific criterion: intra-class encoder state similarity (ESS) [21]. By computing the density matrices resulting from the encoding circuit and measuring quantum distances (e.g., Frobenius norm) between states, the adversary identifies and assigns incorrect labels that maximize intra-class dissimilarity. This manipulation deteriorates the structure of the encoded quantum feature space, thereby disrupting the training process and degrading the generalization capability of the QNN. Such a model degradation attack, achievable without full circuit transparency, exemplifies the threat posed by semi-privileged

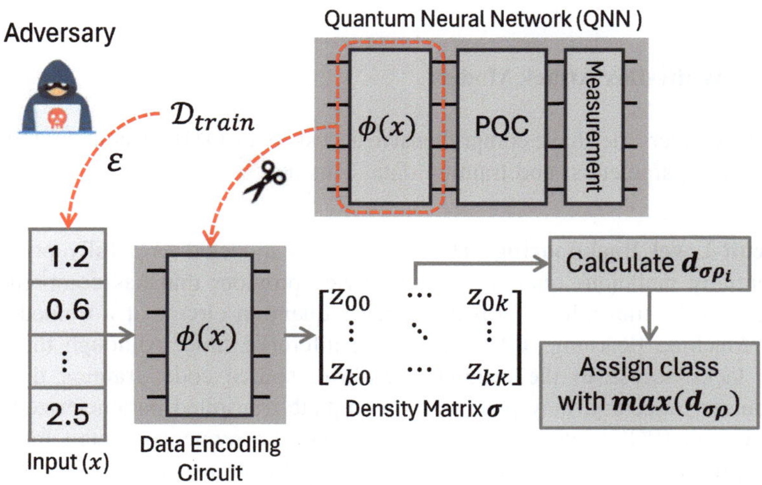

Fig. 3.3 Overview of QUID's label-poisoning technique. Adversary extracts the encoding circuit from the QNN and uses it to compute the output density matrix (σ) for a portion (ε) of the training data $\mathcal{D}_{train}$. It then calculates the matrix distance between σ and the remaining samples in the dataset (ρ_i), assigning the class with the max. distance (max $d_{\sigma\rho}$) [21]

adversaries in practical quantum cloud environments where partial circuit disclosure and data access are common.

- **Vulnerabilities in Hybrid Quantum-Classical Architectures**: The adversary can also be modeled as an untrusted quantum cloud provider with partial but strategically significant visibility into the computation pipeline. Specifically, the adversary does not possess access to the original source code of the QML model, its training data, or the internal architecture design (e.g., parameter initialization, circuit topology, or training dynamics) [22]. However, the attacker does receive access to the compiled, executable quantum circuits submitted for inference, which are composed of a classical-to-quantum data encoder $D(Z)$ and a parameterized quantum model $C(\theta)$. By impersonating a legitimate user and submitting chosen inputs $D(Z)$, the adversary induces the QML service to transmit the composed circuit $C(\theta) \cdot D(Z)$ to the cloud infrastructure. Leveraging the known structure of their own encoder $D(Z)$, the attacker can construct and apply the inverse circuit $D^{-1}(Z)$ to effectively cancel the data-encoding portion, thereby isolating the underlying QML model $C(\theta)$. This exploitation relies on the reversibility of quantum circuits and the transparency of the encoding function, allowing the adversary to reconstruct the core model even without access to its training history or high-level design intent. The attacker's capabilities thus extend beyond a conventional black-box model, given their ability to observe and manipulate low-level quantum instructions. This gray-box attack illustrates a critical vulnerability in centralized QML deployments and motivates the need for distributed execution strategies, such as partitioning the QML model across multiple cloud providers to prevent any single adversary from recovering the complete model.

3.2.1.3 White-Box Attack Models

White-box adversaries have comprehensive access to the QML system, including its architecture, parameters, and training data. This access facilitates the most potent attacks.

- **Circuit-Level Backdooring**: The adversary is modeled as a fully privileged, potentially malicious quantum cloud service provider that has comprehensive access to the transpiled, hardware-specific quantum circuit of a trained Quantum Machine Learning (QML) model at inference time. Although the adversary lacks access to the original high-level source code, training data, and optimizer trajectory, they possess the complete transpiled parameterized quantum circuit (PQC), which includes low-level gate sequences, qubit mappings, and optimized parameter placements. This level of transparency, typical in current QML-as-a-Service (QMLaaS) deployments, allows the attacker to reverse engineer the architecture and parameter values of the original model [11, 23] (Fig. 3.4). The adversary reverse engineers the transpiled circuit by parsing it to extract the gate-level representation based on the backend's basis gate set (e.g., $\{id, X, SX, CNOT, RZ\}$ in IBM's superconducting platforms), the device's

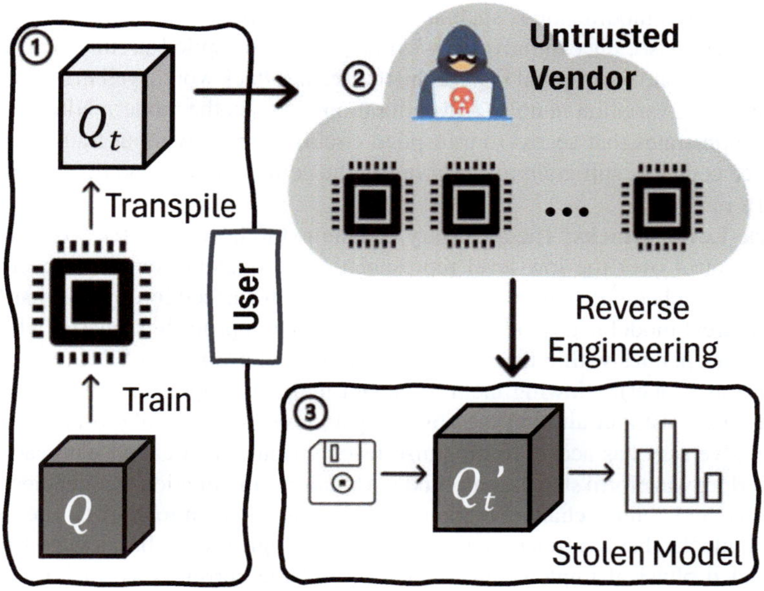

Fig. 3.4 The flow diagram describes reverse engineering of QML parameters by untrusted third-party vendors acting as adversaries. (1) shows the user training and transpiling a QML model Q using non-proprietary quantum hardware and sending the transpiled version of the trained model Q_t to the untrusted vendor for inferencing. (2) and (3) describe the attack model involving the procedure of reverse engineering performed by the untrusted vendor to extract the parameters and steal the IP of the user-designed model [11]

coupling map, and transpilation optimization strategies. The attacker systematically analyzes the gate patterns, especially the sequences of RZ, $CNOT$, and SX gates, using a lookup table (LUT) constructed from known transpilation patterns of common parametric gates (RX, RY, RZ). For example, a transpiled pattern such as $RZ(\pi/2) \cdot SX \cdot RZ(\phi) \cdot SX \cdot RZ(\pi/2)$ can be mapped back to an $RX(\theta)$ rotation gate with an inferred parameter θ derived via decomposition. To recover the entanglement structure, the adversary reverses SWAP insertions and logical-to-physical qubit mappings using the device's coupling constraints. For parameter recovery, a brute-force parameter sweep is conducted over reduced domains (e.g., $\theta \in [-\pi, \pi]$), leveraging access to the same transpiler toolchain and computational resources. The adversary iteratively generates reverse-engineered circuits, transpiles them, and compares their parameterized structure and output statistics against the golden reference circuit using a fidelity metric. The reverse-engineering objective is to minimize the discrepancy between the reverse-engineered and original transpiled circuits, achieving functional equivalence. This capability enables the adversary to reconstruct an approximate but operational copy of the architecture-agnostic PQC, which can then be ported across different quantum backends, recompiled with alternative gate sets, or subjected to unauthorized modifications such as watermark tampering, adversarial retraining, or IP theft. Also, since the adversary works with the transpiled circuits to extract the parameters and the original PQC architecture, the attack works well in the presence of temporal variation in noise and calibration issues in the modern NISQ devices. It demonstrates that access to transpiled circuits alone—without training data or source code—is sufficient to compromise the confidentiality and deployability of QML models.

- **Pulse-Level Attacks**: The adversary is assumed to have complete visibility into and control over the low-level representation of quantum circuits, specifically, the pulse schedules that encode gate operations for execution on real hardware. This threat model is grounded in a supply chain compromise scenario, where the attacker operates within the quantum software toolchain or an SDK (e.g., Qiskit, Amazon Braket), allowing them to embed malicious behavior into custom gate definitions without altering the abstract gate-level logic visible to the user [20]. The adversary has access to the pulse-level attributes associated with each gate, including waveform shapes, amplitudes, durations, modulation frequencies, phase offsets, and control channel mappings. This enables them to craft tailored pulse-level attacks that are functionally correct at the logical level but deviate from the intended physical implementation. In spite of the current state of NISQ quantum hardware, which is affected by the temporal variation of noise and frequent calibration issues, the adversary having access to the control electronics can still launch a pulse-level attack to undermine the execution of the quantum program on the real hardware. Two main classes of attacks are considered: (i) channel-level attacks, which manipulate the routing or scheduling of control pulses (e.g., qubit plundering by leaking operations across channels, blocking pulses to freeze victim qubits, or reordering pulses to disrupt computation), and (ii) pulse-level attacks, which directly alter signal parameters to degrade fidelity, introduce timing shifts, or induce crosstalk. Importantly, these attacks remain stealthy because pulse-level behavior is analog, dynamic (due to frequent calibrations), and largely opaque to users who verify circuits at the gate abstraction layer. The adversary exploits this abstraction gap, embedding payloads that evade detection even under standard circuit validation procedures. Furthermore, the attack model accounts for different classes of victims—ranging from naive users with no verification capability to sophisticated users who conduct unitary fidelity checks—by tailoring the pulse modifications to fall within hardware noise tolerances or mimic expected calibration drift. This comprehensive control over internal representations and execution pathways situates the attacker firmly in the white-box category and reveals a potent, underexplored threat vector in quantum computing: adversarial manipulation at the pulse control layer beneath the visible circuit abstraction.
- **Prompting the need for Quantum Logic Locking (QLL)**: In cases where the adversary is characterized as an honest-but-curious quantum computing server—such as a commercial quantum cloud platform—or an internal malicious actor with full control over the quantum hardware and execution stack. The client, acting as a quantum circuit designer, compiles a high-level quantum program that includes a proprietary quantum oracle and submits the fully decomposed and technology-

mapped physical quantum circuit to the server for execution. The adversary, by virtue of operating the hardware, has unrestricted access to all components of the physical circuit, including detailed gate-level descriptions, qubit mappings, scheduling metadata, and execution results [24]. This level of access enables the adversary to reverse engineer the structure and functionality of the embedded quantum oracle—often used in Grover's search and other oracle-based algorithms—thus compromising intellectual property (IP) and sensitive problem-specific information encoded within the circuit. The attack model assumes that the quantum oracle constitutes proprietary logic encoded as a reversible sub-circuit and that its structure and behavior are critical to the confidentiality of the client's computational objective. The adversary may observe measurement outcomes, simulate alternate input configurations, or perform structural analysis on the gate network to infer the function implemented by the oracle. To counter this, the proposed defense involves locking the oracle via the insertion of additional key-controlled quantum gates—most commonly CNOT gates controlled by dynamically reassigned key qubits—creating multiple functional modes of the oracle indistinguishable to the server. Only the client, who holds the secret key qubit schedule, can identify the correct functional output post-execution. This defense assumes the attacker can access and execute all oracle modes but lacks knowledge of the key qubit values that dictate the correct circuit semantics. As such, this model represents a fully privileged white-box adversary with both read and execution access, capable of structural analysis but unable to discern protected logic without cryptographic or obfuscatory barriers.

- **Obfuscating Phase in Quantum Circuit Compilation**: The adversary is modeled as an untrusted third-party quantum compiler with full access to the pre-compiled quantum circuit supplied by the circuit designer. This model assumes that the quantum circuit, in its algorithmic or logical form, must be transmitted to an external compiler for hardware-specific transpilation and optimization. In doing so, the designer inadvertently exposes the complete structure and functional logic of the circuit, including its gate sequence, entanglement topology, and qubit mappings. The adversary, operating at the compiler level, is thus capable of inspecting every component of the circuit, enabling the extraction of proprietary algorithms (intellectual property theft), the reverse engineering of core functionality, and the malicious alteration of gate logic to embed hardware Trojans or counterfeit modifications. This high-privilege access classifies the model squarely within the white-box paradigm, as the attacker does not need to infer information from indirect observations but instead directly sees and operates on the internal representations of the quantum circuit. The threat is exacerbated by the current reliance on widely available third-party compilers, such as Qiskit, TKET, and Cirq, which may be maintained by external vendors and integrated into diverse cloud-based quantum development environments. These compilers can modify phase relationships, alter gate structures, or introduce stealthy perturbations that are difficult to detect post-compilation [25].

3.3 Countermeasures Against State-of-the-Art Attack Models

The unique characteristics of QML systems necessitate specialized security measures to address their inherent vulnerabilities. Recent research by various scholars has led to the development of targeted countermeasures aimed at fortifying QML models against a spectrum of adversarial threats. This section delineates these strategies, emphasizing their applicability across different stages of the QML pipeline.

3.3.1 Circuit-Level Obfuscation and Logic Locking

Unlike classical logic locking where each key bit is implemented via dedicated key gates or control logic at the transistor level, QLL operates under fundamentally different constraints and opportunities. Quantum gates act on superposed and entangled states, and any direct mapping of classical key-bit-to-control structures is both physically expensive and computationally limiting. To address this, the E-LoQ [24] scheme introduces a key-embedded quantum locking framework, wherein multiple classical key bits are efficiently encoded into a single key qubit. This enables scalable obfuscation of quantum circuits such as those found in modern quantum machine learning models without incurring significant qubit or gate overheads.

The mechanism begins by transforming a quantum circuit $\text{Circ} = G_m \dots G_0$ into a locked variant Circ', via a key-dependent encryption function $\text{Enc}(\text{Circ}, \mathbf{k})$, where $\mathbf{k} \in \{0, 1\}^n$ is a classical key of length n. This transformation involves the addition of a dedicated key qubit q_k, which dynamically progresses through the n key bits via interleaved Hadamard and Pauli-X operations, effectively implementing sequential key indexing within a single quantum wire. Two forms of gate-level modifications are introduced in the circuit. **Functional Gates**: A subset of original gates are replaced with key-controlled versions, where execution is conditional on the state of q_k being $|1\rangle$. This ensures that the circuit performs correctly only when the applied key matches $\mathbf{k}$. **Dummy Gates**: New gates are inserted whose operations are contingent on $q_k = |0\rangle$, effectively acting as no-ops under the correct key and injecting noise under incorrect ones. The key scheduling across the circuit is randomized, and key bit transitions are obscured through "H-masking," which adds Hadamard gates before each key-dependent operation to decorrelate adjacent key bits and resist simplification by transpilers. During the decryption phase $\text{Dec}(\text{Circ}', \mathbf{k})$, the correct sequence of Pauli-X and Hadamard operations is applied to reconstruct the original functionality, and the locked circuit is simplified by removing redundant identity gates and the auxiliary q_k. Critically, the unlocked circuit Circ'' satisfies $\text{Circ}'' = \text{Circ}$ only if $\mathbf{k}$ is correct. When incorrect keys $\mathbf{k}_g \neq \mathbf{k}$ are used, both functional and dummy gates are misaligned, yielding a semantically corrupted circuit. Security analysis of E-LoQ reveals its robustness against key guessing and reverse engineering. The obfuscation affects both the circuit's functional output and structural layout, increasing the difficulty of reverse-engineering attacks. Quantitative metrics such as Total Variation Distance (TVD), Hamming Variation Distance (HVD), and Degree of Functional Corruption (DFC) consistently show significant deviation from the baseline circuit when incorrect keys are used—TVD and HVD often exceed 0.9, and DFC approaches –1 in benchmark evaluations.

Moreover, key guessing is empirically infeasible. The average success rate of randomly guessing the correct key is negligible, and even partial key matches yield circuits with incorrect outputs or high functional divergence. Experiments with various benchmark circuits, including arithmetic units, adders, and comparators, confirm that even short key lengths (e.g., 6 bits) provide sufficient entropy to prevent adversarial inversion, especially when keys are mapped to variable gate positions. E-LoQ further ensures minimal overhead. Post-decryption simplification restores the original circuit depth with fidelity loss constrained below 1%. Unlike earlier schemes that required one qubit per key bit, or retained permanent overhead in the deployed circuit, E-LoQ discards all auxiliary locking logic after compilation, rendering it highly practical for near-term QML circuits that operate under severe qubit and gate budget constraints.

3.3.2 Hardware-Aware Watermarking

In the NSIQ era of quantum computing, noise plays an active role in the quantum circuit design and implementation. Leveraging this property of quantum hardware a passive, hardware-dependent watermark can be embedded in quantum neural networks. The authors in [17] have demonstrated the watermarking scheme on quantum generative adversarial networks (qGANs). Unlike active watermarking strategies, this approach exploits the stochastic variations introduced during quantum circuit execution as a natural means of identifying the quantum backend used for training. These variations, being hardware specific, are encoded into the trained model's parameters and manifest in the output data generated by the qGAN (Fig. 3.5).

During the training phase, the qGAN model is executed on a designated quantum backend. The model architecture consists of multiple sub-generators implemented as PQCs, which operate on latent vectors to produce image patches. Due to the sensitivity of PQCs to hardware noise (e.g., gate error rates, decoherence times, readout errors), the evolution of model parameters inherently reflects the unique noise profile of the training hardware. The generated images carry this noise imprint as a latent watermark, which remains intact even when the model is later inferred on different hardware. To enable watermark extraction, a classical convolutional neural network (CNN) classifier is trained on images generated by qGANs trained on a suite of known hardware. The classifier learns to identify subtle statistical differences arising from the embedded noise, achieving near-perfect classification accuracy for models trained on single hardware and approximately 90% for models trained across sequences of multiple hardware. For validation, the classifier's output is filtered using a confidence threshold M; only predictions exceeding this threshold are accepted as

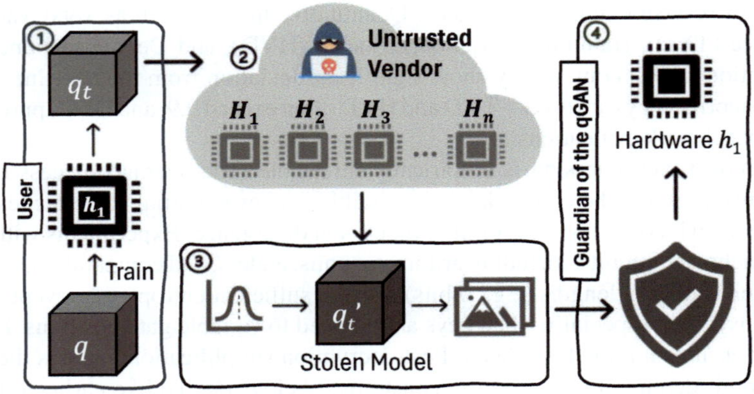

Fig. 3.5 The flow diagram describes our attack model and the proposed security measure. In the figure (1) shows the user training his qGAN, q on hardware h_1 to generate a trained qGAN q_t; (2), (3) describes the threat model of an untrusted quantum hardware vendor where the user sends qt for inferencing (note, the hardware used for inferencing, Hi, could be different than the hardware used for training h_1), from where it gets counterfeited by the untrusted vendor ($q_t^{'}$); (4) is our proposed method of collecting the images generated by $q_t^{'}$ and detecting the hardware where it has been trained using the classifier for proof of ownership [17]

proof of hardware provenance. This enables reliable differentiation between genuine and counterfeit models, even in adversarial scenarios involving unknown or tampered training sources.

Security analysis reveals several strengths of this watermarking strategy. First, the uniqueness of the watermark scales with the size of the available hardware suite. If a model is trained on a sequence of k hardware systems from a set of n, the probability of collision reduces to $\prod_{i=1}^{k} \frac{1}{n-i}$, providing high uniqueness even for moderate values of k. For instance, selecting 5 backends from a pool of 15 yields a collision probability on the order of 10^{-5}. The watermark is resistant to removal because the training process is inherently non-reversible. Even if an adversary has access to the white-box architecture and transpiled quantum circuit, reconstructing the exact noise-induced parameter trajectory is computationally infeasible without knowledge of initial states and training data. Furthermore, attempts to tamper with the watermark—by fine-tuning the stolen model on new hardware—only partially overwrite the original noise signature unless extensive retraining is performed, which reduces the incentive to engage in such tampering. False claims of ownership based on ghost watermarks are similarly deterred by requiring the classifier to match the watermark with high confidence and optionally through multi-hardware signature embedding. By training across an ordered sequence of backends, the watermark becomes both composite and sequence dependent, making it extremely difficult to forge. This approach introduces no architectural overhead, relies solely on existing training dynamics, and is broadly applicable to other QML models where noise-dependent parameter evolution can be captured and later authenticated. The method is particularly suited for cloud-based quantum service environments where trust boundaries are minimal and IP leakage risks are high.

3.3.3 Side-Channel Attack Mitigation

QML workloads, particularly those involving VQCs, are susceptible to crosstalk-induced side-channel attacks when deployed on multi-tenant NISQ hardware. The intrinsic physical interactions among proximate qubits can expose dynamic circuit features, including entangling gate patterns and scheduling information. To safeguard the structural confidentiality of QML circuits—whose layer depth, connectivity, and parameterized entangling operations are critical IP—several crosstalk-aware defense strategies have been developed. These aim to reduce the observability and reproducibility of physical noise signatures that side-channel adversaries may exploit.

- **Qubit Layout Randomization and Dynamic Mapping**: To decouple logical qubit identities from their physical placement, a randomized mapping strategy is employed prior to execution. During transpilation, each QML model instance is compiled using a randomly selected allocation of logical-to-physical qubits, ensuring that sensitive entangling operations (e.g., $CNOT$ or CR gates) do not consistently occupy fixed locations within the hardware topology [15]. This strategy is particularly effective for QML circuits where the placement of entangled qubit pairs conveys model structure, such as patch-wise variational layers in quantum convolutional neural networks or entanglement topologies in quantum graph learning models. By rotating the spatial context of each execution, the correlation between specific victim operations and observer noise signals is weakened, reducing side-channel signal coherence across executions.
- **Noise-Shaping**: A second line of defense introduces execution-time obfuscation through the insertion of padding operations and idle gate shaping [20]. Selective insertion of single-qubit rotations and decoupling gates on both active and idle qubits during known sensitive intervals (e.g., high-density entangling layers) disrupts the crosstalk field's spectral consistency. For QML workloads, this is applied dynamically based on the learned temporal profile of the variational layers. For instance, entangling blocks that dominate the expressivity of the QML model (e.g., in quantum feature maps or variational classifiers) are targeted for padding with dynamically randomized gate sequences. These gates are selected to be unitary equivalent to identity in function, but sufficiently distinct in hardware-level control signals (e.g., shaped microwave pulse envelopes) to introduce temporal and spatial noise decoys.
- **Circuit Obfuscation**: Structural obfuscation of QML models at the transpilation layer is realized through compilation strategies that preserve functional equivalence while altering gate scheduling and topology [15]. Techniques include **Gate Reordering**: Rewriting commutable gate blocks to alter execution timing without affecting output fidelity. **Dummy Gate Insertion**: Adding canceling $CNOT$

pairs or parameterized gates with zero-angle rotations to obscure true operation density. **Template Substitution**: Replacing standard sub-circuits with functionally equivalent but structurally distinct alternatives. These methods preserve the training trajectory and inference behavior of the QML model while masking structural invariants that may be leveraged by adversaries to perform classification or reconstruction attacks on variational models.

- **Randomizing the Execution Scheduling**: To mitigate timing-based leakage exploited through temporal bucketing analysis, a stochastic scheduler introduces non-determinism in circuit queuing and execution windows [20]. Each QML job is fragmented and re-ordered with randomized delays or submitted as batched sequences interleaved with decoy circuits. The scheduler may leverage dummy jobs from a secure controller to increase entropy in system-wide temporal load patterns. This is particularly relevant to QML inference-as-a-service deployments, where repeated invocations of a model on user data must maintain indistinguishability at the execution profile level. Even if the model's layout is consistent, variance in gate timing and circuit duration limits adversarial correlation of output signal deviations to specific circuit epochs.
- **Topology-Aware Device Assignment and Isolation**: Lastly, QML security-aware runtime systems may employ physical isolation by assigning sensitive model components to minimally coupled regions of the quantum processor. Device regions with reduced crosstalk coefficients are selected for model qubits, while known high-coupling zones are reserved for untrusted or third-party tenants [15, 20]. In hybrid QML models where classical and quantum inference are pipelined, this approach allows for staging of the quantum kernel execution into pre-designated hardware zones, minimizing the physical leakage paths accessible to concurrent adversaries.

3.3.4 Secure Model Partitioning and Distribution

To mitigate model theft risks associated with untrusted quantum cloud providers, the QuMoS framework [22] introduces a partitioning-based security strategy for QML workloads. Rather than relying on cryptographic obfuscation or post-training watermarking, QuMoS decomposes the quantum model into a set of computational submodules (nodes), each of which is executed on a different, physically isolated quantum backend. This architectural fragmentation ensures that no single provider has access to the entire quantum circuit, thereby preventing reconstruction of the full QML model from any individual subsystem. Each submodule of the QML model is defined as a quantum computational block $C_i(\theta_i)$, representing a distinct portion of the variational quantum circuit. The blocks are linked via a directed acyclic graph (DAG), with edges denoting data flow and inter-node dependencies. The QuMoS runtime dispatches these blocks for execution on separate quantum cloud providers, with connectivity managed through secure classical post-processing of intermediate measurement results. Security stems from the design constraint that any single submodel (i.e., connected subset of blocks assigned to a single provider) must exhibit low standalone predictive power. That is, no subset executable on a compromised provider should retain a non-trivial approximation of the full model's functionality. To enforce this, the framework defines and optimizes a heuristic metric $\text{SecMec}(M)$, computed as follows:

$$\text{SecMec}(M) = 1 - \max_{q \in \mathcal{Q}_S,\, sm \in \mathcal{S}_M} \frac{\text{ACC}(sm, q)}{\text{ACC}(M)}$$

where $\mathcal{Q}_S$ is the set of quantum devices, $\mathcal{S}_M$ is the set of security submodels (i.e., all maximal subgraphs confined to one provider), and ACC denotes accuracy on a test dataset. This metric quantifies how closely any submodel can approximate the full QML behavior; higher values indicate greater resilience to partial model extraction. To identify optimal partitionings, QuMoS employs a reinforcement learning-based controller. The controller explores the discrete space of node-to-node topologies, quantum architecture mappings, and provider assignments. Each candidate model is evaluated on the basis of its accuracy and $\text{SecMec}(M)$, and controller parameters are updated using a reward signal $R \propto \text{ACC}(M) + \lambda \cdot \text{SecMec}(M)$, where λ modulates the trade-off between performance and security.

Each candidate architecture is trained and validated across a suite of simulated noisy backends (e.g., IBMQ quito, belem, manila), with security submodels identified through topological analysis of the deployment graph. Submodels are then separately evaluated for inference performance on their assigned provider to calculate their leakage potential. Empirical evaluation shows that naive partitioning schemes, even when distributed across multiple providers, often fail to secure the model: some submodels retain disproportionately high accuracy, enabling partial model theft. In contrast, architectures optimized by the QuMoS engine consistently maximize security by ensuring that submodels are entangled in non-trivial data dependencies, reducing their standalone efficacy. Furthermore, QuMoS discovers secure model configurations with minimal loss in accuracy—often within 1–2% of centralized neural architecture search (NAS) baselines, and occasionally outperforming them. Notably, the framework is robust to varying numbers of available providers. Even with only two providers, it can achieve substantial security benefits, and the partitioning strategy dynamically adapts to available infrastructure. Moreover, the system selectively omits unnecessary nodes and providers to optimize the balance between model fidelity and confidentiality.

3.3.5 Hardware-Induced Output Perturbation

To mitigate threats pertaining to the cloning of QML models by repeated querying, authors in [18] introduce a perturbation-based defense mechanism that leverages the intrinsic noise characteristics and hardware heterogeneity of NISQ systems to obfuscate output distributions and degrade the fidelity of cloned QML models. Two concrete techniques are proposed: **Hardware Variation-Induced Perturbation (HVIP)** and **Hardware and Architecture Variation-Induced Perturbation (HAVIP)**. Both exploit naturally occurring device-level variances and architectural diversity across quantum backends to introduce controlled but unpredictable distortions in the QNN outputs.

- **HVIP**: HVIP operates by dynamically varying the execution backend of the cloud-hosted QNN. Rather than consistently executing all inference queries on a single quantum device, the system randomly routes each incoming query to a different hardware instance. Since quantum backends differ significantly in their coupling maps, gate fidelities, basis gate sets, and noise profiles (e.g., amplitude damping, depolarization, readout error), the resulting measurement statistics diverge subtly across devices. This variability introduces stochastic perturbations into the output probability vectors returned to the adversary, even when the same input is queried repeatedly. As a result, the attacker accumulates a training dataset with inconsistent label distributions, thereby impairing the convergence and fidelity of the substitute QNN. HVIP is especially effective when the backend pool includes devices with non-overlapping error characteristics, amplifying divergence in the Softmax output distributions.
- **HAVIP**: HAVIP enhances the defense by jointly introducing architectural heterogeneity into the system. Instead of deploying a single QNN across multiple devices, HAVIP trains multiple, distinct QNN architectures—each optimized and compiled for a different quantum backend. Upon receiving a query, the inference engine selects one of these QNN instances at random to process the input. Since each model is structurally distinct (e.g., differing in circuit depth, entanglement structure, parameter count), and each backend exhibits unique error characteristics, the ensemble produces non-deterministic, architecture-specific outputs. This misaligns the decision boundaries perceived by the adversary and fragments the label distribution in the adversarial dataset. The combination of circuit-level and hardware-level diversity ensures that the attacker is unable to consistently map input features to reliable output labels, leading to noisy, low-fidelity clones even when high-volume queries are issued.

The defense mechanisms were tested on hybrid QML pipelines involving parameterized quantum circuits (PQC-1, PQC-6, PQC-17, PQC-19), using MNIST, Fashion-MNIST, Kuzushiji, and Letters datasets. Clone models trained on Top-1 and Top-k label queries were evaluated across scenarios with and without defense activation. Quantitative results show that HAVIP and HVIP introduce measurable perturbations: up to 15.71% label mismatches in Top-1 queries and 10.2% TVD in Top-k outputs. The cloned models experienced performance drops of up to 13% in test accuracy, particularly when trained with probability vectors returned from the perturbed ensemble. Interestingly, the study also observes that QML models trained in noisy environments (e.g., using SPSA optimizers and mixed-device noise simulations) exhibit higher baseline robustness to minor perturbations. This dual-edged outcome implies that while HVIP and HAVIP degrade attacker success, they must be reinforced with more aggressive or adaptive noise modulation strategies to remain effective as quantum devices scale and standardize.

From a QMLaaS perspective, HVIP and HAVIP represent practical and cost-effective methods to enhance IP protection. These defenses do not require cryptographic encapsulation or quantum authentication but rather exploit the inherent stochasticity of NISQ systems. By treating noise and architectural diversity as security assets rather than liabilities, the system obfuscates functional mappings without compromising user-level inference performance, aligning well with the QML workloads, particularly where small decision boundary shifts or Softmax noise can drastically affect training convergence in adversarial cloning attempts.

3.4 Designing Secure QML Systems

As QML transitions from theory to implementation, ensuring the security of these systems from the ground up is no longer optional—it is essential. Given the hybrid architecture of most QML systems, the high cost of training, and the increasing reliance on cloud-based quantum services, model developers must proactively incorporate security principles throughout the design lifecycle. This section outlines key guidelines for designing QML models with robustness and trustworthiness in mind, informed by state-of-the-art countermeasures and known attack vectors.

3.4.1 Identifying Threat Model

The development of a secure QML pipeline must commence with the formalization of a threat model that characterizes potential adversarial capabilities and objectives. In the context of QML systems, adversaries can be classified by their level of access: black box, gray box, or white box. Black-box adversaries, typically external clients interacting through public QMLaaS APIs, are limited to input-output queries and rely on statistical probing to infer model behavior. Gray-box adversaries may be semi-privileged actors such as cloud providers or software vendors who have partial visibility into compiled circuits, encoders, or training data. White-box adversaries represent the most powerful threat, with full access to model internals, circuit parameters, transpiled gates, and even low-level pulse schedules. Defining the access level and intent—whether model extraction [17], data poisoning [21], or logic subversion [6]—is essential to selecting appropriate defenses and embedding security constraints throughout the QML pipeline.

3.4.2 Data-Encoding Security

The data-encoding stage, which transforms classical inputs into quantum states, is inherently vulnerable due to its exposure at the initial point of interaction. Encoding strategies, such as amplitude or angle encoding, can leak structural information if left unprotected. To mitigate these vulnerabilities, encoding procedures must be designed to introduce entropy and obfuscation, ensuring that the mapping from input space to Hilbert space is not trivially invertible. Encoders can employ randomized gate structures or parameterized transformations whose specifics are concealed from execution environments. Furthermore, encoded data must undergo quantum-aware validation to prevent adversarial manipulation, particularly label-poisoning attacks that exploit the similarity structure of quantum states. Incorporating noise-tolerant and adversarially robust encoding frameworks can prevent perturbation-induced degradation of quantum feature spaces, thus preserving the semantic integrity of the learning process.

3.4.3 Obfuscate Quantum Circuit Structure

Variational quantum circuits, central to QML models, are a primary target for model extraction and reverse-engineering attacks due to their rich representational capacity and high design cost. The structural protection of such circuits requires a combination of logic obfuscation and conditional execution gating. Techniques such as QLL introduce key-dependent behavior by embedding control logic into circuit pathways, ensuring correct output generation only when a secret key is applied. An effective implementation, such as the E-LoQ scheme [24], leverages a single qubit to sequentially encode multiple classical key bits, significantly minimizing overhead while injecting high entropy into the circuit behavior. Additionally, structural diversification through template substitutions, insertion of canceling gate pairs, and reordering of commutable blocks hinders attempts to statically analyze or simulate the original functional logic. These measures are vital in protecting the intellectual property encoded within the circuit and deterring adversarial replication.

3.4.4 Harden the Hybrid Training Loop

The quantum-classical interface in QML training pipelines constitutes a critical attack surface. Variational algorithms rely on classical optimizers to iteratively update quantum parameters based on measured outcomes, creating a bi-directional data flow that can be intercepted or altered. To secure this loop, communication between quantum and classical components must be authenticated and, where feasible, encrypted. Sensitive artifacts such as gradient values or cost functions must be safeguarded against both passive eavesdropping and active manipulation. Furthermore, robust training strategies can be deployed to preemptively inoculate models against poisoning and convergence manipulation. Adversarial training that incorporates quantum-aware perturbations during training, such as those induced by the QUID framework [21], ensures that the model's decision boundaries remain resilient to encoded state-space anomalies. Extensions of differential privacy to quantum optimization may further obfuscate parameter evolution trajectories, mitigating inference risks associated with repeated observations of intermediate states.

3.4.5 Integrate Hardware-Aware Defenses

Quantum hardware, particularly in the NISQ regime, exhibits idiosyncratic noise profiles and physical crosstalk that introduce side channels exploitable by adversaries. To defend against such leakage, hardware-aware mitigations must be integrated at the transpilation and scheduling layers. Logical-to-physical qubit mappings should be randomized across executions to prevent spatial correlation of entangling gates. Gate-level noise shaping, through the insertion of pulse-equivalent identity operations, can disrupt spectral signatures that would otherwise reveal circuit timing or structure. Furthermore, the stochastic fragmentation of execution schedules introduces temporal entropy, preventing timing analysis that correlates inputs with execution duration or gate activity. A complementary approach leverages the inherent noise fingerprint of quantum hardware as a passive watermark, uniquely tying a trained model to the backend used for its development. Such watermarks, when embedded within parameter evolution during training, are non-replicable without access to the original hardware and can be detected through trained forensic classifiers, providing a reliable means of provenance verification.

3.4.6 Partitioning and Distributing Securely

Given the risks associated with centralized deployment, particularly in cloud-hosted environments, secure QML systems must consider partitioning strategies that distribute model functionality across disjoint execution domains. The QuMoS framework exemplifies this approach by decomposing the quantum model into sub-circuits, each executed on a distinct quantum backend. These submodules are designed to be individually non-functional, i.e., their inference accuracy is statistically indistinguishable from random guessing, thereby preventing adversaries from reconstructing meaningful model behavior even in the event of a breach. The orchestration of inter-submodule communication is handled through secure classical channels, often with post-measurement aggregation and normalization. Optimal partitioning is guided by reinforcement learning algorithms that balance task accuracy with the submodels' predictive leakage, quantified via metrics such as $SecMec$. This architectural isolation paradigm significantly raises the difficulty threshold for model reconstruction and enables compliance with data locality and trust boundary requirements in federated quantum computing scenarios.

3.5 Future Directions in Secure QML Design

Future work in secure QML should systematically expand the threat landscape to encompass diverse attack vectors that exploit the unique properties of quantum computation. While current models address threats such as model stealing, data poisoning, and circuit backdooring, more sophisticated adversaries may leverage quantum-specific features, such as entanglement structure, noise profiles, and circuit reversibility, to execute covert or multi-stage attacks. For instance, the potential for adversarial manipulation through entanglement injection, inference-time noise pattern analysis, or optimizer-level gradient hijacking in hybrid setups remains largely unexamined. Additionally, as QML systems increasingly adopt QMLaaS and multi-tenant execution environments, new avenues for cross-user leakage and supply chain compromise need to be analyzed thoroughly for security gaps.

On the defensive front, advancing beyond isolated protections toward integrated and adaptive security frameworks is critical. Existing techniques, such as quantum logic locking, hardware-aware watermarking, and circuit obfuscation, have demonstrated effectiveness but are typically static and context specific. Future defenses should be dynamic, incorporating runtime-aware scheduling, reconfigurable circuit topologies, and adversarially robust training regimes that reflect quantum state-space geometry. The idea of differential privacy also holds promise in the QML landscape for reducing leakage and enhancing confidentiality in untrusted environments. Importantly, the field would benefit from standardized evaluation protocols, threat benchmarking tools, and formal guarantees tailored to quantum computational settings.

Competing Interests The authors have no conflicts of interest to declare that are relevant to the content of this chapter.

Acknowledgements The work is supported in parts by NSF (CNS-2129675, CCF-2210963), gifts from Intel and IBM Quantum Credits.

References

1. Biamonte J, Wittek P, Pancotti N, Rebentrost P, Wiebe N, Lloyd S (2017) Quantum machine learning. Nature 549(7671):195–202. [Online]. https://doi.org/10.1038/nature23474
2. Schuld M, Sinayskiy I, Petruccione F (2015) An introduction to quantum machine learning. Contemp Phys 56(2):172–185. [Online]. https://doi.org/10.1080/00107514.2014.964942
3. Cerezo M, Verdon G, Huang HY et al (2022) Challenges and opportunities in quantum machine learning. Nat Comput Sci 2:567–576. [Online]. https://doi.org/10.1038/s43588-022-00311-3

4. Preskill J (2018) Quantum computing in the NISQ era and beyond. Quantum 2:79. [Online]. https://doi.org/10.22331/q-2018-08-06-79
5. Nguyen HT, Krishnan P, Krishnaswamy D, Usman M, Buyya R (2024) Quantum cloud computing: a review, open problems, and future directions. [Online]. arXiv:2404.11420
6. Farhi E, Neven H (2018) Classification with quantum neural networks on near term processors. [Online]. arXiv:1802.06002
7. Kundu S, Ghosh S (2024) Security concerns in quantum machine learning as a service. [Online]. arXiv:2408.09562
8. Lloyd S, Schuld M, Ijaz A, Izaac J, Killoran N (2020)Quantum embeddings for machine learning. [Online]. arXiv:2001.03622
9. Schuld M, Killoran N (2019) Quantum machine learning in feature Hilbert spaces. Phys Rev Lett 122(4). [Online]. https://doi.org/10.1103/PhysRevLett.122.040504
10. Havlíček V, Córcoles AD, Temme K, Harrow AW, Kandala A, Chow JM, Gambetta JM (2019) Supervised learning with quantum-enhanced feature spaces. Nature 567(7747):209–212. [Online]. https://doi.org/10.1038/s41586-019-0980-2
11. Ghosh A, Ghosh S (2024) The quantum imitation game: reverse engineering of quantum machine learning models. [Online]. arXiv:2407.07237
12. Upadhyay S, Ghosh S (2023) Stealthy swaps: adversarial swap injection in multi-tenant quantum computing. [Online]. arXiv:2310.17426
13. Xiong Y, Hsieh C-J (2020) Improved adversarial training via learned optimizer. [Online]. arXiv:2004.12227
14. Sofer E, Shaked T, Chaux C, Shlezinger N (2025) Unveiling and mitigating adversarial vulnerabilities in iterative optimizers. [Online]. arXiv:2504.19000
15. Choudhury N, Mude CN, Das S, Tikkireddi PC, Tannu S, Basu K (2024) Crosstalk-induced side channel threats in multi-tenant nisq computers. [Online]. arXiv:2412.10507
16. Tan Y, Choudhury N, Basu K, Szefer J (2025) Qubithammer attacks: qubit flipping attacks in multi-tenant superconducting quantum computers. [Online]. arXiv:2504.07875
17. Ghosh A, Kundu D, Chatterjee A, Ghosh S (2024) Guardians of the quantum GAN. [Online]. arXiv:2404.16156
18. Kundu S, Kundu D, Ghosh S (2024) Evaluating efficacy of model stealing attacks and defenses on quantum neural networks. In: Proceedings of the great lakes symposium on VLSI 2024, pp 556–559
19. Fu Z, Zhao L, Zhang X, Xu Y, Huang G, Chen F (2025) CopyQNN: quantum neural network extraction attack under varying quantum noise. [Online]. arXiv:2504.00366
20. Xu C, Szefer J (2024)Security attacks abusing pulse-level quantum circuits. [Online]. arXiv:2406.05941
21. Kundu S, Ghosh S (2025) Adversarial data poisoning attacks on quantum machine learning in the NISQ era. [Online]. arXiv:2411.14412
22. Wang Z, Li J, Hu Z, Gage B, Iwasawa E, Jiang W (2023) Qumos: a framework for preserving security of quantum machine learning model. [Online]. arXiv:2304.11511
23. Ghosh A, Ghosh S (2024) AI-driven reverse engineering of QML models. [Online]. arXiv:2408.16929
24. Saeed SM, Wille R, Karri R (2019) Locking the design of building blocks for quantum circuits. ACM Trans Embed Comput Syst 18(5s). [Online]. https://doi.org/10.1145/3358184
25. Rehman A, Langford V, Liu Y (2025) Opaque: obfuscating phase in quantum circuit compilation for efficient IP protection. [Online]. arXiv:2502.16605

Chapter 4
Quantum Adversarial Machine Learning and Defense Strategies: Challenges and Opportunities

Eric Yocam, Anthony Rizi, Mahesh Kamepalli, Varghese Vaidyan, Yong Wang, and Gurcan Comert

Abstract As quantum computing continues to advance, the development of quantum-secure neural networks is crucial to prevent adversarial attacks. This chapter proposes three principles for quantum-secure design: (1) the use of post-quantum cryptography, (2) the implementation of quantum-resistant neural network architectures, and (3) the assurance of transparent and accountable development and deployment. These principles are supported by various quantum strategies, including quantum data anonymization, quantum-resistant neural networks, and quantum encryption. The chapter also identifies open issues in quantum security, privacy, and trust and recommends exploring adaptive adversarial attacks and auto-adversarial attacks as future directions. The proposed design principles and recommendations provide guidance for the development of quantum-secure neural networks, ensuring the integrity and reliability of machine learning models in the quantum era.

E. Yocam (✉) · M. Kamepalli · V. Vaidyan (✉)
The Beacom College of Computer and Cyber Sciences, Dakota State University, Madison, SD, USA
e-mail: eric.yocam@trojans.dsu.edu

V. Vaidyan
e-mail: varghese.vaidyan@dsu.edu

M. Kamepalli
e-mail: mahesh.kamepalli@trojans.dsu.edu

A. Rizi
Augusta University, Augusta, CA, USA
e-mail: arizi@augusta.edu

Y. Wang
Department of Computer Science, University of Idaho, Moscow, ID, USA
e-mail: stevew@uidaho.edu

G. Comert
Computational Data Science and Engineering Department,
North Carolina A&T State University, Greensboro, NC, USA
e-mail: gcomert@ncat.edu

M. Usman (ed.), *Quantum Robustness in Artificial Intelligence*, Quantum Science and Technology, https://doi.org/10.1007/978-3-032-11153-1_4

4.1 Introduction

A defense against quantum-enabled adversarial attacks requires rethinking conventional machine learning security, privacy, and trust. Quantum computing fundamentally alters the threat landscape of classical computing, expanding the set of possible adversarial attacks. New quantum computing defense strategies will be required to protect quantum machine learning (QML) models against an evolving threat landscape [1].

4.1.1 *Quantum-Evolved Threats*

First, advances in quantum computing hardware and algorithms have advanced dramatically. Second, several major technology companies and research laboratories now operate functional, quantum processors with dozens to over 1,000 qubits (the basic units of quantum information). A qubit is a two-level quantum-mechanical system that can exist in a superposition of the two states at the same time. Third, quantum systems being studied and developed are being prepared for practical use in early QML applications. Fourth, adversarial machine learning has shown, through demonstrated studies, that classical models are inherently vulnerable. Finally, quantum computers are poised to elevate adversarial attacks by executing more sophisticated optimization algorithms capable of efficiently discovering adversarial examples.

4.1.2 *Quantum Emergence*

Quantum computing is becoming increasingly relevant as technology advances. This growth is driven by public and private investment, which continues to spur innovation in quantum hardware and algorithms. IonQ initially brought an 11-qubit quantum computer into public use through the Amazon Braket cloud platform [2]. In October 2020, they introduced a 32-qubit version, marking a breakthrough for commercially available quantum systems. This wave of financial and research support is accelerating the global development and deployment of quantum technologies.

4.1.3 *Quantum-Driven Interest*

An expanding community of researchers, scientists, and engineers who work in quantum computing has translated into an increase in both the quality and diversity of research accomplished in this field. Previous research in quantum computing, specifically QML ([3–5]), has demonstrated considerable potential. However, there is also a potential harmful side to quantum computing, where a wide variety of malicious applications can be created, such as breaking cryptography and even facilitating the discovery of illegal drugs [6].

Recognizing the considerable potential translates into an abundance of applications, which makes it easy to see the multitude of ways quantum computing can and will be relevant—a disruptive technology across various verticals.

Quantum computing has also fostered a global research ecosystem that comprises researchers from different countries and organizations who continuously contribute to and engage in competition with each other, thereby creating a dynamic form of collaboration and international engagement.

4.1.4 Quantum Adversarial Threats

Quantum Adversarial Machine Learning (QAML) uses concepts of quantum mechanics to improve the performance and robustness of machine learning models under adversarial settings. Using the unique properties of quantum mechanics, QAML enables the development of new defenses against adversarial attacks. A quantum adversarial attack can be categorized according to the capability level of the adversary. A quantum attack is made possible by an adversary, such as a malicious organization, that may attempt to attack or exploit quantum algorithms and protocols. The adversarial attack can be categorized into three different levels of threat capability, according to the adversary's capabilities, as summarized in Table 4.1.

At the lowest capability level, an adversary might show basic disruptive behavior. This could involve adding random noise or changes to the quantum system to hurt the performance of quantum algorithms [7]. Another type of low-level threat is measurement attacks, where an adversary interferes with the measurement process to get information about the quantum state [8].

Moving up to an intermediate capability level, an adversary may exhibit a more sophisticated means of interference. The adversary might attempt to manipulate quantum gates to alter the behavior of quantum circuits. Moreover, the adversary at

Table 4.1 Adversarial threat capability levels

Threat capability level	Adversarial behavior
Lowest	Introducing random noise or perturbations into the quantum system, Measurement attacks
Intermediate	Manipulating quantum gates to alter the behavior of quantum circuits, Parameter tampering
Highest	Modifying or developing quantum algorithms to subvert the learning process, Employing advanced quantum techniques like quantum crypto-analysis

Table 4.2 Adversarial intentions and motivation levels

Intentions and motivations	Adversarial actions
Information gathering	Learning sensitive details about quantum state, model parameters, or training data for competitive advantage
Privacy violation	Breaching privacy of training data, Inferring sensitive information about individuals or organizations
Infrastructure compromise	Compromising security of quantum computing infrastructure

this upper level might engage in parameter tampering, which involves determining how to change the parameters in a quantum algorithm to influence the decision.

At the highest capability level, an adversary may have considerable knowledge and resources at the adversary's disposal to carry out attacks. An adversary can study potential designs of quantum algorithms and execute or create a quantum algorithm to prevent learning. The adversary can then employ state-of-the-art quantum methods, such as quantum cryptanalysis, to bypass conventional cryptographic security protocols. This may potentially jeopardize the security of QML systems on a larger scale [9]. In addition to understanding the level of threat capability of an adversary, the intentions and motivations associated with the adversary's actions must be taken into account, as summarized in Table 4.2.

First, an adversary can collect information to learn aspects of the quantum state, model parameters, or training data, thereby gaining a competitive advantage. Second, an adversary may bias the behavior of the quantum model to create bad misclassifications or biased predictions. Third, privacy violations may become an issue, as an adversary can compromise the privacy of training data or infer potentially sensitive information about individuals or organizations from the trained model. Finally, an adversary with significant resources may aim to completely compromise the security of quantum computing infrastructure under the most severe of circumstances.

4.1.5 Quantum-Evolved Threat Mitigation

Quantum-evolved threat mitigation may be employed to counter potential adversarial threats. Potential threat mitigation may include (1) reducing the impact of noise on quantum computations, (2) using randomization, (3) detecting tampering, and (4) applying quantum security computing. An example of mitigating the compromise of security for quantum computing infrastructure is through continuous monitoring and auditing of the quantum computing environment for anomalous activity (or signs of adversarial interference), thereby ensuring the security (along with integrity) of the entire quantum computing infrastructure. As the field of QML continues to evolve, research and best practices will be paramount not only in understanding the adversary's capability levels, intentions, and motivations levels, but also in mitigating adversarial threats and safeguarding quantum-based systems against adversarial attacks.

4.1.6 Quantum Landscape

Quantum computing uses qubits that can represent both 0 and 1 simultaneously due to superposition [10]. Leading companies, such as IBM, Google, D-Wave, and IonQ, are developing specialized quantum processors capable of controlling and manipulating qubits. The number of qubits [11, 12] is an important measure. However, quantum computing also faces a significant challenge pertaining to the number of qubits, such as high error rates, which limit its ability to execute a large number of operations before the error rate accumulates [13]. For example, to address high error rates, researchers are working on error correction codes, though they come with substantial overhead in terms of qubit count [14].

Various physical systems are also being explored for use in specialized quantum processors, including superconducting circuits, trapped ions, topological qubits, and photonics. Many of these systems require specialized cryogenic cooling to operate at temperatures near absolute zero [15].

The quantum software ecosystem has evolved rapidly alongside hardware advances. Several quantum programming languages have emerged for quantum software, including Qiskit, used on IBM systems, Cirq, used on Google systems, and Forest, used on Rigetti systems [16]. Microsoft has also developed comprehensive quantum software tools, including the Quantum Development Kit and other frameworks. Likewise, quantum algorithms have been developed, such as Shor's algorithm for factoring large numbers and Grover's algorithm for unstructured search [17, 18].

Quantum simulators serve as essential tools within the quantum software ecosystem. Quantum simulators enable researchers to experiment with quantum algorithms on classical computers, though they are typically limited to small-scale experiments. Major companies (e.g., IBM, Google, and Rigetti) offer some level of access to their quantum hardware via the cloud, allowing quantum developers and researchers to remotely run experiments without leaving their desktops.

Despite making significant progress, quantum computing still faces major challenges that need to be solved to reach its full potential. These include (1) developing effective error correction techniques that are crucial for building large-scale, fault-tolerant quantum computers; (2) increasing the number of qubits without losing quality; (3) finding practical problems that can benefit from quantum computing; (4) creating practical quantum algorithms; and (5) closing the gap between quantum and classical systems.

A mix of quantum and classical computing presents a hopeful near-term solution to this last challenge. Quantum processors manage certain computational tasks. Meanwhile, classical systems take care of program control, data management, and result processing.

4.1.7 Quantum Datasets

Quantum computing datasets have several inherent qualities that make them suitable for training QML models. These inherited qualities include high-dimensional data, quantum entanglement and superposition, and quantum system simulation [19]. High-dimensional data are complex for quantum computers to represent and attack, since all of the dimensions of the data must be represented. Quantum entanglement and superposition protect datasets against the introduction of noise. Quantum system simulation can be used to generate new QML algorithms (e.g., quantum cryptography and quantum key distribution).

There is no one-size-fits-all solution for protecting quantum computing datasets. However, by curating and preparing quantum computing datasets, the quantum computing dataset can be more resilient to adversarial attacks and preserve the integrity of quantum training data. The quantum computing dataset protections to consider include (1) the use of synthetic data, (2) the addition of noise to the data, and (3) the application of post-processing to the dataset [20].

First, generating artificial data or synthetic data can help create datasets that are more resistant to quantum attacks. For example, by designing synthetic data and adding it to the dataset, it can be tailored to resist known quantum attacks. Second, adding noise to the data, such as just enough to avoid degrading the quality of the dataset, can make it harder for quantum computing to break cryptographic algorithms. Finally, using post-processing techniques like adversarial training can improve the dataset's robustness, increasing its resistance to known quantum attacks.

4.1.8 Quantum Algorithms

Researchers have made important strides in creating quantum algorithms designed for machine learning tasks. The Quantum Support Vector Machine (QSVM) algorithm uses principles of quantum computing to perform tasks like binary classification. QSVM has shown a meaningful quantum advantage over traditional support vector machine algorithms for certain inputs.

Complementing these direct quantum adaptations, variational quantum algorithms have emerged as another promising approach. Variational quantum algorithms typically use parameterized quantum circuits along with a classical optimization routine. The Variable Quantum Eigensolver (VQE), which is a variational algorithm, has been used to efficiently estimate the ground state energy of molecules. This is an important task in quantum chemistry and materials science.

4.1.9 Quantum Challenges

Despite its potential, QML faces several significant challenges that need attention for practical use. Key issues include the limitations of Noisy Intermediate-Scale Quantum (NISQ) devices, a lack of quantum data, optimizing circuits for specific hardware, integrating classical and quantum systems, and broader trust and governance concerns.

First, one of the main challenges is working with NISQ devices. Quantum computers are game changers, but they have noise and are limited in the number of qubits they can use. This can make it difficult to implement and train complex QML models. For instance, traditional quantum error correction is not practical on NISQ devices because of the qubit requirements. This situation calls for different strategies, such as error mitigation techniques and designing noise-resilient algorithms.

Second, obtaining large amounts of high-quality quantum data remains a significant challenge. Classical machine learning typically requires sufficient amounts of data for machine learning to be effective; however, QML faces the additional challenge of a lack of quantum data. For example, in quantum chemistry settings, obtaining high-fidelity experimental data from quantum experiments can be both a resource-intensive and a technically challenging process.

Third, optimizing circuits for hardware is a significant task. For example, any quantum algorithms executed on charged superconducting qubits or trapped ions require work on circuit design, including the choice of gates, and decomposition of circuits involves decisions to obtain circuits that are compatible with the hardware limitations.

Finally, interfacing QML algorithms with classical systems and data processing pipelines remains a challenge due to the need to connect quantum computing to classical computing in a meaningful way, which is crucial for practical applications. For example, in QML workflows, it is essential to determine how to communicate effectively between classical and quantum components and to properly connect them to perform data processing and perform a task.

Beyond technical hurdles, recent research underscores the importance of addressing trust and transparency in quantum-enabled systems. This encompasses issues related to the spread and regulation of quantum computing capabilities [21] and the development of defensible design methodologies to enhance public and institutional trust in intelligent systems [22]. Therefore, QML challenges extend beyond engineering to encompass broader social and ethical considerations.

4.1.10 Quantum-Classical Adversarial Attacks

Quantum-classical adversarial attacks are a significant security issue. Machine learning quantum classifiers, similar to classical classifiers, are vulnerable to specially designed adversarial examples made by adding small disturbances to legitimate input

Table 4.3 Adversarial attack types

Attack type	Adversary's knowledge level
Data poisoning	White-box
Perturbation	White-box, Gray-box, Black-box
Membership inference	Gray-box, Black-box
Evasion	White-box, Gray-box, Black-box
Square	White-box, Gray-box, Black-box

data [23]. Practical defense strategies, like adversarial training, can be used to reduce known weaknesses.

Quantum-classical adversarial machine learning attacks are still evolving, given the limited resources and access available for quantum computing. However, several adversarial attacks are mentioned in classical machine learning studies, which can also be found in quantum-classical neural networks. These adversarial attacks are made against both classical and quantum-classical neural networks, depending on the adversary's knowledge of the neural network.

The levels of knowledge of the adversary are the white-box (with complete understanding of the target classification model), the black-box (without knowledge of the target classification model), and the gray-box (partial understanding of the target classification model) [24]. The differences between white-box, gray-box, and black-box attacks are not always distinctly separated. For example, an adversary may have partial knowledge of a target classification model, allowing them to conduct a gray-box attack on the classification even if they possess neither complete nor full knowledge about the underlying classification architecture and parameters for a quantum, classical, or quantum-classical neural network model.

The adversary's knowledge of either quantum, classical, or quantum-classical neural network models dictates the type of adversarial attack. Data poisoning, perturbation, membership inference, evasion, and squaring are all different types of adversarial attacks depending on the adversary's knowledge level, as shown in Table 4.3.

4.1.11 Quantum-Classical Data Poisoning Attack Type

A quantum-classical data poisoning attack occurs when someone injects harmful or corrupted data into the training dataset. This leads the model to learn incorrect patterns, which harms its performance or behavior [25]. Data poisoning can be a black-box or white-box attack, depending on how much the attacker knows about the model's structure and training process.

4.1.12 Quantum-Classical Perturbation Attack Type

A quantum-classical perturbation attack, or adversarial attack, involves adding small, often hard-to-detect changes to individual input samples. This can lead the model to misclassify those specific inputs [26]. These changes take advantage of the model's decision boundaries without altering the model itself or impacting other samples in the dataset. Perturbation attacks can be categorized as white-box, gray-box, or black-box attacks. In a white-box perturbation attack, the attacker has full access to the target model's structure and parameters. This allows for the creation of very effective adversarial examples.

In contrast, in a gray-box based quantum-classical perturbation attack, the attacker has partial knowledge of the target model, including its architecture or training data distribution. This limited information makes it harder to create effective adversarial examples compared to white-box scenarios, but it is still possible. Black-box perturbation attacks are the most difficult situation. Here, the attacker knows nothing about the model's internal structure and can only see the input-output relationships. Even with this limitation, black-box adversarial attacks can still happen using techniques like query-based optimization or transfer attacks with substitute models.

4.1.13 Quantum-Classical Inference Attack Type

A quantum-classical membership inference attack attempts to find out if a specific data sample was part of the model's training dataset [27]. The attacker takes advantage of the fact that machine learning models usually show different confidence levels or prediction patterns for training data compared to unseen data. To carry out the attack, the attacker collects samples from the same group as the training data and queries the target model to analyze its responses. Often, the attacker builds shadow models that work similarly to the target model. They train these shadow models on known datasets to identify patterns that separate member data from non-member data. Membership inference attacks are usually black-box attacks. They only require query access to the target model and do not need knowledge of its structure or parameters.

4.1.14 Quantum-Classical Evasion Attack Type

A quantum-classical evasion attack occurs when the test data are covertly altered by adding small perturbations to produce adversarial samples. For example, an attacker can add small changes to the input data and measure the impact on an MLP network for two datasets (e.g., CICIDS 2017, TRAbID 2017). By comparing the performance drops in accuracy, precision, recall, and F1 score on adversarial test data with original test data, 21.52% and 29.07% drops in performance may occur, respectively. Evasion attacks were white-box, gray-box, or black-box attacks based on the amount of knowledge that the attacker had regarding the target model.

4.1.15 Quantum-Classical Square Attack Type

A quantum-classical square attack is a black-box adversarial attack method that requires no knowledge of the target model's internal structure. It relies on query-based optimization [28]. The algorithm uses a randomized search strategy that updates perturbations in rounds to reduce the attack objective function. In each round, the algorithm randomly changes parts of the current perturbation, adds these changes to the input, and checks if the modification improves the attack's success. The perturbations must stay within certain norm limits (usually l-infinity or l2) to keep the adversarial examples unnoticed. The process keeps going until a successful adversarial example is found or a maximum number of queries is reached.

4.1.16 Quantum-Classical Inherent Vulnerabilities

Quantum-classical models can exhibit inherent vulnerabilities related to perturbations, accessibility, transferability, and data poisoning [29]. First, models demonstrate sensitivity to small input perturbations, lacking robustness against carefully crafted adversarial examples [30]. Second, accessibility vulnerabilities arise when models are exposed through public APIs and cloud services, providing attackers with query accesses [31]. Third, transferability enables adversarial examples crafted for one model to fool other models, even without direct access to the target system [32]. Finally, data poisoning vulnerabilities occur when attackers compromise training datasets with malicious samples, corrupting the model's learning process [33]. Additionally, models without batch normalization exhibit increased vulnerabilities to adversarial attacks [34].

4.1.17 Quantum-Classical Defense

Quantum-classical defense mechanisms include input transformation, randomization, and adversarial training. Input transformation defenses include pixel deflection, blurring, JPEG compression, dimensionality reduction, and principal component analysis [35]. The use of randomization defenses includes pixel dropout, image resizing, noise injection, model ensemble, and model parameter tuning [36].

Adversarial examples, including quantum-classical defense mechanisms, are designed so that the machine learning model is explicitly trained to increase the robustness of the model against an attack [37]. One such defense mechanism against adversarial examples is the use of JPEG compression during the preprocessing phase of image classification using mathematical models called neural networks. JPEG compression removes adversarial noise by filtering out perturbations typically inaudible or imperceptible to humans when applied to their perception via vision. Adversarial attacks typically consist of the addition of high-frequency artifacts; hence, the use of JPEG compression—via the discrete cosine transform (DCT)—effectively suppresses these distortions through suppression of high-frequency components such as abrupt intensity changes and slight color variations.

For example, in the JPEG compression research study, by retraining the model with different amounts of compression on the images iteratively, using a step size of 10, and repeating the process 9 times, the results demonstrated signs of improvement. Specifically, the CIFAR-10 dataset with adversarial data was included, along with the FGSM attack and a 0.02 perturbation, which improved the accuracy from 28.97% to 79.57%. As with a deep-fool type of attack, the accuracy improved from 27.44% to 82.71%. The GTSRB data set with adversarial data was included, along with the FGSM attack and a 0.08 perturbation, which improved the accuracy from 41.00% to 73.37%. Again, as with the Deep Fool attack, the accuracy improved from 68.19% to 91.70% [38].

4.1.18 Quantum-Classical Protections Applied to Quantum

Several security aspects are involved with protecting quantum-classical machine learning, including post-quantum cryptography, quantum data protection, quantum data privacy, verified quantum models, quantum malware detection, and algorithmic robustness.

4.1.19 Contribution

This chapter surveys QAML by demonstrating sophisticated adversarial attacks (e.g., WTC adversarial attacks) against a quantum-classical neural network model with insights for potential defenses for quantum neural network (QNN) models, reviewing the state of the art in quantum computing, QML, and defense strategies, addressing defense strategy opportunities and challenges associated with security, privacy, and trust, and recommending defense strategy, guidelines, and requirements necessary to construct quantum-secure machine learning models.

The remainder of the chapter is organized as follows. First, a discussion on characteristics of possible attacks against quantum-classical neural networks; second, an empirical study examining attacks against the hybrid model, with results and discussion. Third, a focus on the quantum model, analyzing its characteristics and vulnerabilities to attacks. Fourth, a set of proposed quantum adversarial defense strategies provides potential methods to mitigate risks. Fifth, coverage of crucial aspects of security, privacy, and trust requirements in quantum models. Sixth, quantum-secure recommendations, presenting defense strategies, guidelines, and requirements associated with the security, privacy, and trust of quantum models. Finally, open issues suggest future research directions and opportunities for further development.

4.2 Quantum-Classical Attacks

In this section, a review of a quantum-classical neural network model and adversarial attack characteristics that support topics covered within the quantum-classical empirical study.

4.2.1 Quantum-Classical Model

Exploration of quantum-classical neural network attacks not only provides insight into necessary protections but also suggests guidance for developing quantum neural network protections. A quantum-classical model is a classical model that has been partially quantized [39]. The quantum-classical model requires a defined quantum circuit. A quantum circuit is a computational routine that consists of quantum operations on data concurrent with classical real-time computation. Quantum-classical model is simulated on conventional computer hardware rather than accessing and using costly non-conventional quantum computer hardware. Quantum-classical models also expand the attack surface area for adversaries with new capabilities that quantum-classical models unlock. Careful design is needed to fully secure quantum-classical models against blended classical and quantum threats such as adversarial attacks.

4.2.2 White-Box Versus Black-Box Characteristics

A white-box adversarial attack is when an attacker knows the architecture, parameters, and training dataset of the quantum-classical model for an attack against the model [40]. In contrast, a black-box adversarial attack occurs when an attack does not have internal knowledge of the quantum-classical model. In this case, the attacker must rely on observing input-output patterns to estimate gradients or train substitutes.

4.2.3 Targeted Versus Untargeted Characteristics

A targeted attack is when a classifier within the quantum-classical model has been targeted for misclassification as part of the adversarial attack. In contrast, an untargeted attack is when there is no target classifier within the quantum-classical model specified as part of an adversarial attack [41]. A targeted adversarial attack aims to generate false data to fool the classifiers. For example, image classification utilizes a classifier, which is a set of rules used by machine learning and deep learning models to categorize data. However, the untargeted adversarial attack aims to cause any misclassification in the machine learning model. That is, the objective is to get the quantum-classical model to output something incorrect.

4.2.4 Distinct Versus Compounded Characteristics

Distinct adversarial attacks perturb individual inputs to cause misclassification by the quantum-classical model. In contrast, compounded attacks coordinate perturbations across batches of multiple inputs to bypass the quantum-classical model's defenses that consider each sample in isolation.

4.3 Quantum-Classical Empirical Study

In this section, an empirical study was conducted on a quantum-classical (or hybrid) neural network (HNN) model to show that defenses against new and more powerful adversarial attacks will be effective in QNN models. Likewise, the empirical study demonstrates the need to defend against more sophisticated adversarial attacks, and, by extension, defense strategies and associated design trade-offs to protect quantum neural networks.

Effectiveness and robustness are two different properties in machine learning models. Effectiveness refers to the overall performance of a model in a specific task. Robustness, however, is the performance of a model when presented with input (e.g., adversarial examples). In practice, a trade-off exists between the effectiveness and robustness of machine learning model properties. For example, increasing accuracy at times can lead to a decrease in robustness.

In contrast, improving robustness sometimes leads to a decrease in performance. The effectiveness of a white-box, targeted, compounded (WTC) adversarial attack against an HNN has not yet been studied. In the context of QAML, this type of quantum-classical attack investigates a WTC adversarial attack and its effectiveness, potentially providing insight into defenses for protecting quantum models.

First, this work evaluates white-box, targeted, distinct adversarial attacks through the development of a more complex white-box, targeted adversarial attack. In this case, the effectiveness can be measured for a combination of white-box-targeted distinct adversarial attacks (WTC) against an HNN model. Second, as with other technological advances, the WTC attacks will require additional insight and perspective to generate new defenses with countermeasures to protect an HNN model. Finally, effectiveness can be determined by measuring the accuracy of the prediction after the adversarial attack on the WTC has occurred.

The empirical study bridges an existing research gap through an examination of defense mechanisms that mitigate the impact of WTC adversarial attacks on HNNs. The majority of previous work has overlooked the effectiveness of defenses created with a precise aim of thwarting WTC attacks in classical-quantum models. Through a serious examination of how effective a defense would be after such an attack, this research aims to inform the development of improved and more flexible defense methods for protecting HNNs.

Likewise, the empirical study contribution includes (1) the development of HNN model given that the QNN model is somewhat limited due to the access and usage cost constraints placed on quantum computing hardware is (2) an alternative to access and usage cost constraints using the use of an intermediate neural network that links a classical neural network (CNN) model to a QNN model called the HNN model, (3) a HNN that is simulated on conventional computing hardware, (4) a HNN model, like the CNN model, is vulnerable to a WTC adversarial attack, and (5) an effective defense mechanism can be used to minimize the impact to the HNN model after a WTC adversarial attack.

4.3.1 Development Environment

PyTorch is a popular open-source framework and a preferred framework for deep learning and machine learning research. Cirq is also a popular open-source framework for generic interfaces, used as a simulation package for developing and executing the HNN model. Torchattacks is a PyTorch library with a set of popular adversarial attacks used to generate adversarial examples. The Torchattacks library will be used for the WTC adversarial attack parameters, specifically for the combinations of distinct attacks, including the fast gradient sign method (FGSM), projected gradient descent (PGD), and Carlini and Wagner (CW). These distinct attacks will be combined to produce a compound attack that includes FGSM + PGD, FGSM + CW, and CW + PGD.

FGSM attack injects small perturbations in the gradient direction of the loss function of the model to generate adversarial data [42]. The PGD method runs several iterations, and the displacement of each iteration in the gradient direction is limited in each iteration. By combining FGSM and PGD, the FGSM + PGD attack will have more ability to reduce accuracy. CW attack is similar to the PGD, which runs multiple iterations to find the best perturbation that maximizes the loss function to affect the accuracy. A combination of FGSM and CW as well as CW and PGD will also affect the model more in terms of misclassifying the test data via the FGSM + CW attack and CW + PGD attack.

4.3.2 Design Context

The Python packages used for the construction and evaluation of the HNN model include PyTorch, Cirq, and Torchattacks in Fig. 4.1.

Figure 4.1 shows the design of the HNN model without a defense mechanism. We can see a mismatch in the prediction of the digits in the last step of Fig. 4.1. Whereas Fig. 4.2 has a design context for the HNN model and processing with a

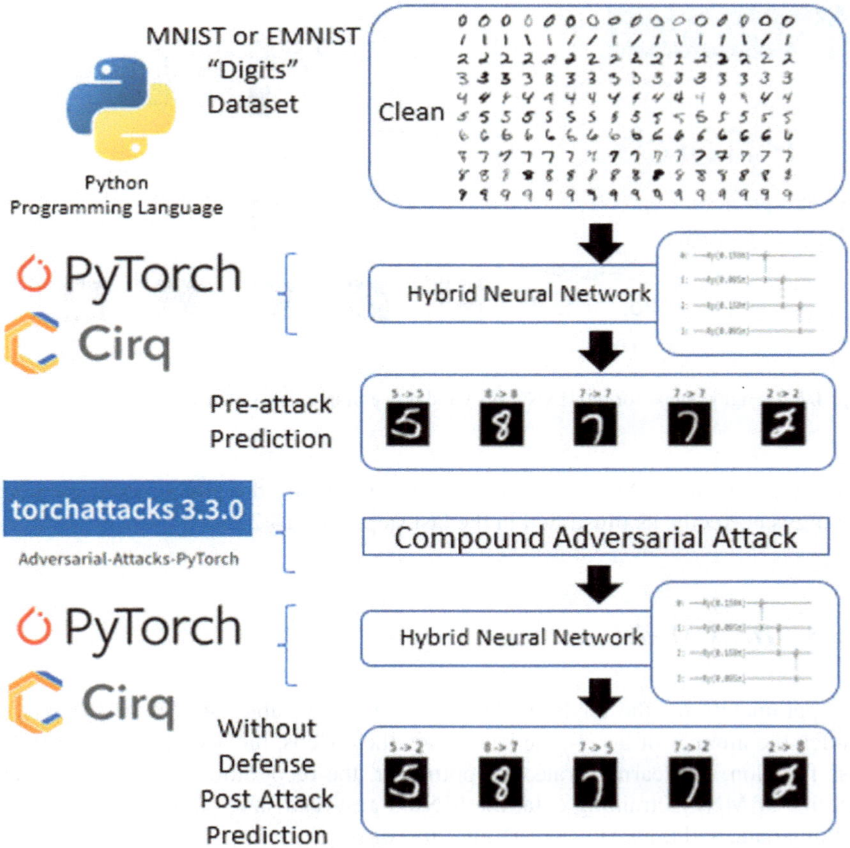

Fig. 4.1 Design context for the HNN model and processing without defense mechanism

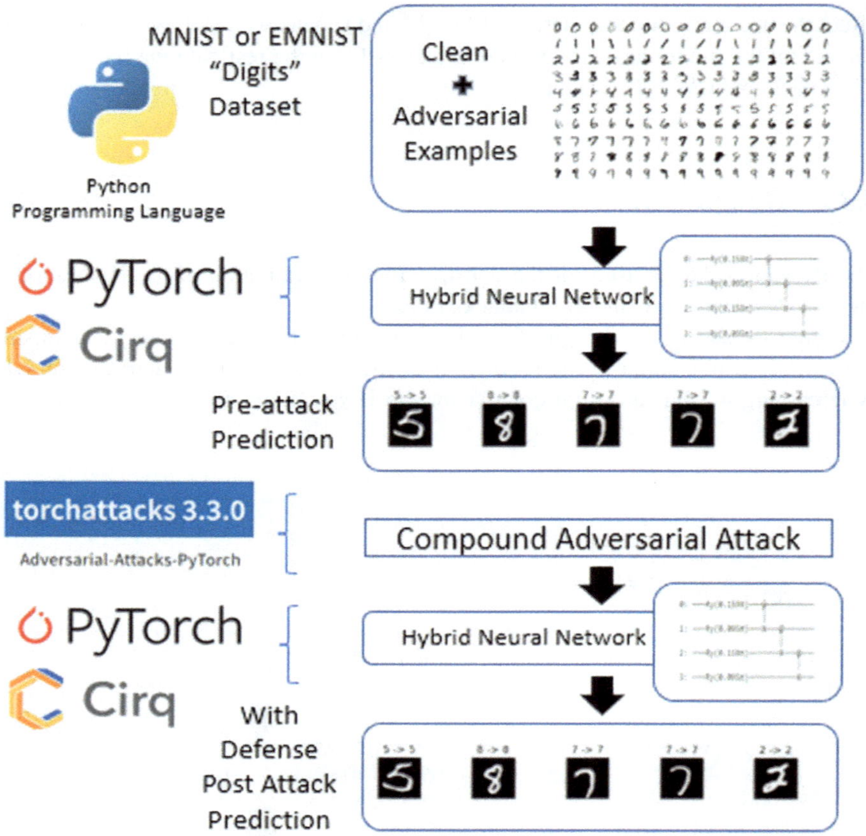

Fig. 4.2 Design context for the HNN model and processing with a defense mechanism

defense mechanism. After using a defense mechanism, the HNN model predicted the digits correctly, as illustrated in the last step of Fig. 4.2.

4.3.3 HNN Model

The parameters for the HNN model summarized in Table 4.4 include the type of model, the number of layers, the batch size, the epochs, the activation function, the loss function, the learning rate, the optimizer, the recognition of digit classifiers, the size of MNIST training data, the MNIST test data size, the EMNIST "Digits" training data, and the EMNIST "Digits" test data.

Table 4.4 Model parameters

No.	Model or data parameter	Value
1	Model type	HNN
2	Number of layers	4
3	Batch size	64
4	Epochs	10
5	Activation functions	Combination ReLU and Log softmax
6	Loss function	NLLoss
7	Learning rate	0.001
8	Optimizer	Adam
9	Digit classifier recognizes	[0, 9]
10	MNIST training data size	12,665
11	MNIST test data size	2,115
12	EMNIST "Digits" training data size	48,000
13	EMNIST "Digits" test data size	8,000

4.3.4 Quantum Circuit

A routine that computes using quantum operations on quantum data (e.g., qubits) represents a quantum circuit. The quantum circuit in Fig. 4.3 is implemented using the Cirq library developed by Google. This quantum circuit consists of two components, called rotation gates and angular gates. Rotation gates (cirq.ry) are applied to each qubit, alternating between the angles theta and phi. Entangling gates (cirq.CNOT) are applied between pairs of adjoint qubits. Rotation gates introduce a single-qubit operation, whereas entangling gates create a correlation among the states of multiple qubits and can manipulate the state of the individual qubits. The number of qubits belonging to the quantum circuit was determined based on the dimension of the model output. The circuit is executed dynamically from the learned parameters theta and phi.

The HNN combines a type of CNN, called a convolutional neural network, with a quantum circuit to enhance the model's performance. The HNN takes the convolutional neural network model as an input parameter and integrates it with the quantum circuit. The HNN initializes trainable parameters (theta and phi) for the quantum circuit, representing the rotation gates' angles. These parameters introduce additional non-linearity and expressiveness to the model. On the forward pass, input features are passed to the convolutional neural network, and the features learned therein are used as inputs to the quantum circuit. The quantum circuit carries out the quantum operations based on the learned parameters to compute the input features. The output of the quantum circuit is further computed and combined with the convolutional neural network model's predictions to obtain the output.

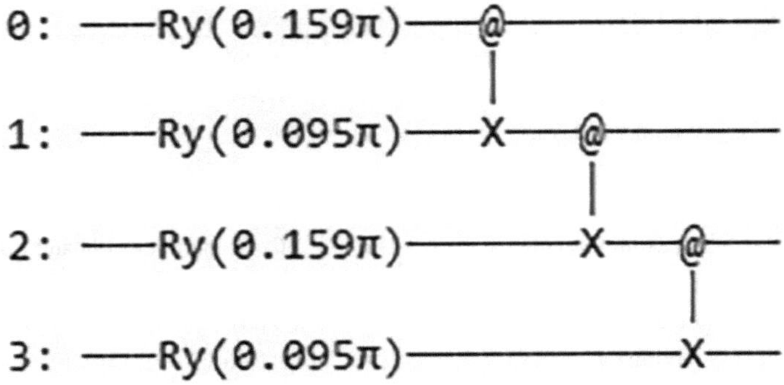

Fig. 4.3 The quantum circuit used for HNN model

4.3.5 Datasets

The HNN model datasets are MNIST and EMNIST "Digits". MNIST is a widely recognized and commonly used dataset for machine learning training and evaluation. The MNIST dataset consists of 60,000 training examples and 10,000 test examples, as shown in Fig. 4.4.

The EMNIST dataset is an extension of the MNIST dataset, consisting of 240,000 training examples and 40,000 test examples. Within EMNIST, there is a subset of the dataset called "Digits", illustrated in Fig. 4.5.

Fig. 4.4 The MNIST dataset sample

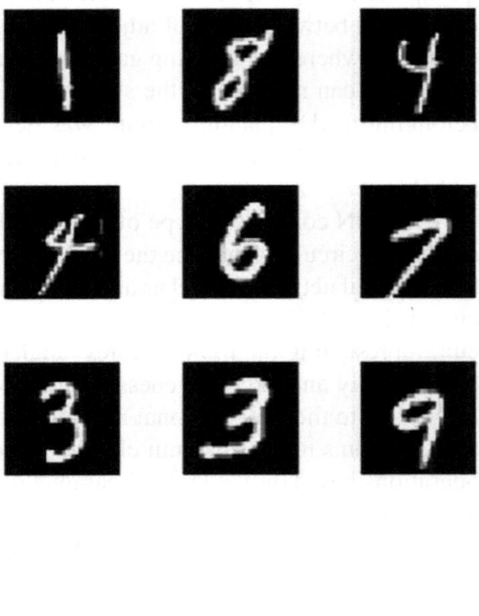

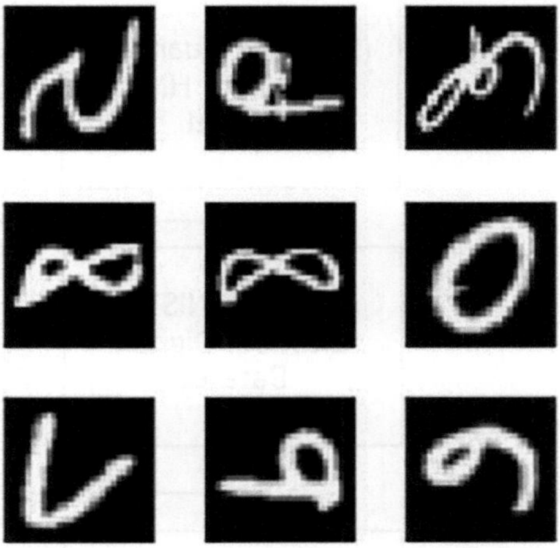

Fig. 4.5 The EMNIST "Digits" dataset sample

4.3.6 Procedures

The empirical study required the development of two procedures with a set of defined steps. An 11-step procedure was created for the HNN model without a defense mechanism, and a 16-step procedure was created for the HNN model with an adversarial training defense mechanism.

4.3.7 Procedure for the Model with No Defense Mechanism

A procedure was developed to take measurements of an HNN model without applying a defense mechanism. Figure 4.6, consists of eleven steps, including (1) load the MNIST or EMNIST "Digits" dataset, (2) define the quantum circuit for the HNN model, (3) create the HNN model and adversarial attacks, (4) fit the HNN model to the lowest loss value, (5) display the loss chart, (6) train the HNN model with no defense mechanism applied, (7) evaluate the HNN model, (8) calculate the compounded pre-attack prediction accuracy value, (9) apply attack against the HNN model, (10) evaluate HNN model, and (11) calculate post-attack prediction accuracy value.

4.3.8 Procedure for Model with Defense Mechanism

A sixteen step procedure was developed, as illustrated in Fig. 4.7, for the model with defense mechanism includes (1) defining the quantum circuit for HNN model, (2) creating HNN model and compounded adversarial attacks, (3) load MNIST or EMNIST "Digits" dataset representing the normal sample, (4) fit HNN model to lowest loss

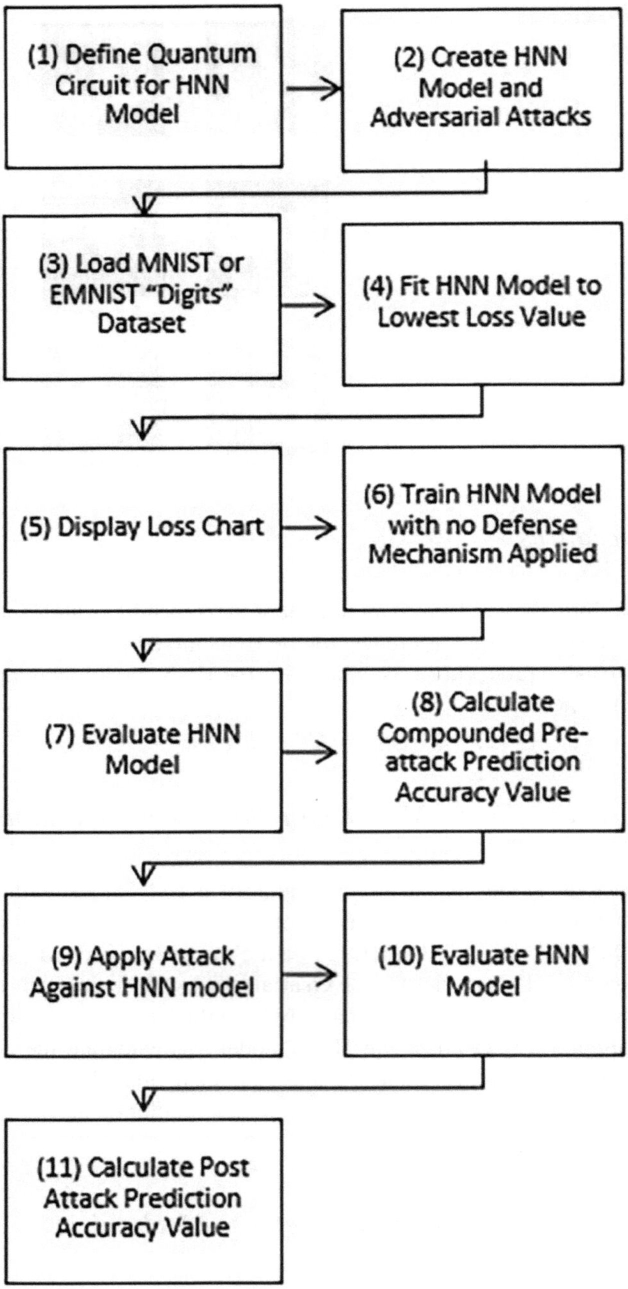

Fig. 4.6 Procedure for HNN model with no defense mechanism

value, (5) display loss chart, (6) train HNN model, (7) apply compounded adversarial attack against HNN model, (8) generate adversarial sample using compounded adversarial attack representing adversarial sample, (9) combine normal sample and adversarial sample into single dataset, (10) load combined dataset, fit HNN model to lowest loss value, and display loss chart, (11) retrain HNN model with defense mechanism applied, (12) evaluate HNN model, (13) calculate pre-attack prediction accuracy value, (14) apply compounded adversarial attack against HNN model, (15) evaluate HNN model, and (16) calculate post attack prediction accuracy value.

Adversarial training is recommended as the best defense against various types of adversarial attacks, including white-box, targeted, and distinct adversarial attacks. The adversarial training process differs from input transformation and randomization methods because it combines the standard sample and the adversarial sample. In contrast, the input transformation process only processes the input dataset, while the randomization process randomizes both the standard sample and the adversarial sample before combining them.

4.3.9 Empirical Results

Accuracy and Loss metrics (see Figs. 4.8 and 4.9) from the results of the HNN model for MNIST and EMNIST datasets with and without adversarial data are illustrated below. Both graphs show three attack scenarios with three types of model techniques. A pre-attack (or clean) accuracy indicates the dataset with no attack trained to a non-defense model. No defense attack accuracy indicates that the attacked dataset is trained with a non-defense model, and defense attack accuracy indicates that the attacked dataset is trained with a model that can resist adversarial attacks.

Tables 4.5 and 4.6 show the exact precision and loss of all metrics of the attacked HNN model after training and evaluating with and without defense models.

The pre-attack (or clean) accuracy for the HNN model with both datasets is almost 100%. Similarly, the loss of the HNN model is much less, nearly negligible. In Tables 4.5 and 4.6, a comparison of pre-attack (or clean), no defense attack, and with defense attack representing accuracy and loss values for the HNN model with both datasets.

The results demonstrate that the compound attacks with FGSM (e.g., FGSM+CW and FGSM+PGD) have a much larger effect on HNN performance than the CW+PGD combination. This is due to the fact that FGSM is a single-step, high-impact attack that adds extreme perturbations, which are compounded when used in combination with CW or PGD. For example, the FGSM+CW attack decreases accuracy on MNIST from 100% to 16% and on EMNIST to 12%, signifying a severe deterioration. In contrast, CW+PGD only creates a minor drop in accuracy (92% on MNIST, 77% on EMNIST), which means that its perturbations are less significant. Furthermore, loss measures support this result: CW+PGD generates much lower attack losses (e.g., 0.134 on MNIST) than FGSM+PGD (7.88) or FGSM+CW (3.93). These results suggest that even though CW+PGD is more stealthy, FGSM-based pairs are more

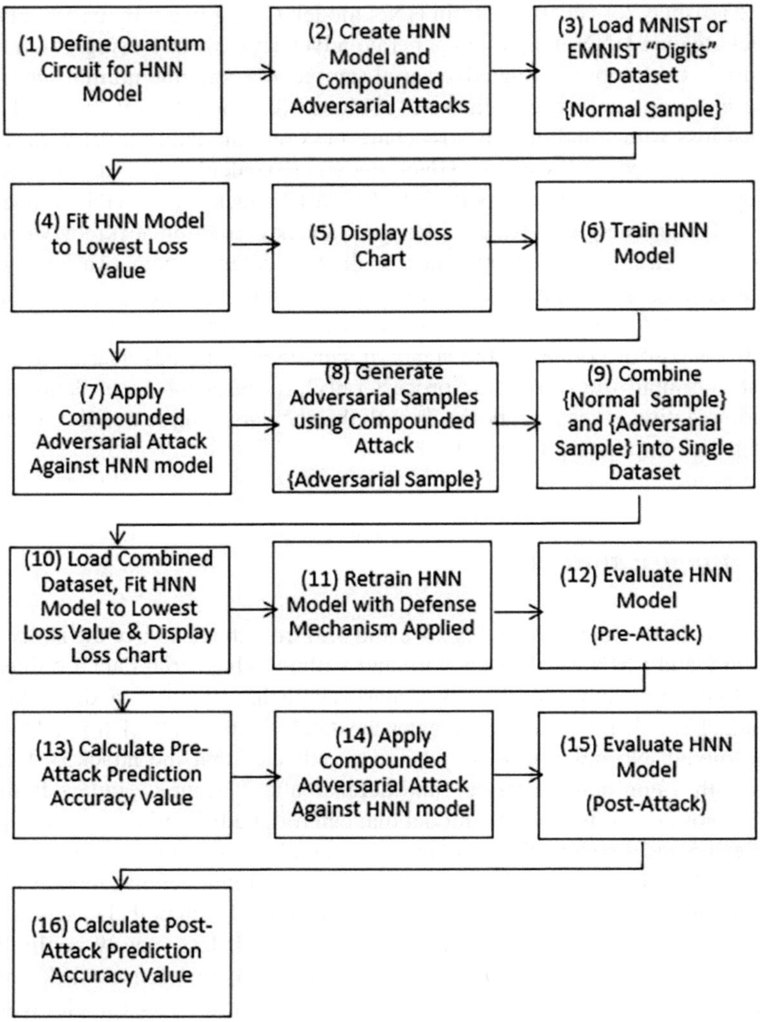

Fig. 4.7 Procedure for HNN model with a defense mechanism

invasive. The best-performing compounding attacks were those involving FGSM, especially combined with CW.

4.4 Quantum Defense Strategies

A QNN is a parameterized quantum model executed on quantum computer hardware [43]. However, access to quantum computer hardware and the associated computational expense of using it (e.g., measured in qubits) constrain the practical use of QNN. Several QNN defense strategies were surveyed, along with their associated opportunities and challenges. The QNN defense strategies include quantum-assisted defense against adversarial artificial intelligence (AI), quantum generative adversarial network, variational quantum generative adversarial network, and quantum error correction.

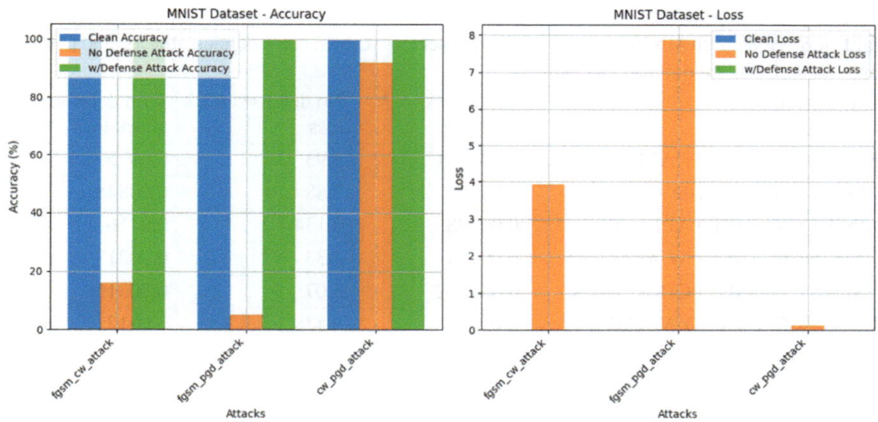

Fig. 4.8 HNN model with MNIST dataset result comparison illustration with three compounded attacks, comparing the accuracy and loss, with and without defense techniques

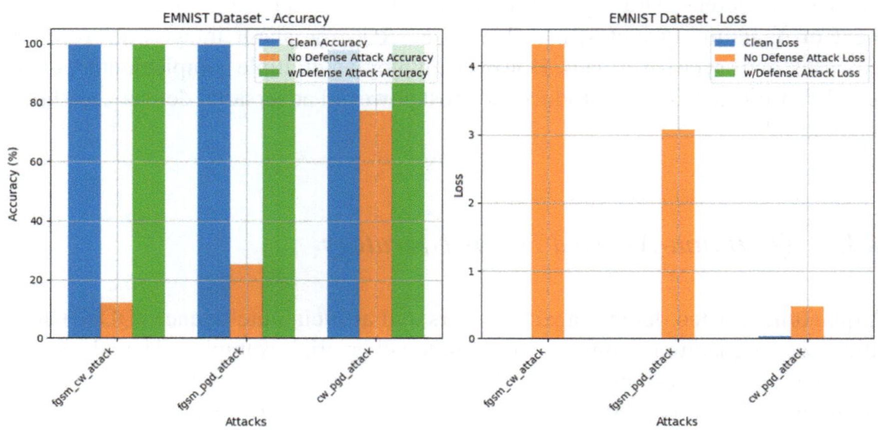

Fig. 4.9 HNN module with EMNIST "Digits" dataset result comparison illustration with three compounded attacks, comparing the accuracy and loss, with and without defense techniques

Table 4.5 HNN Model with MNIST and EMNIST dataset accuracy results, compounded attacks, and comparing with and without defense mechanisms

Attack name	Clean accuracy (%)	No defense attack accuracy (%)	With defense attack accuracy (%)
MNIST FGSM + CW ATTACK	100	16	100
MNIST FGSM + PGD ATTACK	100	5	100
MNIST CW + PGD ATTACK	100	92	100
EMNIST FGSM + CW ATTACK	100	12	100
EMNIST FGSM + PGD ATTACK	100	25	100
EMNIST CW + PGD ATTACK	98	77	100

Table 4.6 HNN with MNIST and EMNIST dataset loss results, compounded attacks, and comparing with and without defense mechanisms

Attack name	Clean loss	No defense attack loss	With defense attack loss
MNIST FGSM + CW ATTACK	0.000641	3.93	0.00759
MNIST FGSM + PGD ATTACK	0.0029	7.88	0.00848
MNIST CW + PGD ATTACK	0.00278	0.134	0.0116
EMNIST FGSM + CW ATTACK	0.00164	4.33	0.00218
EMNIST FGSM + PGD ATTACK	0.000562	3.07	0.00315
EMNIST CW + PGD ATTACK	0.0416	0.478	0.00131

4.4.1 Quantum Defense Strategy Trade-Offs

Every QNN defense strategy within the survey has unique strengths in the deployment of quantum attributes, but also raises concerns about integration, resource expenses, and crypto overloads. However, employing multiple complementary quantum defense strategies can produce the most powerful adversarial defense, as shown in Table 4.7.

4.4.2 Quantum-Assisted Defense Strategy

A quantum-assisted defense against adversarial artificial intelligence (QDAI) algorithm uses a quantum computer to generate adversarial examples. The adversarial examples are used to train the machine learning model to be more robust to adversarial attacks [44]. Opportunities include improving pattern recognition, detecting adversarial examples that classical models miss, and gaining insight through an HNN

Table 4.7 Defense strategy trade-offs

Defense strategy	Opportunities	Challenges
Quantum-assisted defense against artificial intelligence (QDAI)	Enhance pattern recognition, detect adversarial examples, and gain insights	Requires integration with classical systems and noisy quantum processors
Quantum Generative Adversarial Network (QGAN)/Variational Quantum Generative Adversarial Network (VQGAN)	Investigate quantum generative modeling capabilities	Requires significant quantum resources, complex training, and tuning of quantum circuits
Quantum Error Correction (QEC)	Enhance protections against errors/noise, stabilize quantum systems, and develop defense theory	Overhead for encoding/decoding quantum states and trade-offs associated with error correcting codes
Quantum Encryption (QE)	Obscure information about QNN models and data as well as make it difficult for adversaries to interpret or exploit quantum models	Additional cryptography complexity as well as key distribution and management obstacles

model approach. In contrast, the challenges include a quantum circuit that requires significant quantum resources and complex training and tuning.

4.4.3 Quantum GAN Defense Strategy

The defense of the quantum generative adversarial network (QuGAN) is a hybrid generative adversarial network (GAN) that uses a generator implemented in the quantum computer and a discriminator implemented on the classical computer [45]. An alternative defense to a QuGAN is the use of a variational quantum generative adversarial network (VQGAN), where the VQGAN is a variational GAN that utilizes a quantum circuit as its generator. The opportunities include investigating these quantum generative modeling capabilities to synthesize realistic adversarial examples for training, in addition to variational optimization. In contrast, the challenges include the significant quantum resources required, as well as the complex training and tuning of quantum circuits.

4.4.4 Quantum Error Correction Defense Strategy

An adversarial defense involves quantum error correction (QEC) that can potentially protect quantum information from errors. It encodes a single logical qubit into multiple physical qubits. This allows errors on individual physical qubits to be detected and corrected without affecting the logical qubit [46]. The potential lies in increasing protections against errors/noise that can provide attack vectors, stabilizing quantum systems, and advancing the continued evolution of defense theory. On the other hand, the overhead in encoding/decoding quantum states, as well as the trade-offs in error-correcting codes, are the difficulties.

4.4.5 Quantum Encryption Defense Strategy

An adversarial defense using quantum encryption leverages properties such as quantum superposition and entanglement to secure communications and data. Quantum encryption makes it more difficult for adversaries to intercept, measure, or extract usable information. For example, quantum key distribution (QKD) enables the secure sharing of secret keys between parties, allowing them to encrypt/decrypt data. The keys are encoded in quantum states that will be disturbed if improperly accessed [47]. The opportunities involve finding new ways to hide information about neural networks and data. This makes it harder for opponents to understand or use quantum models. On the other hand, the challenges include the added difficulty of cryptography and the issues related to key distribution and management.

4.5 Quantum Security, Privacy, and Trust

This section outlines the requirements for each design guideline related to quantum security, privacy, and trust.

4.5.1 Quantum Security Design Guidelines

The quantum security design is quantum encryption, quantum compartmentalization, and quantum anomaly detection, summarized in Table 4.8.

First, the requirements for the quantum encryption design guideline are post-quantum cryptography (PQC) and QKD. The use of PQC significantly increases the threshold for successful adversarial attacks [48]. The adversary must crack the post-quantum cryptosystem before attempting to analyze the model or construct adversarial inputs [49]. QKD enables the robust encryption and authentication of QML models, algorithms, and data, substantially increasing the difficulty for adversaries attempting to steal, manipulate, or construct adversarial examples without the necessary secret keys [50]. An underlying prerequisite for both PQC and QKD is a robust random number generator, which is a critical component for secure cryptography and key generation, and necessary for encrypting quantum models and data.

Quantum-resistant random number generators provide true randomness rooted in quantum properties to avoid patterns that adversaries could exploit [51].

Second, the requirements for the quantum compartmentalization design guideline, the defense-in-depth approach, are logical segmentation, isolation, and interface minimization [52]. For example, critical quantum model data and parameters can be compartmentalized with only the minimal necessary access, keeping them obscured from adversaries.

Finally, the requirements for the QAD design guideline are advanced quantum analytics and continuous monitoring. For example, tamper detection generates another layer of physical security monitoring to ensure the integrity of quantum processors themselves [53].

4.5.2 Quantum Privacy Design Guidelines

Quantum privacy-preserving computation protects quantum neural networks and is a valid defense mechanism [54]. To date, there has been limited research on how to provide privacy while considering the most effective privacy protection mechanisms. The guidelines for designing quantum privacy are quantum data anonymization, quantum-secure multiparty computation, and quantum access controls, as presented in Table 4.9.

Table 4.8 Quantum security design guidelines and requirements

Design Guidelines	Requirements
Quantum Encryption	Requires post-quantum crypto and QKD
Quantum Compartmentalization	Requires logical segmentation, isolation, and interface minimization
Quantum Anomaly Detection	Requires quantum sensors and measurement, quantum error correction, quantum data analytics, continuous monitoring, and hybrid classical quantum approaches

Table 4.9 Quantum privacy design guidelines and requirements

Design guidelines	Requirements
Quantum data anonymization	Requires quantum scrambling and differential privacy
Quantum-secure multiparty computation	Requires blind quantum computing and federated learning
Quantum access controls	Require attribute-based controls and dynamic policy orchestration

First, the guideline for the design of quantum data anonymization requires quantum scrambling and differential privacy. Differential privacy has been shown to protect user-sensitive information without significantly compromising the model's accuracy [55].

Second, quantum-secure multiparty computation requires blind quantum computing and federated learning. Quantum-secure multiparty computation refers to protocols, such as blind quantum computing, that enable different organizations to collaboratively train neural networks using confidential data in a privacy-preserving manner. Input data are not leaked. QML algorithms have built-in robustness against adversarial examples due to the quantum mechanical effects of superposition and entanglement [56].

Finally, quantum access control requires attribute-based controls and dynamic policy orchestration.

4.5.3 *Quantum Trust Design Guidelines*

The quantum trust design guidelines establish hardware-based trust roots, employ trust-based protocols, and limit trust boundaries, summarized in Table 4.10.

4.6 Quantum-Secure Recommendations

In this section, an overview of related work related to quantum-secure neural networks associated with recommended quantum security, privacy, and trust. Quantum-secure neural networks are designed to resist adversarial attacks. It is important to begin developing defenses against adversarial attacks on quantum neural networks now. As quantum computing continues to evolve, QNN models will become increasingly powerful and widely adopted, making it essential to ensure that they are developed using a specific set of quantum-secure design principles.

Table 4.10 Quantum trust design guidelines and requirements

Design guidelines	Requirements
Quantum hardware-based trust roots	Require quantum physical that cannot be cloned functions for hardware root of trust and tamper-resistant secure enclaves
Quantum trust-based protocols	Require QKD for secure channels and zero-knowledge proofs to validate operations
Quantum trust boundaries	Require least privilege permissions between compartments and validated communication channels

Table 4.11 Quantum-secure design principle no. 1

Quantum strategy	Quantum-secure design principle	Design guideline for quantum security
QDAI, QEC, QE	Use post-quantum cryptography	Quantum-encryption, Quantum-compartmentalization, Quantum-anomaly-detection

4.6.1 *Quantum-Secure Design Principle No. 1*

The first principle of quantum-secure design is to use post-quantum cryptography, summarized in Table 4.11.

QDAI can help protect machine learning models from quantum attacks. QEC can help protect quantum computers from errors. QE can help protect data from unauthorized access. QDAI, QEC, and QE are all crucial for developing quantum-secure systems, as they can help protect against the unique threats that quantum computers pose.

4.6.2 *Quantum-Secure Design Principle No. 2*

The second principle of quantum-secure design is to use the quantum-resistant neural network architecture summarized in Table 4.12.

QDAI can be used to protect the confidentiality of data that is used to train and deploy machine learning models. QGAN and VQGAN can be utilized to safeguard data privacy by generating synthetic data that are indistinguishable from genuine data. Using QDAI, QGAN, and VQGAN, developers can create privacy-preserving systems that protect data privacy.

Table 4.12 Quantum-secure design principle no. 2

Quantum strategy	Quantum-secure design principle	Design guideline for quantum security
QDAI, QGAN/VQGAN	Use quantum-resistant neural network architectures	Quantum-data-anonymization, Quantum-secure-multi-part-computation, Quantum-access-controls

Table 4.13 Quantum-secure design principle no. 3

Quantum strategy	Quantum-secure design principle	Design guideline for quantum security
QDAI, QGAN/VQGAN, QEC, QE	Transparent and accountable development and responsible deployment	Quantum-hardware-based-trust-roots, Quantum-trust-based-protocols, Quantum-trust-boundaries

4.6.3 Quantum-Secure Design Principle No. 3

The third principle of quantum-secure design is transparent and accountable development, and responsible deployment is summarized in Table 4.13.

Transparent and accountable development means that quantum systems should be developed and deployed in a manner that is open, transparent, and allows public scrutiny and oversight.

QDAI can be used to produce more reliable machine learning models. QDAI can protect models from adversarial or perturbation attacks, so as to enhance the accuracy and robustness of machine learning models. QGAN and VQGAN are both strong data synthesizing tools, which can be used in many applications (e.g., training of machine learning models and robustness testing for security systems). Developers can help to ensure that QGAN and VQGAN are used in a transparent and accountable manner to develop secure quantum systems. QEC and QE are both essential for developing quantum-secure systems. QEC can help protect quantum computers from errors, while QE can help protect data from unauthorized access. For example, with QEC and QE, developers can increase the reliability and security of quantum systems (e.g., reducing the risks associated with their deployment).

4.7 Quantum-Secure Open Issues and Future Direction

In this section, several open issues have been identified and recommendations have been made for further investigation in future work.

4.7.1 Quantum-Secure Open Issues

A couple of quantum security, privacy, and trust remain open. First, open issues in quantum security include understanding quantum trade-offs among accuracy, complexity, and security, as well as improving secure multiparty QML techniques and optimizing quantum cryptography schemes for machine learning applications. Second, open issues for quantum privacy are developing end-to-end quantum empirical studies and improving secure multiparty QML techniques. Finally, open issues for quantum trust include developing quantum theories and models that are robust against adversarial attacks and verifying the integrity of quantum data and models.

4.7.2 Quantum-Secure Future Direction

A recommendation for further investigation and future work consists of two potential areas to explore. Quantum-secure defense strategies will continue to evolve for quantum adversarial machine learning. Likewise, a set of comprehensive quantum-secure defense mechanisms has proven to be a challenging task. Quantum-secure defense strategies, not covered within this chapter, that are used to protect against more sophisticated types of QNN adversarial attacks, include adaptive adversarial attacks and auto-adversarial attacks. By exploiting the weak points of each quantum-secure defense strategy, an adversary will develop more sophisticated QNN adversarial attacks, such as adaptive adversarial attacks or auto-adversarial attacks. As a future direction, a recommendation is to explore adaptive adversarial attacks and auto-adversarial attacks since these areas may be worth broadening the investigation into quantum-secure defense mechanisms.

4.8 Conclusion

This chapter aims to help foster multidisciplinary research in the highly relevant domain of QAML. The empirical research has revealed that a QNN model is as susceptible to adversarial attacks as an HNN model. The results of the research, particularly with respect to adversarial attacks, have demonstrated that HNN models are very susceptible, necessitating more efficient defense mechanisms to safeguard QNN models.

In response to this vulnerability, a series of defense strategies, including a suite of QDAI, QGAN, QEC, and QE, was proposed. The trade-offs between the opportunities and challenges of each defense strategy were examined and summarized. These suggested defenses for quantum adversarial machine learning help inform future (long-term) and near-term investigations in this rapidly changing landscape.

The technical novelties of this chapter are three. First, it is an empirical vulnerability comparison of the QNN and HNN models to adversarial attacks. Second, it is a comprehensive set of defense mechanisms tailored to the novel challenges of QAML. Third, it is an in-depth comparison of the trade-offs for each defense approach so that researchers and practitioners can make an informed choice when applying the defenses.

As quantum computing evolves, it is increasingly necessary to design effective and viable countermeasures for preventing adversarial attacks on QML models. This chapter paves the way for future research on the subject, offering sound advice and suggestions for building quantum-secure machine learning protocols.

References

1. Miller DJ, Xiang Z, Kesidis G (2020) Adversarial learning targeting deep neural network classification: a comprehensive review of defenses against attacks. Proc IEEE 108(3):402–433. https://doi.org/10.1109/JPROC.2020.2970615
2. Feder T (2020) Quantum computing ramps up in private sector. Phys Today 73(11):22–25. https://doi.org/10.1063/PT.3.4612
3. Vaidyan VM, Rimal BP (2023) Hybrid quantum artificial intelligence electromagnetic spectrum analysis framework for transportation system security. J Hardw Syst Secur 8(1):1–11. https://doi.org/10.1007/s41635-023-00142-2
4. Vaidyan VM, Tyagi A (2022) Towards quantum artificial intelligence electromagnetic prediction models for ladder logic bombs and faults in programmable logic controllers. In: 2022 international conference on electronic systems and intelligent computing (ICESIC), pp 1–6. https://doi.org/10.1109/ICESIC53714.2022.9783543
5. Vaidyan VM, Tyagi A (2022) Hybrid classical-quantum artificial intelligence models for electromagnetic control system processor fault analysis. In: 2022 IEEE IAS global conference on emerging technologies (GlobConET), pp 798–803. https://doi.org/10.1109/GlobConET53749.2022.9872410
6. Mensa S, Sahin E, Tacchino F, Barkoutsos PK, Tavernelli I (2023) Quantum machine learning framework for virtual screening in drug discovery: a prospective quantum advantage. Mach Learn Sci Technol 4(1):015023. https://doi.org/10.1088/2632-2153/acb900
7. Zhou Y, Zhang P (2023) Noise-resilient quantum machine learning for stability assessment of power systems. IEEE Trans Power Syst 38(1):475–487. https://doi.org/10.1109/TPWRS.2022.3160384
8. Liu N, Wittek P (2020) Vulnerability of quantum classification to adversarial perturbations. Phys Rev A 101(6):062331. https://doi.org/10.1103/PhysRevA.101.062331
9. Akter MS, Shahriar H, Iqbal I, Hossain M, Karim M, Clincy V, Voicu R (2023) Exploring the vulnerabilities of machine learning and quantum machine learning to adversarial attacks using a malware dataset: a comparative analysis. arXiv:2305.19593. https://doi.org/10.48550/arXiv.2305.19593
10. Rietsche R, Dremel C, Bosch S et al (2022) Quantum computing. Electron Mark 32:2525–2536. https://doi.org/10.1007/s12525-022-00570-y
11. Khumalo M, Chieza H, Prag K et al (2022) An investigation of IBM quantum computing device performance on combinatorial optimisation problems. Neural Comput Appl. https://doi.org/10.1007/s00521-022-07438-4
12. Hu F, Lamata L, Sanz M, Chen X, Chen X (2020) Quantum computing cryptography: finding cryptographic Boolean functions with quantum annealing by a 2000 qubit D-wave quantum computer. Phys Lett A 384(10):126214. https://doi.org/10.1016/j.physleta.2019.126214
13. Google Quantum AI (2023) Suppressing quantum errors by scaling a surface code logical qubit. Nature 614:676–681. https://doi.org/10.1038/s41586-023-01607-3
14. Bravyi S, Dial O, Gambetta JM, Gil D, Nazario Z (2022) The future of quantum computing with superconducting qubits. J Appl Phys 132(16):160902. https://doi.org/10.1063/5.0082975
15. Zhao P, Lim YD, Li HY, Guidoni L, Tan CS (2021) Advanced 3D integration technologies in various quantum computing devices. IEEE Open J Nanotechnol 2:101–110. https://doi.org/10.1109/OJNANO.2021.3124363
16. Burgholzer L, Raymond R, Wille R (2020) Verifying results of the IBM Qiskit quantum circuit compilation flow. In: 2020 IEEE international conference on quantum computing and engineering (QCE), pp 356–365. https://doi.org/10.1109/QCE49297.2020.00051

17. Vandersypen LMK, Steffen M, Breyta G, Yannoni CS, Sherwood MH, Chuang IL (2001) Experimental realization of Shor's quantum factoring algorithm using nuclear magnetic resonance. Nature 414:883–887. https://doi.org/10.1038/414883a
18. Cafaro C, Mancini S (2012) On Grover's search algorithm from a quantum information geometry viewpoint. Phys A Stat Mech Appl 391(4):1610–1625. https://doi.org/10.1016/j.physa.2011.09.018
19. Cheng W, Zhang S, Lin Y (2023) Study on the adversarial sample generation algorithm based on adversarial quantum generation adversarial network. In: 2023 3rd international symposium on computer technology and information science (ISCTIS), pp 238–243. https://doi.org/10.1109/ISCTIS58954.2023.10213103
20. Huang C, Zhang S (2023) Enhancing adversarial robustness of quantum neural networks by adding noise layers. New J Phys 25(8):083019
21. Rosch-Grace D, Straub J (2022) Analysis of the likelihood of quantum computing proliferation. Technol Soc 68:101880. https://doi.org/10.1016/j.techsoc.2022.101880
22. Straub J (2022) Increasing trust in artificial intelligence with a defensible AI technique. In: 2022 IEEE applied imagery pattern recognition workshop (AIPR), pp 1–7. https://doi.org/10.1109/AIPR57179.2022.10092226
23. Lu S, Duan L-M, Deng D-L (2020) Quantum adversarial machine learning. Phys Rev Res 2(3):033212. https://doi.org/10.1103/PhysRevResearch.2.033212
24. Ayub MA, Johnson WA, Talbert DA, Siraj A (2020) Model evasion attack on intrusion detection systems using adversarial machine learning. In: 2020 54th annual conference on information sciences and systems (CISS), pp 1–6. https://doi.org/10.1109/CISS48834.2020.1570617116
25. Lin J, Luley R, Xiong K (2022) From adversarial examples to data poisoning instances: utilizing an adversarial attack method to poison a transfer learning model. In: ICC 2022—IEEE international conference on communications, pp 2351–2356. https://doi.org/10.1109/ICC45855.2022.9839219
26. Jagadeesha N (2022) Facial privacy preservation using fgsm and universal perturbation attacks. In: 2022 international conference on machine learning, big data, cloud and parallel computing (COM-IT-CON), vol 1, pp 46–52. https://doi.org/10.1109/COM-IT-CON54601.2022.9850531
27. Chatterjee S, Kungurtsev V (2023) Quantum solutions to the privacy vs. utility tradeoff. arXiv:2307.03118
28. Andriushchenko M, Croce F, Flammarion N, Hein M (2020) Square attack: a query-efficient black-box adversarial attack via random search. arXiv:1912.00049
29. Mehta C, Harniya P, Kamat S (2022) Comprehending and detecting vulnerabilities using adversarial machine learning attacks. In: 2022 2nd international conference on artificial intelligence and signal processing (AISP), pp 1–5. https://doi.org/10.1109/AISP53593.2022.9760580
30. Sadeghi K, Banerjee A, Gupta SKS (2020) A system-driven taxonomy of attacks and defenses in adversarial machine learning. IEEE Trans Emerg Top Comput Intell 4(4):450–467. https://doi.org/10.1109/TETCI.2020.2968933
31. Sadu C, Das PK (2021) A defense method against facial adversarial attacks. In: TENCON 2021–2021 IEEE region 10 conference (TENCON), pp 459–463. https://doi.org/10.1109/TENCON54134.2021.9707433
32. Bortsova G, González-Gonzalo C, Wetstein SC, Dubost F, Katramados I, Hogeweg L, Liefers B, Van Ginneken B, Pluim JP, Veta M, Sánchez CI, De Bruijne M (2021) Adversarial attack vulnerability of medical image analysis systems: unexplored factors. Med Image Anal 73:102141. https://doi.org/10.1016/j.media.2021.102141
33. Hu C, Hu Y-HF (2020) Data poisoning on deep learning models. In: 2020 international conference on computational science and computational intelligence (CSCI), pp 628–632. https://doi.org/10.1109/CSCI51800.2020.00111
34. Kurakin A, Goodfellow IJ, Bengio S (2016) Adversarial machine learning at scale. arXiv:1611.01236
35. Nesti F, Biondi A, Buttazzo G (2023) Detecting adversarial examples by input transformations, defense perturbations, and voting. IEEE Trans Neural Netw Learn Syst 34(3):1329–1341. https://doi.org/10.1109/TNNLS.2021.3105238

36. Wang F, Chai G, Li Q, Wang C (2020) Magnet with randomization defense against adversarial examples. In: 2020 international conference on networking and network applications (NaNA), pp 316–320. https://doi.org/10.1109/NaNA51271.2020.00061
37. Chen X, Zhang N (2020) Layer-wise adversarial training approach to improve adversarial robustness. In: 2020 international joint conference on neural networks (IJCNN), pp 1–8. https://doi.org/10.1109/IJCNN48605.2020.9206760
38. Das N, Shanbhogue M, Chen S-T, Hohman F, Chen L, Kounavis ME, Chau DH (2017) Keeping the bad guys out: protecting and vaccinating deep learning with jpeg compression. arXiv:1705.02900
39. Edwards D, Rawat DB (2020) Quantum adversarial machine learning: status, challenges and perspectives. In: 2020 IEEE international conference on trust, privacy and security in intelligent systems and applications (TPS-ISA), pp 128–133
40. Trochun Y, Stirenko S, Rokovyi O, Alienin O, Pavlov E, Gordienko Y (2021) Hybrid classic-quantum neural networks for image classification. In: 2021 11th IEEE international conference on intelligent data acquisition and advanced computing systems: technology and applications (IDAACS), vol 2, pp 968–972
41. Morovat K, Panda B (2020) A survey of artificial intelligence in cybersecurity. In: 2020 international conference on computational science and computational intelligence (CSCI), pp 109–115
42. Villegas-Ch W, Jaramillo-Alcázar A, Luján-Mora S (2024) Evaluating the robustness of deep learning models against adversarial attacks: an analysis with fgsm, pgd and cw. Big Data Cogn Comput 8(1):8. https://doi.org/10.3390/bdcc8010008
43. Li S, Neupane A, Paul S, Song C, Krishnamurthy SV, Roy-Chowdhury AK, Swami A (2018) Adversarial perturbations against real-time video classification systems. arXiv:1807.00458
44. Gong L, Liu Q (2023) Exploiting quantum noise for adversarial attack in quantum machine learning. Quantum Inf Process 22(1):45. https://doi.org/10.1007/s11128-022-03746-8
45. Stein SA, Baheri B, Chen D, Mao Y, Guan Q, Li A, Fang B, Xu S (2021) Qugan: a quantum state fidelity based generative adversarial network. In: 2021 IEEE international conference on quantum computing and engineering (QCE), pp 71–81. https://doi.org/10.1109/QCE52317.2021.00023
46. Balamurugan C, Singh K, Ganesan G, Rajarajan M (2021) Post-quantum and code-based cryptography: some prospective research directions. Cryptography 5(4). https://doi.org/10.3390/cryptography5040038
47. Tan TG, Zhou J, Sharma V, Mohanty SP (2023) Post-quantum adversarial modeling: a user's perspective. Computer 56(8):58–67. https://doi.org/10.1109/MC.2022.3218046
48. Xu Y, Raja K, Ramachandra R, Busch C (2022) Adversarial attacks on face recognition systems. In: Handbook of digital face manipulation and detection: from deepfakes to morphing attacks, pp 139–161. Springer International Publishing
49. Chen ACH (2023) Post-quantum cryptography neural network. In: 2023 international conference on smart systems for applications in electrical sciences (ICSSES), pp 1–6. https://doi.org/10.1109/ICSSES58299.2023.10201083
50. Hegde SB, Jamuar A, Kulkarni R (2023) Post quantum implications on private and public key cryptography. In: 2023 international conference on smart systems for applications in electrical sciences (ICSSES), pp 1–6. https://doi.org/10.1109/ICSSES58299.2023.10199503
51. Aji A, Jain K, Krishnan P (2021) A survey of quantum key distribution (QKD) network simulation platforms. In: 2021 2nd global conference for advancement in technology (GCAT), pp 1–8. https://doi.org/10.1109/GCAT52182.2021.9587708
52. Arias D, de Guzmán IGR, Rodríguez M, Terres EB, Sanz B, de la Puerta JG, Pastor I, Zubillaga A, Bringas PG (2023) Let's do it right the first time: survey on security concerns in the way to quantum software engineering. Neurocomputing 538:126199. https://doi.org/10.1016/j.neucom.2023.03.060
53. Ash-Saki A, Alam M, Ghosh S (2020) Improving reliability of quantum true random number generator using machine learning. In: 2020 21st international symposium on quality electronic design (ISQED), pp 273–279. https://doi.org/10.1109/ISQED48828.2020.9137054

54. Das S, Ghosh S (2023) Trojannet: detecting trojans in quantum circuits using machine learning. arXiv:2306.16701
55. Liu X, Xie L, Wang Y, Zou J, Xiong J, Ying Z, Vasilakos AV (2021) Privacy and security issues in deep learning: a survey. IEEE Access 9:4566–4593. https://doi.org/10.1109/ACCESS.2020.3045078
56. Watkins W, Chen S, Yoo S (2023) Quantum machine learning with differential privacy. Sci Rep 13. https://doi.org/10.1038/s41598-022-24082-z

Chapter 5
Adversarial Robustness in Quantum Machine Learning

Nicola Franco, Aman Saxena, Tom Wollschläger, Alona Sakhnenko, Stephan Günnemann, and Jeanette Miriam Lorenz

Abstract This chapter provides an advanced introduction to adversarial robustness in quantum machine learning, a rapidly evolving field that intersects quantum computing with adversarial techniques from classical machine learning. Designed for graduate students with a foundational background in quantum computing and machine learning, the material explores the vulnerabilities that quantum learning systems face when subjected to adversarial input manipulations. We begin by examining adversarial attack vectors, and reviewing exact and approximate verification techniques, including methods based on linear relaxation, convex programming, and hypothesis testing. Throughout the chapter, we examine quantum classifiers for classical and quantum data, introducing quantum differential privacy, and detailing noise-based certification. Next, we explore quantum-enhanced formal verification methods that combine mixed-integer programming with decomposition techniques, and present quantum randomized smoothing, which uses quantum amplitude estimation to certify robustness. In doing so, the material provides a comprehensive understanding of the principles underlying quantum adversarial robustness. The chapter concludes with introducing the open research questions that form the frontiers of this interdisciplinary area.

N. Franco (✉) · A. Sakhnenko · J. M. Lorenz
Fraunhofer Institute for Cognitive Systems IKS, Munich, Germany
e-mail: nicola.franco@iks.fraunhofer.de

A. Sakhnenko
e-mail: alona.sakhnenko@iks.fraunhofer.de

J. M. Lorenz
e-mail: jeanette.miriam.lorenz@iks.fraunhofer.de

A. Saxena · T. Wollschläger · S. Günnemann
School of Computation, Information and Technology and Munich Data Science Institute, Technical University of Munich, Garching, Germany
e-mail: aman.saxena@tum.de

T. Wollschläger
e-mail: tom.wollschlaeger@tum.de

S. Günnemann
e-mail: s.guennemann@tum.de

M. Usman (ed.), *Quantum Robustness in Artificial Intelligence*, Quantum Science and Technology, https://doi.org/10.1007/978-3-032-11153-1_5

5.1 Introduction and Motivation

The growing use of machine learning (ML) in both routine activities and high-stakes applications highlights the urgent need for trustworthy and dependable artificial intelligence systems. In classical ML, strong vulnerabilities have been found; for instance, deep-learning perception errors have contributed to autonomous-vehicle crashes [1], and large-language models can be manipulated to bypass safety protocols, potentially generating harmful content such as instructions for bio-weapons [2]. These instances motivate the development of *trustworthy AI*, demanding systems that can provide guarantees about their behavior and ensure that failure rates remain below acceptable risk thresholds.

This need for robustness extends directly into the accelerating field of Quantum Machine Learning (QML). As QML transitions from theoretical exploration to practical application, it will eventually find its way into high-stakes domains too. Quantum classifiers are being prototyped for tasks such as the classification of high-resolution image datasets, with potential applications in areas like medical-image triage and tumor detection [3]. Similarly, pilot projects in finance, chemistry, and cybersecurity are testing QML pipelines, thereby expanding the potential attack surface and the consequences of model failure [4, 5].

Like their classical counterparts, QML models suffer from adversarial fragility. Variational quantum circuits (VQCs), a common QML architecture, have been shown to be vulnerable to tiny, carefully crafted perturbations in input phases or amplitudes, which can flip model predictions with high confidence [6]. Beyond targeted attacks, QML systems must also contend with intrinsic quantum-specific noise, such as decoherence and shot noise. Interestingly, some forms of quantum noise can have complex effects on robustness, as we will see later in this chapter.

Adversarial Goal

In this chapter, we consider a classification model $f : \mathcal{X} \to \mathcal{Y}$, where $\mathcal{X} \subseteq \mathbb{R}^d$ represents the input space and $\mathcal{Y} = \{1, \ldots, K\}$ denotes the set of K possible classes. Given an input x with true label y, the goal of an adversary attacking the model f can be defined as finding a perturbation $\delta^\star$ such that the model predict a chosen target label y_t, formally defined as

$$\delta^\star = \arg\max_{\|\delta\| \leq \varepsilon} \mathbb{1}[f(x + \delta) = y_t], \tag{5.1}$$

where ε bounds the magnitude of the perturbation under a specified threat model. The symbol $\mathbb{1}[\cdot]$ is the indicator function: it returns 1 when the statement inside the brackets is true and 0 otherwise. Put simply, the attacker must work out a systematic way to tweak the input just enough to make the model produce the chosen wrong answer. The most common approaches employ the usage of gradients as they are also used to optimize the model itself.

Because the model is differentiable, the attacker can examine how the loss changes with tiny variations of the input and move the input in the direction that most increases that loss. Using the standard training loss $\mathcal{L}$ (e.g., cross-entropy) but evaluated against the target label y_t, an approximate perturbation can be written as

$$\delta \approx \eta \nabla_x \mathcal{L}\big(f(x), y_t\big), \tag{5.2}$$

where $\tilde{x} = x + \delta$ and $\nabla_x \mathcal{L}(f(x), y_t)$ is the gradient of the loss with respect to the input, $\eta > 0$ is chosen so that $\|\delta\| \leq \varepsilon$. The resulting vector δ is only an approximation to the optimal attack; repeating this gradient step while re-projecting onto the ε-ball often yields a stronger adversarial example.

Gradient-based attacks, such as the Fast Gradient Sign Method (FGSM) [7] and Projected Gradient Descent (PGD) [8], remain strong attacks when model gradients are accessible. In quantum neural networks, adversaries can generally use the same gradient-based attacks as in the classical setting [6]. However, phenomena like barren plateaus can obscure gradients [9], requiring adversaries to pivot to finite-difference approximations or black-box strategies. Representative quantum attack scenarios include *phase-flip attacks*, which apply small controlled phases within the encoding unitary, *parameter-shift attacks* that exploit training queries to leak gradient information, and *data-encoding hacks* that tamper with the classical-to-quantum feature map to cause distinct inputs to collide in Hilbert space [6, 10, 11].

Defense Strategies

In classical ML, heuristic defenses have been widely adopted. Adversarial training, where adversarial examples are generated and used during training, remains a common baseline approach [8]. However, many heuristic defenses have been shown to rely on obfuscated gradients, which can create a false sense of security when evaluated against non-adaptive or weak attacks [12]. A critical limitation of such heuristic approaches is that robustness is typically reported only against a fixed suite of known attacks, offering no guarantees against a more sophisticated, adaptive adversary [12].

These challenges highlight the pressing need for robust verification methods to ensure the reliability and trustworthiness of QML models, especially as they get integrated in critical applications. This chapter will therefore focus on **formal verification techniques** for assessing and guaranteeing adversarial non-robustness. Specifically, we will explore how quantum computing (QC) can be used to enhance verification in the classical ML and examine approaches that are guaranteeing verifiable robustness of QML. We conclude by examining promising future research directions in this rapidly evolving interdisciplinary area.

Formal Verification for Machine Learning

The common use of heuristic defenses against adversarial attacks in ML has highlighted their fundamental weaknesses, as these defenses are often overcome by more advanced attacks. Formal verification offers a strong alternative to overcome these limitations. Rather than depending exclusively on empirical tests, which can identify vulnerabilities but cannot conclusively prove their absence, formal verification seeks to analyze or design a defense mechanism to ensure it is provably secure against a defined threat model. This rigorous formulation allows for a high degree of certainty regarding the security of a model f for a given input x when subjected to a defined threat model $S(x)$. The result of this verification process provides a clear guarantee: either the model is robust or a counterexample, an adversarial attack, is discovered or proven to exist within the boundaries of the threat model. Consequently, this robust methodology can be employed to objectively evaluate and compare different models or defense strategies based on their provable robustness guarantees, moving beyond mere empirical performance against a limited set of known attacks.

A model f is deemed **robust** at a specific input x with respect to a threat model, or perturbation set, $S(x)$ if its prediction remains unaltered for all possible perturbations encompassed by that set. Mathematically, this is expressed as

$$\forall \tilde{x} \in S(x), \quad f(\tilde{x}) = f(x). \tag{5.3}$$

A system, such as a model architecture or a defense mechanism, is defined as being **formally exactly verifiable** if, for any given triple $(f, S(x), x)$, it is possible to algorithmically decide whether or not the model f is robust at input x under the threat model $S(x)$. Achieving this often necessitates solving an optimization problem. If $c = f(x)$ is the original prediction, the central task is to ascertain if any perturbed input $\tilde{x} \in S(x)$ exists such that $f(\tilde{x}) \neq c$. This can be framed as finding the minimum perturbation required to induce a misclassification, or more generally, by checking the sign of a defined robustness margin. Crucially, a precise definition of the **threat model** $S(x)$ is required, as it delineates the set of allowed adversarial perturbations around the input x.

Threat Model

The threat model $S(x)$ must meticulously specify the entire space of permissible perturbed versions $\tilde{x}$ of the original input x that an adversary can generate. It is only when we can ensure that no manipulation $\tilde{x} \in S(x)$ alters the prediction $f(\tilde{x})$ from the original $f(x)$ that we can declare the model provably robust for that specific input and threat model. In classical ML, particularly within the domain of image classification, these threat models frequently define perturbations $\delta = \tilde{x} - x$ that are considered "small" with respect to some L_p-norm (e.g., L_0, L_1, L_2, L_∞). The perturbation set under such a norm is given by

$$S_p(x, \epsilon) = \{\tilde{x} \in \mathcal{X} \mid \|\tilde{x} - x\|_p \leq \epsilon \text{ and } \tilde{x} \text{ is valid (e.g., pixel values in } [0, 1])\}. \tag{5.4}$$

For inputs that are matrices, which can arise in certain quantum contexts or when comparing model parameters, Schatten p-norms may be more relevant. In the field of quantum adversarial robustness, the distinct nature of inputs (quantum states) and operations (unitary transformations) necessitates different types of threat models. For instance, perturbations might be characterized by unitary operators U acting on a quantum state $|\psi\rangle$, such that the perturbed input $\tilde{x}$ corresponds to $U|\psi\rangle$. The operator U is typically constrained, for example, by being close to the identity operator I in some operator norm, or by being restricted to a specific set of quantum gates. The set $S_Q(x)$ would then be defined based on these quantum-specific constraints.

Exact Verification

Exact verification implies that the verification algorithm furnishes a **tight bound** on the robustness of the model. If the verifier certifies a triple $(f, S(x), x)$ as robust, then it is unequivocally guaranteed to be robust. Conversely, if the verifier determines that the model is not robust, this definitively implies that an adversarial example $\tilde{x} \in S(x)$ provably exists, and often, such an example can be explicitly constructed. For a neural network that produces logits $\mathbf{z}(x) = (z_1(x), \ldots, z_K(x))$, where the predicted class is $c = \arg\max_i z_i(x)$, robustness signifies that $z_c(\tilde{x}) > z_j(\tilde{x})$ for all classes $j \neq c$ and for all perturbed inputs $\tilde{x} \in S(x)$. The verification problem can then be formulated as checking whether the following minimum margin is positive:

$$\rho(f, x, S(x)) = \min_{\tilde{x} \in S(x)} \left(\min_{j \neq c} (z_c(\tilde{x}) - z_j(\tilde{x})) \right). \tag{5.5}$$

The model is robust if and only if this margin $\rho(f, x, S(x))$ is greater than zero. Approaches for exact verification often necessitate specific model formulations. For instance, networks composed of piecewise linear activation functions (such as ReLU, Leaky ReLU, and MaxPool) can be analyzed using methods like Mixed Integer Programming (MIP) or satisfiability modulo theories (SMT) solvers. In a MIP formulation, the network's architecture and the perturbation set $S(x)$ are encoded as a system of linear constraints, with integer variables representing the activation status of units like ReLUs. The objective function is then set to find an $\tilde{x} \in S(x)$ that either minimizes $z_c(\tilde{x}) - z_j(\tilde{x})$ for some target class $j \neq c$, or equivalently, maximizes $z_j(\tilde{x}) - z_c(\tilde{x})$. However, a significant challenge is that solving these optimization problems is generally NP-hard. Consequently, for neural networks of even moderate size, such as those with a few hundred neurons or more, exact verification rapidly becomes computationally intractable, thereby limiting its practical application to smaller models or highly specific scenarios.

Approximate Verification

To mitigate the substantial computational expense associated with exact verification, one can **relax** the underlying optimization problem. This relaxation leads to approximate verification methods, which can provide provable robustness guarantees but do not necessarily offer provable non-robustness. These methods compute a *lower bound* $\underline{\rho}(f, x, S(x))$ on the true minimum margin $\rho(f, x, S(x))$ as defined in (5.5), such that

$$\underline{\rho}(f, x, S(x)) \leq \rho(f, x, S(x)). \tag{5.6}$$

If this computed lower bound $\underline{\rho}(f, x, S(x))$ is positive, it logically follows that the true margin $\rho(f, x, S(x))$ must also be positive, and thus the model is provably robust for the given input x and threat model $S(x)$. However, if the computed lower bound $\underline{\rho}(f, x, S(x))$ is less than or equal to zero, the robustness status of the model remains undetermined. In such cases, the true margin ρ could still be positive (meaning the model is robust, but the approximation was too loose to prove it), or ρ could be non-positive (meaning the model is indeed not robust). This ambiguity implies that there can be samples that are genuinely robust for a given threat model, yet the approximate verifier is unable to certify this robustness. This phenomenon results in a **verification gap**, defined as

$$\rho(f, x, S(x)) - \underline{\rho}(f, x, S(x)) \geq 0. \tag{5.7}$$

A primary objective of research in approximate verification is to develop methods that yield tighter (i.e., larger) lower bounds $\underline{\rho}$ while remaining computationally feasible. Several common techniques are employed for approximate verification. These include **Linear Relaxation/Interval Bound Propagation (IBP)**, which over-approximates activation functions with linear bounds or interval arithmetic, offering speed at the cost of potentially loose bounds. Another approach is **Convex Relaxation/Semidefinite Programming (SDP)**, which formulates the verification problem as a convex optimization problem (e.g., an SDP). This often provides tighter bounds than linear relaxation but incurs a higher computational cost. Furthermore, methods based on **Lagrangian Duality/Abstract Interpretation** leverage principles from program analysis and optimization theory to derive sound bounds.

Stochastic Verification

Randomized smoothing [13, 14] is a technique used to construct a new classifier, g, from a base classifier, f, that is certifiably robust to ℓ_2 norm adversarial perturbations. The smoothed classifier $g(x)$ is defined as the class that is most likely to be returned by the base classifier f when its input x is perturbed by isotropic Gaussian noise.

Mathematically, this is expressed as

$$g(x) = \text{argmax}_{c \in \mathcal{Y}} P(f(x + \delta) = c) \tag{5.8}$$

where $\delta \sim \mathcal{N}(0, \sigma^2 I)$, $\mathcal{Y}$ is the set of possible classes, and σ^2 is the variance of the Gaussian noise. The robustness of g can be certified, meaning we can provably guarantee that $g(x)$ will not change its prediction for any perturbation Δ such that $\|\Delta\|_2 < R$, where R depends on σ and the probability $P(f(x + \delta) = c)$ [13, 15, 16].

5.2 Verification for Quantum Machine Learning

In this section, we discuss algorithms that enable the verification of QML models. Specifically, we focus on quantum classifiers operating on either classical or quantum data. Theoretical arguments based on concentration of measure expose the vulnerabilities of quantum classifiers [17, 18] to adversarial manipulation, making their use in safety-critical domains questionable. This highlights the need for developing QML models with provable robustness.

QML models that operate on quantum data typically begin by applying a VQC, followed by measurement. The probability of predicting class k corresponds to the likelihood of observing the k-th measurement outcome. When applied to classical data, the QML pipeline is extended to include a data encoding stage, where classical inputs are mapped to quantum states using a parameterized encoding circuit. The rest of the model, including variational processing and measurement, proceeds as before. Learning involves optimizing the circuit parameters to minimize the empirical loss over training data. Adversaries may attempt to manipulate the prediction of the model by altering either the classical input before encoding or the input quantum state itself.

Quantum Classifier on Quantum Data

Formally, a quantum classifier $F_\theta : \mathcal{D}_n \mapsto \Delta_K$ maps a quantum state $\sigma \in \mathcal{D}_n := \left\{\rho \in \mathbb{C}^{2^n \times 2^n} \,\middle|\, \rho^H = \rho \wedge \mathrm{Tr}(\rho) = 1\right\}$ to the K-dimensional probability simplex $\Delta_K := \{\boldsymbol{y} \in \mathbb{R}^K \mid \forall i : y_i \geq 0 \wedge \sum_{i=1}^K y_i = 1\}$. The classifier first applies a parameterized quantum operation $\mathcal{E}_\theta(\cdot)$ and then measures the output using a positive operator-valued measure (POVM) $\{\Pi_i\}_{i=1}^K$. Let $Y_\theta : \mathcal{D}_n \mapsto \{1, \ldots, K\}$ denote the class assignment, i.e., $Y_\theta(\sigma) = \arg\max_k F_\theta^{(k)}(\sigma)$, and $F_\theta^{(k)}(\sigma) := \Pr[Y_\theta = k \mid \sigma]$ the probability of predicting class k for input σ:

$$F_\theta^{(k)}(\sigma) = \mathrm{Tr}(\Pi_k \mathcal{E}_\theta(\sigma)). \tag{5.9}$$

Quantum Classifier on Classical Data

To perform classification on classical data, we first encode the input into a quantum state using a data encoding scheme, i.e., a quantum circuit parameterized by the input. A data encoding function $\sigma : \mathcal{X} \mapsto \mathcal{D}_n$ maps a data instance $x \in \mathcal{X}$ to the density matrix $\sigma(x) \in \mathcal{D}_n$ of an n-qubit system. The classifier then predicts class k with probability:

$$F_{\theta_1,\theta_2}^{(k)}(x) = \mathrm{Tr}(\Pi_k \mathcal{E}_{\theta_2}(\sigma_{\theta_1}(x))), \tag{5.10}$$

where θ_1 and θ_2 parameterize the data encoding and the variational layers, respectively.

Data-Encoding Methods

Schuld et al. [19] proposed a general representation for near-term quantum data encoding schemes, arguing that many existing strategies can be seen as special cases of this more general framework, often requiring only minor classical preprocessing. This formulation reveals that such encoding schemes naturally induce models that compute partial Fourier series over the input data. This structure provides insight into their expressivity and supports the argument that they can serve as universal function approximators.

Formally, the encoding is described by a parameterized quantum circuit $\rho_\theta : \mathcal{X} \rightarrow \mathcal{D}_n$, mapping classical input $x \in \mathcal{X}$ to a quantum state $\rho_\theta(x) \in \mathcal{D}_n$:

$$\rho_\theta(x) = U_L(x) W_\theta^{(L)} \cdots U_1(x) W_\theta^{(1)} \rho_0 W_\theta^{(1)\dagger} U_1(x)^\dagger \cdots W_\theta^{(L)\dagger} U_L(x)^\dagger. \quad (5.11)$$

Here, $U_l(x)$ are diagonal unitaries that encode the input (typically as $U_l(i,i) := \exp(-i\lambda_i^{(l)} x)$), and $W_\theta^{(l)}$ are parameterized, data-independent unitaries. This architecture captures many widely used encoding strategies, including those based on Hamiltonian eigenvalue encodings and classically intractable feature maps [19–21].

Most robustness guarantees in the literature rely on accurately estimating class probabilities. However, because measurements only reveal sampled outcomes from the quantum state, the true probabilities in Eqs. (5.9) and (5.10) cannot be directly observed. Instead, they are estimated via repeated execution of the quantum circuit. Consequently, although robustness bounds may be derived via deterministic methods, the guarantees themselves are inherently probabilistic.

5.2.1 Noise-Based Verification Against Perturbations in the Quantum Data

A key area of research in certifiable robustness for QML draws inspiration from RS techniques in classical ML [13–15]. Since QML models operate in inherently noisy environments, where noise processes like depolarization can degrade performance, training under noise can be treated analogously to RS, which may enhance model robustness and allow for deriving provable robustness guarantees. Motivated by this connection, recent work has explored incorporating controlled noise into quantum circuits to certify robustness [22–25]. In this section, we discuss several methods developed to achieve this.

We begin by outlining the notion of Differential Privacy (DP), a formal framework for analyzing how effectively an algorithm protects sensitive information contained within its input data. A key feature of DP is the introduction of randomization into the algorithm, which induces a probability distribution over its outputs. The privacy guarantees thus derived aim to ensure that small changes in the input dataset, such as modifying or removing a single entry, result in only minor changes in the corresponding output distribution. Notably, the expected output of a differentially private algorithm, averaged over the introduced randomization, tends to exhibit reduced sensitivity to perturbations in the input. This insight has inspired the development of randomized smoothing as a technique for certifying robustness against adversarial perturbations in machine learning models [26].

Quantum Differential Privacy

Building on the connection between (ϵ, δ)−DP and robustness of classical ML models, recent research has extended this framework to quantum classifiers by formulating analogous privacy definitions for quantum operations [27]. Intuitively, Quantum DP requires that the model's output distribution does not change significantly when the input quantum state is perturbed slightly, as measured via some suitable metrics, such as trace distance. This prevents an adversary from reliably distinguishing between similar inputs based on the classifier's output. Recall the quantum classifier $Y_\theta : \mathcal{D}_n \mapsto \{1, \ldots, K\}$, defined in Eq. (5.9) with corresponding prediction probabilities $F_\theta^{(k)}(\sigma) = \mathrm{Tr}(\Pi_k \mathcal{E}_\theta(\sigma))$. For any measurable subset $S \subseteq \{1, \ldots, K\}$, the probability that the classifier assigns a label in S is given by

$$\mathbf{Pr}(Y_\theta(\sigma) \in S) = \sum_{k \in S} F_\theta^{(k)}(\sigma) = \sum_{k \in S} \mathrm{Tr}(\Pi_k \mathcal{E}_\theta(\sigma)).$$

Further, let $\tau_{\mathcal{D}_n}$ be a metric on $\mathcal{D}_n$. The quantum classifier Y_θ is said to be (ϵ, δ)-differentially private if for all $\sigma, \rho \in \mathcal{D}_n$ with $\tau_{\mathcal{D}_n}(\rho, \sigma) < \tau_0$, and for all measurable subsets $S \subseteq \{1, ..., K\}$:

$$\mathbf{Pr}(Y_\theta(\sigma) \in S) \leq e^\epsilon \mathbf{Pr}(Y_\theta(\rho) \in S) + \delta. \tag{5.12}$$

In the special case of $\delta = 0$, this simplifies to the following condition on class-wise prediction probabilities:

$$\forall k \in \{1, ..., K\} : \quad F_\theta^{(k)}(\sigma) \leq e^\epsilon F_\theta^{(k)}(\rho). \tag{5.13}$$

If a classifier satisfies the above condition for all ρ, σ such that $\tau_{\mathcal{D}_n}(\rho, \sigma) < \tau_0$, then it also satisfies Eq. (5.12) with $\delta = 0$, and vice versa. Moreover, since $\tau_{\mathcal{D}_n}(\cdot, \cdot)$ is symmetric, this yields an equivalent condition:

$$\forall k \in \{1, ..., K\} : \quad e^{-\epsilon} \leq \frac{F_\theta^{(k)}(\sigma)}{F_\theta^{(k)}(\rho)} \leq e^\epsilon. \tag{5.14}$$

This symmetric form captures the idea that for any pair of inputs within distance τ_0, the model's prediction probabilities cannot vary too wildly. In essence, it quantifies the robustness of the classifier to small perturbations in the quantum input.

Robustness Verification through Depolarization Noise

Although depolarization noise is a recognized challenge in QC, evidence suggests it can be exploited to achieve robustness guarantees through DP [22]. Depolarization noise on a single qubit with probability p transforms a state ρ according to:

$$\mathcal{E}(\rho) = p \cdot \frac{I}{N} + (1 - p)\rho, \tag{5.15}$$

where $N = 2^n$ is the Hilbert space dimension. It has been shown in Zhou and Ying [27] that adding depolarization noise (Eq. 5.15) enables the derivation of ϵ−DP guarantees (Eq. 5.14). Let us consider the case of $\delta = 0$, a certified robustness bound linked to depolarization asserts that for a correctly classified input state σ with class probability $F_\theta^{(c)}(\sigma)$ and next highest class probability $F_\theta^{(k)}(\sigma)$ with $k \neq c$, a condition:

$$F_\theta^{(c)}(\sigma) > e^{2\epsilon} F_\theta^{(k)}(\sigma) \tag{5.16}$$

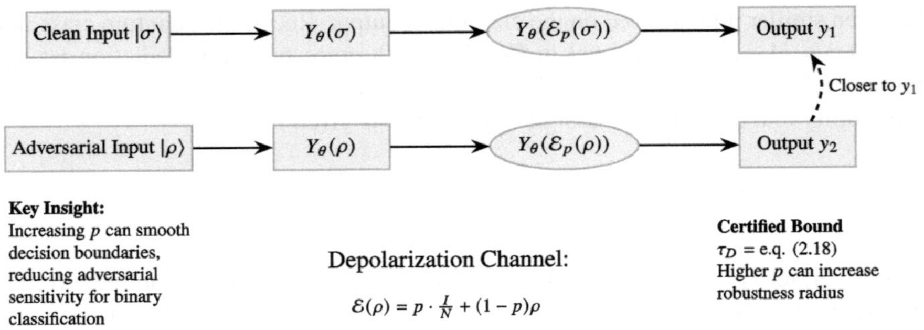

Fig. 5.1 The diagram illustrates how applying the depolarization channel $Y_\theta(\mathcal{E}_p(\sigma))$ to both clean and adversarial inputs causes their outputs to converge, with a higher noise parameter p increasing the certified robustness bound τ_D by smoothing the decision boundaries

must hold [22]. The term ϵ is related to the depolarization probability p and the measurement dimensionality D_{meas} by

$$\epsilon = \ln\left(1 + D_{meas}\frac{(1-p)\tau_D}{p}\right), \tag{5.17}$$

where τ_D is a certified trace distance given as

$$\tau_D = \frac{e^{\epsilon_{min}} - 1}{D_{meas}} \cdot \frac{p}{(1-p)} \quad \text{with} \quad \epsilon_{min} = \frac{1}{2}\ln\left(\frac{F_\theta^{(c)}(\sigma)}{F_\theta^{(k)}(\sigma)}\right). \tag{5.18}$$

Here, p is the depolarization probability defined as $1 - \prod_{i=1}(1-p_i)$, where p_i is the probability of the depolarization noise in the i-the depolarization channel. In this context, p implies a potential for noise to limit the effect of adversarial perturbations (Fig. 5.1).

Let the class prediction be given by $c = \arg\max_k F_\theta^{(k)}(\sigma)$, and choose $R(\sigma)$ such that

$$R(\sigma) < \frac{F_\theta^{(c)}(\sigma)}{\max_{k\neq c} F_\theta^{(k)}(\sigma)}. \tag{5.19}$$

Now, select τ_D in Eq. (5.18) such that

$$\epsilon = \frac{1}{2}\ln R(\sigma). \tag{5.20}$$

Then, for all states ρ satisfying $\tau_{\text{tr}}(\rho, \sigma) < \tau_D$, the following inequalities hold:

$$
\begin{aligned}
\forall c \in \{1, \ldots, K\} : F_\theta^{(c)}(\rho) &\geq e^{-\epsilon} F_\theta^{(c)}(\sigma), && \text{Eq. (2.14)}\\
&> e^{-\epsilon} R(\sigma) F_\theta^{(k)}(\sigma), && \text{Eq. (2.19)}\\
&\geq e^{-2\epsilon} R(\sigma) F_\theta^{(c)}(\sigma), && \text{Eq. (2.16)}\\
&= F_\theta^{(c)}(\sigma). && \text{Eq. (2.20)}
\end{aligned}
$$

Hence, the predicted class remains unchanged under perturbations of the input state within the trace distance τ_D. This establishes the robustness of the classifier against perturbations in the quantum state.

Despite the theoretical promise, experiments on multi-class classification tasks (using datasets like MNIST variants) reveal that adding depolarization noise did not consistently improve adversarial robustness [28]. In fact, for multi-class scenarios, increasing depolarization noise or the number of classes often led to decreased accuracy and robustness. The study concludes that alternative strategies, such as developing QNN architectures that inherently achieve better class separation, are crucial for robust QML, as simply relying on or increasing depolarization noise is not an effective defense in these more complex scenarios.

Constructing Optimal Quantum Noise Channels

In this section, we design noise channels that guarantee DP in quantum systems by limiting state distinguishability (through trace distance contraction) and ensuring minimal measurement probabilities. In particular, we try to construct the best possible noise channel through a unified framework. This unified approach not only protects privacy but also enhances robustness against adversarial attacks by optimizing the channel's parameters via semidefinite programming.

To formalize privacy guarantees using noise, we introduce a new class of quantum noise channels, named (α, γ)-channels [25]. These channels are defined by two conditions. First, the action of the channel on any pair or quantum states must contract the trace distance between them by at most a factor of α, that is:

$$
\tau(\mathcal{E}(\rho), \mathcal{E}(\sigma)) \leq \alpha \cdot \tau(\rho, \sigma). \tag{5.21}
$$

Second, the measurement probability for any outcome must not fall below a minimum threshold γ, such that

$$
\mathrm{Tr}[\Pi_k \mathcal{E}(\sigma)] \geq \gamma. \tag{5.22}
$$

Here, $\tau(\cdot, \cdot)$ denotes the trace distance between density operators, which quantifies how distinguishable two quantum states are.

Given these properties, we prove that any (α, γ)-channel guarantees DP with a privacy parameter ε given by the logarithmic expression:

$$
\varepsilon = \ln\left(1 + \frac{\alpha \cdot \tau_D}{\gamma}\right), \tag{5.23}
$$

where τ_D is an upper bound on the trace distance between the input states. This result generalizes prior known bounds and unifies them under a single framework.

Two notable cases fall under this formulation. First, the depolarizing channel of Eq. (5.15), which introduces isotropic noise into the system. Substituting into the general DP bound, this channel satisfies:

$$\varepsilon = \ln\left(1 + \frac{(1-p)\cdot \tau_D \cdot D}{p}\right), \tag{5.24}$$

where D is the dimension of the measurement space. Second, for channels based on random unitary rotations, such as those induced by random parameters in rotation gates, we can retrieve the bound:

$$\varepsilon = \ln\left(1 + \frac{\tau_D}{t^n}\right), \tag{5.25}$$

where t encapsulates the effective noise level introduced by the random rotations and n is the number of qubits.

An important consequence of the privacy guarantee is its implication for robustness. Rearranging Eq. (5.23) yields a direct way to compute the maximum certifiable distance as

$$\tau_D = \frac{(e^{\varepsilon_{\min}} - 1)\cdot \gamma}{\alpha}, \tag{5.26}$$

with $\varepsilon_{\min}$ as in Eq. (5.23). This expression relates robustness directly to the output confidence of the classifier and the parameters of the noise channel.

Consider a single input state $\sigma(x)$ that is first processed by the pretrained circuit U and afterwards by an unknown noise channel $\mathcal{E}$. For brevity, we denote $\sigma_D = U\sigma U^\dagger$ and abbreviate this by σ. After these two steps, the state is measured with the POVM $\{\Pi_k\}$, giving class–probabilities:

$$F_\theta^{(k)}(\sigma) = \mathrm{Tr}\big[\Pi_k\, \mathcal{E}(U\sigma U^\dagger)\big] \equiv \mathrm{Tr}\big[\Pi_k\, \mathcal{E}(\sigma)\big]. \tag{5.27}$$

For the true label c, we define the robustness ratio:

$$B = \frac{F_\theta^{(c)}(\sigma)}{\max_{k\neq c} F_\theta^{(k)}(\sigma)}. \tag{5.28}$$

Maximizing B under a fixed privacy budget ε is equivalent to the following optimization problem formulation:

$$\begin{aligned} \max \quad & B, \\ \text{s.t.} \quad & \mathcal{E}(.) \text{ is a quantum channel}, \\ & \forall \sigma, \rho \quad \tau(\mathcal{E}(\sigma), \mathcal{E}(\rho)) \leq \alpha \tau(\sigma, \rho), \\ & \forall k \qquad F_\theta^{(k)}(\sigma) \geq \gamma. \end{aligned} \tag{5.29}$$

The first constraint enforces Completely Positive and Trace-Preserving (CPTP), the second encodes trace–distance contractivity, while the third guarantees a minimum confidence for every class. This formulation delivers the optimal noise channel $\mathcal{E}$ that maximizes robustness while strictly respecting all physical and privacy-related requirements.

To find optimal channels that maximize B, we need to reformulate Eq. (5.29) as a semidefinite program (SDP). Because the ratio B leads to a non-convex objective, we instead optimize a tractable surrogate: the correct-class probability $F_\theta^{(c)}(\sigma)$. Additionally, we represent a quantum channel $\mathcal{E}$ via its Choi matrix $J_\mathcal{E}$, a positive semidefinite operator that encodes all actions of $\mathcal{E}$. The optimization maximizes the output probability for the correct class subject to constraints ensuring that the channel remains CPTP. A compact SDP that designs the optimal CPTP noise channel acting after the classifier is defined as

$$\begin{aligned} \max \quad & F_\theta^{(c)}(\sigma) \\ \text{s. t.} \quad & J_\mathcal{E} \succeq 0, \\ & \mathrm{Tr}_{\mathrm{out}}\big[J_\mathcal{E}\big] = I_{\mathrm{in}}, \\ & J_\mathcal{E} - \frac{\alpha}{D} I \succeq 0, \\ & \mathrm{Tr}_{\mathrm{out}}\big[(\Pi_k \otimes I) J_\mathcal{E}\big] \succeq \gamma I \quad \forall k. \end{aligned} \tag{5.30}$$

Here $J_\mathcal{E}$ is the Choi matrix of the channel $\mathcal{E}$, D is the input-output Hilbert-space dimension, α controls contractivity, and γ enforces a minimum confidence on every POVM element. These constraints include the positive semidefiniteness of $J_\mathcal{E}$, the normalization condition $\mathrm{Tr}_{\mathrm{out}}\big[J_\mathcal{E}\big] = I_{\mathrm{in}}$, and linear matrix inequalities encoding the bounds on contractivity and measurement minimums. The SDP provides a principled, implementable method for crafting noise channels that are optimal in terms of privacy and robustness to adversarial attacks.

Classical Robustness via Quantum Noise Channels

So far, we have discussed robustness methods for QML that focus on perturbations within the quantum state space. However, QML models are frequently applied to classical data that is first embedded into quantum states using encoding schemes that are frequently non-linear. The highly non-linear nature of these encodings makes it difficult to translate robustness guarantees expressed in terms of quantum states into meaningful bounds on the original classical inputs.

To address this challenge, we discuss here a framework for certifying robustness to adversarial perturbations in the classical input space [24]. This approach constructs quantum noise channels that, when applied after the data encoding gates in the quantum circuit, effectively emulate the effect of RS on the classical inputs. Essentially, the proposed ansatz for the data encoding scheme enables the interpretation of the resulting QML model as RS applied to some base classifier.

For simplicity, let us consider the effect of a phase damping channel on the RZ encoding of a real-valued input $x \in \mathbb{R}$, i.e., applying the gate $RZ(x)$ to a fixed quantum state ρ:

$$x \mapsto RZ(x)^\dagger \rho RZ(x) := \begin{bmatrix} \exp(\iota\frac{x}{2}) & 0 \\ 0 & \exp(-\iota\frac{x}{2}) \end{bmatrix} \begin{bmatrix} \rho_{11} & \rho_{12} \\ \rho_{12}^* & \rho_{22} \end{bmatrix} \begin{bmatrix} \exp(-\iota\frac{x}{2}) & 0 \\ 0 & \exp(\iota\frac{x}{2}) \end{bmatrix}$$
$$= \begin{bmatrix} \rho_{11} & \rho_{12}\exp(-\iota x) \\ \rho_{12}^*\exp(\iota x) & \rho_{22} \end{bmatrix}.$$

Recall that the phase damping channel $\mathcal{E}_{PD}^{(\lambda)}$ acts on the state ρ as follows:

$$\rho := \begin{bmatrix} \rho_{11} & \rho_{12} \\ \rho_{12}^* & \rho_{22} \end{bmatrix} \mapsto \mathcal{E}_{PD}^{(\lambda)}(\rho) := \begin{bmatrix} \rho_{11} & \rho_{12}\sqrt{1-\lambda} \\ \rho_{12}^*\sqrt{1-\lambda} & \rho_{22} \end{bmatrix}. \tag{5.31}$$

Now, consider the smoothed data encoding block obtained by composing $\mathcal{E}_{PD}^{(\lambda)} \circ RZ$. This composition transforms the input as follows:

$$\begin{aligned} x &\mapsto \mathcal{E}_{PD}^{(\lambda)} \circ RZ(x)(\rho) \\ &= \begin{bmatrix} \rho_{11} & \rho_{12}\sqrt{1-\lambda}\exp(-\iota x) \\ \rho_{12}^*\sqrt{1-\lambda}\exp(\iota x) & \rho_{22} \end{bmatrix} \\ &= \begin{bmatrix} \rho_{11} & \rho_{12}\mathbb{E}_{z\sim\mathcal{N}(0,-\ln(1-\lambda))}[\exp(-\iota(x+z))] \\ \rho_{12}^*\mathbb{E}_{z\sim\mathcal{N}(0,-\ln(1-\lambda))}[\exp(\iota(x+z))] & \rho_{22} \end{bmatrix} \\ &= \mathbb{E}_{z\sim\mathcal{N}(0,-\ln(1-\lambda))}[RZ(x+z)(\rho)]. \end{aligned}$$

For any linear functional $\mathcal{F}:\mathcal{D}_2 \mapsto \mathbb{R}$,

$$\mathcal{F}(\mathcal{E}_{PD}^{(\lambda)} \circ RZ(x)) = \mathcal{F}(\mathbb{E}_{z\sim\mathcal{N}(0,-\ln(1-\lambda))}[RZ(x+z)(\rho)]) \tag{5.32}$$
$$= \mathbb{E}_{z\sim\mathcal{N}(0,-\ln(1-\lambda))}[\mathcal{F}(RZ(x+z)(\rho))]. \tag{5.33}$$

Since any quantum classifier is a linear function on the encoding space of density matrix, $\mathcal{D}_2$, Eq. (5.32) shows that a quantum classifier with a phase damping channel can be expressed as applying RS to the classifier without the noise channel. Saxena et al. [24] extend this idea to find smoothing noise channels corresponding to any smoothing distribution for the general encoding ansatz from Schuld et al. [19] defined in Eq. (5.11).

However, we provide a robustness guarantee for a specific realization of smooth data encoding, as shown in Fig. 5.2. Consider an N-qubit system with L sequential repetitions of data encoding blocks separated by data-independent parameterized circuits $\{W_l(\theta)\}_{l=1}^{L+1}$. Each data encoding block encodes data into N parallel single-qubit subsystems, followed by a phase damping noise channel $\mathcal{E}_{PD}^{(\lambda)}$. We consider this setup as a binary classifier, and therefore, the final measurement process involves only two observables. Denote $y_\theta(x;\lambda)$ as the probability that this classifier predicts class 1 when applied to input x. Suppose the classifier predicts class 1 (class 0) at $x \in \mathbb{R}$, i.e., $y_\theta(x;\lambda) \geq y_0 > 0.5$ ($y_\theta(x;\lambda) \leq y_0 < 0.5$). Then, the prediction of the classifier remains unchanged for all perturbed inputs $\tilde{x} \in \mathbb{R}$ such that,

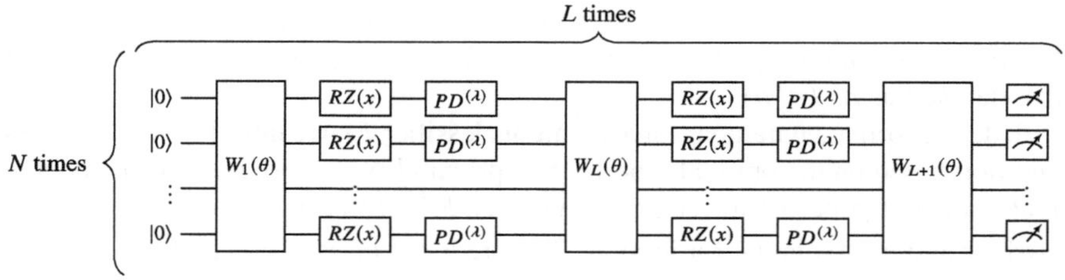

Fig. 5.2 Smooth Quantum Classifier using Phase Damping noise channels. This classifier corresponds to smoothing applied to a special case of near-term encoding proposed by Schuld et al. [19] defined in Eq. 5.11. The robustness guarantees derived using the framework of Saxena et al. [24] are presented in Eq. (5.11)

$$||\tilde{x} - x|| < \frac{1}{\sqrt{NL}}\sqrt{\ln(\frac{1}{1-\lambda})}\Phi^{-1}(y_0)$$

where $\Phi^{-1}(\cdot)$ is the inverse CDF of the standard Gaussian distribution $\mathcal{N}(0, 1)$, i.e., $\Phi(x) := \int_{-\infty}^{x} \frac{1}{\sqrt{2\pi}} \exp(-\frac{z^2}{2})dz$. Importantly, this robustness bound is *black-box*, meaning that it does not depend on the specific design choice of the variational unitaries $\{W_l\}_{l=1}^{L+1}$, their parameters θ, or the details of the measurement (POVM). In practice, we estimate y_0, a high-probability lower bound (respectively, upper bound) of $y_\theta(x; \lambda)$, using Clopper–Pearson confidence intervals for the Binomial distribution. This results in stochastic robustness guarantees.

Additionally, one can derive a non-trivial upper bound on the trace distance between the encoded quantum states corresponding to x and $\tilde{x}$. Let $\rho_{L+1}(x; \lambda)$ denote the density matrix of the quantum state encoding x just before measurement, and let $\tau_{tr}(\cdot)$ denote the trace distance. Then,

$$\tau_{tr}(\rho_{L+1}(x; \lambda), \rho_{L+1}(\tilde{x}; \lambda)) < 2\Phi\left(\frac{\sqrt{NL}}{2\sqrt{\ln\left(\frac{1}{1-\lambda}\right)}}||\tilde{x} - x||\right) - 1.$$

5.2.2 *Hypothesis Testing and SDP-Based Approaches to Robustness*

In contrast to noise-injection methods for robustness verification, another class of approaches seeks to establish formal guarantees using statistical and optimization-theoretic tools. This includes hypothesis testing–based frameworks [29]. In binary quantum hypothesis testing (QHT), a measurement operator M (with $0 \leq M \leq I_d$) is used to reject H_0 in favor of H_1. Given an input σ, the type-I error probability, corresponding to rejecting H_0 when it is true, is defined as $\alpha(M; \sigma) = \text{Tr}[\sigma M]$, while the type-II error probability, corresponding to accepting H_0 when H_1 is true, is given by $\beta(M;\rho) = \mathrm{Tr}[\rho(I_d - M)]$. Under the Neyman-Pearson framework, the objective is to minimize $\beta(M;\rho)$ subject to $\alpha(M;\sigma) \leq \alpha_0$, with the optimal type-II error denoted as $\beta^*_{\alpha_0}(\sigma,\rho)$.

To derive formal guarantees for a quantum classifier F_θ, we introduce a robustness condition based on the optimal type-II error probability, β^*. So, let σ be the benign input state and ρ be the adversarial state. Let $F_\theta^{(k)}(\sigma)$ be the probability that the classifier assigns to class k for input σ. The predicted class is $c = \mathrm{argmax}_k F_\theta^{(k)}(\sigma)$. Suppose we have bounds on the class probabilities for the benign state σ:

(i) The probability of the true class c is at least p_A, i.e., $F_\theta^{(c)}(\sigma) \geq p_A$.
(ii) The probability of any other class is at most p_B, i.e., $\max_{k\neq c} F_\theta^{(k)}(\sigma) \leq p_B$.

If the two conditions hold, then it is guaranteed that

$$\mathrm{argmax}_k F_\theta^{(k)}(\sigma) = \mathrm{argmax}_k F_\theta^{(k)}(\rho) \tag{5.34}$$

for any ρ with:

$$\beta^*_{1-p_A}(\sigma,\rho) + \beta^*_{p_B}(\sigma,\rho) > 1. \tag{5.35}$$

Thus, by verifying that the adversarial state ρ satisfies this inequality, one certifies that the classification decision remains invariant despite the perturbation. This bridge between state discrimination (via quantum hypothesis testing) and classification allows us to provide a formal and operationally meaningful guarantee on the noise tolerance of quantum classifiers.

The QHT Robustness Bound

QHT is traditionally used to optimally distinguish between two quantum states. However, this technique can also be reversed to certify that when two states are sufficiently similar (in terms of trace distance), any classifier will yield the same output for both. This leads to closed-form conditions that are provably optimal in guaranteeing that a quantum classifier remains invariant under small input perturbations.

Given $F_\theta : \mathcal{D}_n \mapsto \{1, \ldots, K\}$ and the benign state σ (or its purification ψ_σ) the classifier outputs the label c with confidence:

$$F_\theta^{(c)}(\sigma) \geq p_A > p_B \geq \max_{k\neq c} F_\theta^{(k)}(\sigma), \tag{5.36}$$

with $p_A + p_B = 1$ in the binary (or dominant class) setting. In the binary setting, robustness is guaranteed if the corresponding optimal type-II error:

$$\beta^*_{1-p_A}(\sigma,\rho) > \tfrac{1}{2}, \tag{5.37}$$

and here we show that for pure states, an equivalent closed-form holds.

In quantum information theory, any mixed state can be viewed as arising from a larger pure state. Given a density operator σ on a Hilbert space $\mathcal{H}$, a purification is a pure state $|\psi_\sigma\rangle$ in an enlarged Hilbert space $\mathcal{H} \otimes \mathcal{A}$ (with $\mathcal{A}$ being an ancillary system) such that

$$\sigma = \mathrm{Tr}_{\mathcal{A}}\left(|\psi_\sigma\rangle\langle\psi_\sigma|\right). \tag{5.38}$$

In practice, this means that any mixed state can be "lifted" to a pure state in a larger space, allowing one to use properties and techniques that are simplest for pure states (for example, relating fidelity to the overlap of state vectors). The choice of purification is not unique, but all purifications of σ yield the same reduced state on $\mathcal{H}$, making this construct especially useful for analyses and proofs in quantum robustness and classification. Namely, if the benign and adversarial states are pure, $\psi_\sigma, \psi_\rho \in \mathcal{H}$, then one obtains the fidelity (squared overlap) bound:

$$\left|\langle\psi_\sigma \mid \psi_\rho\rangle\right|^2 > \frac{1}{2}\left[1 + \sqrt{g(p_A, p_B)}\right], \tag{5.39}$$

where

$$g(p_A, p_B) = 1 - p_B - p_A(1 - 2p_B) + 2\sqrt{p_A\, p_B\,(1 - p_A)(1 - p_B)}. \tag{5.40}$$

In the special case $p_B = 1 - p_A$ this can be written compactly as

$$\left|\langle\psi_\sigma \mid \psi_\rho\rangle\right|^2 > \frac{1}{2} + \sqrt{\frac{p_A(1 - p_A)}{2}}. \tag{5.41}$$

Using the relation between fidelity and the Bures metric [29], one immediately obtains an equivalent robustness condition in terms of the Bures distance d_B:

$$d_B(\psi_\rho, \psi_\sigma) < \left[2 - \sqrt{2\left(1 + \sqrt{g(p_A, p_B)}\right)}\right]^{\frac{1}{2}}. \tag{5.42}$$

Moreover, for pure states the trace distance satisfies:

$$\tau(\psi_\rho, \psi_\sigma)^2 = 1 - \left|\langle\psi_\sigma \mid \psi_\rho\rangle\right|^2, \tag{5.43}$$

so that the robustness condition is equivalently expressed as

$$\tau(\psi_\rho, \psi_\sigma) < \left[\frac{1}{2}\left(1 - \sqrt{g(p_A, p_B)}\right)\right]^{\frac{1}{2}} \tag{5.44}$$

which is equivalent to Eq. (5.35). These closed-form conditions are shown to be both sufficient and necessary (in the regime $p_A + p_B = 1$) for ensuring that an adversarially perturbed input is classified the same as the benign input.

5.3 Quantum-Enhanced Formal Verification

Ensuring the robustness and reliability of ML models is essential for their deployment in safety-critical systems. Traditional formal verification techniques typically use mathematical optimization to model the non-linear behavior of activation functions. Although these methods offer exact guarantees, their scalability is limited due to the rapid growth in computational complexity with network size, motivating the exploration of quantum-enhanced methods.

Hybrid quantum-classical methods have been introduced to accelerate verification by delegating combinatorial subproblems to quantum hardware, while handling continuous components with classical solvers. By integrating quantum optimization algorithms, such as QAOA and quantum annealing, with classical decomposition schemes, these methods aim to leverage near-term quantum devices for scalable and certifiable neural network verification. This chapter explores these quantum-enhanced verification frameworks, highlighting their algorithmic structure, potential advantages, and current limitations.

5.3.1 Mixed-Integer Linear Programming (MILP)

Mixed Integer Linear Programming (MILP) is an optimization framework that combines continuous and discrete variables while maintaining linear objectives and constraints. The standard form is given as

$$\begin{aligned} \min_{x,y} \quad & c^T x + d^T y \\ \text{s.t.} \quad & Ax + By \geq b \end{aligned} \tag{5.45}$$

with $y \in \mathbb{Z}^{n_y}$ for integer variables (or $y \in \{0, 1\}^{n_y}$ for binary variables) and $x \in \mathbb{R}^{n_x}$ for real variables [30]. MILP effectively models discrete decision problems in facility location, scheduling, and logistics [31, 32]. MILP problems are NP-hard due to their discrete nature [33]. To obtain an exact solution, three main approaches exist: Branch and Bound systematically explores the solution space [34], Cutting Planes adds constraints to eliminate fractional solutions [35], and Branch and Cut combines both methods [36].

5.3.2 Decomposition Methods

Decomposition methods address large-scale MILP problems by exploiting their structural properties to break them into smaller, more manageable subproblems. Benders decomposition separates the problem into a master problem containing integer variables and subproblems handling continuous variables, iteratively adding cuts to improve the master problem solution [37]. Dantzig-Wolfe decomposition reformulates the original problem using column generation, where a master problem coordinates decisions while subproblems generate promising solutions [38]. Lagrangian relaxation dualizes complicating constraints, creating a decomposable structure that provides bounds and heuristic solutions [39]. More recent approaches include nested decomposition for multi-stage problems [40] and progressive hedging for stochastic programming applications [41].

Benders Decomposition

Benders decomposition solves MILP problems by separating integer and continuous variables. The method splits Eq. (5.45) into two problems: a master problem that chooses the integer variables y, and a subproblem that optimizes the continuous variables x for a fixed $y = \bar{y}$:

$$\min_{x} \quad c^T x \quad \text{s.t.} \quad Ax \geq b - B\bar{y}. \tag{5.46}$$

The algorithm alternates between solving these two problems. When the subproblem is feasible with optimal value $\theta^{(i)}$, it generates an optimality cut $\theta \geq \theta^{(i)} + \alpha^T(By - B\bar{y})$ that is added to the master problem, where α is an extreme point of the dual feasible region corresponding to the optimal dual solution. When the subproblem is infeasible, it generates a feasibility cut $0 \geq \alpha^T(By - B\bar{y})$ to eliminate the infeasible solution, where α is an extreme ray of the dual feasible region. The dual variables $\alpha \in \mathbb{R}^m$ have dimension equal to the number of constraints in the subproblem. The process continues until convergence. This approach is effective for two-stage problems.

Dantzig-Wolfe Decomposition

Dantzig-Wolfe decomposition solves large-scale linear programming problems by exploiting block structure in the constraint matrix. Consider a problem of the form:

$$\begin{aligned} \min_{x \geq 0} \quad & c^T x \\ \text{s.t.} \quad & Ax = b \\ & Dx = d \end{aligned} \tag{5.47}$$

where A represents linking constraints that couple the variables, and $Dx = d$ represents a block of constraints with special structure. The method reformulates the problem using the fact that any feasible solution can be expressed as a convex combination of extreme points and extreme rays of the constraint set $\{x : Dx = d, x \geq 0\}$. Let $P = \{x^{(1)}, x^{(2)}, \ldots, x^{(K)}\}$ be the set of extreme points and $R = \{r^{(1)}, r^{(2)}, \ldots, r^{(L)}\}$ be the set of extreme rays of the feasible region defined by $Dx = d, x \geq 0$. Any feasible solution can be written as

$$x = \sum_{k=1}^{K} \lambda_k x^{(k)} + \sum_{l=1}^{L} \mu_l r^{(l)} \tag{5.48}$$

where $\sum_{k=1}^{K} \lambda_k = 1$, $\lambda_k \geq 0$, and $\mu_l \geq 0$. This transforms the original problem into a master problem that optimizes over convex combination weights rather than the original variables. The algorithm uses column generation, where a pricing subproblem generates new extreme points or rays by solving optimization problems over the original constraint set $Dx = d$. The subproblem uses dual prices $\alpha \in \mathbb{R}^m$ from the master problem (where m is the number of linking constraints) to determine which new columns to add. The process continues until no improving columns can be found, ensuring optimality. This approach is particularly effective for problems with natural block structures, such as multi-commodity flow problems, vehicle routing, and cutting stock problems.

5.3.3 Hybrid Verification for Machine Learning

Hybrid quantum-classical optimization methods have emerged as a promising approach for addressing the computational challenges inherent in formal verification of neural networks, particularly for ReLU-based architectures. The verification problem can be formulated as a MILP instance, where ReLU non-linearities are encoded using binary variables [42–44]. Recent advances in this field leverage decomposition techniques such as Benders and Dantzig-Wolfe methods to partition the original MILP into smaller subproblems that can be solved iteratively using both classical and quantum hardware [45, 46]. The hybrid approach combines classical linear programming solvers on the one hand with quantum optimization algorithms such as QAOA or quantum annealing for binary subproblems on the other hand, enabling more efficient utilization of current NISQ devices. While these methods currently face limitations in certified accuracy compared to pure classical verifiers [47], they represent a foundational step toward leveraging quantum computing for neural network verification, particularly as quantum hardware continues to mature.

Hybrid Verification with Benders decomposition

Benders decomposes MILP into a mixed-integer master problem and a dual linear programming subproblem, where the master problem (containing binary variables y representing ReLU activations) is transformed into QUBO form and solved using quantum hardware, while the continuous subproblem is solved classically via its dual formulation $\max_{\alpha,\beta}\{\alpha(b - By) + \beta d : \alpha A + \beta C = g^\top\}$ [45].

The iterative process begins with an initial master solution, which is validated by solving the dual subproblem; if infeasible or suboptimal, the algorithm generates Benders cuts in the form of extreme points $(\alpha^{(k)}, \beta^{(k)}) \in \Lambda_p$ or extreme rays $(\alpha^{(k)}, \beta^{(k)}) \in \Lambda_r$, adding constraints $\alpha^{(k)}(b - By) + \beta^{(k)}d \leq \eta$ or $\alpha^{(k)}(b - By) + \beta^{(k)}d \leq 0$ respectively to the master problem.

The QUBO conversion transforms the Benders master problem into a format suitable for quantum hardware by reformulating the constrained binary optimization problem. Following the HQ-CRAN approach [45], the master problem is approximated using binary representations where continuous variables η and auxiliary variables a_k are encoded using two's-complement notation:

$$\eta \approx w_p p = \omega_p \cdot (-2^{n_p-1} p_{n_p-1} + \sum_{i=0}^{n_p-2} 2^i p_i) \tag{5.49}$$

$$a_k \approx w^{(k)} a^{(k)} = \omega_a \cdot \sum_{i=0}^{n_{a_k}-1} 2^i a_i^{(k)} \tag{5.50}$$

where $p \in \{0, 1\}^{n_p}$ and $a^{(k)} \in \{0, 1\}^{n_{a_k}}$ are binary vectors, and ω_p, ω_a are penalty weights. The master problem is then converted to QUBO form:

$$\min_x x^\top Q^{(\tau)} x + q^{(\tau)} x + \kappa^{(\tau)} \tag{5.51}$$

where $x = [p, y, a^{(1)}, \ldots, a^{(\tau)}]$ is the concatenated binary variable vector, and the quadratic matrix $Q^{(\tau)}$ and linear term $q^{(\tau)}$ incorporate both the original objective and Benders cuts accumulated up to iteration τ:

$$Q^{(\tau)} = \sum_{k=1}^{\tau} h^{(k)} \otimes h^{(k)} + \mathrm{diag}(0_{n_p}, 1_{n_y}, 0_{n_{a_1}}, \ldots, 0_{n_{a_\tau}}) \tag{5.52}$$

$$q^{(\tau)} = 2 \sum_{k=1}^{\tau} e_k \cdot h^{(k)} + [w_p, -y^{(t-1)}, 0_{n_{a_1}}, \ldots, 0_{n_{a_\tau}}] \tag{5.53}$$

where $h^{(k)} = [-w_p, -\alpha^{(k)} B, 0, \ldots, w_a^{(k)}, \ldots, 0]$ encodes the k-th Benders cut and $e_k = \alpha^{(k)} b + \beta^{(k)} d$.

The qubit requirements scale with the precision needs: the minimum number of qubits for approximating η is $n_p \geq 1 + \log_2(1 + (u_c + u_t)/\omega_p)$, where u_c, u_t are upper bounds on the predicted and target class outputs. Each accumulated Benders cut requires additional qubits, leading to a total complexity of $n_t = n_p + n_y + \sum_{k=1}^{\tau} n_{a_k}$ qubits for the complete master problem [45] (Fig. 5.3).

The critical limitation of this approach lies in the exponential growth of qubit requirements, as each iteration potentially adds new cuts requiring additional slack variables for QUBO conversion, escalating from $O(n_y + 2 \cdot n_s)$ qubits initially to $O(n_y + 2^{n_y} \cdot n_s)$ qubits in the worst-case scenario when all extreme points are enumerated [45]. Moreover, the heuristic nature of quantum solvers (VQE, QAOA) can produce suboptimal master solutions, leading to invalid cut generation and compromised solution feasibility, as the master solution directly determines the coupling constraints in the robustness verification formulation where $\arg\max_j f(z)$ must remain unchanged for all perturbations $\tilde{z} \in B_\infty^\epsilon(z)$ [45].

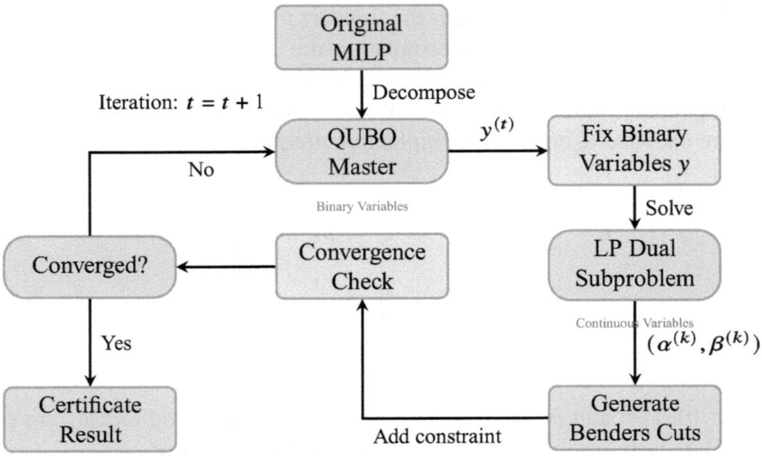

Fig. 5.3 HQ-CRAN hybrid quantum-classical Benders decomposition workflow for neural network robustness verification. The QUBO master problem containing binary ReLU activation variables y is solved on quantum hardware, while the continuous LP dual subproblem is solved classically. Benders cuts $(\alpha^{(k)}, \beta^{(k)})$ are iteratively generated and added to constrain the master problem until convergence

Hybrid Verification with Dantzig-Wolfe Decomposition

The hybrid Dantzig-Wolfe decomposition algorithm for neural network robustness verification operates through a column generation framework that iteratively alternates between a linear programming master problem and two distinct pricing subproblems. The algorithm begins by reformulating the original MILP robustness verification problem Eq. (5.45) into its Dantzig-Wolfe master formulation using extreme point representation, where the feasible region is expressed as convex combinations of extreme points $x^{(i)} \in P_X$ and $y^{(j)} \in P_Y$ through variables λ_i and μ_j respectively [13]. The LP master problem maintains the convexity constraints $\sum_i \lambda_i = 1$ and $\sum_j \mu_j = 1$, while preserving the coupling constraint $Ax + By \geq b$ and the integrality requirement $y \in \{0, 1\}^{n_y}$ through constraint $y = \sum_j y^{(j)} \mu_j$ (Fig. 5.4).

At each iteration t, the algorithm solves the restricted master problem over current extreme point subsets $I' \subseteq I$ and $J' \subseteq J$, obtaining dual multipliers $\alpha^{(t)}$, $\xi^{(t)}$, and $\eta^{(t)}$ corresponding to the coupling, real variable convexity, and integer variable convexity constraints respectively. These dual values drive two parallel pricing subproblems: (1) the classical LP pricing problem $\max\{(c^\top - \alpha^{(t)} A)x^{(i)} - \xi^{(t)} : x^{(i)} \in P_X\}$ solved using traditional linear programming methods, and (2) the quantum QUBO pricing problem $\max\{(d^\top - \alpha^{(t)} B)y^{(j)} - \eta^{(t)} : y^{(j)} \in P_Y\}$ transformed into unconstrained binary optimization form $\min q^\top Qq$ and solved using quantum algorithms such as QAOA or quantum annealing [46].

The key algorithmic advantage lies in the pricing phase: when the LP pricing returns a solution with reduced cost $r < \xi^{(t)}$, the corresponding extreme point $x^{(t)}$ is added to set I'; similarly, when the QUBO pricing yields reduced cost

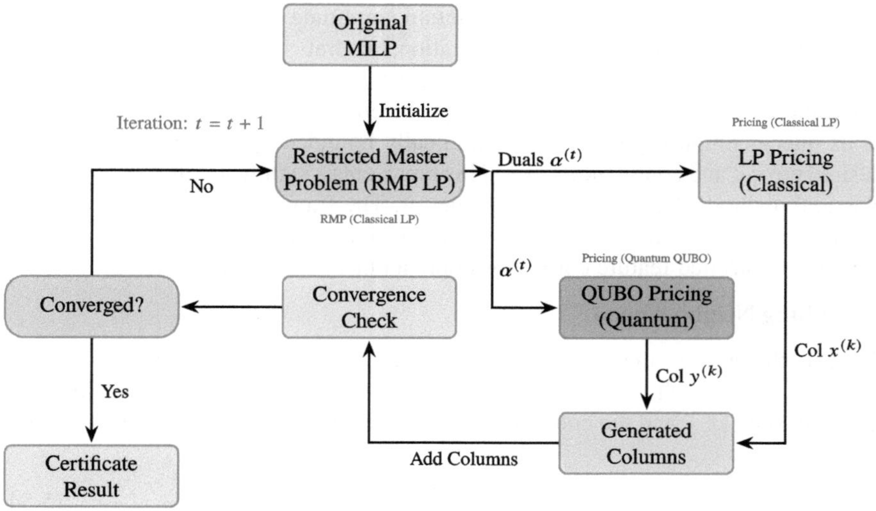

Fig. 5.4 Workflow of the Hybrid Quantum-Classical Dantzig-Wolfe (HQ-CRAN-DW) decomposition. The classical Restricted Master Problem (LP) iteratively provides dual variables $\alpha^{(t)}$ to the pricing subproblems. The continuous variable pricing subproblem is a classical LP, while the integer variable pricing subproblem is a QUBO solved on quantum hardware. Generated columns $x^{(k+1)}$, $y^{(k+1)}$ are added to the RMP, and the process repeats until convergence criteria are met

$p < \eta^{(t)}$, the binary extreme point $y^{(t)}$ is incorporated into set J'. The algorithm terminates when no improving columns are found (i.e., all reduced costs are nonnegative) or when the gap between primal objective ϕ and dual objective $\varphi = \max_{k\in\{0,\ldots,t\}} -\alpha^{(k)}b - \xi^{(k)} - \eta^{(k)}$ falls below threshold θ. This hybrid approach maintains constant qubit requirements $O(n_y + m_y \cdot n_s)$ throughout all iterations, as the binary pricing problem structure remains fixed regardless of the number of generated columns, providing significant quantum resource efficiency compared to exponentially growing alternatives [46].

Formal verification approaches provide theoretical assurance of robustness. These span exact methods, approximate bounds, and stochastic guarantees such as randomized smoothing [48, 49]. Lipschitz continuity estimation plays a key role in bounding model sensitivity [50].

5.3.4 Quantum Randomized Smoothing

In this section, we present a quantum framework for robust classification via RS [13, 14]. The method encodes a classical input into a quantum state and applies a smoothing circuit to create a superposition of perturbed states that form a neighborhood around the input based on a given probability distribution. Quantum techniques, such as amplitude amplification and quantum amplitude estimation, then efficiently measure the likelihood of correct classification, quadratically reducing the number of queries needed compared to classical approaches. We cluster these techniques under the umbrella of quantum RS [48, 49, 51]. The efficiency of quantum RS is evident in its query complexity, which is theoretically bounded by $O\left(\frac{1}{\epsilon}\right)$ for achieving a desired error ϵ and confidence $1-\delta$. This represents a significant improvement over classical methods that require $O\left(\frac{1}{\epsilon^2}\right)$ queries. The framework naturally extends to binary data by encoding bit-flip perturbations and adapts to continuous inputs by mapping controlled feature corruptions onto a binary model [49, 51].

Smoothing Neighborhood

To begin the process, the input vector $x \in \mathbb{N}^d$ is encoded into the basis state denoted as $|\sigma\rangle$. That is, each element x_j of the vector is encoded into its corresponding quantum state $\left|x_j\right\rangle$, and the entire input is represented by the tensor product of these states:

$$|\sigma\rangle = \bigotimes_{j=0}^{d-1} \left|x_j\right\rangle, \tag{5.54}$$

where d is the dimension of the input vector. To create a smoothing neighborhood around the input, we apply a smoothing circuit $U_{p,\lambda}$ that prepares a superposition of perturbed values based on a defined distribution $\mathcal{P}(\cdot;\lambda)$. For a single component x_j, the smoothing operation is given by

$$U_{p,\lambda}|x_j\rangle = \sum_{k=0}^{2^v-1} \sqrt{\mathcal{P}_j(\delta_k;\lambda)}\,|x_j+\delta_k\rangle\,, \tag{5.55}$$

where δ_k represents the k-th possible perturbation, λ is a parameter controlling the smoothing distribution and v is the number of qubits used, which allows for 2^v different perturbation values. When we apply the smoothing circuit to all d input components, the full smoothing operation is expressed as the tensor product over all dimensions:

$$U_{p,\lambda}|\sigma\rangle = \bigotimes_{j=0}^{d-1} U_{p,\lambda_j}|x_j\rangle = \sum_{i=0}^{2^{dv}-1} p(i;\lambda)|i\rangle\,. \tag{5.56}$$

Here, the index i indexes the combined perturbed states from each input dimension and $p(i;\lambda)$ denotes the overall probability amplitude associated with the composite

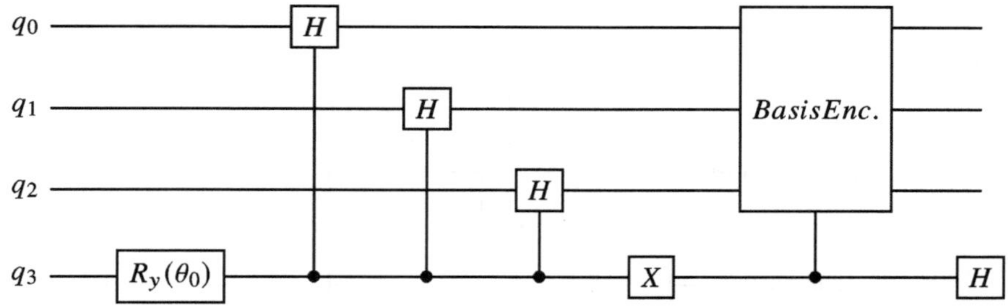

Fig. 5.5 Preparation of a uniform distribution on 4 qubits

state $|i\rangle$. For an uniform distribution every $p(i;\lambda)$ is used with the same weight $\lambda = 1/2^v$. Acting on the basis state $\bigotimes_{j=0}^{d-1}|x_j\rangle$ the circuit produces

$$\sum_{i=0}^{2^{dv}-1} \frac{1}{\sqrt{2^{dv}}} |i\rangle, \tag{5.57}$$

which is the uniform superposition over all allowed perturbations.

In Fig. 5.5, we report an example of a uniform distribution applied on 4 qubits. After the Hadamard layer the offset register already has exactly these equal amplitudes. The circuit reduces to two stages per input component: create the uniform superposition and add the input.

Estimation of Roustness Bounds

Using a QML classifier oracle Y_θ, we can evaluate the smoothed superposition of inputs. The average probability ρ_c of class c is calculated as

$$\rho_c = \mathbb{E}_{\delta\sim\mathcal{N}(0,1)}\left(Y_\theta(x+\delta) = c\right) \tag{5.58}$$

where Y_θ is the output class of the classifier. The Grover diffusion operator G_c is then applied to amplify the probability amplitude of the desired outcomes:

$$G_c = U_{p,\lambda}(2|0\rangle\langle 0| - I)U_{p,\lambda}^\dagger Y^{(c)}. \tag{5.59}$$

The estimation circuit employs Quantum Amplitude Estimation (QAE) to accurately measure ρ_c with high confidence. The measurement of the ancilla qubits yields:

$$\rho_c = \sin^2(\theta/2) \tag{5.60}$$

where θ is the phase estimated through the quantum circuit. The algorithm provides a certified radius for robustness, requiring $M = O(1/\epsilon)$ queries to the base classifier for an error bound ϵ and confidence $1-\alpha$ [48]. This is a significant improvement over classical methods which require $M = O(1/\epsilon^2)$ (Fig. 5.6).

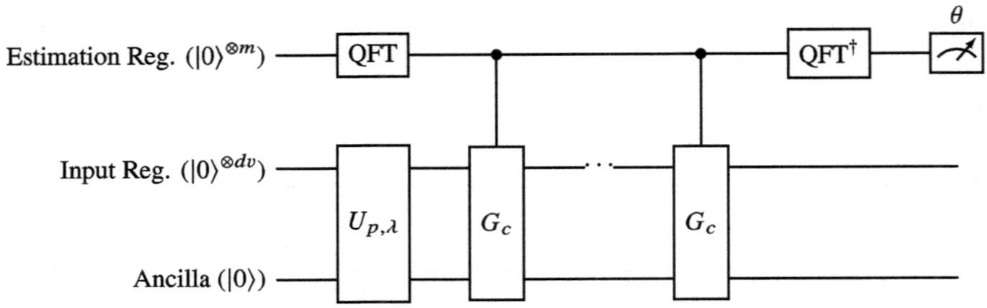

Fig. 5.6 A high-level quantum circuit diagram for the QuAdRo algorithm [48]. The process involves three main stages: (i) preparing a smoothed superposition of the input state via the $U_{p,\lambda}$ operator, (ii) iteratively applying a controlled Grover operator (G_c) to amplify the desired state amplitudes, and (iii) applying an inverse QFT and measuring the estimation register to find the phase θ, from which the robustness probability is calculated via $\rho_c = \sin^2(\theta/2)$

5.3.5 Discrete Quantum Randomized Smoothing

For an input x, typically in a binary space $\mathcal{X}_D^{(N)} := \{0, 1\}^N$, perturbations $\tilde{x}$ are sampled according to a joint PMF $\phi(\tilde{x}|x; p_+, p_-)$ [51]. This PMF is often constructed from data-dependent Bernoulli distributions for each component x_j of the input x. Specifically, each bit $\tilde{x}_j$ of a perturbed sample, given the original input x, is flipped with a probability p_- if $x_j = 1$, or with probability p_+ if $x_j = 0$. This can be expressed as

$$\tilde{x}_j|x \sim \text{Ber}(p_+ \cdot \mathbb{I}(x_j = 0) + (1 - p_-) \cdot \mathbb{I}(x_j = 1)) \tag{5.61}$$

where $\mathbb{I}(\cdot)$ is the indicator function. The joint PMF over all n dimensions is then the product of these individual probabilities (or their complements for non-flips):

$$\phi(\tilde{x}|x; p_+, p_-) = \prod_{j=0}^{n-1} (P_F^{(j)})^{\mathbb{I}(\tilde{x}_j \neq x_j)} (1 - P_F^{(j)})^{\mathbb{I}(\tilde{x}_j = x_j)} \tag{5.62}$$

where $P_F^{(j)} := p_+ \cdot \mathbb{I}(x_j = 0) + p_- \cdot \mathbb{I}(x_j = 1)$ is the probability of a bit-flip for the j-th component.

The smooth classifier $g(x)$ is then defined as the probability that the base classifier f predicts a target class (e.g., class 1) for a perturbed input $\tilde{x}$ drawn from this discrete distribution $\phi(\cdot|x)$:

$$g(x) = \mathbb{P}[f(\tilde{x}) = 1] \text{ for } \tilde{x} \sim \phi(\cdot|x) = \sum_{f(\tilde{x})=1} \phi(\tilde{x}|x). \tag{5.63}$$

Evaluating this sum directly is often intractable due to the 2^n possible perturbations for an n-dimensional binary input, necessitating Monte Carlo (MC) methods in classical approaches.

Quantum Encoding of Discrete Perturbations

The quantum advantage arises from efficiently representing and processing these discrete perturbations. All 2^n possible perturbed states $\tilde{x}_i$ can be encoded into a quantum superposition $|\sigma(x)\rangle$ using n qubits. Crucially, the amplitude of each basis state $|\tilde{x}_i\rangle$ in the superposition is the square root of its transition probability $\sqrt{\phi(\tilde{x}_i|x)}$:

$$|\sigma(x)\rangle := \sum_{i=0}^{2^n-1} \sqrt{\phi(\tilde{x}_i|x)}|\tilde{x}_i\rangle. \tag{5.64}$$

This state preparation can be achieved using a parameterized circuit $U(x, p_-, p_+)$ acting on an initial $|0\rangle^{\otimes n}$ state. Specifically, $U(x, p_-, p_+) := \bigotimes_{m=0}^{n-1} RY(\theta_0 + x_m\theta_D) \circ RX(x_m\pi)$, where $\theta_0(p_+) = 2\arccos(\sqrt{1-p_+})$ and $\theta_D = -2(\arccos(\sqrt{1-p_+}) + \arccos(\sqrt{1-p_-}))$. This yields $U(x, p_-, p_+)|0\rangle = |\sigma(x)\rangle$.

The smooth classifier $g(x)$ is then the sum of probabilities $\phi(\tilde{x}_i|x)$ for which $f(\tilde{x}_i) = 1$. In the quantum state $|\sigma(x)\rangle$, this corresponds to the squared norm of the projection of $|\sigma(x)\rangle$ onto the subspace of "good" perturbations (where $f(\tilde{x}_i) = 1$)[cite: 419]. If we define $|a\rangle$ as the normalized superposition of good states and $|b\rangle$ as that of bad states, then $|\sigma(x)\rangle$ can be written as [51]:

$$|\sigma(x)\rangle = \sqrt{g(x)}|a\rangle + \sqrt{1-g(x)}|b\rangle. \tag{5.65}$$

Here, $g(x)$ is precisely the quantity we want to estimate [51].

Quantum Amplitude Estimation for Discrete Smoothing

QAE is used to find $g(x)$. A Grover-like operator $G = \mathcal{R}_{\sigma(x)}O_f$ is constructed, where O_f is the oracle for the base classifier (e.g., $O_f|\tilde{x}\rangle = (-1)^{f(\tilde{x})}|\tilde{x}\rangle$) and $\mathcal{R}_{\sigma(x)}$ is the reflection about $|\sigma(x)\rangle$ (implementable as $U(x, p_-, p_+)(2|0\rangle\langle 0| - I)U(x, p_-, p_+)^H$). This operator G acts as a rotation in the subspace spanned by $|a\rangle$ and $|b\rangle$ by an angle 2θ, where

$$g(x) = \sin^2(\theta). \tag{5.66}$$

QAE algorithms estimate θ, and thus $g(x)$, with a quadratic speed-up in terms of calls to the oracle O_f compared to classical MC methods. To achieve an estimate of $g(x)$ with accuracy ϵ and confidence $1-\delta$, the quantum algorithm (Smooth Quantum Classifier–SQC) requires $O(\ln(\frac{1}{\delta})\frac{1}{\epsilon})$ oracle calls.

Discrete Perturbations on Continuous Data

A novel extension allows this discrete framework to certify models against discrete perturbations even when the underlying data is continuous. Consider an adversary who can corrupt up to N specific features of a continuous input, where each feature can either be left unchanged or altered to a predefined "corrupted" state. Each such overall perturbation can be uniquely represented by a binary string of length N.

A threat model $\mathcal{B}_r(x)$ defines the set of allowed perturbed inputs $\tilde{x}$ where at most r out of N possible features are changed from their original state x_j to a perturbed state $p(x_j, j)$. The smoothing distribution $\hat{\phi}(\tilde{x}|x; p)$ is defined over these discrete changes:

$$\hat{\phi}(\tilde{x}|x;\, p) = (p)^{\sum\limits_{\tilde{x}_j = p(x_j, j)} 1} (1-p)^{\sum\limits_{\tilde{x}_j = x_j} 1}. \tag{5.67}$$

The smooth classifier $\hat{g}(x)$ is then defined similarly to the purely discrete case:

$$\hat{g}(x) = \sum_{\tilde{x} \in \mathcal{P}_x \text{ and } \hat{f}(\tilde{x}) = 1} \hat{\phi}(\tilde{x}|x;\, p). \tag{5.68}$$

To certify $\hat{g}(x)$, a bijective mapping η_x is established from the set of these 2^N discrete perturbations $\mathcal{P}_x$ onto the binary space $\{0, 1\}^N$. This allows the problem to be mapped back to the discrete binary RS framework described earlier, enabling the use of the same quantum algorithms. The quantum superposition is then created over the N-bit binary representations of the perturbations, and the oracle integrates these with the original continuous data x during evaluation.

Since embedding classical data via basis state encoding makes input perturbations naturally interpretable as bit-flips. We extend our perturbation model to include Hamming distant states [49]. For binary inputs $x \in \{0, 1\}^n$, we design perturbations by constraining the Hamming distance. Define the set of states with exactly k bit-flips from x as

$$D_{n,k}(x) = \left\{\tilde{x} \in \{0, 1\}^n : \|x - \tilde{x}\|_0 = k\right\}. \tag{5.69}$$

To distribute probabilities over these states, we introduce an exponentially decaying weight function. For any Hamming distance i with $0 \le i \le k$, the weight is

$$w(i) = \frac{e^{-i/\xi}}{\sum\limits_{j=0}^{k} e^{-j/\xi}}, \tag{5.70}$$

where $\xi > 0$ controls the decay rate. Since there are $\binom{n}{i}$ states at distance i, each state is assigned a probability:

$$p(i) = \frac{w(i)}{\binom{n}{i}}. \tag{5.71}$$

The overall quantum state that encodes this perturbation distribution is constructed as

$$|\sigma(x)\rangle = \sum_{i=0}^{k} \sqrt{p(i)} \sum_{\tilde{x} \in D_{n,i}(x)} |\tilde{x}\rangle. \tag{5.72}$$

This superposition emphasizes that the original input (with $i = 0$) is most likely, while states with larger Hamming distance occur with diminishing probability.

Realization on a quantum computer is achieved via ancillary qubits and controlled rotational gates. An inverse multi-controlled basis embedding is applied to load the classical probability distribution into the quantum state. This co-design of data encoding and perturbation ensures that the induced noise on the quantum circuit directly corresponds to an interpretable perturbation in the input space, which is essential for obtaining meaningful robustness certificates.

5.4 Research Directions

Despite significant advancements in this field, numerous challenges persist. These range from precision issues that could be addressed by fault-tolerant algorithms to more complex problems such as space complexity and necessity of fairness and privacy constraints, which require further algorithmic innovations. Addressing these challenges will be vital for advancing the practical deployment of quantum technologies in real-world applications.

We begin by examining problems that could benefit from early fault-tolerant quantum computers. Currently, hybrid verification of classical neural networks relies on heuristic quantum subroutines, such as QAOA or VQA, which tend to produce imprecise solutions. However, effective verification routines require exact solutions. To this end, Grover's or Quantum Amplitude Amplification algorithms, which necessitates fault tolerance, are particularly relevant. Is it possible to leverage these algorithms before full fault-tolerance is achieved? In the interim, exploring algorithmic adaptations that utilize imperfect quantum devices for algorithms that deliver exact solution is crucial.

These adaptations may take various forms, e.g., optimizing resource allocation by exploring the noise tolerance of existing algorithms to better align with their specific requirements. Empirical research is increasingly focused on evaluating the performance of QML models in adversarial scenarios, utilizing both real quantum hardware and simulation benchmarks [52, 53]. These studies examine critical aspects such as attack success rates, training methodologies, and the transferability of adversarial examples. Drawing on the insights from this research can refine the resource allocation process, enabling smoother integration and improved performance, even in noisy environments.

One approach to lower the hardware requirements is to explore alternative algorithms, thereby circumventing the challenges associated with the original ones. For example, quantum computing presents significant potential for enhancing stochastic methods by utilizing quantum sampling techniques. These techniques can not only provide convergence benefits [54] but also enable the representation of classically intractable distributions [55]. A promising avenue for future research is to explore how stochastic verification (see Sect. 5.2) can leverage quantum sampling to improve its efficiency.

Not all challenges can be resolved by access to fault-tolerant machines. Estimating the exact Lipschitz constant requires storing the unitary in classical memory, leading to significant space complexity that becomes unfeasible for large QML models. To address similar challenges, some approximate methods have been explored in the literature for different types of QML problems [56, 57], suggesting that pursuing this direction could be advantageous.

References

1. National Transportation Safety Board (2019) Collision between vehicle controlled by developmental automated driving system and pedestrian, Tempe, Arizona, March 18, 2018. Technical Report NTSB/HAR-19/03 PB2019-101216, National Transportation Safety Board, November 2019. https://www.ntsb.gov/investigations/AccidentReports/Reports/HAR1903.pdf
2. OpenAI (2023) Gpt-4 system card. Technical report, OpenAI, March 2023. https://cdn.openai.com/papers/gpt-4-system-card.pdf
3. Khatun A, Usman M (2024) Quantum transfer learning with adversarial robustness for classification of high-resolution image datasets. Adv Quantum Technol 7(2):2400268. https://doi.org/10.1002/qute.202400268
4. Abreu D, Rothenberg CE, Abelém A (2024) Qml-ids: quantum machine learning intrusion detection system. In: 2024 IEEE symposium on computers and communications (ISCC). IEEE, pp 1–6
5. Innan N, Al-Zafar Khan M, Bennai M (2024) Financial fraud detection: a comparative study of quantum machine learning models. Int J Quantum Inf 22(02):2350044
6. Sirui L, Duan L-M, Deng D-L (2020) Quantum adversarial machine learning. Phys Rev Res 2:033212. https://doi.org/10.1103/PhysRevResearch.2.033212
7. Goodfellow IJ, Shlens J, Szegedy C (2015) Explaining and harnessing adversarial examples. In: International conference on learning representations (ICLR). https://arxiv.org/abs/1412.6572
8. Madry A, Makelov A, Schmidt L, Tsipras D, Vladu A (2018) Towards deep learning models resistant to adversarial attacks. In: International conference on learning representations (ICLR). https://arxiv.org/abs/1706.06083
9. McClean JR, Boixo S, Smelyanskiy VN, Babbush R, Neven H (2018) Barren plateaus in quantum neural network training landscapes. Nat Commun 9(1):4812
10. Gong W, Yuan D, Li W, Deng D-L (2024) Enhancing quantum adversarial robustness by randomized encodings. Phys Rev Res 6:023020. https://doi.org/10.1103/PhysRevResearch.6.023020
11. Franco N, Sakhnenko A, Stolpmann L, Thuerck D, Petsch F, Rüll A, Lorenz JM (2024) Predominant aspects on security for quantum machine learning: literature review. In: 2024 IEEE international conference on quantum computing and engineering (QCE), vol 1. IEEE
12. Athalye A, Carlini N, Wagner D (2018) Obfuscated gradients give a false sense of security: circumventing defenses to adversarial examples. In: International conference on machine learning (ICML). PMLR, pp 274–283
13. Cohen J, Rosenfeld E, Kolter Z (2019) Certified adversarial robustness via randomized smoothing. In: International conference on machine learning. PMLR, pp 1310–1320
14. Salman H, Li J, Razenshteyn I, Zhang P, Zhang H, Bubeck S, Yang G (2019) Provably robust deep learning via adversarially trained smoothed classifiers. Adv Neural Inf Process Syst 32
15. Kumar A, Levine A, Feizi S, Goldstein T (2020) Certifying confidence via randomized smoothing
16. Zhang D, Ye M, Gong C, Zhu Z, Liu Q (2020) Black-box certification with randomized smoothing: a functional optimization based framework. https://arxiv.org/abs/2002.09169

17. Liu N, Wittek P (2020) Vulnerability of quantum classification to adversarial perturbations. Phys Rev A 101:062331. https://doi.org/10.1103/PhysRevA.101.062331. https://link.aps.org/doi/10.1103/PhysRevA.101.062331
18. Liao H, Convy I, Huggins WJ, Birgitta Whaley K (2021) Robust in practice: adversarial attacks on quantum machine learning. Phys Rev A 103(4). https://doi.org/10.1103/physreva.103.042427. https://doi.org/10.1103%2Fphysreva.103.042427
19. Schuld M, Sweke R, Meyer JJ (2021) Effect of data encoding on the expressive power of variational quantum-machine-learning models. Phys Rev A 103(3). https://doi.org/10.1103/physreva.103.032430. https://doi.org/10.1103%2Fphysreva.103.032430
20. Kordzanganeh M, Sekatski P, Fedichkin L, Melnikov A (2022) An exponentially-growing family of universal quantum circuits
21. Havlíček V, Córcoles AD, Temme K, Harrow AW, Kandala A, Chow JM, Gambetta JM (2019) Supervised learning with quantum-enhanced feature spaces. Nature 567(7747):209–212. https://doi.org/10.1038/s41586-019-0980-2. https://doi.org/10.1038%2Fs41586-019-0980-2
22. Yuxuan D, Hsieh M-H, Liu T, Tao D, Liu N (2021) Quantum noise protects quantum classifiers against adversaries. Phys Rev Res 3:023153. https://doi.org/10.1103/PhysRevResearch.3.023153
23. Huang J-C, Tsai Y-L, Yang C-HH, Su C-F, Yu C-M, Chen P-Y, Kuo S-Y (2023) Certified robustness of quantum classifiers against adversarial examples through quantum noise
24. Saxena A, Wollschläger T, Franco N, Lorenz JM, Günnemann S (2024) Certifiably robust encoding schemes. https://arxiv.org/abs/2408.01200
25. Winderl D, Franco N, Lorenz JM (2024) Constructing optimal noise channels for enhanced robustness in quantum machine learning. arXiv:2404.16417
26. Lecuyer M, Atlidakis V, Geambasu R, Hsu D, Jana S (2019) Certified robustness to adversarial examples with differential privacy. https://arxiv.org/abs/1802.03471
27. Zhou L, Ying M (2017) Differential privacy in quantum computation. 2017 IEEE 30th computer security foundations symposium (CSF), pp 249–262. https://api.semanticscholar.org/CorpusID:38803461
28. Winderl D, Franco N, Lorenz JM (2024) Quantum neural networks under depolarization noise: exploring white-box attacks and defenses. Quantum Mach Intell 6(2):83
29. Weber M, Liu N, Li B, Zhang C, Zhao Z (2021) Optimal provable robustness of quantum classification via quantum hypothesis testing. npj Quantum Inf 7(1). https://doi.org/10.1038/s41534-021-00410-5. https://doi.org/10.1038%2Fs41534-021-00410-5
30. Wolsey LA (2020) Integer programming. John Wiley & Sons
31. Daskin M (1997) Network and discrete location: models, algorithms and applications. J Oper Res Soc 48(7):763–764
32. Weiss G (1995) Scheduling: theory, algorithms, and systems
33. Hartmanis J (1982) Computers and intractability: a guide to the theory of np-completeness (Michael R. Garey and David S. Johnson). Siam Rev 24(1):90
34. Land AH, Doig AG (2009) An automatic method for solving discrete programming problems. In: 50 years of integer programming 1958-2008: from the early years to the state-of-the-art. Springer, pp 105–132
35. Gomory RE (2009) Outline of an algorithm for integer solutions to linear programs and an algorithm for the mixed integer problem. In: 50 years of integer programming 1958-2008: from the early years to the state-of-the-art. Springer, pp 77–103
36. Fischetti M, González JJS, Toth P (1997) A branch-and-cut algorithm for the symmetric generalized traveling salesman problem. Oper Res 45(3):378–394
37. Benders JF (1962) Partitioning procedures for solving mixed-variables programming problems. Numer Math 4:238–252
38. Dantzig GB, Wolfe P (1960) Decomposition principle for linear programs. Oper Res 8(1):101–111
39. Fisher ML (1981) The Lagrangian relaxation method for solving integer programming problems. Manag Sci 27(1):1–18

40. Birge JR, Louveaux F (2011) Introduction to stochastic programming. Springer Science & Business Media (2011)
41. Tyrrell Rockafellar R, Wets RJ-B (1991) Scenarios and policy aggregation in optimization under uncertainty. Math Oper Res 16(1):119–147
42. Katz G, Barrett C, Dill DL, Julian K, Kochenderfer MJ (2017) Reluplex: an efficient smt solver for verifying deep neural networks. In: Computer aided verification: 29th international conference, CAV 2017, Heidelberg, Germany, July 24–28, 2017, proceedings, part I , vol 30. Springer, pp 97–117
43. Ehlers R (2017) Formal verification of piece-wise linear feed-forward neural networks. In: Automated technology for verification and analysis: 15th international symposium, ATVA 2017, Pune, India, October 3–6, 2017, proceedings, vol 15. Springer, pp 269–286
44. Tjeng V, Xiao KY, Tedrake R (2019). Evaluating robustness of neural networks with mixed integer programming. In: International conference on learning representations
45. Franco N, Wollschläger T, Gao N, Lorenz JM, Günnemann S (2022) Quantum robustness verification: a hybrid quantum-classical neural network certification algorithm. In: 2022 IEEE international conference on quantum computing and engineering (QCE). IEEE, pp 142–153
46. Franco N, Wollschläger T, Poggel B, Günnemann S, Lorenz JM (2023) Efficient milp decomposition in quantum computing for relu network robustness. In: 2023 IEEE international conference on quantum computing and engineering (QCE), vol 1. IEEE, pp 524–534
47. Wang S, Zhang H, Xu K, Lin X, Jana S, Hsieh C-J, Kolter JZ (2021) Beta-crown: efficient bound propagation with per-neuron split constraints for neural network robustness verification. Adv Neural Inf Process Syst 34:29909–29921
48. Sahdev A, Kumar M (2023) Adversarial robustness based on randomized smoothing in quantum machine learning. arXiv:abs/2302.11355
49. Franco N, Kempkes M, Spiegelberg J, Lorenz JM (2024) Quadratic advantage with quantum randomized smoothing applied to time-series analysis. In: 2024 IEEE international conference on quantum computing and engineering (QCE), vol 1. IEEE
50. Berberich J, Fink D, Pranjić D, Tutschku C, Holm C (2024) Training robust and generalizable quantum models. Phys Rev Res 6:043326. https://doi.org/10.1103/PhysRevResearch.6.043326
51. Wollschläger T, Saxena A, Franco N, Lorenz JM, Günnemann S (2024) Discrete randomized smoothing meets quantum computing. arXiv:abs/2408.00895
52. Wendlinger M, Tscharke K, Debus P (2024) A comparative analysis of adversarial robustness for quantum and classical machine learning models. arXiv:abs/2404.16154
53. Montalbano G, Banchi L (2025) Quantum adversarial learning for kernel methods. Quantum Mach Intell 7:15. https://doi.org/10.1007/s42484-025-00238-8
54. Layden D, Mazzola G, Mishmash RV, Motta M, Wocjan P, Kim J-S, Sheldon S (2023) Quantum-enhanced Markov chain monte Carlo. Nature 619(7969):282–287. ISSN 1476-4687. https://doi.org/10.1038/s41586-023-06095-4. http://dx.doi.org/10.1038/s41586-023-06095-4
55. Coyle B, Mills D, Danos V, Kashefi E (2020) The born supremacy: quantum advantage and training of an ising born machine. npj Quantum Inf 6(1). ISSN 2056-6387. https://doi.org/10.1038/s41534-020-00288-9. http://dx.doi.org/10.1038/s41534-020-00288-9
56. Landman J, Thabet S, Dalyac C, Mhiri H, Kashefi E (2022) Classically approximating variational quantum machine learning with random fourier features. https://arxiv.org/abs/2210.13200
57. Huang H-Y, Kueng R, Preskill J (2020) Predicting many properties of a quantum system from very few measurements. Nat Phys 16(10):1050–1057. ISSN 1745-2481. https://doi.org/10.1038/s41567-020-0932-7. http://dx.doi.org/10.1038/s41567-020-0932-7

Chapter 6
Fundamental Questions on Robustness and Accuracy for Classical and Quantum Learning Algorithms

Nana Liu

Abstract This chapter investigates some fundamental questions on the relationship between accuracy and robustness in both classical and quantum classification algorithms under noisy and adversarial conditions. We introduce and clarify various definitions of robustness, including corrupted-instance robustness accuracy and prediction-change robustness, distinguishing them from conventional accuracy measures. Through theoretical analysis and toy models, we establish conditions under which trade-offs between accuracy and robustness accuracy arise and identify scenarios where such trade-offs can be avoided. The framework developed highlights the nuanced interplay between model bias, noise characteristics, and perturbation types, including relevant and irrelevant perturbations. We explore the implications of these results for incompatible noise, adversarial quantum perturbations, the no free lunch theorem, and suggest future methods to examine these problems from the lens of dynamical systems.

6.1 Introduction

Realizing the full potential of quantum-enhanced learning necessitates addressing a fundamental and pervasive challenge: the presence of noise and perturbations in both classical and quantum learning systems. Noise in machine learning arises from multiple sources. On one hand, environmental noise—such as hardware imperfections, decoherence, and stochastic fluctuations—can degrade the fidelity of quantum operations and corrupt input data. On the other hand, adversarial perturbations, often small but intentionally crafted changes to input examples, threaten the reliability and security of learning models by causing misclassifications. While noise is frequently viewed as a detrimental factor limiting performance, it can also play beneficial roles, including regularization to improve generalization, enhancement of privacy, and enabling compression when its parameters are carefully controlled.

N. Liu (✉)
Institute of Natural Sciences, Shanghai Jiao Tong University, Shanghai, China
e-mail: nana.liu@quantumlah.org

M. Usman (ed.), *Quantum Robustness in Artificial Intelligence*, Quantum Science and Technology, https://doi.org/10.1007/978-3-032-11153-1_6

In the context of quantum machine learning, noise plays a dual role: it can be a disruptive influence that impedes algorithmic performance, but also a tool to enhance learning capabilities when harnessed appropriately. This duality raises essential questions about how to evaluate and quantify the robustness of quantum classifiers under realistic noisy conditions. Conventional accuracy measures, defined in noise-free scenarios, fail to capture the stability of model predictions when inputs or models themselves are subject to perturbations. This necessitates the introduction of robustness-specific performance metrics designed for both adversarial and nonadversarial noise environments.

This chapter poses and explores some of these fundamental questions by focusing on classification problems under noisy conditions. We begin by systematically categorizing various notions of robustness, including corrupted-instance robustness accuracy, prediction-change robustness, and error-region robustness. We clarify how these definitions differ conceptually and mathematically, and how they relate to other accuracy measures. We also distinguish between relevant perturbations (which alter the true class of an example), and irrelevant perturbations, which do not change the underlying class but may affect model predictions. From some simple examples, we see the role of relevant and irrelevant perturbations trade-offs that can occur between accuracy and robustness accuracy.

The central theme of the chapter is the exploration of the trade-off relations between accuracy and robustness accuracy. Through theoretical analysis and simplified illustrative models, we demonstrate conditions under which increasing robustness inevitably sacrifices accuracy, and conversely, when both can be simultaneously optimized without compromise. Key results include general relationships for unbiased and biased classifiers, showing how robustness accuracy depends jointly on accuracy and model robustness properties.

We extend the discussion to quantum classification models, examining how quantum noise channels and measurement operators influence these trade-offs. Notably, we analyze noise models, such as bit flip, phase flip, and depolarization, identifying scenarios where trade-offs appear or vanish. We also discuss adversarial perturbations modeled as quantum channels with bounded trace distance, establishing sufficient conditions for maximal trade-offs between accuracy and robustness accuracy. The chapter highlights how the intrinsic geometry of the data and the hypothesis class, rather than model-specific features alone, dictate these trade-offs.

Beyond these core theoretical developments, we explore important practical and conceptual applications of the accuracy-robustness trade-off framework:

- Incompatible Noise: We introduce the notion of incompatible noise, where training a model to be robust against one type of noise can reduce robustness against another. We provide explicit examples involving depolarization and bit-flip noise to illustrate this phenomenon, which has significant implications for model design and training strategies.
- Adversarial Examples: We rigorously characterize adversarial perturbations in the quantum domain, specifying the conditions under which adversarial noise remains irrelevant (does not change true class) but induces prediction errors. This provides

insight into the vulnerability of quantum classifiers and suggests avenues for robust training methods.
- No Free Lunch Theorem: The chapter offers a new perspective on the no free lunch theorem by linking it to robustness accuracy. We interpret the theorem's assertion—that no universally superior model exists—through the lens of perturbations inducing shifts in data distributions, clarifying the relationship between accuracy on one distribution and robustness accuracy on perturbed distributions.

We also outline connections to dynamical systems theory, framing machine learning robustness as a question of stability and convergence under perturbations. We discuss the results and present the open questions in the last section.

6.2 Definition Zoo

In studying the performance of models in the presence of environmental as well as adversarial noise, there are many different characterizations of this performance and the sensitivity to these perturbations. We have various measures of robustness, so it is important to first become clear what these different definitions are, as the literature can often confuse these different definitions and this can even lead to contradictory claims, which are in fact not contradictory if one is careful (Tables 6.1 and 6.2).

Table 6.1 Classification of types of perturbations

	Irrelevant perturbation $\mathcal{N}_{irr}$	Relevant perturbation $\mathcal{N}_{rel}$
Small $D(\sigma, \mathcal{N}(\sigma)) < \epsilon$	Type I	Type II
Unrestricted $D(\sigma, \mathcal{N}(\sigma))$	Type III	Type IV

Table 6.2 Different species of robustness and accuracy

	Corrupted-instance	Prediction-change	Error-region
Robust loss	$\tilde{L}$ Definition 6	L^* Definition 7	$\bar{L}$ Definition 8
Robustness accuracy	$\tilde{A}$ Definition 11	N/A	$\bar{A}$ Definition 13
Robustness	N/A	A^* Definition 12	N/A
Robustness to adversarial perturbations	$\tilde{\Delta}$ Definition 14	Δ^* Definition 15	$\bar{\Delta}$ Definition 16

6.2.1 General Definitions for Classification Problems

Let $\sigma \in \Sigma$ be input states, where the pair (σ, c) is sampled i.i.d from some distribution. The model or hypothesis function $h_{\vec{\theta}} : \Sigma \to C$ makes the prediction for the classification of states $\sigma \in \Sigma$ into K distinct classes, where $C = \{1, ..., K\}$. We denote the ground truth (true class) of σ as $c(\sigma)$. If the model is deterministic, it means that h is a deterministic function. If the model is probabilistic, then h is a probabilistic function, which gives outputs in C with a certain probability.

Definition 1 (*Loss function*) A *local loss function* $l_h(\sigma, c) : (\Sigma, C) \to \Re$ is a penalty function for the state σ when a model h is used. It outputs a small number if h gives the same prediction as c for σ, and a larger number otherwise. A *global loss function* is the expected value of the local cost function over the distribution $(\sigma, c) \sim \mathcal{D}$

$$L_{h,\mathcal{D}} = \mathbf{E}_{(\sigma,c)\sim\mathcal{D}} l_h(\sigma, c). \tag{6.1}$$

If $\mathcal{D} = \mathcal{D}_{train}$ corresponds only to the set of training data, then this global loss is called *empirical loss*. If $\mathcal{D} = \mathcal{D}_{true}$ corresponds to the complete underlying distribution for σ states, then this global loss is the *expected loss* (population risk), but in most cases one does not have direct access to $\mathcal{D}_{true}$.

Definition 2 (*Global loss minimization*) If one found the model that minimized the empirical loss (training process), one gets the *training loss* $\min_h L_{h,\mathcal{D}_{train}}$. The model obtained from this minimization we denote h_{train}

$$h_{train} = \mathrm{argmin}_h L_{h,\mathcal{D}_{train}}. \tag{6.2}$$

One can also instead minimize the expected loss to get $\min_h L_{h,\mathcal{D}_{true}}$. The model obtained from this minimization we denote h_{true}

$$h_{true} = \mathrm{argmin}_h L_{h,\mathcal{D}_{true}} \tag{6.3}$$

where h is optimized over some chosen hypothesis class of functions.

Definition 3 (*Generalization and approximation error*) The *generalization error* (estimation error) G_L is defined as the difference

$$G_L = L_{h_{train},\mathcal{D}_{true}} - L_{h_{true},\mathcal{D}_{true}}. \tag{6.4}$$

Therefore, if the training set better approximates the true distribution (e.g., finite data versus infinite data setting), the generalization error should approach zero. Note that while it is not in general possible to access $L_{h_{train},\mathcal{D}_{true}}$ directly since $\mathcal{D}_{true}$ is generally unknown, it can be approximated by the *testing loss* $L_{h_{train},\mathcal{D}_{test}}$, where

$\mathcal{D}_{test}$ is the distribution corresponding to the testing set.

The model minimizing loss in classification is the Bayes classifier and we can call this value of loss the *Bayes loss* $L_{h_{Bayes},\mathcal{D}_{true}}$. Then the *approximation error* A_L refers to the difference

$$A_L = L_{h_{true},\mathcal{D}_{true}} - L_{h_{Bayes},\mathcal{D}_{true}}. \tag{6.5}$$

The approximation error therefore captures how well the hypothesis class that the model is assumed to belong captures the true classification model.

Definition 4 (*Excess risk*) To assess how good the model h_{train} is, that is derived from the training process, we are interested then in the *excess risk* E_L

$$E_L = L_{h_{train},\mathcal{D}_{true}} - L_{h_{Bayes},\mathcal{D}_{true}} = G_L + A_L. \tag{6.6}$$

Note that since $\mathcal{D}_{true}$ is not generally accessible, the above quantity is approximated by replacing it with a test distribution $\mathcal{D}_{test}$.

6.2.2 *Definitions in Presence of Perturbations*

In the following, we want to deal with loss functions in the presence of perturbations to the input states σ. There are many definitions which must be clearly distinguished and can potentially have very different behavior.

In the presence of a perturbation, the true class itself can change, depending on the type of perturbation on the test state. this leads us to the following definitions where we distinguish perturbations into two types.

Definition 5 (*Relevant and irrelevant perturbations*) A *relevant perturbation* $\mathcal{N}_{rel,\sigma}$ and *irrelevant perturbation* $\mathcal{N}_{irr,\sigma}$ for the state σ are changes to the state such that

$$\begin{aligned} c(\mathcal{N}_{rel,\sigma}(\sigma)) &\neq c(\sigma) \\ c(\mathcal{N}_{irr,\sigma}(\sigma)) &= c(\sigma). \end{aligned} \tag{6.7}$$

We note that if the perturbation is artificially induced by an adversary, it is likely that there is $\sigma-$dependence in the noise $\mathcal{N}$. On the other hand, if the perturbations were induced as a result of the environment, it is more likely that the form of the perturbations themselves are independent of the state σ.

With respect to a specified perturbation, we can also define robust loss, and we identify three kinds.

Definition 6 (*Corrupted-instance robust loss*) Let the perturbation occur so that $\sigma \to \tilde{\sigma}$ for each state $\sigma \in \Sigma$. Then we define the *corrupted-instance robust loss* (CI-robust loss) as the global loss $\tilde{L}$

$$\tilde{L} \equiv \mathbf{E}_{(\sigma,c)\sim\mathcal{D}} l_h(\tilde{\sigma}, c_\sigma). \tag{6.8}$$

Definition 7 (*Prediction-change robust loss*) Let the perturbation occur so that $\sigma \to \tilde{\sigma}$ for each state $\sigma \in \Sigma$. Then we define the *prediction-change robust loss* (PC-robust loss) as the global loss L^*

$$L^* \equiv \mathbf{E}_{(\sigma,c)\sim\mathcal{D}} l_h(\tilde{\sigma}, h(\sigma)), \tag{6.9}$$

where we replace the true class c_σ with the model prediction $h(\sigma)$.

Definition 8 (*Error-region robust loss*) We define the *error-region robust loss* (ER-robust loss) as the global loss $\bar{L}$

$$\bar{L} \equiv \mathbf{E}_{(\sigma,c)\sim\mathcal{D}} l_h(\tilde{\sigma}, c_{\tilde{\sigma}}) \tag{6.10}$$

where we also take into account the change to the true class if the perturbation $\sigma \to \tilde{\sigma}$ occurs.

However, since the true class is usually not accessible (at least those not in the training set), it is not convenient to work with the error-region robust loss, and instead it is more convenient to work with the corrupted-instance robust loss.

6.2.2.1 $1 - 0$ Loss: Accuracy and Robustness

Here we can focus on the different definitions of robustness and accuracy when using $1 - 0$ loss, which is an important loss function for classification problems. Here the local loss $l(\sigma, c)$ is the identity function $\mathbf{1}(h \neq c)$, which takes the value of 1 if $h(\sigma) \neq c$ and is equal to 0 otherwise.

Definition 9 The *error region* $\mathcal{M}$ is defined as the set of samples $\sigma \in \Sigma$ where $h(\sigma) \neq c(\sigma)$.

Definition 10 We define *accuracy* of the model $h_{\vec{\theta}}$ as the probability

$$A_{\vec{\theta},\mathcal{D}} \equiv P_{(\sigma,c)\sim\mathcal{D}}(h_{\vec{\theta}}(\sigma) = c(\sigma)). \tag{6.11}$$

We will often suppress the subscripts $\{\vec{\theta}, \mathcal{D}\}$ in the cases where our model and distribution do not change in our analysis.

Definition 11 We define *corrupted-instance robustness accuracy* by the probability

$$\tilde{A}_{\vec{\theta},\mathcal{D}} \equiv P_{(\sigma,c)\sim\mathcal{D}}(h_{\vec{\theta}}(\tilde{\sigma}) = c(\sigma)) \tag{6.12}$$

where $\tilde{h}$ is a model modified by some perturbation $\mathcal{N}$.

Definition 12 We define *prediction-change robustness* by the probability

$$A^*_{\vec{\theta},\mathcal{D}} \equiv P_{(\sigma,c)\sim\mathcal{D}}(h_{\vec{\theta}}(\tilde{\sigma}) = h(\sigma)). \tag{6.13}$$

Although we will not be using the following definition in our analysis later—since corrupted-instance robustness accuracy and prediction-change robustness are more natural—for completeness we include also the error-region robustness accuracy.

Definition 13 We define *error-region robustness accuracy* by the probability

$$\bar{A}_{\vec{\theta},\mathcal{D}} \equiv P_{(\sigma,c)\sim\mathcal{D}}(h_{\vec{\theta}}(\tilde{\sigma}) = \tilde{c}(\sigma)). \tag{6.14}$$

In the rest of the chapter when using the terminology robustness accuracy, we use this to denote corrupted-instance robustness accuracy unless otherwise specified.

6.2.2.2 Measures of Robustness with Respect to Adversarial Perturbations

Note that all the above measures and definitions require a specification of what the perturbation is. However, in adversarial perturbations where by definition we do not necessarily have full information on the perturbation itself, and in addition these are usually 'small perturbations' according to some definition of "small," it is important to characterize the robustness of a model. In this case, it is more convenient to specify some minimum distance—again different possible definitions of distance are possible—to perturb a test state whereby the relationship between the model prediction and the true prediction changes. Just as there are three types of robust loss, here we have three corresponding types of robustness with respect to adversarial perturbations. We note that depending on the definition of distance one chooses, for example, l_1 or quantum trace distance, we will also obtain different values and can even obtain different behavior.

Definition 14 We define *corrupted-instance robustness to adversarial perturbations* by

$$\tilde{\Delta}(h,\sigma) = \min|\delta|, \quad h(\mathcal{N}_\delta(\sigma)) \neq c(\sigma), \tag{6.15}$$

where $|\delta|$ is some measure of distance between states σ and $\mathcal{N}_\delta(\sigma)$. We can also define the corresponding *expected corrupted-instance robustness to adversarial perturbations* by

$$\langle\tilde{\Delta}(h)\rangle = \mathbf{E}_{(\sigma,c)\sim\mathcal{D}}\Delta(h,\sigma). \tag{6.16}$$

Definition 15 We define *prediction-change robustness to adversarial perturbations* by

$$\Delta^*(h, \sigma) = \min |\delta|, \quad h(\mathcal{N}_\delta(\sigma)) \neq h(\sigma), \tag{6.17}$$

where $|\delta|$ is some measure of distance between states σ and $\mathcal{N}_\delta(\sigma)$. We can also define the corresponding *expected prediction-change robustness to adversarial perturbations* by

$$\langle \Delta^*(h) \rangle = \mathbf{E}_{(\sigma,c)\sim\mathcal{D}} \Delta(h, x). \tag{6.18}$$

Definition 16 We define *error-region robustness to adversarial perturbations* by

$$\bar{\Delta}(h, \sigma) = \min |\delta|, \quad h(\mathcal{N}_\delta(\sigma)) \neq c(\mathcal{N}_\delta(\sigma)), \tag{6.19}$$

where $|\delta|$ is some measure of distance between states σ and $\mathcal{N}_\delta(\sigma)$. We can also define the corresponding *expected error-region robustness to adversarial perturbations* by

$$\langle \bar{\Delta}(h) \rangle = \mathbf{E}_{(\sigma,c)\sim\mathcal{D}} \Delta(h, x). \tag{6.20}$$

6.3 Trade-Offs Between Accuracy and Robustness Accuracy

In this section, we will examine more closely—through simple examples—the relationship between robustness and accuracy, and robustness accuracy from the point of view of classification algorithms at the testing stage. We will use some simplified toy models to illustrate some key observations and to show why it's insufficient to study accuracy of a model on its own. We will see through some simple examples that trade-offs between accuracy and robustness accuracy can exist for both the $1-0$ loss and linear loss. From simple examples as a starting point, we can ask fundamental questions about the requirements we should have for our model which could be subject to noisy perturbations and when we do and do not expect trade-offs between accuracy and robustness accuracy.

For discussions on trade-offs between accuracy and robustness for classification of classical data (where both accuracy and robustness may be defined in various ways), see, for example, [1–5].

6.3.1 $1-0$ *Loss*

From our definition zoo in Sect. 6.2.2 focusing on $1-0$ loss, we see that the main players in characterizing how good the classifier are: (i) accuracy A, (ii) corrupted-instance robustness accuracy $\tilde{A}$, and (iii) prediction-change robustness A^*. A fundamental question is, what is the relationship between these key players?

Briefly, the accuracy denotes the probability that the model h makes the correct description when we are sampling test states from the full distribution $\mathcal{D}$. However, accuracy may not be sufficient if the model is not trained under particular types of noise that is expected to corrupt incoming new states. The corrupted-instance robustness accuracy refers to the accuracy of the model when each test state σ is perturbed by some perturbation $\mathcal{N}$, and we refer this as *robustness accuracy* for short. Since this quantity itself must be related to properties of the model itself, we must also define prediction-change robustness—which we refer as *robustness* for short—which refers to the probability that the model prediction itself does not change when each test state is subject to the same perturbation $\mathcal{N}$.

Here it is possible to identify a fundamental relationship between these quantities when considering K-class classification. Here we look at the simplest case where both the data and the model are unbiased, and we will make some simple observations.

Theorem 1 (Unbiased classifier) *If the data and the model are both unbiased, in the sense that $P(h = i) = 1/K = P(c = j)$ for K-class classification and for any $i, j = 1, ..., K$, then the following holds:*

$$\tilde{A} = A(2A^* - 1) + (1 - A^*) \tag{6.21}$$

$$= A^*(2A - 1) + (1 - A). \tag{6.22}$$

Proof See Appendix A for the proof.

From this simple relationship, we can already make some important observations:

1. It is insufficient to only have a high accuracy A. The aim for a good model is to have high robustness accuracy $\tilde{A}$—which depends just as much on robustness A^* as on accuracy A. Here A and A^* are interchangeable in terms of their effect on $\tilde{A}$.
2. Trade-offs between $\tilde{A}$ and A are possible. For example, if model has low robustness $A^* < 1/2$, higher accuracy leads to lower robustness accuracy, so training more won't help if the model itself is badly chosen. So the choice of the model itself and having in-built invariances to obtain high A^* is very important. Similarly, if accuracy is low $A < 1/2$, increasing robustness alone cannot achieve high robustness accuracy: in fact increasing robustness decreases robustness accuracy.
3. Theorem 1 can be rewritten $|\tilde{A} - A| = |(1 - 2A)(1 - A^*)|$, where it can be seen that the difference between accuracy and robustness accuracy depends independently on A and A^*.

In the above, it is clear that how well the model h is chosen and its own robustness properties is vital to the relationship between accuracy and robustness accuracy. We note that this relationship can also be modified from the above simplest form when different assumptions are included. A simple example we can show explicitly is when the model itself is biased.

Theorem 2 (Biased classifier) *If the data is unbiased, in the sense that* $1/K = P(c = j)$ *for* K*-class classification, but the model is biased. We look at the simplest setting where only one class is biased. Suppose it is class* a *that is biased, so then* $P(h = a) = \alpha$*, and the rest of the classes are unbiased, so* $P(h = i \neq a) = (1 - \alpha)/(K - 1)$ *for any* $i \neq a$*. Then the following holds:*

$$\tilde{A} = \frac{(1-2A)(K-1)}{(1-\alpha)K}\left(A^* + \left(\frac{1-\alpha}{\alpha(K-1)} - 1\right)P(\tilde{h} = h = a)\right) + (1 - A). \tag{6.23}$$

Proof See Appendix B for the proof.

These simple observations show the importance of distinguishing robustness and robustness accuracy and accuracy as three independent considerations that must be studied in their own right, as well as their dependencies upon each other. The fact that trade-offs between these quantities can exist also points out applications of this to investigating whether it is possible to train models under "incompatible noises," where more training under one class of noise can actually adversely affect robustness with respect to another type of noise.

Another application could be to the no free lunch theorem, which is based on the observation that for every distribution for which a model performs well, there should be some distribution for which the same model would perform badly. Although the free lunch theorem guarantees that these "dual distributions" should exist, what is the relationship between these distributions and how do we identify them? These questions are also intimately connected with the trade-off relationships between accuracy, robustness accuracy, and robustness that we see here.

6.3.1.1 Examples of Trade-Off Between Robustness Accuracy and Accuracy

In this section, we provide some examples both in classical and quantum models where trade-offs between accuracy and robustness accuracy can exist. By this, we mean we are provided with two different models h_1 and h_2 and the test states are selected from an identical distribution $\mathcal{D}$. They have corresponding accuracies A_1, A_2; robustness accuracies $\tilde{A}_1$, $\tilde{A}_2$; and robustness values A_1^*, A_2^*. A trade-off between robustness accuracy and accuracy is said to exist if when $A_1 < A_2$ we simultaneously have $\tilde{A}_1 < \tilde{A}_2$. See Appendix C for more details on the individual examples. Example 1 is from [2].

Example 1 Suppose we have the following two models for classical data $x = (x_1, ..., x_{d+1})$ that is selected from a distribution D:

$$\begin{aligned} H_1(x) &= \frac{1}{d}(x_2 + ... + x_{d+1}) \\ H_2(x) &= x_1. \end{aligned} \tag{6.24}$$

This model is related to the true class in the following way. Let $h(x) = \text{sign}(H(x))$. Then $c(x) = h_2(x)$ with probability p (i.e., proportion p of states selected from $x \sim D$ satisfy $c(x) = h_2(x)$). Also if $c(x) = \pm 1$, then $(x_2, ..., x_{d+1}) \sim D_\pm$. Suppose we have a noise model where we have the perturbation $x_2, ..., x_{d+1} \to x'_2, ..., x'_{d+1}$, where if $(x_2, ..., x_{d+1}) \sim D_\pm$ then $(x'_2, ..., x'_{d+1}) \sim D_\mp$, and $x'_1 \to x_1$.

Then when $p > 0.5$, it is possible to have the trade-off $A_1 \gg A_2$ and $\tilde{A}_1 \ll \tilde{A}_2$, so a more accurate model H_1 can lead to worse robustness accuracy.

Example 2 We can generalize Example 1 to a more general example. Suppose we have two models

$$\begin{aligned} H_1(x) &= f(x_1, x_2, ..., x_{d+1}) \\ H_2(x) &= g(x_1), \end{aligned} \tag{6.25}$$

where the perturbation is only on $x_2, ..., x_{d+1}$, leaving x_1 invariant, and they are selected from one of two distributions $D_\pm$ coinciding with $c(x) = \pm$. The effect of the perturbation is to change $D_\pm \to D_\mp$. Here again we have $A_2 = p = \tilde{A}_2$. Then if $A_1 > A_2 = p$, it is always true that $\tilde{A}_1 < \tilde{A}_2$ if $p > 1/2$.

Example 3 Suppose we have two quantum models H_1 and H_2 and a Hamiltonian $h \otimes \mathbf{1}^{\otimes d} + \mathbf{1} \otimes h \otimes \mathbf{1}^{\otimes d-1} + ... + \mathbf{1}^{\otimes d} \otimes h$ with local operator h. Let the input state be of the form $\sigma = \sigma_1 \otimes ... \otimes \sigma_{d+1}$ where $\sigma_{i=1,...,d+1}$ are local states that h acts on and

$$\begin{aligned} H_1(\sigma) &= \frac{1}{d} \sum_{i=1}^{d} \text{Tr}(h\sigma_i) \\ H_2(\sigma) &= \text{Tr}(h\sigma_{d+1}) \end{aligned} \tag{6.26}$$

and each $-1 \leq \text{Tr}(h\sigma_i) \leq 1$. We can define the distribution from which we select σ as D. These models are related to the true class in the following way. Let $h(\sigma) = \text{sign}(H(\sigma))$ such that $c(\sigma) = h_2(\sigma)$ with probability p. Also if $c(\sigma) = \pm 1$, then the values $\text{Tr}(h\sigma_i)|_{i=1,...,d} \sim D_\pm$.

Now we choose a noise model such that the effect of the perturbation is $\text{Tr}(h\sigma_i)|_{i=1,...,d} \to \text{Tr}(h\sigma'_i)|_{i=1,...,d} \sim D_\mp$.

Then it is possible to find a value $p > 0.5$ where we have the trade-off situation that higher accuracy $A_1 > A_2$ implies less robustness accuracy $\tilde{A}_1 < \tilde{A}_2$.

Example 4 We can look at more general models of the following form. We first define

$$f_i(\sigma) = \text{Tr}(\mathcal{O}_i \sigma) \tag{6.27}$$

as the ith *quantum feature with respect to* $\mathcal{O}$ of the state σ, where $\{\mathcal{O}_i\}_{i=1}^{d+1}$ is a set of Hermitian operators. Suppose the elements $\alpha_i = 0, 1$ for simplicity. Then we can compare two models

$$
\begin{aligned}
H_1(\sigma) &= f_1(\sigma) \\
H_2(\sigma) &= \sum_{i=1}^{d+1} \alpha_i f_i(\sigma)
\end{aligned}
\tag{6.28}
$$

and we are selecting states σ in a way such that $f_1(\sigma) = \pm 1$ and $f_{2\leq i\leq d+1}(\sigma)$ has two distributions $D_\pm$ corresponding to $c(\sigma) = \pm$. Suppose we choose perturbations where only $f_2, ..., f_{d+1}$ can change and leave f_1 invariant. Furthermore, the distributions change according to $D_\pm \to D_\mp$. We then define the accuracy A_1 when $P(c = 1) = 1/2 = P(c = -1)$ to be $A_1 = P(h_1 = c) \equiv p$. A similar argument can be followed where when $p > 0.5$, one can achieve $A_2 > A_1$ while $\tilde{A}_2 < \tilde{A}_1$.

Examples 1–4 focus on deterministic classifiers. We can also examine examples with probabilistic classifiers. In the case of a probabilistic classifier, the predicted label $h_s(\sigma)$ is a stochastic variable instead of a deterministic one given σ. For instance, we can consider this label as the outcome of a single-shot quantum measurement. We focus here on binary classifiers which can be straightforwardly generalized to the multi-class scenario.

However, the true label c is still deterministic. This means we are sampling the non-stochastic pair $\{\sigma_i, c_i\} \sim \mathcal{D}$ from some underlying distribution $\mathcal{D}$, where i can also range over continuous values. For simplicity of notation, we consider discrete i and use the summation sign instead of integrals.

Below we will distinguish between two types of probabilities. P_s is the probability associated with the stochastic variable getting a specific outcome given some σ, which we can model with a quantum expectation value as $P_s(h_s(\sigma) = j) = \mathrm{Tr}(\Pi_j \sigma)$, where $\{\Pi_j\}$ are quantum measurement operators, e.g., POVMs. On the other hand, P is the probability associated with averaging over some distribution of input states, for instance, $P(c(\sigma) = 1)$ is the probability of states sampled from $\mathcal{D}$ as having the true label 1. We can define then accuracy A_s of this stochastic classifier h_s as the probability it gives the same outcome as the true label:

$$
A_s \equiv \sum_{(\sigma_i, c_i)} P(\sigma_i, c_i) P_s(h_s(\sigma_i) = c_i). \tag{6.29}
$$

Then we the following lemma for an unbiased binary stochastic classifier. Here we use the definition $\sigma^\pm = \{\sigma | c(\sigma) = \pm 1\}$ and we can define the normalized quantum states:

$$
\Sigma^\pm \equiv \frac{\sum_{\sigma_i^\pm} P(\sigma_i^\pm)\sigma_i^\pm}{\sum_{\sigma_j^\pm} P(\sigma_j^\pm)}. \tag{6.30}
$$

Lemma 1 *We assume an unbiased dataset* $\sum_{\sigma_j^\pm} P(\sigma_j^\pm) = 1/2$. *We assume a noise model that modifies the measurement* $\Pi_{1,-1} \to \tilde{\Pi}_{1,-1} = \sum_l E_l \Pi_{1,-1} E_l^\dagger$ *where* $\{E_l\}$

are Kraus operators and $\Pi_1 = \mathbf{1} - \Pi_{-1}$. Then the difference between the robustness accuracy and accuracy is given by

$$|\tilde{A}_s - A_s| = \frac{1}{2}|Tr(\Pi_1(\tilde{\Sigma}^+ - \Sigma^+ + \Sigma^- - \tilde{\Sigma}^-)|. \tag{6.31}$$

We can also provide an upper bound to this difference

$$|\tilde{A}_s - A_s| \leq \tau(\Sigma_A, \Sigma_B), \tag{6.32}$$

where $\tau(\Sigma_A, \Sigma_B)$ is the trace distance between states $\Sigma_A \equiv (1/2)(\tilde{\Sigma}^+ + \Sigma^-)$ and $\Sigma_B \equiv (1/2)(\Sigma^+ + \tilde{\Sigma}^-)$.

Proof See Appendix D.

Note that furthermore, the projector Π_1 that *maximizes* this upper bound between robustness accuracy and accuracy (for any distribution of states) is achieved by the Helstrom measurement corresponding to the optimal discriminator of Σ_A and Σ_B.

We can now provide examples of cases where trade-offs between robustness accuracy and accuracy occur and when they do not.

Example 5 In the case where the perturbation leaves the states $\Sigma^\pm$ unchanged $\tilde{\Sigma}^\pm = \Sigma^\pm$, then $\tilde{A}_s = A_s$. However, if we have the transformation $\tilde{\Sigma}^\pm \to \Sigma^\mp$, then we have

$$|\tilde{A}_s - A_s| = |1 - 2A_s|. \tag{6.33}$$

So for $A_s > 1/2$, this clearly implies a trade-off between $\tilde{A}_s$ and A_s as A_s grows. We note that even a small perturbation is sufficient to achieve this outcome if the states $\Sigma^\pm$ are near the true decision boundaries and while such a modification could be considered an adversarial example because of the small size of the perturbation. We will see this in the next example. For now, we can test this relation on two simple examples of noise: Pauli noise channel and depolarization noise.

(Pauli noise) Suppose we have the Pauli noise channel $\mathcal{N}_{Pauli}$ for a single qubit σ_i where

$$\mathcal{N}_{Pauli}(\sigma_i) = p_0\sigma_i + p_1\sigma_i + p_2\sigma_i + p_3\sigma_i \tag{6.34}$$

and $p_0 + p_1 + p_2 + p_3 = 1$. Suppose we choose the simplest case where $\Pi_0 = |0\rangle\langle 0|$. Then we can derive

$$\tilde{\Sigma}^\pm = p_0\Sigma^\pm + p_1\sigma_x\Sigma^\pm\sigma_x + p_2\sigma_y\Sigma^\pm\sigma_y + p_3\sigma_z\Sigma^\pm\sigma_z. \tag{6.35}$$

It is straightforward to find

$$\tilde{A}_s = (1 - 2(p_1 + p_2))A_s + \frac{3}{2}(p_1 + p_2). \tag{6.36}$$

Then we have a trade-off if $d\tilde{A}_s/dA_s < 0$, but not if $\tilde{A}_s/dA_s \geq 0$. So we see that the trade-off condition is true when $p_1 + p_2 > 1/2$, but otherwise there is no trade-off. As special examples, for the case of bit-flip error (where only $p_0,\ p_1 \neq 0$), the trade-off condition corresponds to $p_2 > 1/2$, whereas in the case of phase flip (where $p_1.p_2 = 0$) we don't have any trade-off between $\tilde{A}_s$ and A_s.

(Depolarization noise) Another example is depolarization noise $\mathcal{N}_{dep}$ where

$$\mathcal{N}_{dep}(\sigma_i) = (1 - p)\sigma_i + \frac{p}{d}\mathbf{1} \tag{6.37}$$

and

$$\mathcal{N}_{dep}(\Sigma^{\pm}) = (1 - p)\Sigma^{\pm} + \frac{p}{d}\mathbf{1}. \tag{6.38}$$

This leads to

$$\tilde{A}_s = (1 - p)A_s + \frac{p}{2} \tag{6.39}$$

where $d\tilde{A}_s/dA_s = 1 - p > 0$, so there will never be a trade-off between robustness accuracy and accuracy for depolarization noise.

Example 6 We call quantum channels $\mathcal{N}_{adv}$ acting on input quantum states σ *adversarial perturbations* if the trace distance is below some given constant $\epsilon_{max} \ll 1$

$$\tau(\sigma, \mathcal{N}_{adv}(\sigma)) \leq \epsilon_{max} \tag{6.40}$$

for every input state $\sigma \sim \mathcal{D}$. Then it is possible to have a maximal trade-off between accuracy and robustness accuracy $\tilde{A}_s = 1 - A_s$ for adversarial perturbations of the input states if the following sufficient condition is satisfied:

$$\min_i |\frac{p_{-,i}}{p_{+,i}}\mathrm{Tr}(\sigma_i^+\sigma_i^-)| \geq 1 - \epsilon_{max}, \tag{6.41}$$

where $p_{\pm,i} \equiv P(\sigma_i^{\pm})/\sum_{\sigma_j^{\pm}} P(\sigma_j^{\pm})$. This is simple to show using $\mathcal{N}_{Adv}(\Sigma^{\pm}) = \Sigma^{\mp}$, and more specifically $\sigma_i^{\pm} \to \sigma_i^{\mp}$ with normalization constants taken into account. Then using $\tau(\sigma, \rho) \leq \sqrt{1 - \mathcal{F}(\sigma, \rho)}$, the result follows straightforwardly. However, we note that these perturbations are all due to *relevant perturbations* rather than *irrelevant perturbations*, since they change the true class of the states, so they belong to Type II perturbations, rather than Type I. From the above result, we see that the condition of a trade-off depends only on the distances between states of various classes, which is not a property of any particular model in itself, but rather an intrinsic feature of the problem.

We can also look at more general classes of trade-offs between robustness accuracy and accuracy. A more general class of examples of trade-offs between $\tilde{A}_s$ and A_s can come about if we consider small perturbations in the distribution $\mathcal{D}$ itself. Let the initial distribution be $\mathcal{D}_1$ and the distribution after perturbation be $\mathcal{D}_2$. We can define these distributions in the following way.

For example, suppose we have an observable $\mathcal{O}_j$ and we define $F_j(\sigma_i) = \mathrm{SignTr}(\mathcal{O}_j\sigma_i)$ as the sign of the jth *quantum feature* of the state σ_i. Then let us define the sign of one quantum feature F_0 such that

$$\begin{aligned} P(F_0(\sigma_i^{\pm}) &= \pm 1 | c(\sigma_i) = \pm 1) = f_0 \\ P(F_0(\sigma_i^{\mp}) &= \pm 1 | c(\sigma_i) = \mp 1) = 1 - f_0. \end{aligned} \tag{6.42}$$

So if $f_0 > 1/2$, then the feature F_0 is positively correlated with the true label c. We define $\sigma^{f_0^{\pm}} = \{\sigma | f_0(\sigma) = \pm 1\}$ and the following normalized quantum states:

$$\begin{aligned} \Sigma^{++} &= \frac{\sum_{\sigma_i^+} P(\sigma_i^+)\sigma_i^{f_0^+}}{\sum_{\sigma_j^+} P(\sigma_j^+)} \\ \Sigma^{+-} &= \frac{\sum_{\sigma_i^+} P(\sigma_i^+)\sigma_i^{f_0^-}}{\sum_{\sigma_j^+} P(\sigma_j^+)} \\ \Sigma^{-+} &= \frac{\sum_{\sigma_i^-} P(\sigma_i^-)\sigma_i^{f_0^+}}{\sum_{\sigma_j^-} P(\sigma_j^-)} \\ \Sigma^{--} &= \frac{\sum_{\sigma_i^-} P(\sigma_i^-)\sigma_i^{f_0^-}}{\sum_{\sigma_j^-} P(\sigma_j^-)}. \end{aligned} \tag{6.43}$$

Then we have the following general result on when to expect a trade-off between robustness accuracy and accuracy.

Example 7 Now we study a class of perturbations such that

$$\begin{aligned} \tilde{\Sigma}^{++} &= \Sigma^{+-} \\ \tilde{\Sigma}^{+-} &= \Sigma^{++} \\ \tilde{\Sigma}^{--} &= \Sigma^{-+} \\ \tilde{\Sigma}^{-+} &= \Sigma^{--}. \end{aligned} \tag{6.44}$$

Then it can be shown that for any $f_0 > 1/2$ we have the trade-off condition:

$$\tilde{A}_s \leq \frac{f_0}{1-f_0}(1-A_s). \tag{6.45}$$

From the definitions above, these perturbations occur if, for instance, we exchange the following:

$$\sigma_i^{f_0^+} \leftrightarrow \sigma_{i'}^{f_0^-} \tag{6.46}$$

so long as the pair $c_i = c_{i'}$. These are irrelevant perturbations. This means that we can always expect a trade-off between robustness accuracy and accuracy if the perturbations move the state labels according to a quantum feature that is strongly correlated with the true label. This analysis is a generalization of Example 5 where we simply exchange

$$\sigma_i^+ \leftrightarrow \sigma_{i'}^-. \tag{6.47}$$

Furthermore, following straightforwardly from results in Example 6, we see that the above trade-off between robustness accuracy and accuracy for *irrelevant perturbations* can occur if the following sufficient condition is satisfied:

$$\min_i \mathrm{Tr}(\sigma_i^{f_0^+}\sigma_i^{f_0^-}) \geq 1 - \epsilon_{max}. \tag{6.48}$$

Taking into account non-trivial f_0, we can also similarly derive the bound:

$$|\tilde{A}_s - A_s| \leq (2f_0 + 1)\tau(\bar{\Sigma}_A, \bar{\Sigma}_B), \tag{6.49}$$

where

$$\begin{aligned}\bar{\Sigma}_A &\equiv 4f_0(\tilde{\Sigma}^{++} + \tilde{\Sigma}^{-+} + \Sigma^{+-} + \Sigma^{--}) + 2(\tilde{\Sigma}^{+-} + \Sigma^{-+})\\ \bar{\Sigma}_B &\equiv 4f_0(\tilde{\Sigma}^{+-} + \tilde{\Sigma}^{--} + \Sigma^{-+} + \Sigma^{++}) + 2(\tilde{\Sigma}^{-+} + \Sigma^{+-}).\end{aligned} \tag{6.50}$$

6.3.1.2 Absence of Trade-Offs

We know from Theorem 1 that $A^* = 1$ is a *sufficient condition* for there to be *no trade-off* between $\tilde{A}$ and A. For example, see [4]. We can see that this can be a condition on both the input states themselves and a condition on the model h. Note that we only wish for robustness of our model against *irrelevant perturbations*, and we wish for non-robustness against *relevant perturbations*. We suppress indices to distinguish these types of perturbations since all robustness is defined with respect to irrelevant perturbations unless otherwise stated.

To find the circumstances when the model H under a noisy perturbation gives $A^* = 1$, we assume that for all $\sigma \sim D$, we are given $\min_{\sigma\sim D} p_\sigma$, where $p_\sigma = H_k(\sigma)$ (i.e., the probability that σ is assigned to the class that it's in) and σ is in class k. Therefore $\min_{\sigma\sim D} p_\sigma$ is a measure of the closest "distance" of the dataset to the decision boundary for the model. Our purpose is to find a sufficient and necessary condition on the upper bound of $\tau(\sigma, \rho)$ (where $H(\rho) = \tilde{H}(\sigma)$), so that we have prediction-change robustness, i.e., $H(\sigma) = H(\rho) = \tilde{H}(\sigma)$ for all states $\sigma \sim D$, which implies $A^* = 1$.

Locally, i.e., for a specific σ where $H(\sigma)|_{max} \equiv p_\sigma$, we know that a necessary and sufficient condition for prediction-change robustness (i.e., $\tilde{H}(\sigma) = H(\rho) = H(\sigma)$) is

$$\tau(\sigma, \rho) < \delta_\sigma(1 - \sqrt{1 - \delta_\sigma^2}) \equiv \tau_\sigma, \tag{6.51}$$

where $\delta_\sigma = \sqrt{1 - 4p_\sigma(1 - p_\sigma)}/2$. This is a statement about local prediction-change robustness. However, we care about global prediction-change robustness to find the conditions under which $A^* = 1$. To change the above into a global statement for all $\sigma \sim D$, this means we need some upper bound of $\tau(\sigma, \rho)$ for all $\sigma \sim D$. We know that a smaller p_σ should give rise to a larger upper bound to $\tau(\sigma, \rho)$, so we can define the state that gives this smallest probability:

$$\sigma_{min} \equiv \text{argmin}_{\sigma\sim D} p_\sigma. \tag{6.52}$$

Therefore, if we are given no information at all about which states in $\sigma \sim D$ are chosen, then a necessary and sufficient condition for $A^* = 1$ is for the perturbation τ to be below the size

$$\tau < \delta_{\sigma_{min}}(1 - \sqrt{1 - \delta_{\sigma_{min}}^2}) \equiv \tau_{\sigma_{\min}}. \tag{6.53}$$

Later we can compute conditions for other values $0 \leq A^* < 1$.

We can state this as our Proposition 1, which is a condition on the dataset for some given H.

Theorem 3 *For the dataset $\sigma \sim D$ and model H subject to noise where $\tilde{H}(\sigma) = H(\rho)$, then if the trace distance between σ and ρ satisfies the upper bound*

$$\tau(\sigma, \rho) < \delta_{\sigma_{min}}(1 - \sqrt{1 - \delta_{\sigma_{min}}^2}), \tag{6.54}$$

this is a sufficient condition for $A^* = 1$.

We can illustrate this in the following example. We can look at a concrete example of a quantum clustering model where

$$H_k(\sigma) = 1 - \tau(\sigma, C_k), \tag{6.55}$$

where C_k is the centroid state of the cluster K of states for $\sigma \sim D$. For instance, in the case where σ and all C_k are pure states, we have the simple model:

$$H_k(\sigma) = F(\sigma, C_k) = \mathrm{Tr}(|C_k\rangle\langle C_k|\sigma), \tag{6.56}$$

where the classifier is

$$h(\sigma) = \mathrm{argmax}_k H_k(\sigma). \tag{6.57}$$

Clearly, the largest $H_k(\sigma)$ occurs for the state C_k that σ is closest to, hence σ is identified to belong to cluster K. In this case, it is easy to see how the distribution of the sizes of the clusters are sufficient for us to find p_A, since

$$p_A(\sigma) = 1 - \tau(\sigma, C_A). \tag{6.58}$$

Therefore, if we are given the size of the largest cluster, then this gives us $p_{\sigma_{\min}}$, and hence $\tau_{\sigma_{\min}}$. Once inserted into Theorem 3, this would give us the sufficient condition for $A^* = 1$ for this model.

Aside from $A^* = 1$, we see from Theorem 1 that $1 > A^* > 1/2$ is also sufficient for no trade-offs between robustness accuracy and accuracy. Sufficient conditions for this to be satisfied would also be helpful. From Theorem 3, we saw that for any case where $\tau < \tau_{\sigma_{\min}}$, then $A^* = 1$ since $h(\rho) = h(\sigma)$ for all $\sigma \sim D$. However, it may be that we require only $h(\rho) = h(\sigma)$ for some percentage of $\sigma \sim D$ instead of all cases. Suppose we fix $A^* = a$ where $0 \leq a < 1$ is some constant. For the distribution D, suppose we are given also the distribution of p_σ values (i.e., the maximum probability of σ being in a particular class, which will also be the class assigned to σ). Let this distribution be labeled $P_\sigma(p_A)$, distributed over all p_A values for all $\sigma \sim D$. Then we have the following related results.

A sufficient condition for the probability $A^* = a$ to hold is when perturbations of any $\sigma \in D$ satisfy

$$\tau < \delta_{p_a}(1 - \sqrt{1 - \delta_{p_a}^2}) \tag{6.59}$$

where

$$a = A^* = \int_{p_a}^{1} P_\sigma(p_A) dp_A. \tag{6.60}$$

This means that either we can be given some upper bound to τ where we calculate the corresponding p_a and hence the corresponding a value, or that we are given a and identify the corresponding upper bound to τ. Either way, this requires access to the distribution $P_\sigma(p_A)$, which one may or not be able to access.

6.3.1.3 Applications in Studying Trade-Offs

There are at least three main areas where the study of trade-offs between robustness accuracy and accuracy may be helpful for

1. incompatible noise,
2. adversarial perturbations,
3. no free lunch theorem,

and we will briefly discuss these and present open questions.

Incompatible noise

We first present a definition of *incompatible noise*.

Definition 17 Let $\tilde{A}_s^{(1)}$ be the robustness accuracy with respect to noise $\mathcal{N}_1$ and $\tilde{A}_s^{(2)}$ is the robustness accuracy with respect to noise $\mathcal{N}_2$. We call two noise perturbations $\mathcal{N}_1$ and $\mathcal{N}_2$ *incompatible noise* if

$$\frac{d\tilde{A}_s^{(2)}}{d\tilde{A}_s^{(1)}} < 0. \tag{6.61}$$

This means that if we train a model to be more accurate in presence of noise $\mathcal{N}_1$, it becomes less accurate when in presence of noise $\mathcal{N}_2$.

We can show a simple example of this for a quantum classifier.

Example 8 As an illustrative simple example for single qubits with $\Pi_1 = |0\rangle\langle 0|$, we let $\mathcal{N}_1$ be depolarization noise with noise parameter p_1 and $\mathcal{N}_2$ be bit-flip noise with noise parameter p_2, so $\mathcal{N}_2(\Sigma^\pm) = (1 - p_2)\Sigma^\pm + p_2\sigma_x\Sigma^\pm\sigma_x$. Then we can derive

$$\begin{aligned}\tilde{A}_s^{(1)} &= (1 - p_1)A_s + \frac{p_1}{2} \\ \tilde{A}_s^{(2)} &= (1/2 - p_2)A_s + (1/2 + p_2)/2\end{aligned} \tag{6.62}$$

hence

$$\frac{d\tilde{A}_s^{(2)}}{d\tilde{A}_s^{(1)}} = \frac{d\tilde{A}_s^{(2)}}{dA_s}\frac{dA_s}{d\tilde{A}_s^{(1)}} = \frac{1/2 - p_2}{1 - p_1}. \tag{6.63}$$

In this case, since $p_1 < 1$, depolarization noise and bit-flip noise are *incompatible* if and only if $p_2 > 1/2$.

By understanding which perturbations are incompatible with each other, we can better understand the ultimate limitations of our model and to enable better model building.

Adversarial examples

From Example 6, we saw that one way to consider adversarial perturbations are quantum channels $\mathcal{N}_{adv}$ acting on input quantum states σ such that the trace distance is below some given constant $\epsilon_{max} \ll 1$

$$\tau(\sigma, \mathcal{N}_{adv}(\sigma)) \leq \epsilon_{max} \tag{6.64}$$

for every input state $\sigma \sim \mathcal{D}$: these include both Type I and Type II perturbations. However, to get true *adversarial examples for state* σ, we specifically require three conditions:

- $\tau(\sigma, \mathcal{N}_{adv}(\sigma)) \leq \epsilon_{max}, \qquad \epsilon \ll 1$;
- $\mathcal{N}_{adv}$ acting on σ are also *irrelevant perturbations*, i.e., do not change the true label of σ: $c(\sigma) = c(\mathcal{N}_{adv}(\sigma))$;
- $h(\sigma) \neq h(\mathcal{N}_{adv}(\sigma))$.

Examples 1–6 all present cases of trade-offs between robustness accuracy and accuracy for *relevant perturbations*. However, in Example 7, we saw that we could also have these trade-offs for *irrelevant perturbations*, and Example 8 also allows the possibility for irrelevant perturbations. Thus trade-offs might also exist in the case of training to make models more robust to adversarial examples. For instance, in the above three conditions, the last deals with the corrupted-instance robustness of the model A^*, and we already know from Theorem 1 that a low robustness $A^* < 1/2$ to $\mathcal{N}_{adv}$ of Type I perturbations on σ is sufficient to guarantee that increasing accuracy A alone will only lead to worse and worse robustness accuracy. We will see examples of this type of behavior in the next subsection with linear loss.

No free lunch theorem

General trade-off behavior between robustness accuracy and accuracy, to both relevant and irrelevant perturbations, can also be considered from the perspective of the no free lunch theorem and also provide ways of constructing the best models. The no free lunch theorem is the observation that for every distribution $\mathcal{D}$ of states for which a model h performs well (e.g., has high accuracy), there should be some distribution $\tilde{\mathcal{D}}$ of states for which the same model h would perform badly. However, the no free lunch theorem doesn't provide an explicit construction of these distributions for any model, since it's only an existence theorem. We would like to understand what the relationship between one distribution where the particular model h works very well versus a distribution for which the model has low accuracy.

We can connect the setting of the no free lunch theorem to the relationship between accuracy and robustness accuracy in the following way.

We first observe that when we make a perturbation $\mathcal{N}$ on the input state σ for the original distribution $\mathcal{D}_1$, we note that we can instead consider this as the same scenario as when selecting σ from a modified distribution $\mathcal{D}_2$. This means we can rewrite the loss function for the perturbed states $\tilde{\sigma} = \mathcal{N}(\sigma)$ where $\sigma \sim \mathcal{D}$ as identical to the loss function of the unperturbed states selected from a different distribution $\sigma \sim \mathcal{D}_2$

$$\tilde{L}_{\mathcal{D}_1} = \mathbf{E}_{(\sigma,c)\sim\mathcal{D}_1} l_h(\tilde{\sigma}, c_\sigma) = \mathbf{E}_{(\sigma,c_\sigma)\sim\mathcal{D}_2} l_h(\sigma, c_\sigma) = L_{\mathcal{D}_2}. \tag{6.65}$$

This simple rewriting leads to the following interpretation. In the context of $1-0$ loss, the first quantity is the corrupted-instance robustness accuracy with respect to $\mathcal{N}$ when selecting states from $\mathcal{D}_1$. The second quantity is accuracy, when selecting states from a different distribution $\mathcal{D}_2$, where $\mathcal{D}_2$ is simply the selection of states in $\mathcal{D}_1$ but all perturbed by $\mathcal{N}$. This means that it's possible to consider the robustness accuracy of one model with respect to a distribution $\mathcal{D}_1$ as identical to the accuracy of the same model with respect to another distribution $\mathcal{D}_2$.

Thus we can write $\tilde{A}_1 = A_2$. If A^* is the fraction of states σ that satisfy $h(\sigma) = h(\mathcal{N}(\sigma))$, then from Theorem 1 we have

$$A_2 = A_1(2A^* - 1) + (1 - A^*). \tag{6.66}$$

Therefore a trade-off relationship between A_1 and A_2 denotes that the better a model h is with respect to one distribution $\mathcal{D}_1$, the worse the model h is with respect to distribution $\mathcal{D}_2$. We can find the relationship between $\mathcal{D}_1$ and $\mathcal{D}_2$ precisely when we identify the perturbation $\mathcal{N}$ such that $A^* < 1/2$. As a worst-case example, when $A^* = 0$

$$A_2 = 1 - A_1, \tag{6.67}$$

and this means that for the set of states $\sigma \sim \mathcal{D}_1$ and the model h where $h(\sigma) \neq h(\mathcal{N}(\sigma))$ for all states $\sigma \sim \mathcal{D}_1$, when h is the most accurate for $\sigma \sim \mathcal{D}_1$, the same model h is least accurate when the input states are selected from $\mathcal{N}(\sigma)$ where $\sigma \sim \mathcal{D}_1$.

Thus the trade-off relationship between robustness accuracy and accuracy could also provide a window into more constructive ways of examining the no free lunch theorem and to be able to better pinpoint which models work more effectively for which distributions.

6.3.2 *Linear Loss*

We can similarly consider models for classification based on linear loss instead of $1-0$ loss, and examine the relationship between the global loss L and the corrupted-instance robust loss $\tilde{L}$. Note that in the previous section for the $1-0$ loss function, the global loss is termed accuracy A and the corrupted-instance robust loss is termed the corrupted-instance robustness accuracy $\tilde{A}$.

For the linear loss model, we have

$$l_h = -c_i H(\sigma_i) \tag{6.68}$$

where $h(\sigma_i) = \text{sign}H(\sigma_i)$. Suppose we have M training states $\{(\sigma_i, c_i)\}_{i=1}^M$. Then we can write

$$L_\theta = -\frac{1}{M}\sum_{i=1}^{M} c_i H_\theta(\sigma_i)$$
$$\tilde{L}_\theta = -\frac{1}{M}\sum_{i=1}^{M} c_i H_\theta(\tilde{\sigma}_i). \tag{6.69}$$

The relationship between L and $\tilde{L}$ can similarly be applied to identify *incompatible noise*, adversarial perturbations and example, and the no free lunch theorem. We will only provide a few examples here for illustration, since many similar examples already appeared for $1-0$ loss that already motivated some of our fundamental questions.

To understand the relationship between L and $\tilde{L}$, we first study an idealized case, where we assume that there are $K_\theta = K$ states in the training set such that

$$H(\tilde{\sigma}_i) = H(\sigma_i), \tag{6.70}$$

and we denote these as "robust" states σ_{i_R}. The rest of the $M-K$ states satisfy

$$H(\tilde{\sigma}_i) = -H(\sigma_i) \tag{6.71}$$

and we denote these states by $\sigma_{i_{\bar{R}}}$. In this case, $K/M = A^*$. Then we can decompose the loss functions in the following way:

$$\begin{aligned} L &= -\frac{1}{M}\left(\sum_{i_{R=1}}^{K} c_i H(\sigma_{i_R}) + \sum_{i_{\bar{R}}=1}^{M-K} c_i H(\sigma_{i_{\bar{R}}})\right) \\ &= \frac{K}{M}L_R + \frac{M-K}{M}L_{\bar{R}}, \end{aligned} \tag{6.72}$$

where L_R and $L_{\bar{R}}$ denote the "robust" and "non-robust" parts of the loss function and are defined as

$$L_R = -\frac{1}{K}\sum_{i_{R=1}}^{K} c_i H(\sigma_{i_R}) \tag{6.73}$$

$$L_{\bar{R}} = -\frac{M-K}{M}\sum_{i_{\bar{R}}=1}^{M-K} c_i H(\sigma_{i_{\bar{R}}}). \tag{6.74}$$

Similarly we can rewrite

$$\tilde{L} = -\frac{1}{M}\left(\sum_{i_R=1}^{K} c_i H(\tilde{\sigma}_{i_R}) + \sum_{i_{\bar{R}}=1}^{M-K} c_i H(\tilde{\sigma}_{i_{\bar{R}}})\right)$$
$$= \frac{K}{M}L_R - \frac{M-K}{M}L_{\bar{R}}. \tag{6.75}$$

Comparing L and $\tilde{L}$, it becomes clear where any possible trade-off between L and $\tilde{L}$ comes from: the non-robust $L_{\bar{R}}$ part of the cost function and a small A^*, since

$$\frac{L-\tilde{L}}{2} = \frac{M-K}{M}L_{\bar{R}}. \tag{6.76}$$

We observe from the above expression that if the initial good performance of the model—measured according to L—is mostly due to the "non-robust" parts $L_{\bar{R}}$, then when it comes to robustness training using $\tilde{L}$, the model will perform poorly. Although $L_{\bar{R}}$ is not quite the prediction-change robust loss L^*, this case is still analogous to the case of $1-0$ loss, where if robustness is poor $A^* < 1/2$, even if A is very high, $\tilde{A}$ will be low.

We can then examine the relationship between L and $\tilde{L}$ for quantum models.

Theorem 4 *Suppose we look at quantum models where*

$$H(\sigma) = Tr(\mathcal{M}\sigma). \tag{6.77}$$

Then for binary classification with unbiased data we have the difference

$$L - \tilde{L} = Tr(\mathcal{M}(T^- - T^+)) \tag{6.78}$$

where $T^\pm \equiv (\Sigma^\pm + \tilde{\Sigma}^\mp)/2$ *and* $\sigma_i^\pm = \{\sigma_i | c_i = \pm 1\}$ *and* $\Sigma^\pm = (2/M)\sum_{i=1}^{M/2}\sigma_i^\pm$, $\tilde{\Sigma}^\pm = (2/M)\sum_{i=1}^{M/2}\tilde{\sigma}_i^\pm$.

Proof This follows easily from the definitions

$$L = -\frac{1}{M}\left(\sum_{i^+=1}^{M/2}\mathrm{Tr}(\mathcal{M}\sigma_i^+) - \sum_{i^-=1}^{M/2}\mathrm{Tr}(\mathcal{M}\sigma_i^-)\right) = -\frac{1}{2}(\mathrm{Tr}(\mathcal{M}\Sigma^+) - \mathrm{Tr}(\mathcal{M}\Sigma^-)), \tag{6.79}$$

$$\tilde{L} = -\frac{1}{2}(\mathrm{Tr}(\mathcal{M}\tilde{\Sigma}^+) - \mathrm{Tr}(\mathcal{M}\tilde{\Sigma}^-)). \tag{6.80}$$

Suppose we have two irrelevant perturbations $\mathcal{N}_1$ and $\mathcal{N}_2$. In the idealized case there are K states that obey

$$H(\mathcal{N}_1(\sigma_i)) = H(\mathcal{N}_2(\sigma_i)) \tag{6.81}$$

and $M - K$ states obeying

$$H(\mathcal{N}_1(\sigma_i)) = -H(\mathcal{N}_2(\sigma_i)), \tag{6.82}$$

where the former states are "robust" states, whereas the latter are considered "non-robust." Then the relationship between the L_1 and L_2 is given by

$$L_1 - L_2 = 2\frac{M-K}{M} L_{\bar{R}} \tag{6.83}$$

where

$$L_{\bar{R}} = -\frac{1}{M} \sum_{i_{\bar{R}}}^{M-K} c_i H(\sigma_{i_{\bar{R}}}) \tag{6.84}$$

where $\{\sigma_{i_{\bar{R}}}\}$ are the set of states of the non-robust states. In the more realistic quantum case, we have

$$L_1 - L_2 = \mathrm{Tr}(\mathcal{M}(N^- - N^+)) \tag{6.85}$$

where $N^\pm \equiv (\Omega_1^\pm + \Omega_2^\mp)/2$ with $\Omega_1^\pm = (2/M)\sum_{i=1}^{M/2} \mathcal{N}_1(\sigma_i^\pm)$ and $\Omega_2^\pm = (2/M)\sum_{i=1}^{M/2} \mathcal{N}_2(\sigma_i^\pm)$.

We see that it's also possible to have trade-offs between robust loss and loss for those adversarial perturbations that are in addition irrelevant perturbations.

Example 9 Here we have an explicit example for a single-qubit model where an irrelevant perturbation gives rise to a trade-off between the loss and the perturbed loss. For instance, let us have the input states from the training set:

$$\begin{aligned}
\sigma_1^+ &= |0\rangle\langle 0| \\
\sigma_2^+ &= |1\rangle\langle 1| \\
\sigma_1^- &= \frac{5}{6}|0\rangle\langle 0| + \frac{1}{6}|1\rangle\langle 1| \\
\sigma_2^- &= \frac{1}{3}|0\rangle\langle 0| + \frac{2}{3}|1\rangle\langle 1|.
\end{aligned} \tag{6.86}$$

Suppose under an irrelevant perturbation $\mathcal{N}_{irr}$ we have the following transition:

$$\begin{aligned}
\tilde{\sigma}_1^+ &= |1\rangle\langle 1| \\
\tilde{\sigma}_2^+ &= \frac{1}{4}|0\rangle\langle 0| + \frac{3}{4}|1\rangle\langle 1| \\
\tilde{\sigma}_1^- &= \frac{5}{6}|0\rangle\langle 0| + \frac{1}{6}|1\rangle\langle 1| \\
\tilde{\sigma}_2^- &= \frac{4}{5}|0\rangle\langle 0| + \frac{1}{5}|1\rangle\langle 1|.
\end{aligned} \tag{6.87}$$

In this case, we see that

$$\begin{aligned}
\Sigma^+ &= \frac{1}{2}(|0\rangle\langle 0| + |1\rangle\langle 1|) \\
\Sigma^- &= \frac{1}{36}(21|0\rangle\langle 0| + 15|1\rangle\langle 1| \\
\tilde{\Sigma}^+ &= \frac{1}{8}(|0\rangle\langle 0| + 7|1\rangle\langle 1|) \\
\tilde{\Sigma}^- &= \frac{1}{60}(49|0\rangle\langle 0| + 11|1\rangle\langle 1|).
\end{aligned} \tag{6.88}$$

Here we see that

$$\begin{aligned}
L &= \frac{1}{12}\mathrm{Tr}(\mathcal{M}(|0\rangle\langle 0| - |1\rangle\langle 1|)) \\
\tilde{L} &= -\frac{83}{120}\mathrm{Tr}(\mathcal{M}(|0\rangle\langle 0| - |1\rangle\langle 1|)).
\end{aligned} \tag{6.89}$$

This means that L and $\tilde{L}$ cannot be simultaneously minimized. In fact, minimizing L leads to $\mathcal{M} = |1\rangle\langle 1| - |0\rangle\langle 0|$, which would in fact maximize $\tilde{L}$.

We note that there are fundamental reasons why linear loss with linear quantum models are not always sufficient to contain the symmetries of the true underlying models. For instance, we know (from previous theorems [ ]) that if $\tau(\sigma, \rho) < p_\sigma - 1/2$, then it is the case that $h(\sigma) = h(\rho)$ for the linear quantum model h. However, suppose we have $\sigma = \sigma^+$ and $\rho = \mathcal{N}(\sigma^-)$, which means that $c(\rho) = -c(\sigma)$. If in addition $\tau(\sigma^+, \tilde{\sigma}^-) < p_\sigma - 1/2$, this means that $h(\rho) = h(\sigma)$, which cannot represent c exactly. Therefore, if enough states σ satisfy $\tau(\sigma^+, \tilde{\sigma}^-) < p_\sigma - 1/2$, then the ability of h to represent c is low. We can also consider the following example.

Example 10 We have the training states $\{\sigma_1^+, \sigma_2^+, \sigma_3^-, \sigma_4^-\}$ and the states modified by irrelevant perturbations, so that the test states are $\{\tilde{\sigma}_1^+, \tilde{\sigma}_2^+, \tilde{\sigma}_3^-, \tilde{\sigma}_4^-\}$, where by definition

$$\begin{aligned} c(\sigma_1^+) &= c(\sigma_2^+) = c(\tilde{\sigma}_1^+) = c(\tilde{\sigma}_2^+) = 1 \\ &= -c(\sigma_3^-) = -c(\sigma_4^-) = -c(\tilde{\sigma}_3^-) = -c(\tilde{\sigma}_4^-). \end{aligned} \tag{6.90}$$

In this example, we see that when L_{train} is optimized there is a large trade-off with L_{test}. Suppose we use the linear loss using $h(\sigma) = \mathrm{sign} H(\sigma)$ so that $L = -\sum_i c_i h(\sigma_i)$. Now we see that the condition

$$\tau(\sigma_i^\pm, \mathcal{N}_{irr}(\sigma_i^\mp)) \leq p_{\sigma_i} - 1/2 \tag{6.91}$$

for $i = 1, 2, 3, 4$ is sufficient to make $L = -\tilde{L}$ when L is optimized. For instance, the above condition is sufficient to ensure that h does *not* preserve the same symmetries as c, while still optimizing L, since

$$c(\sigma_i^\pm) = h(\sigma_i^\pm) = \pm 1 \tag{6.92}$$

while

$$c(\tilde{\sigma}_i^\pm) = -h(\tilde{\sigma}_i^\pm) = \pm 1. \tag{6.93}$$

Then we see that while $L = -4$, we get $\tilde{L} = 4$. In this case, we see this to be a consequence of the states $\mathcal{N}_{irr}(\sigma^\pm)$ being too close to $\sigma^\mp$ in trace distance, while they are actually differently classified.

Note that this is *not* necessarily saying that $\mathcal{N}_{irr}$ is an adversarial perturbation, since it is possible for $\tau(\sigma_i^\pm, \mathcal{N}_{irr}(\sigma_i^\mp))$ to be small, while $\tau(\sigma_i^\pm, \mathcal{N}_{irr}(\sigma_i^\pm))$ not to be small (hence not adversarial).

From these examples, it is clear that it is necessary in some cases it is better that the symmetries of the true classification be explicitly included.

6.3.3 Role of Relevant and Irrelevant Features

We know from previous examples that standard training (optimizing accuracy) and robustness training (optimizing robustness accuracy) for states selected from the *same* distribution can give rise to different outcomes in terms of how well they ultimately perform, and this is because different properties or *features* of the input states are being learned or prioritized. Although we might have an intuitive feeling what a feature might mean (e.g., the roundness of a ladybug, the cuteness level of a cat, the toxicity of a death-cap mushroom, the purity of a quantum state, etc.), we must begin by formally defining what we might mean by a feature.

Definition 18 We can define a *feature* f_Λ with respect to operator Λ acting on a quantum state σ as $f_\Lambda : \sigma \to \mathbb{R}$, where

$$f_\Lambda(\sigma) = \text{Tr}(\Lambda(\sigma)). \tag{6.94}$$

Then we can call a feature robust or non-robust depending on how well correlated this feature is to the true label $c(\sigma)$ of σ. In principle, Λ can also be nonlinear (e.g., $\Lambda(\sigma) = \sigma^2$, as in the case of obtaining the purity of a quantum state).

Definition 19 We call a feature f_Λ *k-useful* for a given distribution $\mathcal{D}$ if

$$\mathbf{E}_{(\sigma,c)\sim\mathcal{D}}(c(\sigma) \cdot f_\Lambda(\sigma)) \geq k. \tag{6.95}$$

Definition 20 We say f_Λ is a *robust feature* with respect to $(\mathcal{D}, \mathcal{N}_\delta, k, \gamma)$ if it is k-useful and that under perturbations $\mathcal{N}_\delta$ it remains γ-useful, so

$$\mathbf{E}_{(\sigma,c)\sim\mathcal{D}}(\inf_\delta c(\sigma) \cdot f_\Lambda(\mathcal{N}_\delta(\sigma))) \geq \gamma. \tag{6.96}$$

Then a useful but non-robust feature is a feature which is k-useful for some $k > 0$ but is not γ-robust for any $\gamma \geq 0$. Note that the above robust feature definition can also be extended to be defined with respect to different loss functions and having different versions of robust features in the same way we have different robustnesses in Sect. 2.

A working hypothesis is that, given $\mathcal{D}$ and hypothesis class $\mathcal{H}$, if the (standard) accuracy relies heavily on k-useful but non-robust features, then the accuracy can be high while the robustness accuracy remains low. This would identify the cause of these trade-off relationships and provide a means to more efficiently resolve these, if possible.

To express this mathematically, we have the following conjecture.

Conjecture 1 Let the feature f_Λ be $k-$useful, so let $\mathbf{E}_{(\sigma,c)\sim\mathcal{D}}(c(\sigma) \cdot f_\Lambda(\sigma)) \geq k_\Lambda$ and it is non-robust, so $\mathbf{E}_{(\sigma,c)\sim\mathcal{D}}(\inf_\delta c(\sigma) \cdot f_\Lambda(\mathcal{N}_\delta(\sigma))) \leq \gamma_\Lambda$. Then if the hypothesis function *h* has high standard accuracy so $A > a > 1/2$, then it is possible to show that this provides an upper bound to the robustness accuracy $\tilde{A} < g(k_\Lambda, \gamma_\Lambda, a)$, where larger k_Λ, small γ, and larger a lead to smaller $\tilde{A}$.

There are some models which demonstrate this intuition and we discuss simple cases.

Example 11 We can give a concrete demonstration of a quantum example, where we have the product state $\sigma = \sigma_0 \otimes \sigma_1 \otimes ... \otimes \sigma_d$. We can also consider this a map $x_i \to \sigma_i$. Other possibilities exist, but we start with a simple example.

The distribution we are selecting from $(\sigma, c(\sigma)) \sim \mathcal{D}$ is specified. So we let $\{p_i, \sigma_i\}$ to be the σ states we are selecting from, where p_i are the probabilities in which density matrices σ_i are sampled. Note that this means the full density matrix describing the distribution $\bar{\sigma} = \sum_i p_i \sigma_i$ is *not* part of the distribution $\mathcal{D}$. *Only* the states ρ_i are included.

Suppose each $\sigma_i = |\sigma_i\rangle\langle\sigma_i|$ are single pure qubits and $|\sigma_0\rangle = \cos(\theta_0/2)|0\rangle + \exp(i\phi)\sin(\theta_0/2)|1\rangle$. Let the true class $c(\sigma)$ be *defined* as whether the qubit σ_0 is in the upper hemisphere of the Bloch sphere or not, so

$$c(\sigma) = \text{sgn}(\text{Tr}(|0\rangle\langle 0|\sigma_0) - 1/2) = \text{sgn}(\cos(\theta_0^2/2) - 1/2). \tag{6.97}$$

This means if σ_0 is in the upper hemisphere, or $\theta_0 \in [0, \pi/2)$, then $c(\sigma) = 1$ and otherwise $\theta_0 \in [\pi/2, \pi]$ so $c(\sigma) = -1$.

We can rewrite $c(\sigma) = \text{sgnTr}((\lambda_0 \otimes \mathbf{1}^{\otimes d})\sigma)$ where $\lambda_0 = |0\rangle\langle 0| - (1/2)\mathbf{1} = (1/2)(|0\rangle\langle 0| - |1\rangle\langle 1|)$. However, the classifier $h(\sigma)$ need not be perfect. Suppose it misclassifies a fraction p of states in the upper hemisphere but classifies all the states in the lower hemisphere correctly. Then we can write such a hypothesis as

$$h_0(\sigma) = \text{sgn}(\text{Tr}((\lambda_0' \otimes \mathbf{1}^{\otimes d})\sigma) \tag{6.98}$$

where $\lambda_0' = |0\rangle\langle 0| - (1/2 + \eta)\mathbf{1} = (1/2 - \eta)|0\rangle\langle 0| - (1/2 + \eta)|1\rangle\langle 1|$.

Suppose in our distribution half of the states σ_0 are sampled from the upper hemisphere and half the states are sampled from the lower hemisphere, so $P(c(\sigma) = 1) = 1/2 = P(c(\sigma) = -1)$. By construction of our hypothesis, we define that $P(c(\sigma) = 1 = -h_0(\sigma)) = p$ and $P(c(\sigma) = -1 = h_0(\sigma)) = 1/2$. If we use this hypothesis, then we know that the accuracy is the probability

$$\begin{aligned} P(c(\sigma) = h_0(\sigma)) &= P(c(\sigma) = 1 = h_0(\sigma)) + P(c(\sigma) = -1 \\ &= h_0(\sigma)) = (1/2 - p) + 1/2 = 1 - p. \end{aligned} \tag{6.99}$$

This classifier $h_0(\sigma)$ is quite robust in the sense that the distribution of σ_0 (originally evenly distributed over the whole Bloch sphere) must be modified substantially before the accuracy decreases (i.e., moving σ_0 from the top hemisphere all toward the equator around which the fraction of p states are).

However, this $h_0(\sigma)$ is not the only hypothesis. Suppose we have observables on the other states σ_1, ..., σ_d, but these are more weakly correlated with $c(\sigma)$. For instance, let the distribution of $\sigma_{l\neq 0}$ be such that the feature

$$f_l(\sigma) = \text{Tr}(\Lambda_l \sigma) = \text{Tr}(\lambda_l \sigma_l) \tag{6.100}$$

has a probability distribution centered on $\pm|\xi| << 1$ for $c(\sigma) = \pm 1$ and we call these distributions G_+ and G_-, respectively. The local operators $\lambda_l = (1/2)(|0\rangle\langle 0| - |1\rangle\langle 1|$ and the details of the distributions $G_\pm$ come from the choice in states σ_1, ..., σ_d (e.g., unlike for σ_0 which are chosen randomly across the entire Bloch sphere, $\sigma_{l\neq 0}$ are chosen to lie only around the equator of the Bloch sphere where the decision boundary is.

Suppose for these distributions $P_{G_+}(f_l(\sigma) = -1 = -c(\sigma)) = \bar{p} = P_{G_-}(f_l(\sigma) = 1 = -c(\sigma))$ where $\bar{p} < 1/2$. Since $|\xi|$ is very small, we expect $\bar{p}$ to be quite close to $1/2$, which is quite large. This means a large probability of misclassification. For instance, if we used just the hypothesis function $h_l(\sigma) = \text{sgn} f_l(\sigma)$, then the accuracy is

$$P(c(\sigma) = h_l(\sigma)) = 1 - \bar{p}. \tag{6.101}$$

However, suppose we added all of the $l \neq 0$ features together. So each feature on its own is only weakly correlated with $c(\sigma)$, but together they could make a better classifier in the following way. Let

$$h(\sigma) \equiv (1/d)\text{sgn}(\sum_{l=1}^{d} f_l(\sigma)). \tag{6.102}$$

Now, as d grows, we know that the standard deviation the sum of the $G_\pm$ distributions decreases (central limit theorem). In the limit $d \to \infty$, the *sum* of the $G_\pm$ distributions should be delta functions centered on $\pm|\xi|$. This means that the probability the whole sum should be of the opposite sign goes from $\bar{p}$ (for a single l) to a probability $\bar{p}' << \bar{p}$, which drastically increases the accuracy $P(c(\sigma) = h(\sigma)$, even though $h(\sigma)$ consists only of features that are weakly correlated with the true class.

These weakly correlated features are also non-robust features, we can perturb the distributions of $\sigma_1 \otimes ... \otimes \sigma_d$ slightly so that $G_\pm \leftrightarrow G_{mp}$. One possibility is that one can write

$$\begin{aligned} P(h(\sigma') = c(\sigma)) &= \frac{1}{2}(P_{G_+}(h(\sigma') = 1|c(\sigma) = 1) + P_{G_-}(h(\sigma') = -1|c(\sigma) = -1)) \\ &= \frac{1}{2}(P_{G_-}(h(\sigma) = 1|c(\sigma) = 1) + P_{G_+}(h(\sigma) = -1|c(\sigma) = -1)). \end{aligned} \tag{6.103}$$

However, in this form in the last line it cannot be compared directly to the original standard accuracy

$$P(h(\sigma) = c(\sigma)) = \frac{1}{2}(P_{G_+}(h(\sigma) = 1|c(\sigma) = 1) + P_{G_-}(h(\sigma) = -1|c(\sigma) = -1)). \tag{6.104}$$

One method to compare these terms is to break up $P(h = 1|c = 1)$ terms into $P(h = 1|f_0 = 1)P(f_0 = c) + P(h = 1|f_0 = -1)P(f_0 = -c)$, so just like in the classical example above. The proof then follows in exactly the same way. Then it is clear how the trade-off appears in the quantum case, that one can train classifiers to be very accurate, but they will not be robust.

Example 12 We can extend to more general setups, so instead of making the $(x_0, x_1, ..., x_d) \to (\sigma_0, \sigma_1, ..., \sigma_d)$ correspondence, we can also map the amplitudes of a d-dimensional state σ, or some other variant.

So we adopt the following ansatz, where $h(\sigma) = \text{sgn}H(\sigma)$

$$H(\sigma) = \sum_l w_l f_l(\sigma) = \sum_l w_l \text{Tr}(\Lambda_l \sigma). \tag{6.105}$$

We want to generalize the notion of what is a robust and what is a non-robust feature. So in the previous example, we saw clearly how projection of $|0\rangle\langle 0|$ onto σ_0 gives robust features, but projection onto $\sigma_{l\neq 0}$ does not. Of course, the robustness properties are dependent very much on the distributions from which the states σ are selected. The key goal is to identify tests on the individual features to test for both their contribution to accuracy and their contribution to the accuracy when a perturbation is present. The details of the analysis will differ depending the on the loss functions chosen.

For example, for the $1 - 0$ loss function, the interesting quantum quantities are

$$P(c(\sigma) = w_l f_l(\sigma)) \tag{6.106}$$

and

$$P(c(\sigma) = w_l f_l(\sigma')). \tag{6.107}$$

If the former is high and the latter is even lower, then know this lth feature is a feature that contributes a lot to the standard accuracy and robustness accuracy trade-offs.

For other loss functions, it's some other correlation between $c(\sigma)$ and $w_l f_l(\sigma)$.

In the end, what we want is a decomposition of $A - \tilde{A}$ in terms of the difference between the accuracies for every individual feature f_l so we can identify the features most responsible for the trade-off and try to eliminate those non-robust features and to inject more robust features. So the procedure is

$$\text{Choose loss function} \to A(l) - \tilde{A}_\delta(l) \to A - \tilde{A}_\delta, \tag{6.108}$$

where the subscript δ denotes that the robustness accuracy is defined with respect to the perturbation $\mathcal{N}_\delta$. The first arrow depends on the loss function chosen. However, the last arrow may also be difficult because the total accuracy may not be a linear sum of the individual components.

However, we can first examine a case where such a linear relationship exists. So when $H(\sigma) = \sum_l H_l(\sigma)$ and $h_l(\sigma) = \text{sgn}H_l(\sigma)$, let the *linear loss function* be written as

$$\mathbf{E}_{(\sigma,c)\sim D}\mathcal{L}(\sigma, c(\sigma), H(\sigma)) = \mathbf{E}_{(\sigma,c)\sim D}\sum_l \mathcal{L}(\sigma, c(\sigma), H_l(\sigma)). \tag{6.109}$$

In this case

$$A - \tilde{A}_\delta = \sum_l (A(l) - \tilde{A}_\delta(l)) \tag{6.110}$$

and in this way the dominant l components can be identified. These linear loss functions will then be linear functions of $H(\sigma)$, which is in turn linear in the features $f_l(\sigma)$.

For these models, we can separate robust and the most non-robust features in the following way. By labeling robust features by R and non-robust features by $\bar{R}$, we can write

$$\begin{aligned} H_{l\in R}(\mathcal{N}_\delta(\sigma)) &= H_{l\in R}(\sigma) \\ H_{l\in \bar{R}}(\mathcal{N}_\delta(\sigma)) &= -H_{l\in \bar{R}}(\sigma). \end{aligned} \tag{6.111}$$

Of course, the exact relationships are simplified in the above example for an idealized case, but it can be generalized. Then if $\mathcal{L}(\sigma, c(\sigma), H_l(\sigma)) = -c(\sigma)H_l(\sigma)$ we have

$$\begin{aligned} |A - \tilde{A}_\delta| &= |\mathbf{E}_{(\sigma,c)\sim D} c(\sigma)(\sum_l^L - \sum_{l\in R}) H_l(\sigma) + \sum_{l\in \bar{R}} H_l(\sigma)| \\ &= 2|\mathbf{E}_{(\sigma,c)\sim D} c(\sigma) \sum_{l\in \bar{R}} H_l(\sigma)| \end{aligned} \tag{6.112}$$

where we used $\sum_l^L = \sum_{l\in R} + \sum_{l\in \bar{R}}$. From this analysis, we see that the trade-off is due entirely to the non-robust features, and if the non-robust features were to disappear, then the trade-offs would disappear.

What about *nonlinear* loss functions in the features? For instance, if we have a nonlinear function $g(\cdot)$ as a function of $-c(\sigma)H(\sigma)$, then

$$\begin{aligned} &|A - \tilde{A}_\delta| = \\ &|\mathbf{E}_{(\sigma,c)\sim D}(g(c(\sigma)(\sum_{l\in R} H_l(\sigma) + \sum_{l\in \bar{R}} H_l(\sigma))) - g(c(\sigma)(\sum_{l\in R} H_l(\sigma) - \sum_{l\in \bar{R}} H_l(\sigma))))|. \end{aligned} \tag{6.113}$$

So if $g(\cdot)$ is a K-Lipschitz continuous function, then

$$|A - \tilde{A}_\delta| \le 2K|\mathbf{E}_{(\sigma,c)\sim D} c(\sigma) \sum_{l\in \bar{R}} H_l(\sigma)|. \tag{6.114}$$

This means that even for a very accurate classifier that relies on non-robust features, the amount of trade-off is dependent at most on the non-robust elements only.

The above analysis points out that the key quantities to figure out are the non-robust elements: what they are and how they are related to the distribution $\mathcal{D}$ and the perturbation $\mathcal{N}_\delta$. Note that the really problematic features are the ones that are *both non-robust and useful*, where the latter means they contribute to a high A.

So the key question is: how to figure out which features are both useful and non-robust, hence one needs to get rid of them in one's classifier? Furthermore, once they are eliminated, how to replace them by robust features so that the accuracy is not compromised?

Note that in the above inequality, the right-hand side is related to the usefulness of the non-robust features to the standard accuracy. So if those terms are low, then the difference between the accuracy and robustness accuracy must likewise be low. Otherwise, they can be large.

One can also apply spectral analysis to identify better those features that are both useful and robust, for example, see techniques in [6] for computer vision.

6.4 Connection Between Robustness in Quantum Machine Learning and Dynamical Systems

Robustness in machine learning (both classical and quantum) is not only a "static" property of a trained model, as has been considered thus far. In fact, closely related concepts to robustness—numerous forms of stability, for instance—emerges from the study of dynamical processes that can govern both the training stage and inference stage in continuous-time analog models. The notion of stability comes in many forms and is heavily studied in dynamical systems, including Lyapunov stability, asymptotic stability, stability in optimization, and stability in signal processing and control theory (e.g., bounded input-bounded output BIBO). While stability is chiefly concerned with how a system reacts to small perturbations in initial conditions or parameters, often focusing on the system's ability to return to an equilibrium point, different notion of robustness refers to good performance in the presence of uncertainty and perturbations.

By framing learning as a dynamical system, we can gain access to powerful mathematical tools—from ordinary and partial differential equations, stochastic calculus, optimal control to quantum mechanics, which can both sharpen and broaden our theoretical perspective to unlock new architectures and training paradigms that can perform better in the presence of uncertainty, noise and adversarial perturbations. We just mention a few possibilities here.

Gradient-based methods for training can be considered in terms of dynamical systems. Training a machine learning model using gradient descent can be understood as evolving a parameter vector $\theta(t)$ under the influence of a force field derived from the loss function $\mathcal{L}(\theta(t))$. In the continuous-time limit, this evolution satisfies the gradient flow differential equation

$$\frac{d\theta(t)}{dt} = -\nabla_\theta \mathcal{L}(\theta(t)) \tag{6.115}$$

which is a deterministic dynamical system. The trajectories of this flow dictate how the model evolves toward local minima. Studying the stability of these flows, particularly through the Lyapunov framework, allows us to assess convergence properties and sensitivity to initialization. The robustness of the resulting model is thus linked to the geometry of the loss landscape and the conditioning of the flow—flatter minima often correspond to more robust solutions.

Machine learning algorithms can also be built on continuous-time forms of residual neural networks (both classical and quantum), also called neural ordinary and neural partial differential equations (NODEs and NPDEs). For example, the model h evolves with continuous depth t as

$$\frac{dh(t)}{dt} = f(h(t), t; \theta). \tag{6.116}$$

Stability in dynamical systems refers to the behavior of a system under small perturbations to its internal state or initial conditions, and here the stability of the solutions is governed by the Jacobian $J = \partial f / \partial h$. If all eigenvalues of the Jacobian have non-positive real parts, the trajectories of $h(t)$ are contractive or non-expansive, implying that small perturbations in initial conditions $h(0)$ do not grow uncontrollably with time. This stability ensures that the system behaves predictably and is essential for numerical integration and convergence.

In these dynamical systems, the concepts of stability and robustness, while closely related, play distinct roles and are governed by different mathematical principles. Robustness, on the other hand, is concerned with the model's output sensitivity to perturbations in its input, including noise, distribution shifts, or adversarial attacks. While a stable NODE might suppress the effects of internal perturbations, robustness assesses the model's functional generalization—its ability to maintain output performance despite changes or uncertainty in the input state σ. If in addition $\|f(h_1, t) - f(h_2, t)\| \leq L\|h_1 - h_2\| \; \|f(\mathcal{N}_\delta(\sigma)) - f(\sigma)\|$, the system is Lipschitz continuous, and the solution is stable under bounded perturbations. This property translates to robustness against adversarial perturbations, since the model resists sharp changes in output due to small input noise.

In summary, while stability is a mathematical property of the internal learning dynamics, and often a prerequisite for training convergence and numerical solvability, robustness relates to the reliability of a model's predictions under external disturbances. The two are connected: stability in neural ODEs and PDEs often enhances robustness, but the reverse is not guaranteed. A model can be stable but still vulnerable to specific structured input attacks. Understanding both are critical in designing machine learning systems that are dependable in real-world, noisy, or adversarial settings.

On a related note, robustness can also be framed through feedback control, where training adapts to perturbations using corrective signals. Consider the controlled learning dynamics

$$\frac{d\theta(t)}{dt} = -\nabla_\theta \mathcal{L}(\theta(t)) + K(\sigma, \delta) \tag{6.117}$$

where K is a feedback term based on perturbation $\mathcal{N}_\delta$ with "size" δ—however one chooses to define the size according to a specific metric. By designing K using robust control techniques (e.g., H-infinity control), one can stabilize the learning process even under worst-case data corruptions. This control-theoretic lens provides a new way to consider for adversarial robustness and other types of robustness from the viewpoint of dynamical systems, for both classical and quantum learning.

The language of stability, more familiar in the study of ODEs and PDEs, can be relevant in other ways to notions of robustness of both classical and quantum machine learning. For example, adversarial training solves a minimax optimization problem:

$$\min_\theta \mathbf{E}_{(\sigma,c)} \max_\delta \mathcal{L}(\mathcal{N}_\delta(\sigma), c, \theta). \tag{6.118}$$

This can be interpreted through the Hamilton-Jacobi-Bellman (HJB) equation, which governs optimal control under adversarial dynamics. In this formulation, the adversary acts as a controller that perturbs the input, and training seeks a policy (parameter θ) that minimizes the worst-case loss. Viewing adversarial training through this lens facilitates PDE-based approximations that lead to robust classifier architectures. Here the differences between classical and quantum cases would lie in different measures of distance, different structure of the loss function, different classes of "natural" input states, etc.

There are also machine learning algorithms that are designed for the learning of different dynamical systems, e.g., physically informed neural networks (PINNs), and in these cases it's clear that there is a potential rich interplay of the various notions of stability and robustness for these applications.

6.5 Discussion and Open Questions

From the theorems and examples we've explored in this chapter, we can see that it is very important to be careful of the definitions that are used, for example, robustness, accuracy, and robustness accuracy. Even from simple examples, these suggest some preliminary conclusions:

– Accuracy and robustness accuracy trade-offs can occur due to relevant perturbations.

- Accuracy and robustness accuracy trade-offs can also occur due to irrelevant perturbations, for example, when the model has low accuracy (below a certain threshold) as well as a high model robustness. Alternatively, it can also occur when the model has high enough accuracy (above a threshold), but the model has low model robustness. This can be due to the hypothesis class for the model not allowing for the symmetry corresponding to the irrelevant perturbation. Here a higher accuracy gives a lower robust accuracy. This in turn shows that robustness accuracy can be improved with higher model robustness against irrelevant perturbations if accuracy is already high enough.
- It is possible to find sufficient conditions where accuracy and robustness accuracy trade-offs do not occur at all.
- When training states include the same symmetry corresponding to the irrelevant perturbations, the trade-off condition can be shown to improve and even disappear altogether under fairly general conditions.
- To distinguish the effect of relevant and irrelevant perturbations, one can more effectively improve the models by explicitly including known symmetries.
- It is possible that when a model is trained under *incompatible noise*, better accuracy under training with one type of perturbation can induce worse accuracy when training with another type of perturbation. This is particularly important when requiring robustness of quantum models, to explore especially when one trains models under quantum noise.
- The robustness accuracy and accuracy trade-off relationship can inform a constructive form of the no free lunch theorem, which can help suggest better inductive biases for the distributions under which the model works well.

We see that even simple examples in this chapter already suggest many open questions, for example:

1. **General Trade-off Characterization**: Can a complete taxonomy of trade-off types be formulated, especially in higher-dimensional quantum systems with non-trivial entanglement and noise correlations? What are the chief differences between classical and quantum examples?
2. **Incompatible noise**: Can a taxonomy be developed more systematically to identify incompatible quantum noise, and what are the implications for the training of quantum models?
3. **No free lunch theorem**: How can better understanding of trade-off relationships between different forms of robustness and accuracy provide a more constructive form of the no free lunch theorem, to suggest better inductive biases? What are the differences between classical and quantum cases?
4. **Robust Training Strategies**: What are the optimal training algorithms and cost functions that prioritize robustness without excessively compromising accuracy? Can quantum-specific regularization strategies be developed?
5. **Robustness in Regression and Unsupervised Learning**: This chapter focused on classification problems. Can these definitions and trade-off principles be extended rigorously to regression tasks or quantum unsupervised learning frameworks?

6. **Adversarial Quantum Examples**: Is it always possible to construct quantum adversarial examples efficiently where no trade-offs between accuracy and robustness accuracy exist, and are there practical defenses against them in near-term quantum hardware?
7. **Model Selection and Hypothesis Class Constraints**: Are there universal bounds on robustness accuracy for specific classes of quantum models?
8. **Connections to Dynamical Systems**: We noted parallels with dynamical stability and control theory. Can ideas from Lyapunov stability or Hamiltonian systems yield new insights into robustness in quantum learning, in a more system way?

Acknowledgements This book chapter is dedicated to the memory of a good colleague and friend Peter Wittek, with whom the author wrote the first quantum paper studying the robustness of quantum classification algorithms against adversarial perturbations.
NL acknowledges funding from the Science and Technology Commission of Shanghai Municipality (STCSM) grant no. 24LZ1401200 (21JC1402900), NSFC grant no.12471411 and no. 12341104, the Shanghai Jiao Tong University 2030 Initiative, the Shanghai Science and Technology Innovation Action Plan (24LZ1401200) and the Fundamental Research Funds for the Central Universities.

Appendix A: Proof of Theorem 1

Let C_σ be the class label output of our noiseless model for input state σ. However, this model does not necessarily coincide with the "ground truth" for σ, which we present by T_σ. Thus the model is correct for σ if $T_\sigma = C_\sigma$ and is incorrect for σ if $C_\sigma \neq T_\sigma$. For both our training and our test states, let us sample from the states σ from some (usually unknown) distribution $\mathcal{D}$. Then by *accuracy in the absence of noise* we mean the probability that $T_\sigma = C_\sigma$ when $\sigma \sim \mathcal{D}$, which can be denoted

$$A \equiv P_{\sigma\sim\mathcal{D}}(C_\sigma = T_\sigma) \equiv P(C = T) \tag{6.119}$$

where we have dropped the subscripts for convenience. Now let us include noise in our model so $C_\sigma \to \tilde{C}_\sigma$. Then the *accuracy in the presence of noise* can denoted by the probability

$$\tilde{A} \equiv P_{\sigma\sim\mathcal{D}}(\tilde{C}_\sigma = T_\sigma) \equiv P(\tilde{C} = T). \tag{6.120}$$

There is *another* type of accuracy which we can call *robustness accuracy*, which refers to the probability that the model itself gives rise to the same prediction after adding noise, irrespective of the relationship to the ground truth. This robustness accuracy we can denote by

$$A^* = P_{\sigma\sim\mathcal{D}}(\tilde{C}_\sigma = C_\sigma) \equiv P(\tilde{C} = C). \tag{6.121}$$

Usually it is A and $\tilde{A}$ that is of interest for generalization performance, so by relating $\tilde{A}$ with A, we can determine by what amount accuracy degrades in the presence of noise. However, we will see that this relationship will also depend on A^* and it is through A^* that we can include information about the type of noise that is added.

Let there be K classes, so each of C, $\tilde{C}$, T can take values 0, 1, ..., $K-1$. So now we can rewrite $\tilde{A}$ as

$$\tilde{A} \equiv P(\tilde{C}=T) = \sum_{j=0}^{K-1} P(\tilde{C}=T=j) = \sum_j P(\tilde{C}=j|T=j)P(T=j) = \frac{1}{K}\sum_j P(\tilde{C}=j|T=j)$$
$$= \frac{1}{K}\left(\sum_j P(\tilde{C}=j|C=j)P(C=T) + \sum_j\sum_{k\neq j} P(\tilde{C}=j|C=k\neq j)P(C\neq T)\right)$$
$$= \frac{1}{K}\left(A\sum_j P_{jj} + (1-A)\sum_j\sum_{k\neq j} P_{jk}\right), \tag{6.122}$$

where $P_{ij} \equiv P(\tilde{C}=i|C=j)$. We also used the generic assumption in the first line that the samples we have are "unbiased," in the sense that there are as many states of one class as the other determined by the ground truth, so $P(T=j)=1/K$ for any j.

From normalization of probability, we can write $P_{kk} = 1-\sum_{j\neq k} P_{jk}$, which allows us to rewrite $\sum_j\sum_{k\neq j} P_{jk} = \sum_k(\sum_{j\neq k} P_{jk}) = \sum_k(1-P_{kk}) = K - \sum_j P_{jj}$. Now inserting this into Eq. (6.122) we find

$$\tilde{A} = \frac{1}{K}(A\sum_j P_{jj} + (1-A)(K-\sum_j P_{jj}) = \frac{1}{K}(1-2A)\sum_j P_{jj} + (1-A). \tag{6.123}$$

Now using the generic assumption in the second line that the samples we have are "unbiased" also with respect to the mode, so $P(C=j)=1/K$, we see that

$$\sum_j P_{jj} = \sum_j \frac{P(\tilde{C}=C=j)}{P(C=j)} = KP(\tilde{C}=C) = KA^*. \tag{6.124}$$

Therefore, we can write the general relation for *any* noise type as

$$\tilde{A} = A^*(2A-1) + (1-A). \tag{6.125}$$

So now we see that the source of the decrease in accuracy $\tilde{A}$ due to noise itself is due *only* to $A^* = P(\tilde{C}=C)$. The rest of the accuracy dependence is on the accuracy in the noiseless case A which is independent of noise.

It is through the behavior of A^* that makes depolarization noise quite special, since in the *infinite sampling* limit $N \to \infty$, then $\tilde{C}_\sigma = C_\sigma$ for all states σ, which gives $A^* = 1$. This was the result already proved in Lemma 1, in the manuscript with details in Appendix A. Therefore, if we allow $N \to \infty$ sampling of the quantum circuit, inserting $A^* = 1$ into Eq. (6.125) gives

$$\tilde{A} = (2A - 1) + (1 - A) = A, \tag{6.126}$$

so the accuracy in the absence of noise is the same as the accuracy in the presence of depolarization noise! Note that this is special to depolarization noise having the property $\tilde{C} = C$ and is not true for other noises. However, there are also other classes of noises that can help the model remain robust. Some other robustness properties have been investigated in [44] for other types of noises and their impact on accuracy can be found through Eq. (6.125).

However, in the *finite sampling* limit, $A^* \neq 1$ for depolarization noise. We already proved in Lemma 2 of the manuscript that $A^* > 1 - 2\exp(-2N\zeta^2)$ where N is the number of samples we take from the quantum circuit and ζ is the precision to which we determine the quantity $\tilde{\mathbf{y}}_0(\sigma)$. Therefore, we see that in the finite sampling limit, the accuracy in the presence of noise degrades as

$$\tilde{A} > (1 - 2\exp(-2N\zeta^2))(2A - 1) + (1 - A), \tag{6.127}$$

where $\tilde{A} \to A$ exponentially quickly as N grows, so accuracy in the presence of depolarization noise is not in fact compromised very much. Thus, all effects of depolarization noise on accuracy can be remedied by taking more measurements (efficiently).

Appendix B: Proof of Theorem 2

The argument runs in the same way as for Theorem 1. Here we look at K-class classification where the data is unbiased, but the model is biased. However, we look at the simplest setting where only one class is biased. Suppose it is class A that is biased, so then $P(h = a) = \alpha$, and the rest of the classes are unbiased, so $P(h = i \neq a) = (1 - \alpha)/(K - 1) \equiv \alpha'$. In this case, we just need the formula for $\sum_j P_{jj} = P(\tilde{h} = 1|h = 1) + ... + P(\tilde{h} = K|h = K)$, since in the multi-class case we have derived:

$$\tilde{A} = \frac{1}{K}(1 - 2A)\sum_j P_{jj} + (1 - A). \tag{6.128}$$

So we begin with

$$
\begin{aligned}
A^* &= P(\tilde{h} = h = 1) + \ldots + P(\tilde{h} = h = K) \\
&= \alpha P(\tilde{h} = a | h = a) + \alpha' \sum_{i \neq A} P(\tilde{h} = i | h = i) \\
&= \alpha \sum_j P_{jj} + (\alpha' - \alpha)(\sum_j P_{jj} - P(\tilde{h} = a | h = a)) \\
&= \alpha' \sum_j P_{jj} - (\alpha'/\alpha - 1) P(\tilde{h} = h = a).
\end{aligned} \tag{6.129}
$$

We can rearrange to find

$$
\sum_j P_{jj} = \frac{1}{\alpha'}(A^* + (\alpha'/\alpha - 1) P(\tilde{h} = h = a)) \tag{6.130}
$$

where $\alpha' = (1 - \alpha)/(K - 1)$. Finally we have the relation

$$
\tilde{A} = \frac{(1 - 2A)(K - 1)}{(1 - \alpha)K} \left(A^* + \left(\frac{1 - \alpha}{\alpha(K - 1)} - 1 \right) P(\tilde{h} = h = a) \right) + (1 - A), \tag{6.131}
$$

which reduces to Theorem 1 expression when $\alpha' = \alpha = 1/K$.

Appendix C: Details of Examples

Example 1

We now have a noise model where we have the perturbation $x_2, \ldots, x_{d+1} \rightarrow x'_2, \ldots, x'_{d+1}$, where if $(x_2, \ldots, x_{d+1}) \sim D_{\pm}$ then $(x'_2, \ldots, x'_{d+1}) \sim D_{\mp}$.

Under this noise model, we have the transformation

$$
\begin{aligned}
\tilde{H}_1(x) &= \frac{1}{d}(x'_2 + \ldots + x'_{d+1}) \\
\tilde{H}_2(x) &= H_2(x),
\end{aligned} \tag{6.132}
$$

where we see that $\tilde{H}_2 = H_2$ but $\tilde{H}_1 \neq H_1$. We can easily see that $A_2 = p$ and $A^*_2 = 1$. Assuming that $P(c = 1) = 1/2 = P(c = -1)$, A_1 can be computed using

$$
A_1 = \frac{1}{2}(P(h = 1 | c = 1) + P(h = -1 | c = -1)). \tag{6.133}
$$

This accuracy can be increased to arbitrarily high values. For example, if $D_\pm$ are Gaussian distributions centered on $\pm\eta c$ for some $\eta > 0$ with standard deviation 1 and for large enough d for the central limit theorem to apply, then $A_1 = P(h = c)$ corresponds to the probability that the value of a Gaussian distribution centered on η with standard deviation $1/d$ has a value larger than 0. This means that so long as η is larger than $1/\sqrt{d}$, we can have a high A_1. We also find that

$$\begin{aligned} A_1^* &= P_{x_2,\ldots x_{d+1}\sim D_+}(h_1(x) = \tilde{h}_1(x)) \\ &+ P_{x_2,\ldots x_{d+1}\sim D_-}(h_1(x) = \tilde{h}_1(x)) = 0 \end{aligned} \tag{6.134}$$

since we can imagine a perturbation where $x_{i\geq 2} \to -x_{i\geq 2}$ to give $D_\pm \to D_\mp$. This means that $h_1(x) = -\tilde{h}_1(x)$ for all x selected from the distribution $D_+ \bigcup D_-$. To compute the robustness accuracy, we note that the noise perturbation makes the following change in distribution $D_\pm \to D_\mp$, which means that for the Gaussian distribution example we have $\tilde{A}_1 = 1 - A_1$, which from Eq. (6.21) is also consistent with $A_1^* = 0$.

In the special case of the Gaussian distributions, we have

$$\begin{aligned} &A_1 = 1 - \mathrm{Erf}(\eta\sqrt{d}),\, A_1^* = 0,\, \tilde{A}_1 = 1 - A_1 \\ &A_2 = p,\, A_2^* = 1,\, \tilde{A}_2 = A_2. \end{aligned} \tag{6.135}$$

This means that we can have the trade-off scenario where $A_1 >> A_2$ and $\tilde{A}_1 = 1 - A_1 << 1 - A_2 < A_2 = \tilde{A}_2$ when $A_2 = p > 0.5$. Therefore, in the scenario $p > 0.5$, we have a trade-off condition where a better model in accuracy $H_2 \to H_1$ can lead to a less robust model.

Example 2

This means we have the following relationships between probabilities:

$$\begin{aligned} &p_{++} \equiv P_{\{x_1,x_2,\ldots,x_{d+1}\}\sim D_+}(f = 1\, g = 1) \\ &= P_{\{x_1,x_2,\ldots,x_{d+1}\}\sim D_+}(f = 1\, g = -1) \equiv p_{-+} \\ &p_{+-} \equiv P_{\{x_1,x_2,\ldots,x_{d+1}\}\sim D_-}(f = 1\, g = 1) \\ &= P_{\{x_1,x_2,\ldots,x_{d+1}\}\sim D_-}(f = 1\, g = -1) \equiv p_{--} \\ &p'_{++} \equiv P_{\{x_1,x_2,\ldots,x_{d+1}\}\sim D_+}(g = 1\, g = 1) \\ &= P_{\{x_1,x_2,\ldots,x_{d+1}\}\sim D_-}(g = 1\, g = 1) \equiv p'_{+-} = 1 \\ &p'_{-+} \equiv P_{\{x_1,x_2,\ldots,x_{d+1}\}\sim D_+}(g = 1\, g = -1) \\ &= P_{\{x_1,x_2,\ldots,x_{d+1}\}\sim D_-}(g = 1\, g = -1) \equiv p'_{--} = 0. \end{aligned} \tag{6.136}$$

Then it is possible to compute

$$\begin{aligned} A_1 &= 1 - \frac{1}{2}pa - \frac{1}{2}(1-p)b \\ \tilde{A}_1 &= \frac{1}{2}(1-p)a + \frac{1}{2}pb, \end{aligned} \tag{6.137}$$

where $a \equiv 1 - p_{++} + p_{--}$ and $b \equiv 1 - p_{-+} + p_{+-}$. From this we can prove that if $A_1 > A_2 = p$, then it is always true that $\tilde{A}_1 < \tilde{A}_2$ if $p > 1/2$. We see that $A_1 > p$ implies $1/2pa < (1-p)(1-1/2b)$. $p > 1/2$ is equivalent to $1 - p < p$, which implies $1/2pa < p(1-1/2b)$. Inserting this inequality into $\tilde{A}_1 = 1/2(1-p)a + 1/2pb < 1/2pa + 1/2pb < p(1-1/2b) + 1/2pb = p = \tilde{A}_2$ gives our result.

Example 3

If the distributions $D_\pm$ are the same Gaussian distributions in Example 1, then the argument proceeds in the same way, so that if $1 - \mathrm{Erf}(\eta\sqrt{d}) > p > 0.5$, then we have a trade-off situation that $A_1 > A_2$ gives $\tilde{A}_1 < \tilde{A}_2$.

This means that if we are training a model

$$H(\sigma) = \sum_{i=1}^{d+1} \mathrm{Tr}(h\sigma_i) = \alpha_1 H_1(\sigma) + \alpha_2 H_2(\sigma) \tag{6.138}$$

where we begin with $\alpha_1 = 0, \alpha_2 = 1$ and after more training we get $\alpha_0 = 1, \alpha_2 = 0$.

Example 4

Similar to the classical scenario, we can make the following definitions for $j = 1, 2$:

$$\begin{aligned} p_{++}^{(j)} &\equiv P_{\{f_2,\ldots,f_{d+1}\}\sim D_+}(H_j = 1|H_1 = 1) \\ p_{+-}^{(j)} &\equiv P_{\{f_2,\ldots,f_{d+1}\}\sim D_-}(H_j = 1|H_1 = 1) \\ p_{-+}^{(j)} &\equiv P_{\{f_2,\ldots,f_{d+1}\}\sim D_+}(H_j = 1|H_1 = -1) \\ p_{--}^{(j)} &\equiv P_{\{f_2,\ldots,f_{d+1}\}\sim D_-}(H_j = 1|H_1 = -1). \end{aligned} \tag{6.139}$$

We then define the accuracy A_1 when $P(c = 1) = 1/2 = P(c = -1)$ to be

$$A_1 = P(h_1 = c) \equiv p. \tag{6.140}$$

The accuracy for the second model is then

$$A_2 = 1 - \frac{1}{2}pa_2 - \frac{1}{2}(1-p)b_2 \tag{6.141}$$

where $a_2 = 1 - p^{(2)}_{++} + p^{(2)}_{--}$ and $b_2 = 1 - p^{(2)}_{-+} + p^{(2)}_{+-}$. Suppose we choose perturbations where only $f_2, ..., f_{d+1}$ can change and leave f_1 invariant. Furthermore, the distributions change according to $D_\pm \to D_\mp$.

Suppose the perturbation leaves H_1 intact, so then

$$\tilde{A}_1 = A_1 = p, \tag{6.142}$$

but model H_2 has a robustness accuracy different to its accuracy

$$\tilde{A}_2 = \frac{1}{2}(1-p)a_2 + \frac{1}{2}pb_2. \tag{6.143}$$

Then a similar trade-off argument follows where if $p > 1/2$, then if we train values of α_i beginning from $\alpha_1 = 1, \alpha_{i\neq 0} = 0$ to more general $\{\alpha_i\}$ we can achieve

$$A_2 > A_1 \tag{6.144}$$

but this would lead to

$$\tilde{A}_2 < \tilde{A}_1. \tag{6.145}$$

Example 7

A more general class of examples of trade-offs between $\tilde{A}_s$ and A_s can come about if we consider small perturbations in the distribution $\mathcal{D}$ itself. Let the initial distribution be $\mathcal{D}_1$ and the distribution after perturbation be $\mathcal{D}_2$. We can define these distributions in the following way.

For example, suppose we have an observable $\mathcal{M}_j$ and we define $F_j(\sigma_i) = \text{SignTr}(\mathcal{M}_j\sigma_i)$ as the sign of the jth *quantum feature* of the state σ_i. Then let us define the sign of one quantum feature F_0 such that

$$P(F_0(\sigma_i^\pm) = \pm 1 | c(\sigma_i) = \pm 1) = f_0$$
$$P(F_0(\sigma_i^\mp) = \pm 1 | c(\sigma_i) = \mp 1) = 1 - f_0.$$

So if $f_0 > 1/2$, then the feature F_0 is positively correlated with the true label c. With this definition, we can write

$$\begin{aligned}\sum_{\sigma_i^+} P_s(h_s(\sigma_i^+) = 1) &= \sum_{\sigma_i} P_s(h_s(\sigma_i)|c_i) \\ &= \sum_{\sigma_i} (P_s(h_s(\sigma_i) = 1|f_0(\sigma_i) = 1)\mathbf{1}(f_0(\sigma_i) = c_i) \\ &+ P_s(h_s(\sigma_i) = 1|f_0(\sigma_i) = -1)\mathbf{1}(f_0(\sigma_i) = -c_i)).\end{aligned} \tag{6.146}$$

This can be included in the definition of the accuracy where we define $\sigma^{f_0^\pm} = \{\sigma | f_0(\sigma) = \pm 1\}$.

$$\begin{aligned}A_s &= \sum_{\sigma_i} P(\sigma_i, c_i = 1)((\mathbf{1}(f_0(\sigma_i) = c_i)\mathrm{Tr}(\Pi_1 \sigma_i^{f_0^+}) + \mathbf{1}(f_0(\sigma_i) = -c_i)\mathrm{Tr}(\Pi_1 \sigma_i^{f_0^-}))) \\ &+ \sum_{\sigma_i} P(\sigma_i, c_i = -1)((\mathbf{1}(f_0(\sigma_i) = c_i)\mathrm{Tr}(\Pi_{-1} \sigma_i^{f_0^-}) + \mathbf{1}(f_0(\sigma_i) = -c_i)\mathrm{Tr}(\Pi_{-1} \sigma_i^{f_0^-}))) \\ &= \sum_{\sigma_i^+} \mathrm{Tr}(\Pi_1 \sigma_i^{f_0^+}) P(\sigma_i^+, f_0(\sigma_i) = c_i = 1) + \sum_{\sigma_i^+} \mathrm{Tr}(\Pi_1 \sigma_i^{f_0^-}) P(\sigma_i^+, f_0(\sigma_i) = -c_i = -1) \\ &+ \sum_{\sigma_i^-} \mathrm{Tr}(\Pi_{-1} \sigma_i^{f_0^-}) P(\sigma_i^-, f_0(\sigma_i) = c_i = -1) + \sum_{\sigma_i^-} \mathrm{Tr}(\Pi_{-1} \sigma_i^{f_0^+}) P(\sigma_i^-, f_0(\sigma_i) = -c_i = 1).\end{aligned} \tag{6.147}$$

We observe that

$$\begin{aligned}P(\sigma_i^\pm, f_0(\sigma_i) = \pm 1) &= P(f_0(\sigma_i) = \pm 1 | \sigma_i^\pm) = f_0 P(\sigma_i^\pm) \\ P(\sigma_i^\pm, f_0(\sigma_i) = \mp 1) &= P(f_0(\sigma_i) = \mp 1 | \sigma_i^\pm) = (1 - f_0) P(\sigma_i^\pm).\end{aligned} \tag{6.148}$$

Inserting this into the above expression for A_s gives

$$\begin{aligned}A_s &= \sum_{\sigma_i^+} \mathrm{Tr}(\Pi_1 \sigma_i^{f_0^+}) P(\sigma_i^+) f_0 + \sum_{\sigma_i^+} \mathrm{Tr}(\Pi_1 \sigma_i^{f_0^-}) P(\sigma_i^+)(1 - f_0) \\ &+ \sum_{\sigma_i^-} (1 - \mathrm{Tr}(\Pi_1 \sigma_i^{f_0^-})) P(\sigma_i^-) f_0 + \sum_{\sigma_i^-} (1 - \mathrm{Tr}(\Pi_1 \sigma_i^{f_0^+})) P(\sigma_i^-)(1 - f_0).\end{aligned} \tag{6.149}$$

We can define the normalized states

$$\Sigma^{++} = \frac{\sum_{\sigma_i^+} P(\sigma_i^+)\sigma_i^{f_0^+}}{\sum_{\sigma_j^+} P(\sigma_j^+)}$$
$$\Sigma^{+-} = \frac{\sum_{\sigma_i^+} P(\sigma_i^+)\sigma_i^{f_0^-}}{\sum_{\sigma_j^+} P(\sigma_j^+)}$$
$$\Sigma^{-+} = \frac{\sum_{\sigma_i^-} P(\sigma_i^-)\sigma_i^{f_0^+}}{\sum_{\sigma_j^-} P(\sigma_j^-)}$$
$$\Sigma^{--} = \frac{\sum_{\sigma_i^-} P(\sigma_i^-)\sigma_i^{f_0^-}}{\sum_{\sigma_j^-} P(\sigma_j^-)}. \tag{6.150}$$

If we use the unbiased dataset assumption, then we can simplify

$$A_s = \frac{1}{2} f_0 \mathrm{Tr}(\Pi_1 \Sigma^{++}) + \frac{1}{2}(1 - f_0)\mathrm{Tr}(\Pi_1 \Sigma^{+-})$$
$$+ \frac{1}{2} f_0 (1 - \mathrm{Tr}(\Pi_1 \Sigma^{--})) + \frac{1}{2}(1 - f_0)(1 - \mathrm{Tr}(\Pi_1 \Sigma^{-+})). \tag{6.151}$$

Then a perturbed classifier leads to

$$\tilde{A}_s = \frac{1}{2} f_0 \mathrm{Tr}(\Pi_1 \tilde{\Sigma}^{++}) + \frac{1}{2}(1 - f_0)\mathrm{Tr}(\Pi_1 \tilde{\Sigma}^{+-})$$
$$+ \frac{1}{2} f_0 (1 - \mathrm{Tr}(\Pi_1 \tilde{\Sigma}^{--})) + \frac{1}{2}(1 - f_0)(1 - \mathrm{Tr}(\Pi_1 \tilde{\Sigma}^{-+})). \tag{6.152}$$

Now we study a class of perturbations such that

$$\tilde{\Sigma}^{++} = \Sigma^{+-}$$
$$\tilde{\Sigma}^{+-} = \Sigma^{++}$$
$$\tilde{\Sigma}^{--} = \Sigma^{-+}$$
$$\tilde{\Sigma}^{-+} = \Sigma^{--}. \tag{6.153}$$

Then it is straightforward to show that for any $f_0 > 1/2$ we have the trade-off condition:

$$\tilde{A}_s \leq \frac{f_0}{1 - f_0}(1 - A_s). \tag{6.154}$$

Appendix D: Proof of Lemma 1

Below we will distinguish between two types of probabilities. P_s is the probability associated with the stochastic variable getting a specific outcome given some σ, which we can model with a quantum expectation value as $P_s(h_s(\sigma) = j) = \text{Tr}(\Pi_j \sigma)$. On the other hand, P is the probability associated with averaging over some distribution of input states, for instance, $P(c(\sigma) = 1)$ is the probability of states sampled from $\mathcal{D}$ as having the true label 1. We can define then accuracy A_s of this stochastic classifier h_s as the probability it gives the same outcome as the true label

$$A_s \equiv \sum_{\sigma_i} \sum_{c_i} P(\sigma_i, c_i) P_s(h_s(\sigma_i) = c_i). \tag{6.155}$$

Using the definition $\sigma^{\pm} = \{\sigma | c(\sigma) = \pm 1\}$ we can then rewrite A_s

$$\begin{aligned} A_s &= \sum_{\sigma_i} \sum_{c_i} P(\sigma_i, c_i) P_s(h_s(\sigma_i) = c_i) \\ &= \sum_{\sigma_i^+} P(\sigma_i^+, c_i = 1) P_s(h_s(\sigma_i^+) = 1) + \sum_{\sigma_i^-} P(\sigma_i^-, c_i = -1) P_s(h_s(\sigma_i^-) = -1) \\ &= \sum_{\sigma_i^+} P(\sigma_i^+, c_i = 1) \text{Tr}(\Pi_1 \sigma_i^+) + \sum_{\sigma_i^-} P(\sigma_i^-, c_i = -1) \text{Tr}(\Pi_{-1} \sigma_i^-) \\ &= \text{Tr}(\Pi_1 \sum_{\sigma_i^+} P(\sigma_i^+) \sigma_i^+) + \text{Tr}(\Pi_{-1} \sum_{\sigma_i^-} P(\sigma_i^-) \sigma_i^-) \\ &= (\sum_{\sigma_i^+} P(\sigma_i^+)) \text{Tr}(\Pi_1 \Sigma^+) + (\sum_{\sigma_i^-} P(\sigma_i^-)) \text{Tr}(\Pi_{-1} \Sigma^-) \end{aligned} \tag{6.156}$$

where in the third equality we used $P(\sigma_i^{\pm}, c_i = \pm 1) = P(c_i = - \pm 1 | \sigma_i^{\pm}) P(\sigma_i^{\pm}) = P(\sigma_i^{\pm})$ and we define the normalized states

$$\Sigma^{\pm} \equiv \frac{\sum_{\sigma_i^{\pm}} P(\sigma_i^{\pm}) \sigma_i^{\pm}}{\sum_{\sigma_j^{\pm}} P(\sigma_j^{\pm})}. \tag{6.157}$$

For a unbiased dataset we have $\sum_{\sigma_j^{\pm}} P(\sigma_j^{\pm}) = 1/2$. Then using the relation $\Pi_1 = \mathbf{1} - \Pi_{-1}$ we can write

$$A_s = \frac{1}{2} \text{Tr}(\Pi_1 (\Sigma^+ - \Sigma^-)) + \frac{1}{2}. \tag{6.158}$$

We will work using this unbiased assumption unless otherwise stated. Suppose we modify only the stochastic classifier, so we change the measurement $\Pi \rightarrow \tilde{\Pi} =$

$\sum_l E_l \Pi E_l^\dagger$ where $\{E_l\}$ are Kraus operators. Then the robustness accuracy under this perturbation becomes

$$\begin{aligned}\tilde{A}_s &= \frac{1}{2}\mathrm{Tr}(\tilde{\Pi}_1(\Sigma^+ - \Sigma^-)) + \frac{1}{2} \\ &= \frac{1}{2}\mathrm{Tr}(\Pi_1(\tilde{\Sigma}^+ - \tilde{\Sigma}^-)) + \frac{1}{2}\end{aligned} \tag{6.159}$$

where $\tilde{\Sigma}^\pm = \sum_l E_l^\dagger \Sigma^\pm E_l$.

Then we can derive the difference

$$|\tilde{A}_s - A_s| = \frac{1}{2}|\mathrm{Tr}(\Pi_1(\tilde{\Sigma}^+ - \Sigma^+ + \Sigma^- - \tilde{\Sigma}^-)|. \tag{6.160}$$

We can rearrange Eq. (6.160) to get

$$\begin{aligned}|\tilde{A}_s - A_s| &= \frac{1}{2}|\mathrm{Tr}(\Pi_1(\tilde{\Sigma}^+ + \Sigma^- - (\Sigma^+ + \tilde{\Sigma}^-)| \\ &= |\mathrm{Tr}(\Pi_1(\Sigma_A - \Sigma_B))| \leq \tau(\Sigma_A, \Sigma_B)\end{aligned} \tag{6.161}$$

where $\Sigma_A \equiv (1/2)(\tilde{\Sigma}^+ + \Sigma^-)$ and $\Sigma_B \equiv (1/2)(\Sigma^+ + \tilde{\Sigma}^-)$. As a very quick check, when the noise is the identity operation, we see that the upper bound is saturated, since $\tau(\Sigma_A, \Sigma_B = \Sigma_A) = 0$.

References

1. Dobriban E, Hassani H, Hong D, Robey A (2023) Provable tradeoffs in adversarially robust classification. IEEE Trans Inf Theory 69(12):7793–7822
2. Tsipras D, Santurkar S, Engstrom L, Turner A, Madry A (2018) Robustness may be at odds with accuracy. arXiv:1805.12152
3. Xu H, Mannor S (2012) Robustness and generalization. Mach Learn 86:391–423
4. Yang Y-Y, Rashtchian C, Zhang H, Salakhutdinov RR, Chaudhuri K (2020) A closer look at accuracy vs. robustness. Adv Neural Inf Process Syst 33:8588–8601
5. Zhang H, Yu Y, Jiao J, Xing E, El Ghaoui L, Jordan M (2019) Theoretically principled trade-off between robustness and accuracy. In: International conference on machine learning. PMLR, pp 7472–7482
6. Yin D, Gontijo Lopes R, Shlens J, Cubuk ED, Gilmer J (2019) A fourier perspective on model robustness in computer vision. Adv Neural Inf Process Syst 32

Chapter 7
Adversarial Robustness Guarantees for Quantum Classifiers

Maxwell West, Neil Dowling, Angus Southwell, Azar C. Nakhl, Martin Sevior, Muhammad Usman, and Kavan Modi

Abstract Despite their ever more widespread deployment throughout society, machine learning algorithms remain critically vulnerable to being spoofed by subtle adversarial tampering with their input data. The prospect of near-term quantum computers being capable of running quantum machine learning (QML) algorithms has therefore generated intense interest in their adversarial vulnerability. Here we show that quantum properties of QML algorithms can confer fundamental protections against such attacks, in certain scenarios guaranteeing robustness against classically armed adversaries. We leverage tools from many-body physics to identify the quantum sources of this protection. Our results offer a theoretical underpinning of recent evidence which suggest quantum advantages in the search for adversarial robustness. In particular, we prove that quantum classifiers are: (i) protected against weak perturbations of data drawn from the trained distribution, (ii) protected against local attacks if they are insufficiently scrambling, and (iii) show evidence that they are protected against universal adversarial attacks if they are sufficiently chaotic. Our analytic results are supported by numerical evidence demonstrating the applicability

M. West and N. Dowling: Equal contribution.

M. West (✉) · A. C. Nakhl · M. Sevior · M. Usman
School of Physics, The University of Melbourne, Parkville, VIC, Australia
e-mail: westm2@student.unimelb.edu.au

N. Dowling
Institut für Theoretische Physik, Universität zu Köln, Köln, Germany

N. Dowling · K. Modi
School of Physics and Astronomy, Monash University, Clayton, VIC, Australia

A. Southwell
Centre for Quantum Technology, Transport for New South Wales, Sydney, NSW, Australia

M. Usman
Data61, CSIRO, Clayton, VIC, Australia

K. Modi
Science, Mathematics and Technology Cluster, Singapore University of Technology and Design, Singapore, Singapore

Quantum for New South Wales, Sydney, NSW, Australia

M. Usman (ed.), *Quantum Robustness in Artificial Intelligence*, Quantum Science and Technology, https://doi.org/10.1007/978-3-032-11153-1_7

of our theorems and the resulting robustness of a quantum classifier in practice. This line of inquiry constitutes a concrete pathway to advantage in QML, orthogonal to the usually sought improvements in model speed or accuracy.

7.1 Introduction

A dozen years on from their initial discovery [1–3], adversarial attacks remain a surprisingly successful technique for deceiving even highly sophisticated machine learning (ML) models [4]. Remarkably, for example, large-scale image classifiers can be fooled by carefully chosen perturbations which are almost invisible to a human eye [5], or even by changing (if non-invisibly) the value of a single pixel [6]. Due to the accelerating delegation of important tasks to ML, and the tendency of empirical defense strategies to be later bypassed [7], the need for provable guarantees against such spoofing attempts is only growing [8, 9].

Concurrently, the increasing capabilities of quantum computers have generated significant research to determine whether quantum advantage may be expected in machine learning [10–13]. However, but the extent to which they can be expected to deliver direct speed-ups has been brought into question, largely due to the provable difficulty in training large randomly-initialized variational algorithms: the phenomenon of *barren plateaus* [13–22]. It is therefore an opportune moment to search for a different kind of advantage in QML [23]. In fact, the field of quantum adversarial machine learning has generated considerable interest [23–37]. Notably, in a series of recent papers, QML models were studied that indicated significantly increased adversarial robustness against classical adversaries [32–35] (Fig. 7.1a). However, these results are largely empirical, lacking a foundational understanding of the source of the advantage.

This chapter[1] addresses that gap by establishing provable robustness guarantees for QML models under broad and practically relevant scenarios. These guarantees stem from distinct features of quantum data-encoding methods and the dynamical complexity of quantum circuits. The results presented include analytical theorems that leverage fundamental quantum mechanical properties, yielding robustness guarantees unattainable by classical ML models. Additionally, probabilistic bounds and numerical validations are provided for realistic quantum classifier architectures.

Importantly, these guarantees refine prior existence proofs of adversarial vulnerabilities in QML (see Ref. [25, 29]) by focusing on physically realistic scenarios where adversaries are classical and their perturbations are constrained by the encoding strategy used by the quantum model. Specifically, the chapter examines robustness against:

1. weak targeted perturbations designed to induce misclassification on specific inputs,

[1] Based on our preprint [38].

2. strong universal perturbations (see, e.g., Refs. [39, 40]) aimed at misclassifying all inputs simultaneously, and
3. strong local perturbations targeting specific input states.

A summary of these results is provided in Table 7.1.

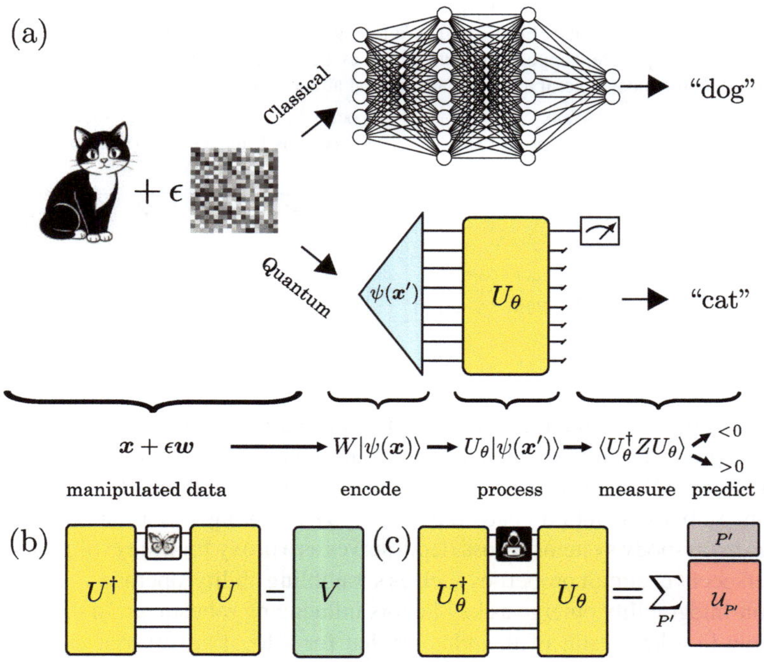

Fig. 7.1 Schematic of adversarial machine learning setting. a Machine learning models are generally highly susceptible to extremely subtle adversarial tampering with their input data, but quantum models have been empirically found to be robust to attacks by classical adversaries [33]. In the general quantum machine learning setting, a classical data string $\boldsymbol{x}$ is encoded in a state $|\psi(\boldsymbol{x})\rangle$, a (trained) quantum algorithm U_θ is applied before measurement of some few-qubit operator Z. An adversarial attack can then be modeled by some change to the initial bit string $\boldsymbol{x} \to \boldsymbol{x} + \epsilon \boldsymbol{w}$, which is equivalent to the action of a unitary W on the encoded state, $|\boldsymbol{x}'\rangle = W|\boldsymbol{x}\rangle$. **b** Chaotic unitaries scramble information throughout quantum degrees of freedom in a many-body system. **c** It is difficult for an adversary to carefully manipulate a chaotic circuit in the precise way needed to induce misclassification. Here, P' is some (spoofed) Pauli string which flips the measurement outcome, and $\mathcal{U}_{P'}$ is some unitary on the subsystem which is not measured

Table 7.1 Summary of robustness guarantees. The applicability of our theorems, which depend on both the attack strategy and the form of data-encoding, $x \in \mathbb{R}^N \mapsto \psi(x) = \mathcal{E}(x) |\mathbf{0}\rangle \langle \mathbf{0}| \mathcal{E}^\dagger(x)$. ϵ denotes the ℓ_∞ norm of the adversarial perturbation. In some cases our results apply unconditionally (denoted by a tick) while in others there is a specified dependence on the details of the encoding. Non-applicability is denoted by a dash. In the bottom row we record the property of the model (qualitatively) responsible for the guarantee: "Quantum" refers to the contractive nature of any quantum classifier (e.g., a unitary circuit), "Scrambling" refers to a quickly decaying out-of-time-ordered correlator (OTOC) [Eq. (7.12)], while by "Chaotic" we mean a linearly-growing local-operator entanglement (LOE) [Eq. (7.14)]. We also note that the ticks in the right-hand column are based on a conjecture, supported by numerical evidence and analytic results under a stronger condition than the most general universal adversarial attack (see Eqs. (7.13) and (7.16))

	Weak (Theorem 1)	Local (Theorems 1 and 2)	Universal (Theorem 3)
Amplitude	✓	✓	–
Angle	$\epsilon \lesssim 1/\sqrt{N}$	OTOC $\ll 1$	✓
Dense	$\epsilon \lesssim 1/\sqrt{N}$		✓
Arbitrary	$\epsilon \lesssim \lvert\Delta x/\Delta\psi\rvert$		–
	Quantum	Scrambling	Chaotic

These robustness guarantees fall into two primary categories. The first (item one above) exploits the unitarity of quantum circuits combined with the statistical properties of typical ML training data. The second (items two and three) draws on concepts from many-body quantum chaos theory [41–57], treating variational quantum circuits as many-body systems whose depth serves as a proxy for time evolution. Within this perspective, circuit properties such as scrambling ability, operator entanglement, and non-integrability emerge as key factors influencing robustness. In this direction, we show in a key result of this chapter that for a local adversarial attack to succeed, the variational quantum circuit must exhibit scrambling behavior. Scrambling is characterized by the out-of-time-ordered correlator (OTOC), which quantifies how information contained in an initially local operator spreads under time evolution in the Heisenberg picture [44–47, 49, 56, 58–60]. Previous studies have observed that certain QML architectures, particularly those involving backward evolution, possess scrambling properties [61, 62], implying potential vulnerability to local attacks. However, contrasting this necessity of scrambling, we show that quantum chaos provides protection against universal adversarial attacks. Here, "quantum chaos" is defined via the linear growth of local-operator entanglement (LOE) [41, 42]. LOE measures the complexity of representing a Heisenberg-evolved operator as a matrix product operator. In locally interacting many-body systems, LOE has been observed to grow maximally—i.e., linearly with time—only for non-integrable dynamics [43, 48, 50–53, 56]. This defines a stricter condition than scrambling alone [56].

Consequently, a variational algorithm exhibiting high operator entanglement possesses an intrinsic robustness against universal adversarial attacks. Intuitively, chaotic unitaries disperse perturbations uncontrollably throughout the system (Fig. 7.1b), preventing an adversary from applying the precise manipulations required to mislead the model (Fig. 7.1c). Before detailing these results, we first describe the general setup and necessary background knowledge in adversarial QML.

The general QML models we consider in this chapter follow a standard three-step architecture consisting respectively of data-encoding, data processing, and measurement.

In the first step, a vector $\boldsymbol{x}$ of classical data is loaded into the quantum computer by means of some encoding method,

$$\mathcal{E}(\boldsymbol{x})\,|\mathbf{0}\rangle = |\psi(\boldsymbol{x})\rangle\,, \tag{7.1}$$

where $|\mathbf{0}\rangle := |0\rangle^{\otimes n}$, with the number of qubits $n = n_{\mathcal{E}}(N)$ being some encoding-dependent function of the size of the classical data, $N := \text{len}(\boldsymbol{x})$. It will often be necessary to consider the corresponding density matrix, $\psi(\boldsymbol{x}) := |\psi(\boldsymbol{x})\rangle\langle\psi(\boldsymbol{x})|$. We will explore three of the most natural and popular encoding methods [63], amplitude encoding, angle encoding, and dense encoding:

$$\mathcal{E}_{\text{amp}}(\boldsymbol{x})\,|\mathbf{0}\rangle = \frac{1}{\|\boldsymbol{x}\|_2}\sum_{j=1}^{N} x_j\,|j\rangle\,, \tag{7.2}$$

$$\mathcal{E}_{\text{angle}}(\boldsymbol{x})\,|\mathbf{0}\rangle = \bigotimes_{j=1}^{N} e^{-ix_j\sigma_y}\,|\mathbf{0}\rangle\,, \tag{7.3}$$

$$\mathcal{E}_{\text{dense}}(\boldsymbol{x})\,|\mathbf{0}\rangle = \bigotimes_{j=1}^{N/2} e^{-ix_{2j}\sigma_z}e^{-ix_{2j-1}\sigma_y}\,|\mathbf{0}\rangle\,. \tag{7.4}$$

Each of these encodings has different advantages, for example with respect to expressibility, learnability, and resource costs [63]. For instance, amplitude encoding is drastically more space efficient, requiring a number of qubits only logarithmic in the dimension of the data, but in general requires exponentially deep circuits [34, 63], as opposed to the constant-depth but extensive-size angle encoding circuits.

In the second and third steps, the encoded state is acted upon by a trainable variational unitary U_θ, following which a local measurement is made. Without loss of generality, we choose this measurement to be of the Pauli-Z operator on the first k qubits. In the binary classification case, upon which for simplicity we focus (although the generalization to multiple classes is straightforward) we take the prediction $\hat{y}$ of the model on the input $\boldsymbol{x}$ to be the sign of the final measurement,

$$y_\theta(\boldsymbol{x}) := \langle\psi(\boldsymbol{x})|\,U_\theta^\dagger Z U_\theta\,|\psi(\boldsymbol{x})\rangle\,, \tag{7.5}$$

choosing $Z := \sigma_z^{\otimes k} \otimes I^{\otimes(n-k)}$ as the operator to be measured. We will also use the notation $Z_U := U_\theta^\dagger Z U_\theta$ to denote an operator Heisenberg-evolved by the circuit. During training, the parameters θ are optimized to minimize a loss function $\ell_\theta(\boldsymbol{x}, y)$, for example of the form

$$\ell_\theta(\boldsymbol{x}, y) = -y(\boldsymbol{x})y_\theta(\boldsymbol{x}) \tag{7.6}$$

over a training set S, where we denote the true label of the datapoint $\boldsymbol{x}$ as $y(\boldsymbol{x}) \in \{\pm 1\}$.

An *adversarial attack* is a vector $\boldsymbol{w}$ which perturbs the input data as $\boldsymbol{x} \mapsto \boldsymbol{x}' := \boldsymbol{x} + \boldsymbol{w}$, intended to change the prediction of the model. At the quantum circuit level, after encoding the classical perturbation induces a unitary W satisfying $|\psi(\boldsymbol{x}')\rangle := W|\psi(\boldsymbol{x})\rangle$. An attack is deemed successful if this perturbation changes the prediction of the model,

$$\text{sgn}[y_\theta(\boldsymbol{x}')] = -\text{sgn}[y_\theta(\boldsymbol{x})]. \tag{7.7}$$

The properties of W will be heavily influenced by the choice of data-encoding map. In the case of angle or dense encoding (Eq. (7.3), (7.4)) for example, it will take the form of a product of local unitaries $W = \bigotimes_i W_i$, but this is not generally true for amplitude encoding (Eq. (7.2)).

The existence of adversarial attacks which can spoof QML models (equivalently, the existence of nearby pairs of states classified differently) has already been established [25, 29], and seems to indicate a significant vulnerability of quantum classifiers. What is less well understood, however, and the focus of this chapter, is the extent to which it is possible to construct and implement these attacks in practice. A key contribution of this chapter is to examine this adversarial setting in terms of concrete scenarios, identifying distinct (dis)advantages of the various encoding methods described above under different types of adversarial attack (see Table 7.1). We will derive fundamental robustness guarantees in three distinct situations: (i) tailored but weak attacks, (ii) strong, local attacks, and (iii) universal attacks which spoof all images with a single attack $\boldsymbol{w}$. To derive these guarantees, we will leverage the contractiveness of quantum dynamics, the dynamical complexity of the trained circuit U_θ, and nature of the encoding in (Eqs. (7.3)–(7.4)). Remarkably, we will argue that given an encoding, and broad class of attack, our theorems hold in full generality. This means that they will apply to all future quantum adversarial scenarios that fit within one of these settings.

7.2 Results

At a high level our results (summarized in Table 7.1) can be split into two categories: statements about the strength of the perturbation required to induce a misclassification, relying only on the unitarity/contractiveness the model (the first column of Table 7.1), and statements about the impossibility of carrying out certain classes of attacks, regardless of the strength of the perturbation (the second and third columns of Table 7.1). While in the first case we show a robustness for all quantum classifiers, the latter category relies on some notions from the theory of many-body chaos. We present our results based on progressively strongly requirements on the dynamical properties of the trained circuit U_θ, as summarized in the bottom row of Table 7.1.

7.2.1 Weak Attacks

We first consider the simplest, and arguably the most realistic, threat model: an input-specific perturbation as weak as possible, so as to maximize the probability that the tampering is not detected. In this case, and in contrast to the generally highly non-linear nature of classical neural networks [64], we can use the unitarity of isolated quantum circuits to arrive at the following result:

Theorem 1 *Given an input state* $|\psi(\boldsymbol{x})\rangle$*, a quantum model as defined in Eq.* (7.5) *will classify all states within a 1-norm ball of* $|\psi(\boldsymbol{x})\rangle$ *of radius* $|y_\theta(\boldsymbol{x})|$ *identically.*

A proof of Theorem 1 may be found in Ref. [38]. In fact, this result extends beyond unitary circuits U_θ to entirely general quantum dynamics. This means, for example, that this robustness guarantee holds in the presence of noise (which effectively weakens the adversary's control of the situation; e.g., see Ref. [26]). The practical usefulness of Theorem 1 depends on two factors: the extent to which changes in the classical data vector $\boldsymbol{x}$ translate to 1-norm changes in $\psi(\boldsymbol{x})$, and how big one can expect the magnitude $|y_\theta(\boldsymbol{x})|$ of the unperturbed output to be.

Investigating the first point, which will depend on the choice of data-encoding scheme, in Ref. [38] we estimate $\Delta\psi = \|\psi(\boldsymbol{x}) - \psi(\boldsymbol{x}')\|_1$ under an adversarial attack $\boldsymbol{x} \mapsto \boldsymbol{x}'$ with $\Delta\boldsymbol{x} = \max_i |x_i - x_i'| = \epsilon \ll 1$ for our considered encoding schemes (Eqs. (7.2)–(7.4)). In the archetypal example of image classification, this would correspond to changing each pixel value by no more than ϵ. We find (see Ref. [38]), for a classical data vector of length N,

$$\Delta\psi_{\text{angle}} \sim \sqrt{N}\epsilon + \mathcal{O}\left(\epsilon^2\right), \tag{7.8}$$

$$\Delta\psi_{\text{dense}} \sim \sqrt{N}\epsilon + \mathcal{O}\left(\epsilon^2\right), \tag{7.9}$$

$$\Delta\psi_{\text{amp}} \sim \epsilon + \mathcal{O}\left(\epsilon^2\right). \tag{7.10}$$

So for large N, when using angle or dense encoding changing each value of the input vector by a small amount can induce a large change in the corresponding quantum states, and effectively weakens the applicability of Theorem 1 to perturbations with $|\epsilon| \lesssim 1/\sqrt{N}$. In the case of amplitude encoding, on the other hand, the resulting change is independent of N, implying that weakly perturbed data will be mapped close to the original irrespective of the dimension of the classical input data. This will therefore impart a strong robustness guarantee on the quantum classifier if $|y_\theta(\boldsymbol{x})|$ is of appreciable magnitude. A similar analysis could be carried out for other encoding strategies; in general, the relevant quantity is the magnitude $|\Delta\boldsymbol{x}/\Delta\psi|$ of the change in the encoded state as a function of the change in the input classical data vector (see Table 7.1).

At first glance, however, it is unclear that one should expect $|y_\theta(\boldsymbol{x})|$ to be reasonably large, as due to standard concentration effects the measurement values for a large quantum circuit will concentrate strongly around zero, and thus a small perturbation will be sufficient to change the sign of most (typical) measurement results, independent of the data-encoding strategy. Indeed, this phenomenon has been used to conclude that in general quantum classifiers will possess extreme adversarial vulnerability, with perturbations of magnitude $\mathcal{O}(2^{-n})$, exponentially falling with the number of qubits, capable of changing the prediction of a model [25]. However, and as has previously been recognized [29], the distribution of states in which one is interested in practice in ML is typically highly non-uniform, which can lead to a merely polynomially vanishing minimum perturbation size for a successful adversarial attack. Yet further progress can be made, eliminating entirely the dependence on the number of qubits, if one assumes that (modulo a potential adversarial perturbation) the test time samples are being drawn from the same distribution μ as was the training set S over which the loss function ℓ_θ (e.g., Eq. (7.6)) was minimized. In this case, one has with high probability that the difference between the expected risk $R(\boldsymbol{\theta}) = \mathbb{E}_{(x,y)\sim\mu}\, \ell_\theta(x, y)$ and the empirical risk $\hat{R}_S(\boldsymbol{\theta}) = \frac{1}{|S|}\sum_{(x_i,y_i)\in S} \ell_\theta(x_i, y_i)$ is bounded by $\mathcal{O}\left(\sqrt{\frac{T\log T}{|S|}}\right)$ where T is the number of trainable 2-qubit unitaries [65]. So if one trains until (say) $\hat{R}_S(\boldsymbol{\theta}) < -1/2$ on a training set S with $|S| \gtrsim T \log T$ then (with high probability) attacked states with $\|\psi(\boldsymbol{x}) - \psi(\boldsymbol{x}')\|_1 < 1/4$ and $(\boldsymbol{x}, y(\boldsymbol{x})) \sim \mu$ will be classified correctly, requiring an adversary to implement perturbations of magnitude $\mathcal{O}(1)$. We conclude that a well-trained QML model will be drastically more adversarially robust on encoded states of data drawn from μ than it will be on (Haar) random states, which, conveniently, are exactly the states that we care about the most.

A useful consequence of the above construction is that it also automatically implies a robustness to non-adversarial noise. In particular, Theorem 1 immediately also includes a robustness in terms of the strength ϵ of the noise in *any* perturbation $\boldsymbol{x} \mapsto \boldsymbol{x} + \epsilon \boldsymbol{w}$. An adversarial perturbation can be seen as a "worst case" scenario, where $\boldsymbol{w}$ is picked to optimize changing the prediction in Eq. (7.5), compared to noise where $\boldsymbol{w}$ would be sampled from some distribution. This can also be viewed as encompassing noise in the quantum algorithm itself. That is, for sufficient training, any weak, coherent noise will not rotate the state vector outside of the 1-norm ball of Theorem 1. Moreover, Theorem 1 readily extends also to incoherent noise; see Ref. [38].

We note that in Ref. [37], in a conceptually similar argument to Theorem 1, Lipschitz bounds are employed to show how certain variational circuits can be trained to constrain (in our notation) $|y_\theta(\boldsymbol{x}') - y_\theta(\boldsymbol{x})|$ as a function of $\|\boldsymbol{x}' - \boldsymbol{x}\|$, and the resulting relation between expressivity and robustness is explored. The present chapter, however, relaxes an assumption made in Ref. [37] on the form of data-encoding, with Theorem 1 being encoding-agnostic.

The results discussed so far depend on the strength of the attack being weak—the isometric nature of unitary maps does not give as useful a bound when the attack is strong. In the search for guarantees even in the face of strong perturbations, then, we need to turn to a different property of the models. As we will see in the next two sections, covering local and universal attacks, it will turn out that different degrees of dynamical complexity in U_θ can either safeguard or jeopardize its robustness against strong adversarial attacks.

7.2.2 Local Attacks

The second attack scenario that we consider is the case of strong, local attacks, where the assumption we make on W is that it acts only on a few qubits. That is, instead of a weak attack in Theorem 1, we take $\boldsymbol{x}' = \boldsymbol{x} + \boldsymbol{w}$, where $w_i \neq 0$ only for some small number $k \ll N$ bits. This threat model is inspired by the surprising result that certain classical neural networks can be successfully spoofed even if the attacker can change only a single pixel [6].

As a first result, we note that a strongly (but still locally) perturbed classical vector $\boldsymbol{x}'$ need not lead to strongly perturbed state $|\psi(\boldsymbol{x}')\rangle$ after encoding. In particular, in the case of amplitude encoding (7.2), changing only a small number of bits necessarily leads to a weak attack, $\|\psi(\boldsymbol{x}) - \psi(\boldsymbol{x}')\|_1 \ll 1$. Intuitively, this is because for a large quantum state with many non-zero amplitudes, changing only a small fraction of these cannot significantly change the global state. The results of Theorem 1 then directly apply, as we prove in Ref. [38]. Therefore, for amplitude encoding, QML circuits are robust against local attacks in addition to weak attacks (first cell of the middle column in Table 7.1). For other forms of encoding, however, this will not be true in general. In the cases of angle and dense encoding, for example, one can orthogonalise a pair of encoded states by changing only a single element of the corresponding classical data vector. Nonetheless, we can make progress by considering the scrambling characteristic of U_θ.

In contrast to the previous results, we now make the simplifying assumption that initial state is sampled from a unitary 2-design, a mild relaxation of the approach of, e.g., Ref. [25]. This is admittedly a strengthening of the previous assumptions that will not hold in general; when it does, however, we have.

Theorem 2 *If the state $|\psi(\boldsymbol{x})\rangle$ representing the classical data $\boldsymbol{x}$ is sampled from a 2-design $\mathcal{E}$ over the uniform measure on states, then for any attacked state $|\psi(\boldsymbol{x}')\rangle = W\,|\psi(\boldsymbol{x})\rangle$ and for any $\delta > 0$,*

$$\Pr_{|\psi(\boldsymbol{x})\rangle\sim\mathcal{E}} \left\{|y_\theta(\boldsymbol{x}) - y_\theta(\boldsymbol{x}')| \geq \delta\right\} \leq \frac{\langle [Z_U, W]^2\rangle}{(d+1)\delta^2}, \tag{7.11}$$

where the prediction $y_\theta(\boldsymbol{x})$ is defined in Eq. (7.5), and the expectation value on the r.h.s. is over a maximally mixed state, $\langle [Z_U, W]^2\rangle = (1/d)\,\mathrm{tr}([Z_U, W]^2)$.

This theorem is entirely independent of the form of the attack and holds generally for any W. Similar encoding-specific bounds could be obtained by instead averaging over states attainable from a specific encoding scheme. We explain this in Ref. [38], and supply there a proof of the above theorem. When W is a local attack, the numerator of the r.h.s. can be interpreted as an out-of-time-order correlator (OTOC), which probes how scrambling the process U_θ is. We first interpret the theorem for a generic W and then analyze it in terms of the OTOC.

To understand this result for a general W, we note that the form of Theorem 2 is not exactly surprising. If the attack operator W commutes with the circuit-evolved measurement $Z_U = U_\theta^\dagger Z U_\theta$, then in the expectation value $y_\theta(\boldsymbol{x})$ it will have no influence.

Similarly, if the commutator is small, $(1/d)\|[Z_U, W]\|_2 \ll 1$, then the adversary can only affect the outcomes of the Z measurement weakly on average. That is, for an adversary to strongly affect a measurement, the strength of the commutator needs to be large. Theorem 2 quantifies this intuition.

In the above, we note that Haar random quantum states typically lead to a concentrated expectation value $|y_\theta(\boldsymbol{x})| \sim \frac{1}{\sqrt{d}}$ [25]. This means that δ in Theorem 2 needs to be at least larger than $\frac{1}{\sqrt{d}}$ for a successful attack, in such a typical situation. Substituting this in, we see that the dependence on d drops out from the r.h.s. of the probability bound Eq. (7.11). This means that even with concentration of measure effects, in this random-state setting the OTOC mediates the viability of an adversarial attack.

Now we return to the case where W is a local attack. In many-body physics, the scaling of the OTOC $\langle [A_t, B]^2 \rangle$ diagnoses quantum information scrambling, where A_t is a time-evolving Heisenberg operator of an initially local operator A, and where B is also a local operator. An early-time exponential growth of an OTOC indicates a fast-scrambling property of the dynamics [45, 49, 58], and the OTOC has played an important role in studies of the black hole information paradox [44], and in understanding feature of quantum chaos without a classical analogue [46, 47, 56, 59]. Converting this to our setting of variational quantum circuits, for a scrambling circuit U_θ we can conclude that

$$\langle [Z_U, W]^2 \rangle \sim \exp\left[\lambda \, \mathrm{depth}(U_\theta)\right] \leq 1, \tag{7.12}$$

for some $\lambda > 0$. On the other hand, for a circuit that does not (quickly) scramble quantum information, this quantity scales slowly with the depth of the circuit, meaning that adversarial attacks are impossible according to Theorem 2.

There is evidence in the literature that trained QML architectures tend to be scrambling quantum circuits [61, 62], and furthermore that scrambling generally impacts trainability [66]. This again highlights the question of whether trained QML circuits are always vulnerable to attack? The above results depend only on the scrambling characteristic of the circuit. This concept is strictly independent of classical simulability [67], for instance. In the following, we will investigate the robustness of a quantum classifier against attacks when U_θ instead is genuinely (quantum) chaotic, a distinct notion [56].

7.2.3 Universal Attacks

We now turn to the setting of universal adversarial attacks, i.e., a perturbation that changes the prediction of the model when applied to *any* input state. While the existence of such perturbations is far from obvious, remarkably they have been shown to exist in both the classical [68] and quantum [40] case. In our framework, from Eqs. (7.5) and (7.7) a universal attack is when for all data $\boldsymbol{x}$,

$$\mathrm{sgn}(\langle\psi(\boldsymbol{x})| Z_U |\psi(\boldsymbol{x})\rangle) = -\mathrm{sgn}(\langle\psi'(\boldsymbol{x})| Z_U |\psi'(\boldsymbol{x})\rangle), \tag{7.13}$$

where $|\psi'(\boldsymbol{x})\rangle = W_{\mathrm{univ}} |\psi(\boldsymbol{x})\rangle$ for some unitary W_{univ}, which is independent of $\boldsymbol{x}$. The key point is that the allowable form of W_{univ} is restricted based on the encoding method, under our assumption that the adversary may only spoof the classical data $\boldsymbol{x}$. We will assume in this section that the adversary may only apply product unitaries: this is the case for angle and dense encoding. Then, we conjecture that if Z_U is sufficiently *complex*, then no product W_{univ} satisfying Eq. (7.13) exists. By complex, we formally mean that it has high operator entanglement, as measured by the local-operator entanglement (LOE). LOE is a dynamical signature of chaos defined in terms of a Heisenberg operator. Through the Choi-Jamiołkowski isomorphism, an operator $X_U = U_\theta^\dagger X U_\theta \in \mathcal{B}(\mathcal{H})$ has a quantum state representation

$$|X_U\rangle := (X_U \otimes \mathbb{1}) |\phi^+\rangle, \tag{7.14}$$

where $|\phi^+\rangle \in \mathcal{H} \otimes \mathcal{H}$ is a maximally entangled state across a doubled Hilbert space. The LOE is then defined as the entanglement of this state, across some bipartition. This is usually measured by a Rényi entropy,

$$S^{(\alpha)}(\omega) := \lim_{\beta\to\alpha} \frac{1}{1-\beta} \log(\mathrm{tr}[\omega^\beta]), \tag{7.15}$$

for $\alpha \geq 0$, and where, e.g., $S^{(1)}$ is the von Neumann entropy.

In the remainder of this section, we will study the conjecture that operator entanglement precludes the existence of a universal adversarial attack, both analytically (Theorem 3 and Corollary 4) and numerically (see Fig. 7.2).

First, to make analytic progress, we will make a simplifying assumption that the universal spoof satisfies

$$Z_U = -W_{\mathrm{univ}}^\dagger Z_U W_{\mathrm{univ}}. \tag{7.16}$$

We call a W_{univ} satisfying this condition a strong universal attack. While the condition (7.16) is sufficient to satisfy Eq. (7.13), it is not necessary. This is because Eq. (7.16) implies the stronger condition that not only the sign is flipped, but also that the magnitude of all expectation values is preserved. When seen in conjunction with the finite-sized numerics of a realistic model classifier (note that in the numerical setting we do not make the assumption of Eq. (7.16)) we will argue that there is a deep relationship between LOE and the existence of a universal adversarial attack (see Fig. 7.2). A surprising facet of this setting is that even if the adversary can solve Eq. (7.16) and determine W_{univ}, a priori it is not clear whether it will be implementable, given the restrictions enforced by the data-encoding (here, locality), but even having dropped the assumption that the perturbation is weak. With this in hand, we present the following result.

Theorem 3 *The distance D between a product of local unitary channels $W = \bigotimes_i W_i$ and a strong universal adversarial attack W_{univ} (satisfying Eq. (7.16)) is*

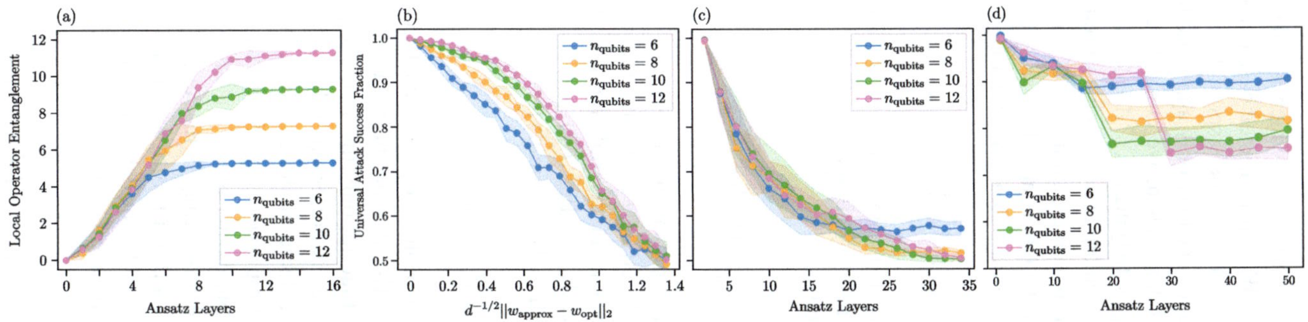

Fig. 7.2 Numerical results for a common architecture. **a** Local operator entanglement (LOE) growth in standard quantum machine learning (QML) models consisting of hardware efficient layers of single qubit rotations and nearest neighbor CNOTs. The initial linear growth of the LOE indicates that these models are implementing chaotic quantum dynamics [56]. **b** The fraction of states successfully spoofed by an approximation to a universal adversarial attack. The attack is carried out by random unitaries with various 2-norm distance from the ideal strong attack W_{univ} (satisfying Eq. (7.16)). For each distance we generate ten circuits, each with five attacks constructed by randomly rotating away from the ideal attack (see Eqs. (7.20) and (7.21)). The mean success fraction is plotted, with the regions within one standard deviation shaded. **c** Here the attack is carried out by optimized local unitary operators on each qubit for random models of increasing circuit depth. For each choice of layer number we generate 20 circuits and train the adversary on 32, 000 training datapoints, and evaluate it on 10, 000 test datapoints. We plot the mean attack success fraction for up to 34 layers, by which point both the LOE in the circuit and the attack success fraction have plateaued. **d** Similar to part (**c**), but employing a model trained to classify images of handwritten digits [69]. While the adversary enjoys improved performance compared to the random case, we nonetheless observe the emergence of increasing robustness with both circuits depth and qubit count. "Universal Attack Success Fraction" is a common vertical axis for parts (**b–d**)

bounded as

$$1 - e^{-\frac{1}{2}S^{(2)}(\nu)} \leq D \leq 1 - e^{-nS^{(2)}(\nu)}, \tag{7.17}$$

where $S^{(2)}(\nu)$ is the Rényi 2-entropy of the reduced Choi state of a backwards circuit-evolved flip operator, $\nu := \mathrm{tr}_A[W_{\mathrm{univ}} |\phi^+\rangle\langle\phi^+| W_{\mathrm{univ}}^\dagger]$, maximized over all congruent bipartitions of the Hilbert space; $\mathcal{H} = \mathcal{H}_A \otimes \mathcal{H}_{\bar{A}}$. Explicitly,

$$D := \inf_{W^{(i)}, 1 \leq i \leq n} \frac{1}{2d} \| \mathcal{W}_{\mathrm{univ}} - \bigotimes_i \mathcal{W}^{(i)} \|_2^2, \tag{7.18}$$

where $\mathcal{W} := W \otimes W^*$ denotes the (superoperator representation of the) quantum map for unitary W.

Here, an optimal spoof W_{univ} acts as a "flip operator" on the k qubits which are measured (in the σ_z basis) and so has only an odd number of σ_x and σ_y in its Pauli decomposition on these qubits, therefore flipping all of the measured expectation values $y_\theta(\boldsymbol{x})$. A proof of Theorem 3 may be found in Ref. [38].

Theorem 3 applies directly to the case of dense angle encoding (7.4), as the classical adversary can effectively only apply a tensor product of local unitaries. In particular, examining the two extreme cases of Eq. (7.17), we see that

$$D \approx \begin{cases} 1, & \text{if } LOE \gg 0 \\ 0, & \text{if } LOE \approx 0. \end{cases} \tag{7.19}$$

This says that a close approximation to a universal attack is possible for a circuit with low LOE, when $D \approx 0$, in the strong sense of Eq. (7.16). The converse case, when the bounds of Theorem 3 are looser, is investigated numerically below. As a simple application of Theorem 3, we consider the case where the model U_θ can be implemented by a Clifford circuit. We have:

Corollary 4 *If the variational quantum circuit U_θ can be implemented using only Clifford gates, then for a local data-encoding map, a universal adversary exists.*

Here, Theorem 3 predicts the existence of universal adversarial attacks composed of local unitaries: the flip operator $F = \sigma_x \otimes \mathbb{1}^{n-1}$ has zero LOE, and under conjugation with a Clifford unitary $U_\theta = C$, so does the backwards evolved $CFC^\dagger$. This operation can therefore in principle be applied by an adversary through an attack on the classical data. In the case of Clifford dynamics we can also see this directly: belonging to the normaliser of the Pauli group, C maps the Pauli string F to the Pauli string $CFC^\dagger =: W_{\mathrm{univ}}$, the required local universal attack.

Corollary 4 applies, for example, to angle encoding, modulo the subtlety that depending on the precise details of the angle encoding, the classical adversary may not be able to apply arbitrary local unitaries. In the formulation of Eq. (7.3), for example, only linear combinations of I and X gates are actually realizable. Nonetheless, with high probability one of the local universal adversarial attacks will be of this form, which we discuss in detail in Ref. [38].

We note that the behavior of the LOE of the model is distinct from the standard QML assumption of a 2−design [14, 19, 20]. For example, the Cliffords generate a 2−design, but as we argue above, LOE is constant for any Clifford dynamics. In Fig. 7.2a we plot the scaling of the LOE in a typical hardware efficient ansatz, finding an extensive growth indicative of quantum chaos [41–43, 48, 50–53, 56]. This is consistent with previous work arguing that effective QML models are efficient scramblers [61, 62], a necessary condition for chaos [56, 60].

We stress again that a fast growing OTOC, the subject of Theorem 2, is not equivalent to a chaotic circuit [56]. This means that the combination of Theorems 2 and 3 dictates that a trained QML circuit U_θ is vulnerable to a universal adversarial attack if it is sufficiently scrambling in terms of the OTOC (Theorem 2), yet not chaotic according to the LOE (Theorem 3).

Finally, we also prove in Ref. [38] that an ϵ-approximation to a universal attack (in 2-norm distance) is itself a 2ϵ-approximate universal adversarial attack, in the sense $\left|\langle\psi|\left(Z_U + W^\dagger Z_U W\right)|\psi\rangle\right| \leq 2\epsilon$ where W is an ϵ-approximate universal adversarial attack. In turn, this means that any state $\boldsymbol{x}$ where $|y_\theta(x)| > 2\epsilon$ will be misclassified after applying W. This implies that the range of ϵ for which an ϵ-approximate universal adversarial example correctly spoofs the entire training set S grows as the empirical risk $\hat{R}_S(\boldsymbol{\theta})$ of U_θ falls (see discussion below Theorem 1).

7.2.4 Numerical Results

We now complement our analytic results with supporting numerical calculations, supplied in Fig. 7.2. Although Theorem 3 gives a bound between the distance from a perfect universal attack to the subset of attacks which can actually be realized by a classical adversary (i.e., consisting only of single qubit rotations) enforced by the LOE of the circuit, its operational meaning remains unclear. For example, if W_{univ} is a universal attack, then so is $-W_{\text{univ}}$, but $\|W_{\text{univ}} - (-W_{\text{univ}})\|_2 = 2d \gg 0$, so being 2-norm far from a given universal attack does not guarantee that the adversary will be unsuccessful. We therefore now seek to connect Theorem 3 to a metric more transparently relevant in practice: the fraction of states that are misclassified following a specific attack. We investigate two scenarios: attacks a given distance in 2-norm away from a perfect attack, and optimized local attacks for QML models of varying circuit depth.

We begin by investigating the relationship between the distance of a given unitary from a perfect universal attack and its efficacy in practice. For concreteness, we generate approximations W_{approx} to W_{univ} satisfying

$$\|W_{\text{approx}} - W_{\text{univ}}\|_2 = \epsilon\sqrt{d} \tag{7.20}$$

for various choices of ϵ by rotating away from W by a unitary generated by a random Pauli string P, i.e.,

$$W_{\text{approx}} = e^{-itP} W_{\text{univ}} e^{itP}, \tag{7.21}$$

with t chosen so as to satisfy Eq. (7.20). The average fraction of states successfully spoofed by the resulting attacks is plotted in Fig. 7.2b. We find a clear dependence between the success of the adversarial attack and the 2-norm distance between the corresponding unitary and W_{univ}. For $\epsilon\sqrt{d} \approx \sqrt{2}$ (i.e., $\langle W_{\text{univ}} | W_{\text{approx}} \rangle_{HS} \approx 0$) the success probability drops to 1/2, with the adversarial attack faring no better than a (strong) random perturbation. In Ref. [38] we also prove that, for small ϵ, we have in this setup that the fraction (with respect to the Haar measure) of states that are misclassified after an attack by W_{approx} is given approximately by

$$\Pr_{|\psi\rangle \sim \mu_{\mathbb{U}}} \left[\text{sgn}(y_\theta(|\psi\rangle)) \neq \text{sgn}(y_\theta(W_{\text{approx}} |\psi\rangle)) \right] \approx 1 - \frac{2\epsilon}{\pi} \tag{7.22}$$

showing that a close approximation in 2-norm distance is a sufficient condition for a successful universal attack, as argued more generally in Ref. [38].

We next consider optimized local attacks, i.e., consisting of a tensor product of parameterised single-qubit unitaries. This is a scenario that an attacker would face in practice if they had the ability to tamper with inputs before they were fed to a quantum classifier employing a local data-encoding map, e.g., dense angle encoding (Eq. (7.4)). The attack is optimized using the `ADAM` optimiser [70], and tested on a QML model consisting of two-qubit unitary operations laid out in a brick-like fashion (see Ref. [38]). The simulations are performed using the matrix product state (MPS) simulator `quimb` [71] in which the bond dimension of the simulation is tracked throughout. The bond dimension of the MPS can be directly related to the entanglement entropy of the resulting state, and is maximized after $\sim n$ layers of the circuit. In Fig. 7.2c we consider a set of 20 random classifiers of varying circuit depth and plot the fraction of states for which the adversary is able to learn to induce a misclassification, finding a sharp decrease as the length of the circuits (and the entanglement present) increase. For $n > 6$ the adversary fails to outperform a random perturbation long before the circuits become maximally expressible, which occurs at a depth of $\sim 2^n$ layers. In the 6 qubit case the limiting behavior of the adversary is to maintain a success probability greater than 0.5. Indeed, as the system becomes small, the set of local unitaries becomes a larger portion of the total unitary group $\mathbb{U}(2^n)$, with the restrictions on the adversary relatively lessened. In Fig. 7.2d we consider a simple classification task where we classify the "0" and "1" handwritten digits from the MNIST dataset [69]. The input images were preprocessed using principal component analysis [72] to compress them to vectors of length n_{qubits}, so that they may be angle encoded. In determining the universal attack success fraction we take the average performance of 20 adversaries trained using the same procedure as for the random case. Quantitatively, we find that the success probability increases compared to the random case, with a distinct transition at approximately $2N$ layers wherein the adversary goes from being moderately successful to having noticeably weakened performance. Intriguingly, we also find that in the large-depth limit the performance of the adversary degrades with increasing system size.[2]

7.3 Discussion

Despite considerable excitement and intense research activity, the prospect of quantum advantage in machine learning has remained, to put it mildly, questionable. Perhaps chief among the difficulties has been the discovery of barren plateaus in the training landscapes of generic variational quantum models, with a long sequence of papers [13–20] ruling out their trainability under various reasonable assumptions. With this phenomenon seemingly tied to the ability of the models to implement classically-intractable operations, and known techniques for avoiding barren plateaus [73–76] resulting in classically-*simulable* circuits [22], searches for alternate sources of advantage in QML are increasingly timely [23]. Within this context, the adversarial vulnerability of quantum models is a natural place to look, and has indeed recently attracted significant attention [23–37].

While it is certainly not a priori clear why one would expect quantum dynamics to be suited to implementing machine learning algorithms more efficiently than classical methods, guarantees against spoofing are far more in line with capabilities that naturally arise in, for example, quantum communication [77]. Similarly, our results do not depend on the details of the model being secret, and are guaranteed purely by its quantum mechanical nature. In the universal adversarial attack case with local data-encoding, for example, even if the classical adversary knows exactly what W_{univ} is, they are unable to apply any attack close to it when the corresponding LOE is high enough.

The theoretical framework we have developed in this chapter [38] attempts to examine the vulnerability of quantum classifiers in the context of a practically relevant threat model: that of an adversary who can manipulate classical data before it is sent to the quantum computer for encoding. Given the various existence proofs of adversarial examples for quantum models which can classify arbitrary states [25] or states smoothly generated from a Gaussian latent space [29], it is an interesting open question to characterize exactly which classes of states, and exactly which variational circuits, lead to quantum models with provable robustness guarantees. While our focus has been on the impossibility of implementing adversarial perturbations given only access to the classical data, showing for example that encoding schemes which do not produce entanglement yield provable guarantees against universal adversarial attacks, also of interest is the computational cost of finding such attacks in the first place. Future searches for robustness guarantees could seek to connect the difficulty of

[2] We do however note that for this simple classification task, one can train a highly accurate classifier with very few layers, obviating the need for large and deep classifiers to begin with.

spoofing with the difficulty of simulating the classifier itself, or that of understanding the data that is being classified.

In the current era of noisy intermediate scale (NISQ) quantum computers, it is also interesting to investigate the validity of our results in the presence of uncontrolled external noise. In this case, one would have to instead take $U_\theta \mapsto \mathcal{L}_\theta$, a completely positive trace preserving (CPTP) map, which generally describes open quantum system evolution [78]. Our results in the first two columns of Table 7.1 readily hold also in this paradigm. In particular, the bound of Theorem 1 immediately holds also for CPTP map classifiers $\mathcal{L}_\theta$, as we show in Ref. [38]. In fact, as trace preserving maps cannot increase the distance between quantum states [79], with the distance remaining constant if and only if the dynamics are unitary, Theorem 1 is in fact strengthened by the presence of noise-induced non-unitarity. A similar point was made in Ref. [26] in the context of depolarisation noise, the strength of which was linked to differential privacy, a measure of the insensitivity of a map to changes in its input [80]. The key difficulty with such guarantees, strengthened though they are by increasing noise, is separating the (non-perturbed) predictions from zero in the first place. Further, in terms of Theorem 2, the only quantity in Eq. (7.11) depending on the circuit is the OTOC, which also tends to grow fast in noisy, scrambling systems [81, 82]. For our results in the final column of Table 7.1, it is less clear as the LOE has not been well studied in the open systems setting. Finally, we note that in Ref. [83] it is shown that noisy circuits are generally only be as useful as shallow circuits for computing expectation values of Pauli operators, which is what QML is largely concerned with. This suggests that noise-tolerance will be a necessary ingredient of QML advantage, before one needs to consider adversarial robustness.

On this note, it remains to be seen to what extent the chaos-based quantum guarantees of adversarial robustness are consistent with the trainability of the safeguarded model in the first place. Highly entangling operators can suffer from entanglement-induced barren plateaus [84] (although this is not guaranteed [19]), as does the related problem of trying to learn a given scrambling operator [85]. As the theory of the trainability of variational QML models continues to advance [19, 20], the search for models that maximize trainability and robustness while minimizing classical simulability remains an important research direction.

References

1. Szegedy C, Zaremba W, Sutskever I, Bruna J, Erhan D, Goodfellow I, Fergus R (2013) Intriguing properties of neural networks. arXiv:1312.6199
2. Biggio B, Corona I, Maiorca D, Nelson B, Šrndić N, Laskov P, Giacinto G, Roli F (2013) Evasion attacks against machine learning at test time. In: Joint European conference on machine learning and knowledge discovery in databases. Springer, pp 387–402. https://doi.org/10.48550/arXiv.1708.06131
3. Goodfellow IJ, Shlens J, Szegedy C (2014) Explaining and harnessing adversarial examples. arXiv:1412.6572

4. Chakraborty A, Alam M, Dey V, Chattopadhyay A, Mukhopadhyay D (2021) A survey on adversarial attacks and defences. CAAI Trans Intell Technol 6:25. https://doi.org/10.1049/cit2.12028
5. Ilyas A, Santurkar S, Tsipras D, Engstrom L, Tran B, Madry A (2019) Adversarial examples are not bugs, they are features. Adv Neural Inf Process Syst 32. https://proceedings.neurips.cc/paper_files/paper/2019/file/e2c420d928d4bf8ce0ff2ec19b371514-Paper.pdf
6. Su J, Vargas DV, Sakurai K (2019) One pixel attack for fooling deep neural networks. IEEE Trans Evol Comput 23:828. https://doi.org/10.1109/TEVC.2019.2890858
7. Athalye A, Carlini N, Wagner D (2018) Obfuscated gradients give a false sense of security: circumventing defenses to adversarial examples. In: International conference on machine learning. PMLR, pp 274–283. https://proceedings.mlr.press/v80/athalye18a/athalye18a.pdf
8. Cohen J, Rosenfeld E, Kolter Z (2019) Certified adversarial robustness via randomized smoothing. In: International conference on machine learning. PMLR, pp 1310–1320. https://proceedings.mlr.press/v97/cohen19c.html
9. Lecuyer M, Atlidakis V, Geambasu R, Hsu D, Jana S (2019) Certified robustness to adversarial examples with differential privacy. In: 2019 IEEE symposium on security and privacy (SP). IEEE, pp 656–672. https://doi.org/10.1109/SP.2019.00044
10. Biamonte J, Wittek P, Pancotti N, Rebentrost P, Wiebe N, Lloyd S (2017) Quantum machine learning. Nature 549:195. https://doi.org/10.1038/nature23474
11. Abbas A, Sutter D, Zoufal C, Lucchi A, Figalli A, Woerner S (2021) The power of quantum neural networks. Nat Comput Sci 1:403. https://www.nature.com/articles/s43588-021-00084-1
12. Liu Y, Arunachalam S, Temme K (2021) A rigorous and robust quantum speed-up in supervised machine learning. Nat Phys 17:1013. https://www.nature.com/articles/s41567-021-01287-z
13. Holmes Z, Sharma K, Cerezo M, Coles PJ (2022) Connecting ansatz expressibility to gradient magnitudes and barren plateaus. PRX Quantum 3:010313. https://doi.org/10.1103/PRXQuantum.3.010313
14. McClean JR, Boixo S, Smelyanskiy VN, Babbush R, Neven H (2018) Barren plateaus in quantum neural network training landscapes. Nat Commun 9:4812. https://doi.org/10.1038/s41467-018-07090-4
15. Wang S, Fontana E, Cerezo M, Sharma K, Sone A, Cincio L, Coles PJ (2021) Noise-induced barren plateaus in variational quantum algorithms. Nat Commun 12:6961. https://www.nature.com/articles/s41467-021-27045-6
16. Cerezo M, Sone A, Volkoff T, Cincio L, Coles PJ (2021) Cost function dependent barren plateaus in shallow parametrized quantum circuits. Nat Commun 12:1791. https://www.nature.com/articles/s41467-021-21728-w
17. Larocca M, Czarnik P, Sharma K, Muraleedharan G, Coles PJ, Cerezo M (2022) Diagnosing barren plateaus with tools from quantum optimal control. Quantum 6:824. https://doi.org/10.22331/q-2022-09-29-824
18. Patti TL, Najafi K, Gao X, Yelin SF (2021) Entanglement devised barren plateau mitigation. Phys Rev Res 3:033090. https://doi.org/10.1103/PhysRevResearch.3.033090
19. Ragone M, Bakalov BN, Sauvage F, Kemper AF, Ortiz Marrero C, Larocca M, Cerezo M (2024) A lie algebraic theory of barren plateaus for deep parameterized quantum circuits. Nat Commun 15:7172. https://doi.org/10.1038/s41467-024-49909-3
20. Fontana E, Herman D, Chakrabarti S, Kumar N, Yalovetzky R, Heredge J, Sureshbabu SH, Pistoia M (2023) The adjoint is all you need: characterizing barren plateaus in quantum ansätze. arXiv:2309.07902
21. Larocca M, Thanasilp S, Wang S, Sharma K, Biamonte J, Coles PJ, Cincio L, McClean JR, Holmes Z, Cerezo M (2024) A review of barren plateaus in variational quantum computing. arXiv:2405.00781
22. Cerezo M, Larocca M, García-Martín D, Diaz NL, Braccia P, Fontana E, Rudolph MS, Bermejo P, Ijaz A, Thanasilp S et al (2023) Does provable absence of barren plateaus imply classical simulability? or, why we need to rethink variational quantum computing. arXiv:2312.09121

23. West MT, Tsang SL, Low JS, Hill CD, Leckie C, Hollenberg LCL, Erfani SM, Usman M (2023) Towards quantum enhanced adversarial robustness in machine learning. Nat Mach Intell 5:581. https://doi.org/10.1038/s42256-023-00661-1
24. Lu S, Duan L-M, Deng D-L (2020) Quantum adversarial machine learning. Phys Rev Res 2:033212. https://doi.org/10.1103/PhysRevResearch.2.033212
25. Liu N, Wittek P (2020) Vulnerability of quantum classification to adversarial perturbations. Phys Rev A 101:062331. https://doi.org/10.1103/PhysRevA.101.062331
26. Du Y, Hsieh M-H, Liu T, Tao D, Liu N (2021) Quantum noise protects quantum classifiers against adversaries. Phys Rev Res 3:023153. https://doi.org/10.1103/PhysRevResearch.3.023153
27. Guan J, Fang W, Ying M (2021) Robustness verification of quantum classifiers. In: International conference on computer aided verification. Springer, pp 151–174. https://doi.org/10.1007/978-3-030-81685-8_7
28. Weber M, Liu N, Li B, Zhang C, Zhao Z (2021) Optimal provable robustness of quantum classification via quantum hypothesis testing. npj Quantum Inf 7:1.https://doi.org/10.1038/s41534-021-00410-5
29. Liao H, Convy I, Huggins WJ, Whaley KB (2021) Robust in practice: adversarial attacks on quantum machine learning. Phys Rev A 103:042427. https://doi.org/10.1103/PhysRevA.103.042427
30. Kehoe A, Wittek P, Xue Y, Pozas-Kerstjens A (2021) Defence against adversarial attacks using classical and quantum-enhanced Boltzmann machines. Mach Learn Sci Technol 2:045006. https://doi.org/10.1088/2632-2153/abf834
31. Ren W, Li W, Xu S, Wang K, Jiang W, Jin F, Zhu X, Chen J, Song Z, Zhang P, et al (2022) Experimental quantum adversarial learning with programmable superconducting qubits. Nat Comput Sci 2:711. https://www.nature.com/articles/s43588-022-00351-9
32. Wu Y, Adermann E, Thapa C, Camtepe S, Suzuki H, Usman M (2023) Radio signal classification by adversarially robust quantum machine learning. arXiv:2312.07821
33. West MT, Erfani SM, Leckie C, Sevior M, Hollenberg LCL, Usman M (2023) Benchmarking adversarially robust quantum machine learning at scale. Phys Rev Res 5:023186. https://doi.org/10.1103/PhysRevResearch.5.023186
34. West MT, Nakhl AC, Heredge J, Creevey FM, Hollenberg LC, Sevior M, Usman M (2023) Drastic circuit depth reductions with preserved adversarial robustness by approximate encoding for quantum machine learning. arXiv:2309.09424
35. Khatun A, Usman M (2024) Quantum transfer learning with adversarial robustness for classification of high-resolution image datasets. arXiv:2401.17009
36. Winderl D, Franco N, Lorenz JM (2024) Constructing optimal noise channels for enhanced robustness in quantum machine learning. arXiv:2404.16417
37. Berberich J, Fink D, Pranjić D, Tutschku C, Holm C (2023) Training robust and generalizable quantum models. arXiv:2311.11871
38. Dowling N, West MT, Southwell A, Nakhl AC, Sevior M, Usman M, Modi K (2024) Adversarial robustness guarantees for quantum classifiers. arXiv:2405.10360
39. Hendrik Metzen J, Chaithanya Kumar M, Brox T, Fischer V (2017) Universal adversarial perturbations against semantic image segmentation. In: Proceedings of the IEEE international conference on computer vision, pp 2755–2764
40. Gong W, Deng D-L (2022) Universal adversarial examples and perturbations for quantum classifiers. Natl Sci Rev 9:nwab130. https://doi.org/10.1093/nsr/nwab130
41. Prosen T, Znidarič M (2007) Is the efficiency of classical simulations of quantum dynamics related to integrability? Phys Rev E 75:015202(R). https://doi.org/10.1103/PhysRevE.75.015202
42. Prosen T, Pižorn I (2007) Operator space entanglement entropy in a transverse ising chain. Phys Rev A 76:032316. https://doi.org/10.1103/PhysRevA.76.032316
43. Pižorn I, Prosen T (2009) Operator space entanglement entropy in xy spin chains. Phys Rev B 79:184416. https://doi.org/10.1103/PhysRevB.79.184416

44. Shenker SH, Stanford D (2014) Black holes and the butterfly effect. J High Energy Phys 2014:67. https://doi.org/10.1007/JHEP03(2014)067
45. Maldacena J, Shenker SH, Stanford D (2016) A bound on chaos. J High Energy Phys 2016:106. https://doi.org/10.1007/JHEP08(2016)106
46. Swingle B, Bentsen G, Schleier-Smith M, Hayden P (2016) Measuring the scrambling of quantum information. Phys Rev A 94:040302. https://doi.org/10.1103/PhysRevA.94.040302
47. Roberts DA, Swingle B (2016) Lieb-robinson bound and the butterfly effect in quantum field theories. Phys Rev Lett 117:091602. https://doi.org/10.1103/PhysRevLett.117.091602
48. Dubail J (2017) Entanglement scaling of operators: a conformal field theory approach, with a glimpse of simulability of long-time dynamics in 1..+..1d. J Phys A Math Theoret 50:234001. https://doi.org/10.1088/1751-8121/aa6f38
49. Swingle B (2018) Unscrambling the physics of out-of-time-order correlators. Nat Phys 14:988. https://doi.org/10.1038/s41567-018-0295-5
50. Jonay C, Huse DA, Nahum A (2018) Coarse-grained dynamics of operator and state entanglement. arXiv:1803.00089
51. Alba V, Dubail J, Medenjak M (2019) Operator entanglement in interacting integrable quantum systems: the case of the rule 54 chain. Phys Rev Lett 122:250603. https://doi.org/10.1103/PhysRevLett.122.250603
52. Bertini B, Kos P, Prosen T (2020) Operator entanglement in local quantum circuits i: chaotic dual-unitary circuits. Sci Post Phys 8:067. https://doi.org/10.21468/SciPostPhys.8.4.067
53. Alba V (2021) Diffusion and operator entanglement spreading. Phys Rev B 104:094410. https://doi.org/10.1103/PhysRevB.104.094410
54. Anand N, Styliaris G, Kumari M, Zanardi P (2021) Quantum coherence as a signature of chaos. Phys Rev Res 3:023214. https://doi.org/10.1103/PhysRevResearch.3.023214
55. Kim J, Murugan J, Olle J, Rosa D (2022) Operator delocalization in quantum networks. Phys Rev A 105:L010201. https://doi.org/10.1103/PhysRevA.105.L010201
56. Dowling N, Kos P, Modi K (2023) Scrambling is necessary but not sufficient for chaos. Phys Rev Lett 131:180403. https://doi.org/10.1103/PhysRevLett.131.180403
57. Dowling N, Modi K (2024) Operational metric for quantum chaos and the corresponding spatiotemporal-entanglement structure. PRX Quantum 5:010314. https://doi.org/10.1103/PRXQuantum.5.010314
58. Sekino Y, Susskind L (2008) Fast scramblers J High Energy Phys 2008:065. https://doi.org/10.1088/1126-6708/2008/10/065
59. Foini L, Kurchan J (2019) Eigenstate thermalization hypothesis and out of time order correlators. Phys Rev E 99:042139. https://doi.org/10.1103/PhysRevE.99.042139
60. Xu T, Scaffidi T, Cao X (2020) Does scrambling equal chaos? Phys Rev Lett 124:140602. https://doi.org/10.1103/PhysRevLett.124.140602
61. Shen H, Zhang P, You Y-Z, Zhai H (2020) Information scrambling in quantum neural networks. Phys Rev Lett 124:200504. https://doi.org/10.1103/PhysRevLett.124.200504
62. Wu Y, Zhang P, Zhai H (2021) Scrambling ability of quantum neural network architectures. Phys Rev Res 3:L032057. https://doi.org/10.1103/PhysRevResearch.3.L032057
63. LaRose R, Coyle B (2020) Robust data encodings for quantum classifiers. Phys Rev A 102:032420. https://doi.org/10.1103/PhysRevA.102.032420
64. Agarap AF (2018) Deep learning using rectified linear units (relu). arXiv:1803.08375
65. Caro MC, Huang H-Y, Cerezo M, Sharma K, Sornborger A, Cincio L, Coles PJ (2022) Generalization in quantum machine learning from few training data. Nat Commun 13:4919. https://www.nature.com/articles/s41467-022-32550-3
66. Garcia RJ, Bu K, Jaffe A (2022) Quantifying scrambling in quantum neural networks. J High Energy Phys 2022:27. https://doi.org/10.1007/JHEP03(2022)027
67. Xu S, Swingle B (2020) Accessing scrambling using matrix product operators. Nat Phys 16:199. https://doi.org/10.1038/s41567-019-0712-4
68. Moosavi-Dezfooli S-M, Fawzi A, Fawzi O, Frossard P (2017) Universal adversarial perturbations. In: Proceedings of the IEEE conference on computer vision and pattern recognition, pp 1765–1773. https://openaccess.thecvf.com/content_cvpr_2017/html/Moosavi-Dezfooli_Universal_Adversarial_Perturbations_CVPR_2017_paper.html

69. Lecun Y, Bottou L, Bengio Y, Haffner P (1998) Gradient-based learning applied to document recognition. Proc IEEE 86:2278. https://doi.org/10.1109/5.726791
70. Kingma DP, Ba J (2014) Adam: a method for stochastic optimization. arXiv:1412.6980
71. Gray J (2018) quimb: a python package for quantum information and many-body calculations. J Open Source Soft 3:819. https://doi.org/10.21105/joss.00819
72. Jolliffe IT, Cadima J (2016) Principal component analysis: a review and recent developments. Philos Trans R Soc A Math Phys Eng Sci 374:20150202
73. Pesah A, Cerezo M, Wang S, Volkoff T, Sornborger AT, Coles PJ (2021) Absence of barren plateaus in quantum convolutional neural networks. Phys Rev X 11:041011. https://doi.org/10.1103/PhysRevX.11.041011
74. Schatzki L, Larocca M, Nguyen QT, Sauvage F, Cerezo M (2024) Theoretical guarantees for permutation-equivariant quantum neural networks. npj Quantum Inf 10:12. https://doi.org/10.1038/s41534-024-00804-1
75. Wang Y, Qi B, Ferrie C, Dong D (2023) Trainability enhancement of parameterized quantum circuits via reduced-domain parameter initialization. arXiv:2302.06858
76. West MT, Heredge J, Sevior M, Usman M (2023) Provably trainable rotationally equivariant quantum machine learning. arXiv:2311.05873
77. Bennett CH, Brassard G (2014) Quantum cryptography: public key distribution and coin tossing. Theoret Comput Sci 560:7. https://doi.org/10.1016/j.tcs.2014.05.025
78. Wilde MM (2013) Quantum information theory. Cambridge University Press. https://doi.org/10.1017/CBO9781139525343
79. Nielsen MA, Chuang IL (2010) Quantum computation and quantum information. Cambridge University Press. https://www.cambridge.org/highereducation/books/quantum-computation-and-quantum-information/01E10196D0A682A6AEFFEA52D53BE9AE#overview
80. Zhou L, Ying M, Differential privacy in quantum computation. In: 2017 IEEE 30th computer security foundations symposium (CSF). IEEE, pp 249–262. https://ieeexplore.ieee.org/document/8049724
81. Zhang Y-L, Huang Y, Chen X (2019) Information scrambling in chaotic systems with dissipation. Phys Rev B 99:014303. https://doi.org/10.1103/PhysRevB.99.014303
82. Schuster T, Yao NY (2023) Operator growth in open quantum systems. Phys Rev Lett 131:160402. https://doi.org/10.1103/PhysRevLett.131.160402
83. Mele AA, Angrisani A, Ghosh S, Khatri S, Eisert J, Frana DS, Quek Y (2024) Noise-induced shallow circuits and absence of barren plateaus. arXiv:2403.13927
84. Marrero CO, Kieferová M, Wiebe N (2021) Entanglement-induced barren plateaus. PRX Quantum 2:040316. https://doi.org/10.1103/PRXQuantum.2.040316
85. Holmes Z, Arrasmith A, Yan B, Coles PJ, Albrecht A, Sornborger AT (2021) Barren plateaus preclude learning scramblers. Phys Rev Lett 126:190501. https://doi.org/10.1103/PhysRevLett.126.190501

Chapter 8
Understanding the Generalization of Adversarially Trained Quantum Classifiers

Petros Georgiou and Sharu Theresa Jose

Abstract Quantum classifiers have recently been found to be vulnerable to adversarial attacks. These attacks perturb the input data to mislead the classifier to make incorrect predictions. Adversarial training has since emerged as a promising defence strategy for quantum classifiers. In adversarial training, the training set consists of both clean and adversarially perturbed data, where the latter are generated assuming a known adversarial attack. Thus, the classifier is trained to minimize an adversarial loss, which quantifies the worst loss suffered by the classifier due to adversarial perturbations. Although quantum classifiers trained in this way incur lower adversarial loss on the training dataset, there is still limited understanding of how well these classifiers generalize to new, previously unseen, adversarially perturbed data. This chapter reviews recent research focused on understanding the generalization performance of adversarially trained quantum classifiers. To this end, we first introduce the learning-theoretic concept of Rademacher complexity and demonstrate how it can quantify the generalization performance of standard, non-adversarially trained, quantum classifiers. We then extend this framework to the adversarial setting considering different types of adversarial attacks, including classical adversarial attacks perturbing the classical input data before it is embedded into a quantum state, and quantum adversarial attacks perturbing the quantum states directly. For classical attacks, we also discuss the impact of the quantum embedding on the generalization performance of adversarially trained quantum classifiers.

8.1 Introduction

Recent studies have shown that quantum classifiers are vulnerable to small, adversarial perturbations in the input data, leading to incorrect predictions [1, 2]. For instance, a quantum classifier based on variational quantum circuits and implemented on programmable transmon qubits—achieving test accuracy up to 99% on medical MRI

P. Georgiou · S. T. Jose (✉)
University of Birmingham, Birmingham, United Kingdom
e-mail: s.t.jose@bham.ac.uk

M. Usman (ed.), *Quantum Robustness in Artificial Intelligence*, Quantum Science and Technology, https://doi.org/10.1007/978-3-032-11153-1_8

images—was shown to be easily deceived by adversarial perturbations imperceptible to human eyes [2]. These vulnerabilities not only threaten any potential quantum advantage, but also limits the applicability of quantum classifiers to safety-critical industries.

In this context, *adversarial training* has emerged as a promising defense strategy for quantum classifiers, demonstrating improved robustness to various attacks [1–3]. This approach assumes prior knowledge of the adversarial attack—its type and strength—to design an *adversarial loss* function, that captures the worst-case performance under adversarial attack. The goal of adversarial training is to find the robust classifier that minimizes the adversarial loss function over the observed training dataset. This results in a challenging min-max optimization problem, with the learner trying to minimize the worst-case loss incurred under adversarial attacks.

While much prior work has focused on the optimization challenges of adversarial training, this chapter shifts the focus toward understanding the *adversarial robust generalization*—the ability of an adversarially trained classifier to perform well on new, previously unseen, adversarially perturbed test data. A classifier that achieves low adversarial training loss, yet performs poorly on adversarial test data, suffers from poor adversarial generalization due to overfitting on the training data. Adversarial robust generalization has been extensively studied in the literature for classical machine learning models including neural network models [4–6]. The key insight is that adversarial robust generalization is significantly harder than standard generalization, often requiring substantially more training data.

Building on insights from classical machine learning, this chapter investigates the adversarial robust generalization of quantum classifiers. Unlike classical models, quantum classifiers present unique characteristics in the adversarial setting: Attacks can target either the classical data space or the quantum state space, and the data encoding layers in quantum classifiers may confer additional robustness. These distinct features prompt a careful reevaluation of adversarial robust generalization in the quantum context. To this end, we aim to address the following fundamental questions:

1. *How well does an adversarially trained quantum classifier generalize to similarly perturbed test data?*
2. *What is the sample complexity required to guarantee adversarial robust generalization in a probably approximately correct sense?*
3. *What factors affect the generalization capacity of adversarially trained quantum classifiers?*

We address these questions systematically in the next few sections.

8.2 Generalization Error of Quantum Classifiers

We start by defining quantum classifier models, and then discuss the generalization analysis of standard trained quantum classifiers.

Throughout this chapter, we consider input data of the form (x, y) where $x \in \mathcal{X}$ denotes the feature input taking values in the space $\mathcal{X}$ and y denotes the corresponding label taking values in $\mathcal{Y}$. For simplicity, we focus on binary classification with label space $\mathcal{Y} = \{+1, -1\}$. The feature input x can denote purely classical data (e.g., images of a cat or dog) or can denote a quantum state $\rho(x)$ indexed by a classical parameter $x \in \mathcal{X}$. A quantum classifier $g : \mathcal{X} \to \mathcal{Y}$ then models the unknown relationship between feature input and corresponding label.

When the input data is classical, the quantum classifier first encodes the classical data x to a quantum state $\rho(x)$ via a quantum feature map: $\rho : x \mapsto \rho(x)$. This step can be omitted if the data is quantum. The *quantum feature map* can be implemented using various strategies—amplitude embedding, rotation embedding, repeated embedding, etc. [7] (see Sect. 8.2.2 for more details). Following the embedding stage, the quantum classifier measures the expected value of a Hermitian observable H given the quantum state $\rho(x)$. The expected value of the observable is given by $f(x) = \mathrm{Tr}(H\rho(x))$, where $\mathrm{Tr}(\cdot)$ denotes the trace operation. The binary classifier then assigns the label depending on the sign of $f(x)$ as $g(x) = \mathrm{sgn}(f(x)) = \mathrm{sgn}(\mathrm{Tr}(H\rho(x)))$, where $\mathrm{sgn}(\cdot)$ denotes the sign function.

We can then define the function class of quantum binary classifiers as

$$\mathcal{G}_r = \{x \mapsto g(x) = \mathrm{sgn}(\mathrm{Tr}(H\rho(x))) : H \in \mathcal{H}_r\}, \tag{8.1}$$

where

$$\mathcal{H}_r = \{H : H = H^\dagger, \|H\|_r \leq B\} \tag{8.2}$$

denotes the model class of Hermitian observables with B-bounded r-Schatten norm for $B > 0$. Note that for $r \in [1, \infty)$, the r-Schatten norm of an observable A is defined as $\|A\|_r = \left(\mathrm{Tr}(|A|^r)\right)^{1/r}$ with $|A| = \sqrt{AA^\dagger}$. In the limiting case of $r = \infty$, $\|A\|_\infty = \max\{\lambda(|A|)\}$, where $\{\lambda(|A|)\}$ denotes the set of eigenvalues of $|A|$.

To train the above classifier, we first define a loss function $\ell(g, \rho(x), y)$ that measures the performance of the classifier g on a data point (x, y). For binary classification, following standard classical machine learning literature [6], we use

$$\ell(g, \rho(x), y) = \phi(yf(x)), \tag{8.3}$$

where $\phi(\cdot)$ is a $[0, M]$-bounded, monotonically non-increasing, η-Lipschitz continuous function.

The goal of learning is to find the classifier that minimizes the *population loss* (or test loss),

$$L(g) = \mathbb{E}_{P(x,y)}[\ell(g, \rho(x), y)], \tag{8.4}$$

which is the expected loss with respect to the true, unknown data distribution $P(x, y)$. However, finding this classifier is impossible, since the data distribution $P(x, y)$ is unknown to the learner. Instead, the learner uses the observed training set $\mathcal{D} = \{(x^1, y^1), \ldots, (x^N, y^N)\}$ of N samples, that are independently and identically generated from the data distribution $P(x, y)$, to define an empirical *training loss*,

$$\hat{L}(g) = \frac{1}{N} \sum_{i=1}^{N} \ell(g, \rho(x^i), y^i), \tag{8.5}$$

and finds the classifier $g \in \mathcal{G}_r$ that minimizes the above training loss.

The difference between the population loss and training loss of a classifier g defines the *generalization error*,

$$\Delta L(g) = L(g) - \hat{L}(g).$$

Intuitively, generalization error quantifies how well the performance of the classifier on a finite training set indicates its performance on test data. A larger gap indicates larger mismatch between training and testing, and thus poor generalization (or large generalization error). For instance, a classifier that memorizes the training data can incur zero training loss, but non-zero test loss on new, previously unseen data. The generalization error quantifies the amount of overfitting of the classifier on the training set.

8.2.1 Bounds on Generalization Error

Understanding and characterizing the generalization error of classifiers has long been a central problem in statistical learning theory [8]. Several theoretical frameworks have been developed to quantify the generalization error in terms of the *complexity* of the classifier function class, the number of training samples, or the properties of the training algorithm. This includes the probably approximately correct (PAC) framework, the PAC-Bayesian framework (see [9] for a nice introduction), the algorithmic-stability framework (see [10]), and information-theoretic framework (see [11]).

In this chapter, we will adopt the probably approximately correct framework to analyze the generalization error. The PAC framework gives a data-distribution independent, high probability upper bound on the generalization error, that holds uniformly for all models in the function class. Specifically, in the PAC setting, we determine the minimum number $N(\epsilon, \delta)$ of training samples to ensure, with probability at least $1-\delta$ (for $\delta \in (0, 1)$), that the generalization error $\Delta L(g)$ of any classifier $g \in \mathcal{G}_r$ is at most ϵ [12]. The minimum number $N(\epsilon, \delta)$ is called the *sample complexity* required to (ϵ, δ)-PAC learn the function class $\mathcal{G}_r$. This, in turn, depends on the complexity of the function class measured using various metrics like Vapnik-Chervonenkis (VC) dimension, Rademacher complexity, etc.

In particular, for $[0, M]$-bounded loss function $\ell(g, x, y)$, the following upper bound on the generalization error holds with probability at least $1-\delta$, with respect to the random sampling of training data, for all $g \in \mathcal{G}_r$ [12]:

$$\Delta L(g) \leq 2\mathcal{R}(\ell \circ \mathcal{G}_r) + 3M\sqrt{\frac{1}{2N}\log\frac{2}{\delta}}, \tag{8.6}$$

where

$$\mathcal{R}(\ell \circ \mathcal{G}_r) = \mathbb{E}_{\boldsymbol{\sigma}}\Big[\sup_{g\in\mathcal{G}_r}\frac{1}{N}\sum_{i=1}^{N}\sigma_i \ell(g, \rho(x^i), y^i)\Big] \tag{8.7}$$

is the empirical *Rademacher complexity* of the composed function class $\ell \circ \mathcal{G}_r$. Here, the expectation is with respect to the vector $\boldsymbol{\sigma} = (\sigma_1, \ldots, \sigma_N)$ of N i.i.d Rademacher variables, each taking values $\sigma_i = \pm 1$ with equal probability. Intuitively, Rademacher complexity measures the richness of a function class by its ability to fit random noise: a class capable of fitting random noise well is more complex.

Following (8.6), an upper bound on the generalization error $\Delta L(g)$ of any classifier g belonging to the class $\mathcal{G}_r$ can be derived by analyzing its Rademacher complexity. To this end, we have the following upper bound.

Theorem 8.1 *For an η-Lipschitz loss function $\ell(g, \rho(x), y)$, the empirical Rademacher complexity $\mathcal{R}(\ell \circ \mathcal{G}_r)$ evaluated on an N-sample training data $\mathcal{D}$ is upper bounded as*

$$\mathcal{R}(\ell \circ \mathcal{G}_r) \leq \eta\mathcal{R}(\mathcal{F}_r) \leq \frac{\eta B}{N}\begin{cases}\sqrt{2\ln(d_H)\Big\|\sum_{i=1}^{N}\rho(x^i)^2\Big\|_\infty}, & r = 1\\[2ex] T_{\frac{r}{r-1}}\|\sqrt{\sum_{i=1}^{N}\rho(x^i)^2}\|_{\frac{r}{r-1}}, & 2 \leq r < \infty\\[2ex] \|\sqrt{\sum_{i=1}^{N}\rho(x^i)^2}\|_1 & , r = \infty\end{cases} \tag{8.8}$$

where

$$\mathcal{R}(\mathcal{F}_r) = \mathbb{E}_{\boldsymbol{\sigma}}\Big[\sup_{f\in\mathcal{F}_r}\frac{1}{N}\sum_{i=1}^{N}\sigma_i y^i f(x^i)\Big] \tag{8.9}$$

is the Rademacher complexity of the function class $\mathcal{F}_r = \{x \mapsto f(x) = \mathrm{Tr}(H_r\rho(x)) : H \in \mathcal{H}_r\}$*,* $T_\beta = 2^{-\frac{1}{4}}\sqrt{\frac{\pi\beta}{e}}$ *and* d_H *is the dimension of quantum Hilbert space.*

Proof The proof follows in two steps. In the first step, since the loss function is η-Lipschitz, one can apply Talagrand's contraction lemma (see Appendix 8.6) to get that $\mathcal{R}(\ell \circ \mathcal{G}_r) \leq \eta\mathcal{R}(\mathcal{F}_r)$, whereby the required Rademacher complexity is upper bounded by the Rademacher complexity of the function class $\mathcal{F}_r$. Note that the function class $\mathcal{F}_r$ consists of functions linear in the observables. Hence, in the second step, one can use Hölder's inequality to derive a further upper bound on $\mathcal{R}(\mathcal{F}_r)$ that is proportional to the term

$$\mathbb{E}_\sigma\left[\left\|\frac{1}{N}\Sigma_{i=1}^N\sigma_i\rho(x^i)\right\|_{\frac{r}{r-1}}\right].$$

Depending on the value of r, this term can be further upper bounded using either the operator Khintchine inequalities or Tropp's inequalities (see Appendix 8.6) to get the desired upper bound. We refer the readers to [13] for more details. □

From Theorem 8.1, we see that Rademacher complexity scales as $\mathcal{R}(\ell \circ \mathcal{G}_r) = O(\frac{\mathcal{B}_r}{N})$, with $\mathcal{B}_r(\cdot)$ depending on appropriate norm of the sum $\sum_{i=1}^N \rho(x^i)^2$ of encoded quantum states. This, in turn, depends on the specific quantum embedding scheme adopted. In fact, we can see that $\mathcal{B}_r = O(\sqrt{N})$, resulting in the Rademacher complexity scaling as $1/\sqrt{N}$ with the number of training samples. This shows that classifier classes with high Rademacher complexity require large number of training samples to ensure low generalization error.

Remark 8.1 Throughout this chapter, we consider function class $\mathcal{G}_r$ determined by the space of observables $\mathcal{H}_r$ with norm constraints. In practice, such classifiers are implemented through parameterized observables $H_\theta = U_\theta^\dagger H U_\theta$, where U_θ is a unitary gate whose parameters θ are optimized during the learning process, and H is some fixed Hermitian matrix with bounded norm. Such parameterized observables yield functions of the form $f(x) = \mathrm{Tr}(H_\theta\rho(x)) = \mathrm{Tr}(U_\theta^\dagger H U_\theta\rho(x))$, where the latter can be equivalently written as $f(x) = \mathrm{Tr}(HU_\theta\rho(x)U_\theta^\dagger) = \mathrm{Tr}(H\rho(\theta, x))$ with $\rho(\theta, x) = U_\theta\rho(x)U_\theta^\dagger$. There exist several works in the literature that study the generalization error of such parameterized quantum classifiers. In particular, Ref. [14] shows that the generalization error of such classifiers scales as $O(\sqrt{\frac{T}{N}})$, where T denotes the number of parameterized unitary gates considered in the model. Alternative upper bounds that explicitly take into account the impact of quantum embedding schemes can be found in [15], where the generalization error is upper bounded as $O(\sqrt{\frac{|\Omega|}{N}})$, where $|\Omega|$ denotes the cardinality of the frequency spectrum of the quantum feature map.

8.2.2 *Quantum Embeddings and Impact on Generalization Error*

We now discuss three commonly used quantum embedding schemes that map the input data $x = (x_1, \ldots, x_d) \in \mathbb{R}^d$ to a pure quantum state $|\psi(x)\rangle$ with corresponding density matrix $\rho(x) = |\psi(x)\rangle\langle\psi(x)|$.

Amplitude embedding encodes the normalized feature entries to the amplitudes of the computational basis states $|i\rangle$ for $i = 1, \ldots d$, as

$$x \mapsto |\psi(x)\rangle = \sum_{i=1}^{d} \frac{x_i}{\|x\|_2}|i\rangle.$$

Implementing this requires $n = \lceil \log_2 d \rceil$ qubits, resulting in a Hilbert space dimension of $d_H = 2^n$.

Angle rotation embedding uses $n = d$ qubits to encode the d-dimensional input x. Specifically, the i-th entry of the feature vector x is encoded into the i-th qubit via a Pauli rotation. Using Pauli-Y rotations, this can be achieved via applying a d-qubit unitary gate $U(x) = \bigotimes_{j=1}^{d} e^{-ix_j\sigma_Y}$, with σ_Y denoting the Pauli-Y matrix, on an initial state $|0\rangle$ as $|\psi(x)\rangle = U(x)|0\rangle$. The resulting embedded quantum state is obtained as

$$x \mapsto |\psi(x)\rangle = U(x)|0\rangle.$$

Dense rotation embedding uses $n = d/2$ qubits to encode two feature entries into each qubit. Specifically, this is achieved via the unitary matrix $U(x) = \bigotimes_{j=1}^{d/2} e^{-ix_{2j}\sigma_Z} e^{-ix_{2j-1}\sigma_Y}$ with σ_Z denoting Pauli-Z matrix [16]. The resulting embedded quantum state is then obtained as $x \mapsto |\psi(x)\rangle = U(x)|0\rangle$.

Finally, L**-layer rotation embedding** schemes obtain the quantum state $|\psi(x)\rangle$ by applying the rotation unitary $U(x)$ followed by a fixed unitary $V^{(l)}$ L-times. This results in

$$|\psi(x)\rangle = V^{(L)}U(x) \ldots V^{(2)}U(x)V^{(1)}U(x)|0\rangle.$$

Such repeated encoding schemes facilitate data re-uploading, and are known to yield highly expressive quantum models [7].

In nutshell, each quantum embedding maps classical data to a quantum state within a Hilbert space, the dimension of which is determined by the specific embedding scheme. For instance, amplitude embedding maps a d-dimensional input to a Hilbert space of dimension $d_H = d$, whereas angle rotation embedding maps it to an exponentially large $d_H = 2^d$-dimensional Hilbert space. The choice of the quantum embedding impacts the Rademacher complexity, and thus the generalization error.

8.3 Adversarial Attacks on Quantum Classifiers and Adversarial Robust Generalization

We now discuss two categories of *white-box* adversarial attacks on quantum classifiers, assuming the adversary has full access to the loss function and the quantum classifier (see Fig. 8.1). The first category encompasses the (p, ϵ)-*classical adversarial attacks*, where the adversary perturbs the feature input $x \in \mathcal{X}$ within an l_p-ball of radius $\epsilon > 0$ to maximize the loss incurred. Specifically, a (p, ϵ)-classical attack on input x returns an adversarial input,

$$\tilde{x}^* = \arg \max_{\tilde{x} \in \mathcal{X}: \|x - \tilde{x}\|_p \leq \epsilon} \ell(g, \rho(x), y). \tag{8.10}$$

The corresponding *adversarial loss* function under classical (p, ϵ)-attack is defined as

$$\ell^C_{p,\epsilon}(g, \rho(x), y) = \max_{\tilde{x} \in \mathcal{X}: \|x - \tilde{x}\|_p \leq \epsilon} \ell(g, \rho(\tilde{x}), y), \tag{8.11}$$

which quantifies the worst-case effect of adversarial attack on quantum classifier.

Beyond classical attacks, we can also consider the class of (p, ϵ)-*quantum attacks*, where the adversary directly perturbs the embedded quantum state $\rho(x)$ within a p-

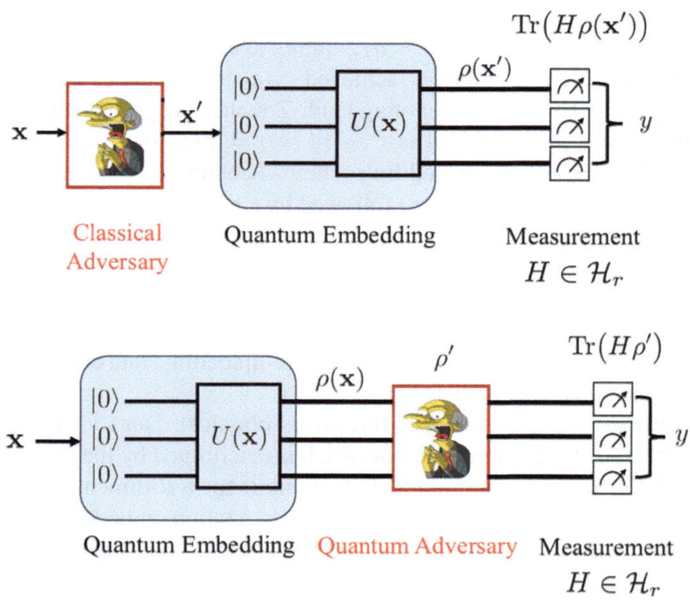

Fig. 8.1 Illustration of a quantum classifier being subject to classical (top) and quantum (bottom) adversarial attacks

Schatten norm ball of radius $\epsilon > 0$ to maximize the loss incurred. Concretely, this attack returns a perturbed quantum state $\tilde{\rho}$ that maximizes the adversarial loss,

$$\ell^{Q}_{p,\epsilon}(g, \rho(x), y) = \max_{\tilde{\rho}:\|\rho(x)-\tilde{\rho}\|_p \leq \epsilon} \ell(g, \tilde{\rho}, y). \quad (8.12)$$

under (p, ϵ)-quantum attack.

Notably, quantum attacks require access to the exact quantum state $\rho(x)$ to perturb it. Furthermore, the optimization problem in (8.12) to generate quantum attacks is more challenging than its classical counterpart in (8.10), as it must ensure that the perturbed state $\tilde{\rho}$ remains a valid quantum state—a positive semi-definite, unit trace, Hermitian matrix.

Remark 8.2 Note that in practice, the inner maximization in (8.10) and (8.12) are analytically intractable for general loss functions. Consequently, practical adversarial attacks typically rely on approximate solutions to the inner maximization problem. For instance, under $p = \infty$, common attack methods include the Projected Gradient Descent (PGD), Basic Iterative Method (BIM), and Fast Gradient Sign Method (FGSM) (see [17]).

Assuming that the attack is known, adversarial robust learning aims to find the quantum classifier that minimizes the *adversarial population loss*

$$L^{C/Q}_{p,\epsilon}(g) = \mathbb{E}_{P(x,y)}[\ell^{C/Q}_{p,\epsilon}(g, \rho(x), y)] \quad (8.13)$$

under classical (C) or quantum (Q) attacks. However, this cannot be achieved in practice, since the data distribution $P(x, y)$ is unknown. Instead, the learner resorts to *adversarial training*, using an empirically evaluated *adversarial training loss*,

$$\hat{L}^{C/Q}_{p,\epsilon}(g) = \frac{1}{N}\sum_{i=1}^{N} \ell^{C/Q}_{p,\epsilon}(g, \rho(x^i), y^i). \quad (8.14)$$

Thus, adversarial training of a quantum classifier entails minimizing the empirical average loss evaluated on adversarially perturbed data inputs. This process, in turn, requires identifying a sufficiently expressive quantum classifier capable of maintaining high performance across all data points within a perturbation ball centered around each original data point.

The difference between adversarial population loss and adversarial training loss quantifies the *adversarial robust generalization error*,

$$\Delta L^{C/Q}_{p,\epsilon}(g) = L^{C/Q}_{p,\epsilon}(g) - \hat{L}^{C/Q}_{p,\epsilon}(g). \quad (8.15)$$

Intuitively, this measures how well the adversarial training loss of a classifier approximates its performance on previously unseen, similarly perturbed data. In the next section, we discuss upper bounds on adversarial robust generalization error through Rademacher complexity.

Before proceeding, let us consider the following example from [13] to study the adversarial generalization error of an adversarially trained quantum classifier and compare it with its conventional generalization error.

Example 8.1 Consider a binary classification problem with equiprobable labels. Assume that the feature input x is generated from the multi-variate Gaussian distribution $P(x|y = y') = \mathcal{N}(\mu_{y'}, \mathbb{I}_d)$ with identity covariance matrix $\mathbb{I}_d$ of dimension d and mean $\mu_{y'} \in \mathbb{R}^d$ taken as

$$\mu_{y'} = \begin{cases} \frac{\pi}{4} I_d & \text{if } y' = +1 \\ \frac{\pi}{4}\left(I_{\lfloor d/2 \rfloor} \oplus (-I_{d-\lfloor d/2 \rfloor})\right) & \text{if } y' = -1, \end{cases}$$

where I_n denotes a vector of ones in $\mathbb{R}^n$. Consider the following class of binary quantum classifiers: $f(x) = \mathrm{Tr}(H_\theta \rho(x))$, where $\rho(x)$ is obtained via rotation angle embedding and $H_\theta = U_\theta^\dagger H U_\theta$ denotes a parameterized observable obtained via the application of a unitary operator U_θ on a fixed Hermitian H with bounded $r = \infty$-norm. In practice, H_θ is implemented via a quantum neural network that has four strongly entangling layers, followed by a Pauli-Z measurement on all qubits. The resulting function class of quantum classifiers is $\mathcal{G}_\infty = \{x \mapsto g(x) = \mathrm{sgn}(\mathrm{Tr}(H_\theta \rho(x))) : \theta \in \Theta\}$. Consider a Lipschitz-continuous, monotonically increasing loss function $\ell(g, \rho(x), y) = \frac{1}{1+e^{\alpha f(x)}}$, with $\alpha = 10$, and assume a classical Fast Gradient Sign Method (FGSM) attack—an approximate $p = \infty$-norm attack—on the quantum classifier.

Figure 8.2 shows the adversarial training loss and adversarial test loss incurred by the quantum classifier during adversarial training, as a function of the training epochs. Also shown is the conventional training and test loss of the adversarially trained quantum classifier at each epoch. Note that the adversarial training and testing assume FGSM attack with perturbation strength $\epsilon = 0.3$. The adversarial generalization error (i.e., the difference between adversarial training and test loss) of the model obtained at the end of 20 training epochs is highlighted in the figure. It is easy to see that the adversarial generalization error of this model is much higher than its standard generalization error.

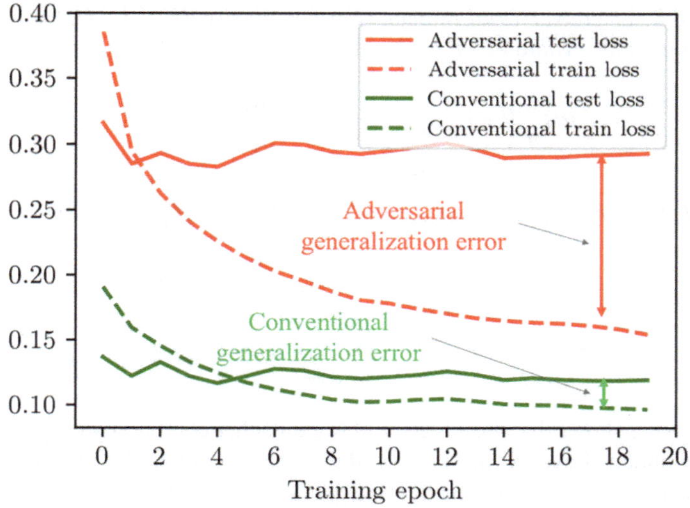

Fig. 8.2 Adversarial and conventional generalization errors plotted against number of training samples for an *adversarially trained* angle embedding based classifier. Figure adopted from [13]

8.3.1 Upper Bounds on Adversarial Robust Generalization Error

To derive an upper bound on the adversarial robust generalization error in (8.15), we can extend the Rademacher complexity based analysis in (8.6) to the adversarial setting. This yields that with probability at least $1-\delta$ with respect to the random draws of the dataset $\mathcal{D}$,

$$\Delta L_{p,\epsilon}^{C/Q}(g) \leq 2\mathcal{R}(\ell_{p,\epsilon}^{C/Q} \circ \mathcal{G}_r) + 3M\sqrt{\frac{1}{2N}\log\frac{2}{\delta}}, \tag{8.16}$$

where

$$\mathcal{R}(\ell_{p,\epsilon}^{C/Q} \circ \mathcal{G}_r) = \mathbb{E}_{\sigma}\Big[\sup_{g\in\mathcal{G}_r} \frac{1}{N}\sum_{i=1}^{N} \sigma_i \ell_{p,\epsilon}^{C/Q}(g, \rho(x^i), y^i)\Big] \tag{8.17}$$

is the *adversarial Rademacher complexity* of the composed function class $\ell_{p,\epsilon}^{C/Q} \circ \mathcal{G}_r$.

The following theorem presents an upper bound on the adversarial Rademacher complexity above in terms of the non-adversarial Rademacher complexity.

Theorem 8.2 *The adversarial Rademacher complexity $\mathcal{R}(\ell^{C/Q}_{p,\epsilon} \circ \mathcal{G}_r)$ under classical/quantum attack evaluated on an N-sample dataset can be upper bounded as*

$$\mathcal{R}(\ell^{C/Q}_{p,\epsilon} \circ \mathcal{G}_r) \leq \eta \mathcal{R}(\mathcal{F}_r) + \frac{\eta B J(r)}{\sqrt{N}} \mathcal{S}^{C/Q}_{r,p,\epsilon}, \tag{8.18}$$

where $\mathcal{R}(\mathcal{F}_r)$ can be upper bounded as in (8.6), $J(r) = O(\sqrt{1 + \frac{1}{r}})$ *and*

$$\mathcal{S}^{C}_{r,p,\epsilon} = d_H \sup_{x,x': \|x-x'\|_p \leq \epsilon} \|\rho(x) - \rho(x')\|_{\frac{r}{r-1}}, \tag{8.19}$$

$$\mathcal{S}^{Q}_{r,p,\epsilon} = d_H \sup_{\rho(x),\rho': \|\rho(x)-\rho'\|_p \leq \epsilon} \|\rho(x) - \rho'\|_{\frac{r}{r-1}}. \tag{8.20}$$

Theorem 8.2 shows that the adversarial Rademacher complexity of quantum classifiers is bounded above by the standard Rademacher complexity plus an additional term that scales as $1/\sqrt{N}$ with the training data size. This additional term depends on a sensitivity quantity denoted by $\mathcal{S}^{C/Q}_{r,p,\epsilon}$. This sensitivity term vanishes in the absence of adversarial attacks, i.e., when $\epsilon = 0$.

We now examine this sensitivity term under both classical and quantum attacks. In the classical case, the sensitivity term $\mathcal{S}^{C}_{r,p,\epsilon}$ in (8.19) can be interpreted as a measure of the Lipschitz-continuity of the quantum embedding $x \mapsto \rho(x)$. For instance, assume that the embedding map $x \mapsto \rho(x)$ is L-Lipschitz continuous everywhere with respect to 1-Schatten norm, i.e., $\|\rho(x) - \rho(x')\|_1 \leq L\|x - x'\|_p$ for all $x, x' \in \mathcal{X}$. Then, the sensitivity term can be upper bounded as $\mathcal{S}^{C}_{r=\infty,p,\epsilon} \leq L\epsilon$. Therefore, a quantum embedding with small Lipchitz constant L exhibits reduced sensitivity to adversarial perturbations, thereby enhancing robustness and improving adversarial robust generalization. In the next section, we will analyze the sensitivity term for different types of commonly employed quantum embeddings.

In contrast to classical attacks, the sensitivity term under quantum attacks can be analyzed more directly as the adversary directly perturbs the embedded quantum state. Specifically, the following relationship holds:

$$\mathcal{S}^{Q}_{r,p,\epsilon} \leq \epsilon \max\{d_H, d_H^{2-1/r-1/p}\}, \tag{8.21}$$

which captures the sensitivity explicitly in terms of the perturbation norm p, the perturbation budget ϵ, the Hilbert space dimension d_H, and the r-norm of the measurement observable. For fixed r, this term scales linearly with Hilbert space dimension for $p < \frac{r}{2r-1}$-type attack, i.e.,

$$\mathcal{S}^{Q}_{r,p,\epsilon} = O(\epsilon d_H).$$

In contrast, when $p > \frac{r}{2r-1}$, the dependence on the Hilbert space dimension becomes worse. This highlights that quantum attacks are particularly sensitive to the dimension of the Hilbert space they act on: larger d_H leads to higher sensitivity, reduced robustness, and consequently resulting in larger adversarial robust generalization error.

8.3.1.1 Proof of Theorem 8.2

Deriving an upper bound on the adversarial Rademacher complexity presents more technical challenges compared to the non-adversarial case addressed in Theorem 8.1. This increased difficulty arises primarily from the additional inner maximization inherent in the adversarial loss function.

To address this, [13] employs *covering number*-based arguments, which are well-established tools in classical statistical learning theory, to upper bound the adversarial Rademacher complexity. For a comprehensive introduction to these tools, please see Appendix. Intuitively, a set B is said to be a δ-cover of a set $A \subset \mathbb{R}^n$ with respect to the Euclidean norm $\|\cdot\|_2$ if, for all $a \in A$, there exists a $b \in B$ such that $\|a - b\|_2 \leq \delta$. The covering number, denoted as $C(A, \|\cdot\|, \delta)$, is defined as the smallest cardinality of any δ-cover of the set A. This in turn relates to Rademacher complexity via Dudley entropy integral bound [18] (see Appendix). For detailed derivation of Theorem 8.2 using the aforementioned techniques, we refer the readers to [13].

8.4 Impact of Quantum Embeddings on Adversarial Robust Generalization

In this section, we investigate how different quantum embeddings impact the sensitivity term $\mathcal{S}^C_{r,p,\epsilon}$ under classical attacks, and thus contribute to the adversarial robust generalization error. We start by considering the standard embedding schemes described in Sect. 8.2.2. The following theorem [13] summarizes the main findings.

Theorem 8.3 *The sensitivity term $\mathcal{S}^C_{r,p,\epsilon}$ under classical attacks can be upper bounded for amplitude embedding as*

$$\mathcal{S}^C_{r,p,\epsilon} \leq 2d \min\left\{\frac{\epsilon \max\{1, d^{1/2-1/p}\}}{\min_{x\in\mathcal{X}:\|x\|_2>0}\|x\|_2}, 1\right\}, \tag{8.22}$$

for L-layered angle-rotation embedding as

$$\mathcal{S}^C_{r,p,\epsilon} \leq 2L(2\epsilon)^d d^{-d/p}, \tag{8.23}$$

and for L-layered dense rotation embedding as

$$\mathcal{S}^C_{r,p,\epsilon} \leq 2L(2\sqrt{2}\epsilon)^{d/2} \max\{d^{-d/4}, d^{-d/2p}\}. \tag{8.24}$$

Proof We refer the readers to [13] for detailed derivations. □

Theorem 8.3 highlights the following key points:

- Under amplitude embedding, the sensitivity term scales at most linearly with data dimension d. This dependence on d is intrinsic to the embedding, and, therefore, generally unavoidable.
- In contrast, for (repeated) angle rotation embedding, the sensitivity exhibits markedly different behavior. When the data dimension d is sufficiently large, i.e., when $d \gg e^{\max\{2,p\}\ln(2\sqrt{2}\epsilon)-1}$, the sensitivity term has negligible dependence on d. In fact, the sensitivity term decays at least exponentially as $d \to \infty$. This result suggests that repeated angle encoding enables the representation of high-dimensional data within a quantum classifier, without incurring a significant penalty in adversarial robustness. Consequently, one can exploit the high expressivity of such embeddings to maintain adversarially robust generation that is comparable to standard generalization, with only a negligible increase in the number of required training samples.

It is important to note that the behavior explained above pertains only to the sensitive term. As established in Theorem 8.2, the adversarial robust generalization error also depends on the standard, non-adversarial Rademacher complexity studied in Sect. 8.2. By jointly considering the behavior of both the standard Rademacher complexity and the sensitivity term, we can arrive at the following conclusions in the high-dimensional data regime:

- **Amplitude Embedding**: The adversarial robust generalization error is predominantly determined by the sensitivity term (under $r = \infty$), leading to a significant deviation from the standard generalization error.
- **Repeated Angle Rotation Embedding** The adversarial robust generalization is predominantly influenced by the standard Rademacher complexity term, resulting in a behavior that closely aligns with the standard generalization error.

Beyond the standard embedding schemes described in Sect. 8.2.2, the sensitivity term can be analyzed for *Lipschitz-continuous quantum embedding schemes*. For instance, consider the following embedding scheme: $\rho(x) = S(x)\rho_0 S(x)^\dagger$, where $S(x) = \prod_{l=1}^{L}\prod_{k=1}^{d} S_{lk}(x_k)$ with $S_{lk}(\alpha) = \exp(-\frac{i}{2}\alpha G_{lk})$. Let $\lambda = \max_{k,l} \|G_{lk}\|_\infty$. Then, the quantum embedding map $\rho : x \mapsto \rho(x)$ is Lipschitz-continuous with respect to ∞-Schatten norm as [19]

$$\|\rho(x) - \rho(x')\|_\infty \leq \sqrt{dL}\lambda\|x - x'\|_2. \tag{8.25}$$

The Lipschitz-continuity of the above quantum embedding scheme implies that the sensitivity term $\mathcal{S}^C_{r,p,\epsilon}$, for $r = 1$ and $p = 2$, can be upper bounded as $\mathcal{S}^C_{1,2,\epsilon} \leq \sqrt{dL}\lambda\epsilon$.

8.5 Conclusion

This chapter provided an introduction to the statistical learning-theoretic analysis of adversarial robust generalization of quantum classifiers. Given the early stage of research in this area, the discussion was primarily informed by the limited existing literature [13, 20]. This stands in contrast to the extensive body of work focused on understanding the generalization of standard, non-adversarially trained, quantum machine learning models [14, 15, 21]. The majority of these studies have employed the PAC framework to derive generalization error bounds that hold uniformly over all models, without incorporating the properties of the training algorithm. As highlighted in this chapter, such bounds tend to be loose.

Recent developments in the generalization analysis of quantum models have begun to explore the algorithmic stability of quantum neural network models trained using stochastic gradient descent [22]. These approaches explicitly account the properties of the training algorithm. A potential interesting line of future research would be to investigate the impact of quantum resources such as shot noise and entanglement on the stability of standard and adversarial training algorithms. Additionally, an information-theoretic approach to characterizing generalization error has been investigated in [23]. However, these methods have not yet been applied to the analysis of adversarial robust generalization of quantum classifiers.

8.6 Appendices

8.6.1 Useful Inequalities, and Results from Statistical Learning Theory

In this section we present several useful inequalities to bound the Rademacher complexity:

Firstly, Talagrand's contraction lemma allows us to bound the Rademacher complexity of the loss function class $\ell \circ \mathcal{G}_r$.

Lemma 8.1 (Talagrand's contraction lemma [24]) *Let $\phi : A \mapsto B$ be an η-Lipschitz function. Then the following inequality holds:*

$$\mathcal{R}(\phi \circ A) \leq \eta \mathcal{R}(A).$$

The operator Khintchine's inequality allows us to evaluate expectations with respect to Rademacher variables.

Lemma 8.2 (Operator Khintchine's inequality [25]) *Let X_i be Hermitian operators and $\sigma_i = \pm 1$ be Rademacher variables, then the following inequality holds:*

$$\left(\mathbb{E}_\sigma\left[\left\|\sum_i \sigma_i X_i\right\|_\alpha^\alpha\right]\right)^{1/\alpha} \leq \begin{cases} \left\|\sqrt{\sum_i X_i^2}\right\|_1 & \textit{for } \alpha = 1 \\ 2^{-1/4}\sqrt{\pi\alpha/e}\left\|\sqrt{\sum_i X_i^2}\right\|_\alpha & \textit{for } 2 \leq \alpha \leq \infty \end{cases}.$$

In the case of $\alpha = \infty$, Khintchine's inequality is not applicable. Instead, one can use the following Tropp's inequality.

Lemma 8.3 (Tropp's inequality [25]) *Let X_i be Hermitian operators of dimensions d_H and $\sigma_i = \pm 1$ be Rademacher variables, then the following inequality holds:*

$$\mathbb{E}_\sigma\left[\left\|\sum_i \sigma_i X_i\right\|_\infty\right] \leq \sqrt{2\log d_H \left\|\sum_i X_i^2\right\|_\infty}.$$

Let us now recall some standard definitions and results which we used in the proofs in the main body of the text. In particular these were used to upper bound the adversarial Rademacher complexity in Theorem 8.2. We start by defining what a covering set is.

Definition 8.1 (*Covering Set*)
Let $A \subseteq \mathbb{R}^N$. The set B is a δ covering set of A with respect to $\|\cdot\|$ (denoted as $\mathcal{B}(A, \|\cdot\|, \delta)$) if

$$\forall a \in A \; \exists \; b \in B : \|a - b\| \leq \delta.$$

The space of δ-covering sets of a set A defines the δ-covering number for the set, as explained below.

Definition 8.2 (*Covering Number*)
Let $A \subseteq \mathbb{R}^m$. The δ-covering number $C(A, \|\cdot\|, \delta)$ of A with respect to the norm $\|\cdot\|$ is the smallest cardinality of a δ-covering set of A, i.e.,

$$C(A, \|\cdot\|, \delta) = \inf_{B:\, \forall a \in A \; \exists \; b \in B: \|a-b\| \leq \delta} |B|.$$

The covering number is a useful tool which can be used to upper bound the Rademacher complexity of a set via the Dudley entropy integral bound below [26].

Theorem 8.4 (Dudley Entropy Integral Bound [18])
Let $A \subseteq \mathbb{R}^N$ denote a set of N-dimensional vectors. Let $B = \max_{a \in A} \|a\|_2$. The Rademacher complexity $\mathcal{R}(A) = \mathbb{E}_\sigma[\sup_{a \in A} \sum_{i=1}^N \frac{1}{N} \sigma_i a_i]$ can be upper bounded as

$$\mathcal{R}(A) \leq 12 \int_0^{B/2} \frac{\sqrt{\log(C(A, \|\cdot\|_2, \alpha))}}{N} d\alpha.$$

References

1. Lu S, Duan L-M, Deng D-L (2020) Quantum adversarial machine learning. Phys Rev Res 2
2. Ren W, Li W, Xu S, Wang K, Jiang W, Jin F, Zhu X, Chen J, Song Z, Zhang P et al (2022) Experimental quantum adversarial learning with programmable superconducting qubits. Nat Comput Sci 2(11):711–717
3. West MT, Erfani SM, Leckie C, Sevior M, Hollenberg LC, Usman M (2023) Benchmarking adversarially robust quantum machine learning at scale. Phys Rev Res 5(2):023186
4. Schmidt L, Santurkar S, Tsipras D, Talwar K, Madry A (2018) Adversarially robust generalization requires more data. In: Advances in neural information processing systems, vol 31
5. Xiao J, Fan Y, Sun R, Luo Z-Q (2022) Adversarial rademacher complexity of deep neural networks. arXiv:2211.14966
6. Yin D, Kannan R, Bartlett P (2019) Rademacher complexity for adversarially robust generalization. In: International conference on machine learning. PMLR, pp 7085–7094
7. Schuld M, Sweke R, Meyer JJ (2021) Effect of data encoding on the expressive power of variational quantum-machine-learning models. Phys Rev A 103(3):032430
8. Vapnik V (2013) The nature of statistical learning theory. Springer Science & Business Media
9. Guedj B (2019) A primer on pac-Bayesian learning. arXiv:1901.05353
10. Bousquet O, Elisseeff A (2000) Algorithmic stability and generalization performance. In: Advances in neural information processing systems, vol 13
11. Hellström F, Durisi G, Guedj B, Raginsky M, et al (2025) Generalization bounds: perspectives from information theory and pac-bayes. In: Foundations and trends® in machine learning, vol 18, no 1, pp 1–223
12. Shalev-Shwartz S, Ben-David S (2014) Understanding machine learning: from theory to algorithms. Cambridge University Press
13. Georgiou P, Thomas AM, Jose ST, Simeone O (2025) On the generalization of adversarially trained quantum classifiers. arXiv:2504.17690
14. Caro MC, Huang H-Y, Cerezo M, Sharma K, Sornborger A, Cincio L, Coles PJ (2022) Generalization in quantum machine learning from few training data. Nat Commun 13(1):4919
15. Caro MC, Gil-Fuster E, Meyer JJ, Eisert J, Sweke R (2021) Encoding-dependent generalization bounds for parametrized quantum circuits. Quantum 5:582
16. Dowling N, West MT, Southwell A, Nakhl AC, Sevior M, Usman M, Modi K (2026) Adversarial robustness guarantees in quantum classifiers. npj Quantum Information 12:16
17. Liu N, Wittek P (2020) Vulnerability of quantum classification to adversarial perturbations. Phys Rev A 101:062331 Jun
18. Vershynin R (2018) High-dimensional probability: an introduction with applications in data science, vol 47. Cambridge University Press
19. Kempkes M, Ijaz A, Gil-Fuster E, Bravo-Prieto C, Spiegelberg J, van Nieuwenburg E, Dunjko V (2025) Double descent in quantum machine learning. arXiv:2501.10077
20. Georgiou P, Jose ST, Simeone O (2024) Adversarial quantum machine learning: an information-theoretic generalization analysis. In: 2024 IEEE international symposium on information theory (ISIT), pp 789–794
21. Bu K, Koh DE, Li L, Luo Q, Zhang Y (2022) Statistical complexity of quantum circuits. Phys Rev A 105(6):062431
22. Yang J, Xie W, Xu X (2025) Stability and generalization of quantum neural networks. arXiv:2501.12737
23. Caro MC, Gur T, Rouzé C, Franca DS, Subramanian S (2024) Information-theoretic generalization bounds for learning from quantum data. In: The thirty seventh annual conference on learning theory. PMLR, pp 775–839
24. Michel Ledoux MT (2013) Probability in Banach spaces: isoperimetry and processes. Springer
25. Banchi L, Pereira JL, Jose ST, Simeone O (2023) Statistical complexity of quantum learning. arXiv:2309.11617
26. Wainwright MJ (2019) High-dimensional statistics: a non-asymptotic viewpoint, vol 48. Cambridge University Press

Chapter 9
The Interplay of Robustness and Generalization in Quantum Machine Learning

Julian Berberich, Tobias Fellner, and Christian Holm

Abstract While adversarial robustness and generalization have individually received substantial attention in the recent literature on quantum machine learning, their interplay is much less explored. In this chapter, we address this interplay for variational quantum models, which were recently proposed as function approximators in supervised learning. We discuss recent results quantifying both robustness and generalization via Lipschitz bounds, which explicitly depend on model parameters. Thus, they give rise to a regularization-based training approach for robust and generalizable quantum models, highlighting the importance of trainable data-encoding strategies. The practical implications of the theoretical results are demonstrated with an application to time series analysis.

9.1 Introduction

Adversarial attacks are data perturbations which aim to cause failures in machine learning algorithms. It is well known in classical machine learning that even adversarial attacks of very small size can lead to misclassification [1, 2]. Therefore, significant efforts were made to develop robust training methods which improve adversarial robustness and, thereby, the practical reliability of machine learning algorithms [3–5]. Beyond its protection against adversarial attacks, the importance of robustness stems from its close connection to generalization. The generalization performance of a machine learning algorithm refers to its ability of extrapolating from the train-

J. Berberich (✉)
Institute for Systems Theory and Automatic Control and Center for Integrated Quantum Science and Technology (IQST), University of Stuttgart, Stuttgart, Germany
e-mail: julian.berberich@ist.uni-stuttgart.de

T. Fellner · C. Holm
Institute for Computational Physics, University of Stuttgart, Stuttgart, Germany
e-mail: tobias.fellner@icp.uni-stuttgart.de

C. Holm
e-mail: holm@icp.uni-stuttgart.de

M. Usman (ed.), *Quantum Robustness in Artificial Intelligence*, Quantum Science and Technology, https://doi.org/10.1007/978-3-032-11153-1_9

ing dataset to (unseen) test data. The interplay of robustness and generalization is well explored in the classical machine learning literature [1, 2, 6–8]. In particular, enhanced robustness typically improves generalization performance, which can be explicitly quantified via generalization bounds, i.e., bounds on the difference between the expected testing loss and the empirical training loss [7].

In this chapter, we study the interplay of robustness and generalization in quantum machine learning (QML). QML is a recently emerging field at the intersection of classical machine learning and quantum computing [9, 10]. This chapter focuses on the usage of quantum computers as machine learning models. Various recent works have highlighted the importance of adversarial robustness in this context, demonstrating the sensitivity of quantum models w.r.t. adversarial attacks and proposing training strategies for improved robustness [11–21]. Generalization is another key property of quantum models. In particular, a substantial body of literature has developed bounds on the generalization performance [21–29]. However, most of these bounds are uniform and do not involve model parameters, which leads to substantial limitations [30].

In this chapter, we focus on a particular QML approach which uses variational quantum algorithms (VQAs) for supervised learning [31–33]. A VQA combines a parameterized quantum circuit with a classical optimization routine, see [34] for an overview. We discuss the robustness of variational quantum models against adversarial attacks using Lipschitz bounds [21], which are a powerful tool from classical machine learning and control theory [35–37]. Further, we emphasize the strong links between robustness and generalization in QML by stating a generalization bound from [21] which, in contrast to alternative results, depends explicitly on model parameters and, therefore, overcomes their limitations. We also discuss the implications of these results on the data encoding in QML, highlighting the important benefits of trainable encoding strategies. With novel numerical results, we demonstrate that the considered robustness perspective not only brings theoretical insights but also relevant practical advantages.

The chapter is structured as follows. After introducing the considered quantum model class in Sect. 9.2, we separately address robustness and generalization in Sects. 9.3 and 9.4, respectively. Section 9.5 contains our numerical results on time series analysis, and we conclude the chapter in Sect. 9.6.

9.2 Variational Quantum Models

This chapter focuses on variational quantum models, which rely on parameterized unitary operators

$$U_{j,\Theta_j}(x) = e^{-i(w_j^\top x+\theta_j)H_j} \tag{9.1}$$

for $j = 1, \ldots N$. Here, $x \in \mathbb{R}^d$ denotes the classical data that are considered, e.g., for classification tasks. Moreover, $\Theta_j = \{w_j, \theta_j\}$ are parameters that are updated during training using an optimization strategy introduced below in more detail. Finally, the H_j's are Hermitian generators, i.e., they satisfy $H_j = H_j^\dagger$, and they may act on one or multiple qubits. Based on the operators $U_{j,\Theta_j}(x)$, we define the parameterized quantum circuit

$$U_\Theta(x) = U_{N,\Theta_N}(x) \cdots U_{1,\Theta_1}(x). \tag{9.2}$$

For any data sample x, a set of trainable parameters $\Theta = \{\Theta_1, \ldots, \Theta_N\}$, and Hermitian generators H_j, the expression (9.2) constitutes a valid quantum circuit that can be implemented on a quantum computer (after an additional compilation step, if necessary). For example, a common choice for H_j are Pauli operators acting on individual qubits

$$I \otimes \cdots \otimes P \otimes \cdots \otimes I \tag{9.3}$$

with $P \in \{I, X, Y, Z\}$, in which case $U_{j,\Theta_j}(x)$ is a rotation gate with angle $w_j^\top x + \theta_j$. The circuit may also include constant (parameter-independent) gates such as entangling CNOT gates, for which $w_j = 0$ and $\theta_j = 1$.

Finally, to obtain an output of the quantum circuit, we perform a measurement. In particular, repeatedly running $U_\Theta(x)$ for the initial state $|0\rangle$ and measuring w.r.t. the observable $\mathcal{M}$, one can retrieve the expectation value

$$f_\Theta(x) = \langle 0 | U_\Theta(x)^\dagger \mathcal{M} U_\Theta(x) | 0 \rangle. \tag{9.4}$$

For any choice of parameters Θ, f_Θ is a real-valued function $f_\Theta : \mathbb{R}^d \to \mathbb{R}$ whose value can be evaluated using a quantum computer. In Fig. 9.1, the corresponding

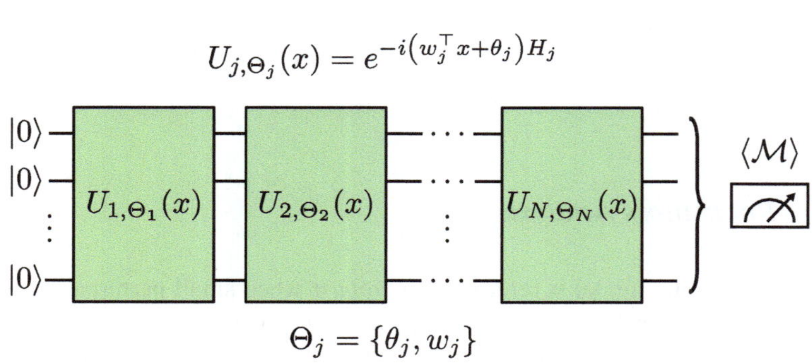

Fig. 9.1 Quantum circuit diagram corresponding to the function $f_\Theta(x)$ in Eq. (9.4). Unitaries of the form (9.1) are applied consecutively. The expectation value of an observable $\mathcal{M}$ is measured at the end of the quantum circuit

quantum circuit diagram is illustrated. In variational QML, this function f_Θ is used as a function approximator to perform learning tasks, e.g., classification or regression.

Note that the considered quantum model includes the full data vector x in each of the N unitaries $U_{j,\Theta_j}(x)$. This strategy is commonly referred to as data re-uploading and it gives rise to a universal classifier already for a single qubit [38]. In contrast, the majority of existing variational QML approaches employs an alternating structure between data and parameters, i.e., a parameterized quantum circuit of the form $U(\theta)V(x)$ which first encodes the data in $V(x)$ and then acts with a trainable unitary operator $U(\theta)$, where

$$U(\theta) = e^{-i\theta_N G_N} \cdots e^{-i\theta_1 G_1}, \quad V(x) = e^{-ix_1 L_1} \cdots e^{-ix_d L_d} \tag{9.5}$$

for Hermitian generators G_j, L_j. Clearly, the alternating structure $U(\theta)V(x)$ is a special case of $U_\Theta(x)$ defined in (9.2), i.e., they are equal for suitable choices of the parameters Θ_j. Since the encoding of the data x in $V(x)$ is fixed a priori via the choice of L_j and cannot be adapted online, we refer to it as *fixed encoding*. Conversely, we refer to $U_\Theta(x)$ as a circuit with *trainable encoding* since the weights w_j multiplying the data are adapted during training. It was demonstrated in [38, 39] that the additional flexibility provided by a trainable encoding substantially improves the expressivity. Later in this chapter, we will see that trainable encoding models also admit crucial advantages over fixed-encoding models in terms of robustness and generalization.

9.3 Robustness of Quantum Models

This section focuses on the robustness of the quantum model (9.4) using Lipschitz bounds, which are frequently used in classical machine learning and control theory. In Sect. 9.3.1, we analyze the robustness of a given quantum model by computing a Lipschitz bound involving the model parameters. Next, in Sect. 9.3.2, we propose a regularization-based training strategy to train models which are inherently more robust.

9.3.1 Robustness Analysis

A machine learning model is referred to as *robust* when small perturbations of the input data do not lead to excessive changes of the output. Lipschitz bounds are a powerful mathematical tool for characterizing robustness. For a given quantum model f_Θ, we say that $L_\Theta > 0$ is a Lipschitz bound if[1]

---

[1] Throughout this chapter, all involves matrix and vector norms are (induced) 2-norms, but the results can be adapted to different norms.

$$\|f_\Theta(x+\varepsilon) - f_\Theta(x)\| \leq L_\Theta \|\varepsilon\| \tag{9.6}$$

for any $x, \varepsilon \in \mathbb{R}^d$. Intuitively, L_Θ bounds the worst-case change of the function f_Θ when perturbing its input by ε. Thus, smaller values of L_Θ imply an improved robustness of f_Θ. The following result from [21] provides a Lipschitz bound for the variational quantum model f_Θ.

Theorem 9.1 ([21]) *The following is a Lipschitz bound for* f_Θ

$$L_\Theta = 2\|\mathcal{M}\| \sum_{j=1}^{N} \|w_j\| \|H_j\|. \tag{9.7}$$

The Lipschitz bound (9.7) depends on the norm of the observable $\mathcal{M}$, the Hermitian generators H_j, and the trainable parameters w_j. In particular, if any of these quantities decreases, then the Lipschitz bound decreases and, therefore, the guaranteed robustness of f_Θ improves. Lipschitz bounds have found increasing usage in the recent quantum computing literature. In particular, related Lipschitz bounds were derived for establishing convergence of VQA training [40], analyzing robustness of quantum algorithms [41], studying quantum models with single shots [42], and investigating double-descent properties of QML [43].

While the quantities $\mathcal{M}$ and H_j appearing in the Lipschitz bound (9.7) are fixed a priori, the parameters w_j are adapted during training. As we show in Sect. 9.3.2, this allows us to systematically influence the robustness of the quantum model during training by penalizing the norm of w_j.

9.3.2 Training Robust Quantum Models

In the following, we employ Theorem 9.1 to derive a robust training strategy for variational quantum models. To define the training problem, suppose we are given a set of data points $\{x_k, y_k\}_{k=1}^n$ from a domain $\mathcal{X} \times \mathcal{Y}$. Based on the data, we want to fit a quantum model f_Θ by minimizing a loss function ℓ, i.e., by solving

$$\min_\Theta \frac{1}{n} \sum_{k=1}^{n} \ell(f_\Theta(x_k), y_k). \tag{9.8}$$

Theorem 9.1 shows that the norm of the weights w_j influences the Lipschitz bound and, thereby, the robustness of the quantum model. With this motivation, instead of the original training problem (9.8), it was proposed by [21] to solve the modified problem

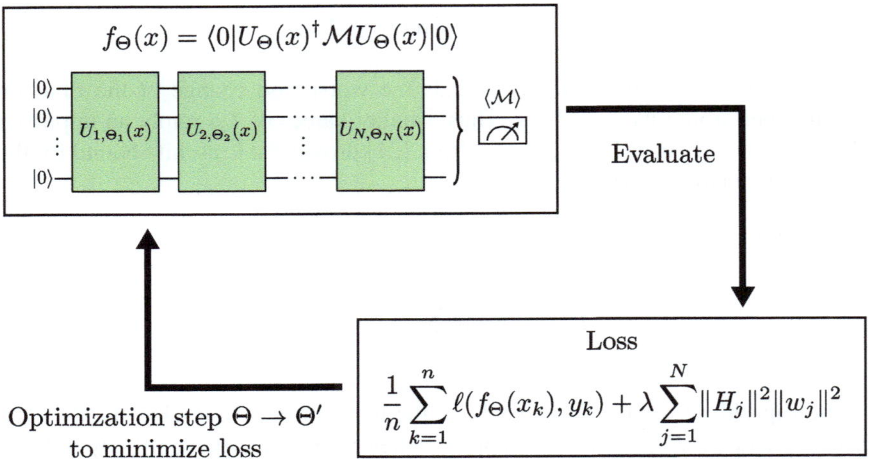

Fig. 9.2 Training scheme of the VQA. The function $f_\Theta(x)$ is evaluated by executing the corresponding quantum circuit. The resulting output, along with the norm of the encoding weights $\|\mathbf{w}_j\|$, is used to compute a regularized loss function. This loss is minimized iteratively by optimizing all trainable parameters Θ

$$\min_{\Theta} \frac{1}{n} \sum_{k=1}^{n} \ell(f_\Theta(x_k), y_k) + \lambda \sum_{j=1}^{N} \|w_j\|^2 \|H_j\|^2. \tag{9.9}$$

The variable $\lambda > 0$ can be used to trade off the two competing objectives of robustness and training error, and it is a hyperparameter that can be tuned via cross validation. The added regularization encourages robustness of the trained quantum model, and it is analogous to ridge regression in classical machine learning [44]. Figure 9.2 shows the iterative update algorithm of the VQA.

Note that the Lipschitz bound in Theorem 9.1 is independent of the parameters θ_j. For fixed-encoding quantum models, the parameters θ_j are the only trainable parts of the model and, therefore, the Lipschitz bound (9.7) is fully determined by offline design choices and cannot be modified during training. In particular, adding a regularization of the parameters θ_j to the cost of (9.9) does not have a direct effect on robustness and, hence, on generalization, compare Sect. 9.4. Indeed, it was observed in the literature that fixed-encoding models are not overly sensitive w.r.t. perturbed data [45, 46]. However, this also means that robustness cannot be influenced systematically in quantum models with fixed encoding, which leads to important practical limitations, see Sect. 9.5 for details.

9.4 Generalization of Quantum Models

It is desirable to obtain a quantum model that not only performs well on the training data (i.e., minimizes the training loss in (9.8)), but that also leads to a small loss on test data samples that were not included in the training. In the following, we show that the Lipschitz bound (9.7) can be used to quantify the ability of the quantum model (9.4) to generalize. To state the formal result, let us assume that the data (x, y) are sampled according to some distribution $\mathbb{P}$. The central goal of the considered supervised learning task can be formulated as follows: Find a set of parameters Θ that minimize the *expected risk* over all possible data samples, i.e.:

$$\mathcal{R}_{\text{exp}}(\Theta) = \mathbb{E}_{(x,y)\sim\mathbb{P}}\left[\ell(f_\Theta(x), y)\right]. \tag{9.10}$$

However, $\mathcal{R}_{\text{exp}}(\Theta)$ cannot be evaluated since the distribution $\mathbb{P}$ is typically unknown and we only have access to the training data $\{x_k, y_k\}_{k=1}^n$. Hence, we use the data to define a proxy for the expected risk: the *empirical risk*. The latter is defined as the cost in the training problem (9.8), i.e.:

$$\mathcal{R}_{\text{emp}}(\Theta) = \frac{1}{n}\sum_{k=1}^{n} \ell(f_\Theta(x_k), y_k). \tag{9.11}$$

The generalization performance of a model describes how well insights obtained from the training data transfer to the unseen test data. Mathematically, the ability of a model to generalize can be quantified via the *generalization error* $\mathcal{E}(\Theta)$, which bounds the difference between the expected risk and the empirical risk

$$\mathcal{E}(\Theta) = |\mathcal{R}_{\text{exp}}(\Theta) - \mathcal{R}_{\text{emp}}(\Theta)|. \tag{9.12}$$

Intuitively, one can expect that a small Lipschitz bound implies good generalization: When small data perturbations lead to small output changes, then the model is less affected by overfitting and the insights from finitely many data points in $\mathcal{R}_{\text{emp}}(\Theta)$ can be more easily transferred to the (infinitely many) data points corresponding to $\mathcal{R}_{\text{exp}}(\Theta)$. This is demonstrated with two simple scalar functions in Fig. 9.3.

The following result[2] shown in [21] formalizes this intuition by bounding the generalization error $\mathcal{E}(\Theta)$ in terms of the Lipschitz bound (9.7).

Theorem 9.2 ([21], Informal) *The generalization error* (9.12) *is bounded as*

$$\mathcal{E}(\Theta) \leq C_1 \|\mathcal{M}\| \sum_{j=1}^{N} \|w_j\| \|H_j\| + \frac{C_2}{\sqrt{n}} \tag{9.13}$$

[2] We only provide the informal version from [21, Theorem IV.1] and refer to [21, Theorem B.1] for the detailed technical statement.

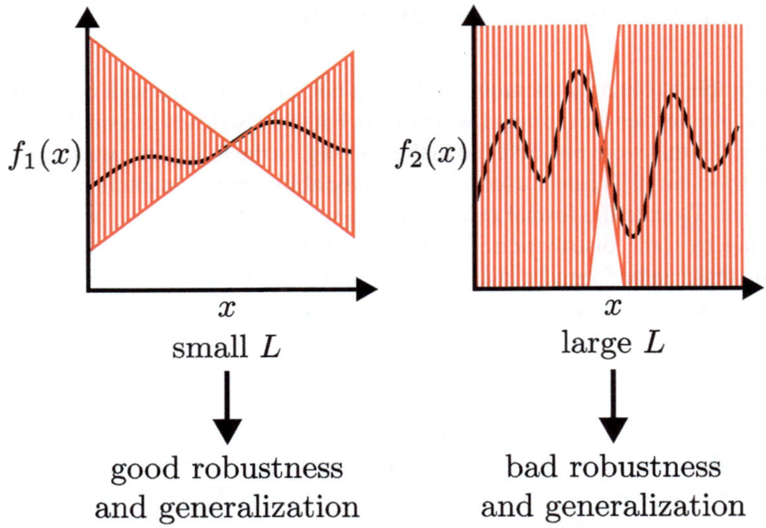

Fig. 9.3 The function on the left-hand side ($f_1(x)$) has a smaller Lipschitz bound than the function on the right-hand side ($f_2(x)$). This means that incremental changes from any point to left or right are confined to a smaller range of possible values, indicated via the conic regions shaded in red. As a result, the function $f_1(x)$ is smoother, whereas the function $f_2(x)$ can be interpreted as the result of overfitting and can, therefore, be expected to have worse generalization performance

for some $C_1, C_2 > 0$.

The generalization error bound (9.13) depends on the Lipschitz bound (9.7) as well as on the data length n. This shows that a small Lipschitz bound as encouraged via the robust training problem (9.9) not only improves robustness but also generalization. One can show that the bound (9.13) vanishes in the infinite data limit $n \to \infty$, see [21, Appendix B] for details. Further, the generalization error bound is minimized and the best generalization is achieved for $w_j = 0$, $j = 1, \ldots, N$, i.e., when the Lipschitz bound is zero. Of course, this does not imply that the expected risk is satisfactory since a reduced Lipschitz bound can increase the empirical risk, which may outweigh the reduced generalization error. In practice, there typically is a sweet spot for which the expected risk can be reduced via the Lipschitz bound by substantially decreasing the generalization error while only mildly increasing the empirical risk (see Sect. 9.5).

Note that the bound (9.13) explicitly depends on the parameters Θ of the model (9.4). In contrast, the majority of the existing QML generalization bounds in the literature, e.g., [23–29] hold uniformly over all parameters in the model class, with the exception of [22] that relies on the effective dimension of a quantum model. The parameter-dependent generalization bound in Theorem 9.2 has the advantage of leading to targeted training strategies for improved generalization, e.g., via regularization as in (9.9). A more detailed analysis of the limitations of uniform generalization bounds is provided in [30].

Finally, we note that, like the Lipschitz bound in Theorem 9.1, also the generalization bound (9.13) is independent of the parameters θ_j. As a result, for fixed-encoding models, it cannot be influenced systematically during training, which severely limits the possibilities for controlling overfitting (compare Sect. 9.5 for the practical implications).

9.5 Numerical Results

To validate our theoretical findings, we train variational quantum machine learning models with the proposed regularization strategy and assess their robustness and generalization capabilities. We begin by motivating and outlining the learning problem used to verify the theoretical statements from Sects. 9.3 and 9.4 (Sect. 9.5.1), followed by a detailed description of the quantum model and training procedure (Sect. 9.5.2). Finally, we present results on robustness and generalization in Sects. 9.5.3 and 9.5.4, respectively. The code to reproduce the numerical results of our study is publicly available on GitHub [47].

9.5.1 Problem Statement

Extracting informative features from time series data is a central challenge across numerous disciplines, including medicine, meteorology, and finance. Given the critical role of sequential data in many real-world applications, ensuring that models designed for feature extraction are robust to input perturbations is essential.

We consider a learning task based on the logistic map [48], defined by

$$x_t = r x_{t-1}(1 - x_{t-1}), \tag{9.14}$$

where the parameter $r \in \mathbb{R}$ governs the dynamics of the system. For $r \in [0, 4]$, the resulting time series remain bounded within the interval $[0, 1]$. The behavior of the system varies significantly with r: for $0 \leq r \leq 3$, the system converges to a fixed point; for $3 \leq r \lesssim 3.5$, the dynamics become periodic, oscillating between a finite set of values, and for $3.5 \lesssim r \leq 4$, the system exhibits chaotic behavior, characterized by extreme sensitivity to initial conditions [49]. These different behaviors are visualized in the bifurcation diagram in Fig. 9.4, where we plot the first 50 elements of the sequence $\{x_i\}$ for various values of r. Throughout this study, we fix the initial condition to $x_1 = 0.5$.

Chaotic dynamics are of particular importance in modeling complex phenomena such as climate systems and financial markets [50, 51]. Given the practical relevance of learning from chaotic time series, we here consider the task of inferring the system parameter $r \in [3.5, 4]$ from a finite sequence $\{x_i\}_1^l$ generated according to Eq. (9.14).

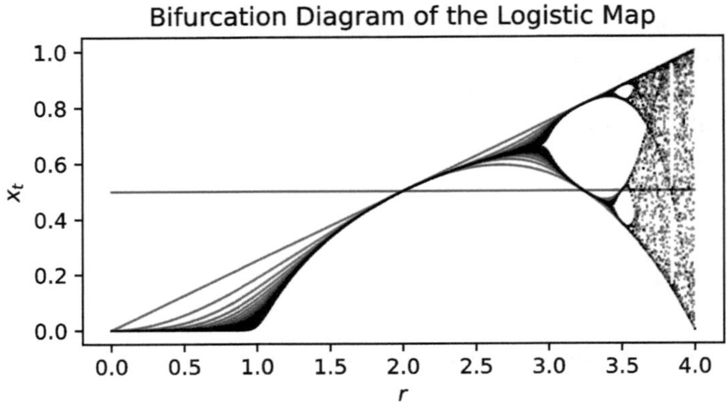

Fig. 9.4 Bifurcation diagram of the logistic map. The first 50 iterations of the logistic map are plotted as a function of the control parameter r. For $0 \leq r \leq 3$, the system converges to a single fixed point. In the range $3 \leq r \lesssim 3.5$, the dynamics become periodic, cycling through a finite set of values. For $3.5 \lesssim r \leq 4$, the system enters a chaotic regime, characterized by aperiodicity and sensitive dependence on initial conditions

Formally, the variational quantum machine learning model is trained to approximate the function:

$$f_\Theta(\{x_i\}_1^l) = r\,. \tag{9.15}$$

We generate a dataset comprising 1000 sequences, each constructed according to Eq. (9.14), with values of the parameter r chosen equidistantly from the interval $[3.5, 4]$. We fix the sequence length to $l = 12$. From the resulting dataset of tuples $(\{x_i\}_1^l, r)$, we randomly allocate 200 samples for training and reserve the remaining 800 for testing purposes.

9.5.2 *Model and Training Procedure*

The variational quantum model is designed such that the time series is processed in a sequential manner. The parameterized quantum circuit used in this work is depicted in Fig. 9.5. Its design is inspired by a quantum model developed for benchmarking variational quantum machine learning methods on time series prediction tasks [52].

Each data point x_i from the time series is encoded into the quantum circuit via Pauli-Z and Pauli-Y rotation gates, with rotation angles defined as

$$\alpha_{i,j} = w_{i,j}^{(1)} x_i + \theta_{i,j}^{(1)}\,, \tag{9.16}$$

$$\beta_{i,j} = w_{i,j}^{(2)} x_i + \theta_{i,j}^{(2)}\,, \tag{9.17}$$

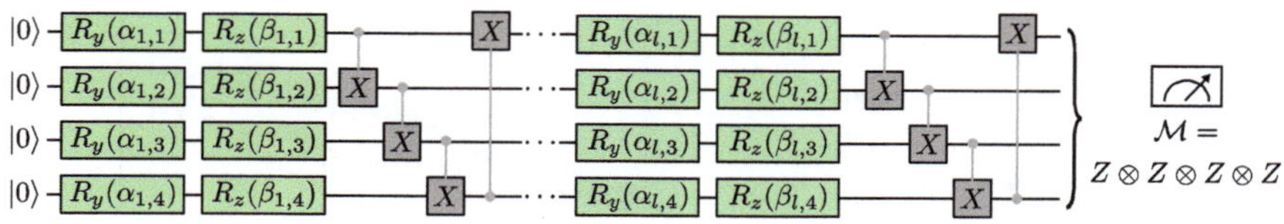

Fig. 9.5 Circuit diagram of the variational quantum circuit employed in this study. The input data sequence is sequentially encoded using Pauli-Z and Pauli-Y rotation gates with angles $\alpha_{i,j}$ and $\beta_{i,j}$, respectively, as defined in Eq. 9.16. Following each encoding block, controlled Pauli-X (CNOT) gates are applied in a nearest-neighbor pattern to introduce entanglement between adjacent qubits. At the end of the circuit, the observable $\mathcal{M} = Z \otimes Z \otimes Z \otimes Z$ is measured to obtain the output expectation value

where $w_{i,j}^{(k)}$ and $\theta_{i,j}^{(k)}$ are trainable parameters, and the subscripts i and j index the time step and the qubit, respectively. These rotations are applied on qubit j to embed the input feature x_i. To introduce entanglement between the qubits, we apply a layer of controlled Pauli-X (CNOT) gates in a nearest-neighbor configuration following each data-encoding layer. For our experiments, the quantum circuit operates on four qubits. Finally, at the end of the circuit, we perform a measurement of the multi-qubit observable $\mathcal{M} = Z \otimes Z \otimes Z \otimes Z$ to determine the output of the quantum model. Since the output domain of the quantum model is $[-1, 1]$ (the range of possible expectation values when measuring $\mathcal{M}$), we linearly scale the output to the target domain $[3.5, 4]$.

The models are simulated using the Python-based quantum machine learning library `PennyLane` [53]. Training is carried out using the Adam optimizer [54] with a learning rate of 0.01 and proceeds for a total of 2000 epochs.

9.5.3 *Evaluating Model Robustness Under Data Perturbations*

To empirically validate the theoretical findings on robustness, we train three quantum models using different values of the regularization parameter, $\lambda \in \{0, 0.004, 0.03\}$. In addition, we include a variant with a fixed encoding, in which the parameters $w_{i,j}^{(k)}$ are not optimized but instead held constant throughout training. All trainable parameters are initialized by sampling from a uniform distribution over the interval $[-\pi/2, \pi/2]$.

To evaluate the robustness of the models against input perturbations, we add random noise to the test sequences. Specifically, noise is sampled uniformly from $[-\epsilon, \epsilon]^l$ and added to each input sequence, where ϵ denotes the noise level. For each value of ϵ, we compute the worst-case mean squared error (MSE) between the predicted and true values of the parameter r over 100 independently perturbed versions of the test set. The results of these experiments are presented in Fig. 9.6,

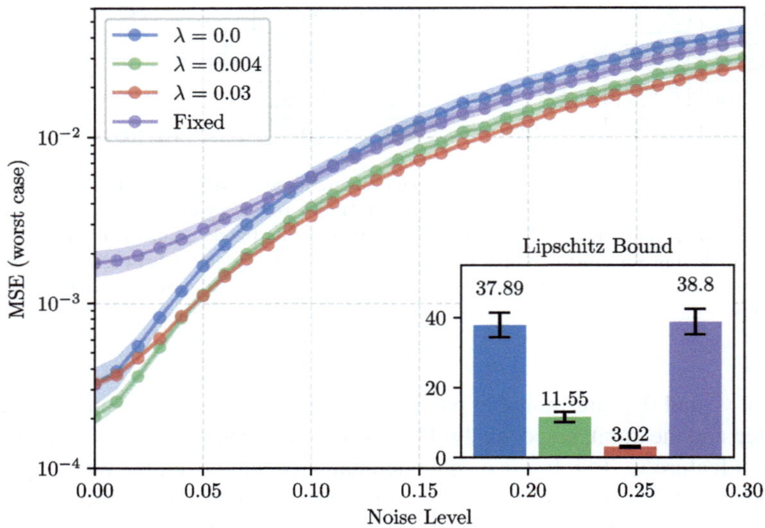

Fig. 9.6 The plot shows the prediction accuracy on the test set as a function of the noise level ϵ for various trained models. For each data point, 50 models are trained with different random weight initializations. Each model is evaluated on 100 noise-perturbed versions of the test sequences, where the noise is sampled uniformly from the box $[-\epsilon, \epsilon]^l$. From these, the worst-case accuracy for each model is recorded. The plotted values represent the average worst-case accuracy across the 50 trained models, with the shaded regions indicating the standard deviation. We compare four training scenarios: three with trainable encoding using different regularization strengths λ, and one with a fixed encoding. The inset shows the Lipschitz bounds of the trained models

illustrating how different regularization strategies influence the robustness of the quantum models under noisy input conditions.

As anticipated, the prediction error increases across all models as the noise level ϵ rises. In the high-noise regime, models trained with a larger regularization parameter λ consistently exhibit improved performance. This observation aligns with the theoretical expectations derived from the Lipschitz bound L_Θ.

Specifically, increasing the regularization strength imposes a stronger penalty on the magnitude of the encoding weights during training. As indicated by Eq. (9.7), smaller encoding weights correspond to a reduced Lipschitz constant. A smaller Lipschitz bound, in turn, implies increased robustness to input perturbations, thereby explaining the enhanced predictive stability of more heavily regularized models under noisy conditions.

The effect of regularization on the Lipschitz bound is illustrated in the inset of Fig. 9.6, which shows a decrease in L_Θ for models with trainable encodings as λ increases. In contrast, the model with fixed-encoding parameters cannot reduce the norm of the encoding weights through optimization. Consequently, its Lipschitz bound remains comparatively large, which contributes to its inferior robustness performance in the presence of noise.

Notably, for small noise levels, there exists a significant performance gap between the model with fixed encoding and the models with trainable encoding. This discrepancy can be attributed to the fact that the model with fixed encoding lacks sufficient expressivity (the ability to approximate a function from the encoded data) due to the reduced number of trainable parameters. Consequently, models with trainable encodings are better equipped to approximate the training data, which translates into improved accuracy on the test set. However, as the noise level increases, the model with fixed encoding demonstrates superior predictive accuracy compared to the model with trainable encoding and regularization parameter $\lambda = 0$. This can be explained by the increased expressivity of the latter model, which, despite its enhanced capacity to fit the training data, may also be more susceptible to overfitting. As a result, it exhibits poorer generalization performance and reduced robustness against data perturbations.

The connection between robustness and generalization also becomes evident when comparing models with trainable encoding in the low-noise regime. Among these, the model trained with $\lambda = 0.004$ achieves the highest predictive accuracy, whereas models with $\lambda = 0$ and $\lambda = 0.03$ perform less accurately. This observation suggests that an intermediate level of regularization can provide an optimal trade-off between expressivity and generalization. In the following section, we will examine generalization performance in more detail.

9.5.4 Evaluating Model Generalization

To study generalization, we train a series of models over a range of regularization parameters $\lambda \in [0, 0.03]$. The corresponding results are shown in Fig. 9.7, where we examine how variations in λ affect both the prediction accuracy and the Lipschitz bound.

We begin by analyzing the prediction accuracy on the training dataset. As λ increases, accuracy steadily declines, which can be attributed to reduced model expressivity resulting from smaller encoding weights. In contrast, on the test dataset, we observe that the prediction error reaches a minimum at $\lambda = 0.004$. Relative to the case without regularization this corresponds to an improvement in performance of approximately 50%. This suggests that moderate regularization enhances the model's ability to generalize from the training data to unseen data. However, for larger values of λ, performance deteriorates again, as the model's expressivity is overly constrained, leading to reduced accuracy on both the training and test datasets.

To gain deeper insight into the model's generalization capability, we examine the generalization gap defined as

$$\mathrm{MSE}_{\mathrm{test}} - \mathrm{MSE}_{\mathrm{train}} \, . \tag{9.18}$$

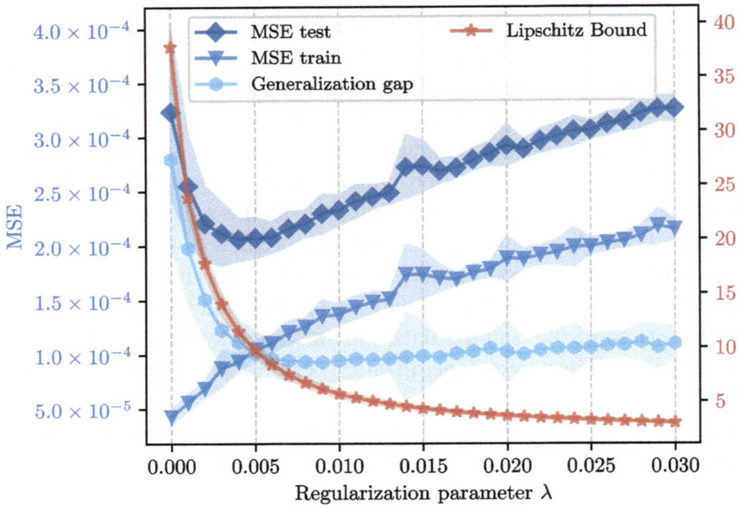

Fig. 9.7 Evaluation of model generalization for varying regularization strengths $\lambda \in [0, 0.03]$. The blue plots show the mean squared error (MSE) on the training and test datasets, as well as the generalization gap. These are affiliated with the left y-axis. The Lipschitz bound is displayed in red and is associated with the right y-axis. All displayed values are the mean over 50 random weight initializations. The shaded areas correspond to the standard deviation

The generalization gap decreases rapidly for small values of λ, and then stabilizes at an approximately constant level for larger values. This behavior indicates that the regularization approach described in Eq. (9.9) effectively improves generalization.

In addition, we observe that the Lipschitz bound decreases monotonically with increasing λ, consistent with the original motivation for incorporating the regularization term, namely, to constrain the model's Lipschitz constant. Taken together, these findings support the theoretical connection between generalization and the Lipschitz bound as stated in Theorem 9.2.

To illustrate the generalization performance, we plot the predicted versus true values r for both training and test data in Fig. 9.8. For the trainable encoding with $\lambda = 0$ and $\lambda = 0.004$, the training points are well fitted. On the test set, performance improves noticeably near $r = 4$ when using regularization with $\lambda = 0.004$. In contrast, fixing the encoding parameters leads to less accurate predictions, consistent with the high MSE observed for this model in the low-noise regime, see Fig. 9.6. Notably, all models struggle to predict values near $r = 3.5$ and $r = 4$, likely due to edge effects in measuring the expectation value $\langle \mathcal{M} \rangle \in [-1, 1]$, which is linearly mapped to $[3.5, 4]$. The set of quantum states yielding a given expectation value is largest around $\langle \mathcal{M} \rangle = 0$ and shrinks toward the extremes $\langle \mathcal{M} \rangle = \pm 1$, making it harder for the model to learn states producing such extreme values.

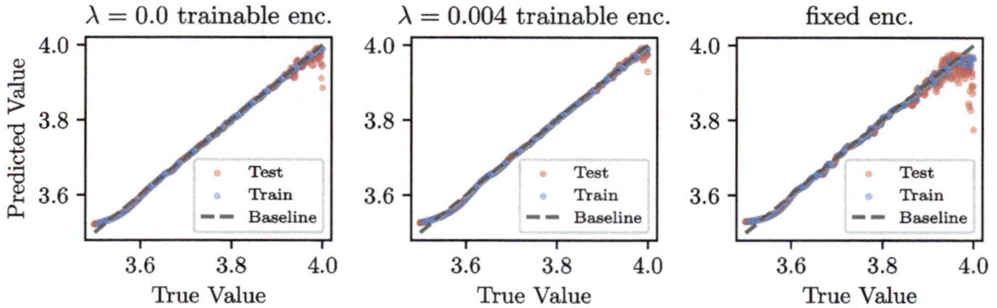

Fig. 9.8 The figure shows the predicted versus true values of r for the training (blue) and test (red) data. The black dashed line indicates perfect prediction. Left and center: Models with trainable encoding without ($\lambda = 0$) and with regularization ($\lambda = 0.004$), respectively. Right: Model with fixed (non-trainable) encoding parameters. The results shown correspond to the median mean squared error on the test data of all 50 random weight initializations

9.5.5 Related Numerical Results from the Literature

We note that the literature contains further numerical results demonstrating the practicality of Lipschitz bounds in variational QML. In particular, in [21], the results shown in Sects. 9.3 and 9.4 are applied in supervised learning for a classification problem. Moreover, [55] compares the robustness of classical models and quantum models, where the latter are analyzed via Theorem 9.1 and trained via (9.9). Finally, [56] studies robustness and generalization properties of variational quantum models as in (9.4) in the context of quantum reinforcement learning. The results shown in the present section constitute the first application of Lipschitz-based regularization in QML for time series analysis.

9.6 Conclusion

In this chapter, we reviewed the interplay of robustness and generalization in variational QML. We introduced theoretical robustness and generalization results from [21] based on Lipschitz bounds. These results can be used to design regularization strategies for robust training of quantum models. We also highlighted the role of trainability in the data encoding. With a numerical application to a time series problem, we demonstrated the practical impact of the presented theoretical results.

While this chapter focuses on the use of Lipschitz bounds and regularization, we note that alternative approaches for handling adversarial robustness in QML exist. For example, the robustness of a quantum model can be improved during training by including adversarial examples into the training data, see [19] for details and references. Further approaches for enhancing robustness during training include adding quantum noise [14] or solving a min-max optimization problem [13], for which tailored generalization bounds can be derived [57, 58]. The key benefit of the robust regularization-based training strategy (9.9) is that it relies on a rigorous theoretical foundation (Theorem 9.1) and does not increase the computational complexity. Studying connections between the robust training approaches from the literature and the presented Lipschitz-based strategy is a promising direction for future research, potentially improving the understanding of robustness and generalization in QML.

References

1. Goodfellow IJ, Shlens J, Szegedy C (2014) Explaining and harnessing adversarial examples. arXiv:1412.6572
2. Szegedy C, Zaremba C, Sutskever I, Bruna J, Erhan D, Goodfellow I, Fergus R (2014) Intriguing properties of neural networks. arXiv:1312.6199
3. Wong E, Kolter Z (2018) Provable defenses against adversarial examples via the convex outer adversarial polytope. In: Proceedings of international conference on machine learning, pp 5283–5292. https://proceedings.mlr.press/v80/wong18a/wong18a.pdf
4. Tsuzuku Y, Sato I, Sugiyama M (2018) Lipschitz-margin training: scalable certification of perturbation invariance for deep neural networks. In: Proceedings of advances in neural information processing systems, pp 6541–6550. https://proceedings.neurips.cc/paper_files/paper/2018/file/485843481a7edacbfce101ecb1e4d2a8-Paper.pdf
5. Madry A, Makelov A, Schmidt L, Tsipras D, Vladu A (2019) Towards deep learning models resistant to adversarial attacks. arXiv:1706.06083
6. Krogh A, Hertz J (1991) A simple weight decay can improve generalization. In: Advances in neural information processing systems, vol 4. https://proceedings.neurips.cc/paper/1991/file/8eefcfdf5990e441f0fb6f3fad709e21-Paper.pdf
7. Xu H, Mannor S (2012) Robustness and generalization. Mach Learn 86(3):391–423
8. Papernot N, McDaniel P, Wu X, Jha S, Swami A (2016) Distillation as a defense to adversarial perturbations against deep neural networks. In: Proceedings of IEEE symposium on security and privacy (SP), pp 582–597
9. Biamonte J, Wittek P, Pancotti N, Rebentrost P, Wiebe N, Lloyd S (2017) Quantum machine learning. Nature 549(7671):195–202
10. Cerezo M, Verdon G, Huang H-Y, Cincio L, Coles PJ (2022) Challenges and opportunities in quantum machine learning. Nat Comput Sci 2:567–576
11. Edwards D, Rawat DB (2020) Quantum adversarial machine learning: status, challenges and perspectives. In: Proceedings 2nd IEEE international conference on trust, privacy and security in intelligent systems and applications (TPS-ISA), pp 128–133
12. Liu N, Wittek P (2020) Vulnerability of quantum classification to adversarial perturbations. Phys Rev A 101:062331
13. Lu S, Duan L-M, Deng D-L (2020) Quantum adversarial machine learning. Phys Rev Res 2:033212
14. Du Y, Hsieh M-H, Liu T, Tao D, Liu N (2021) Quantum noise protects quantum classifiers against adversaries. Phys Rev Res 3:023153
15. Guan J, Fang W, Ying M (2021) Robustness verification of quantum classifiers. In: Proceedings of international conference on computer aided verification, pp 151–174
16. Liao H, Convy I, Huggins WJ, Whaley KB (2021) Robust in practice: adversarial attacks on quantum machine learning. Phys Rev A 103:042427
17. Weber M, Liu N, Li B, Zhang C, Zhao Z (2021) Optimal provable robustness of quantum classification via quantum hypothesis testing. Quantum Inf 7:76
18. Gong W, Deng D-L (2022) Universal adversarial examples and perturbations for quantum classifiers. Natl Sci Rev 9(6):nwab130

19. West MT, Tsang S-L, Low JS, Hill CD, Leckie C, Hollenberg LCL, Erfani SM, Usman M (2023) Towards quantum enhanced adversarial robustness in machine learning. Nat Mach Intell 5:581–589
20. West MT, Erfani SM, Leckie C, Sevior M, Hollenberg LCL, Usman M (2023) Benchmarking adversarially robust quantum machine learning at scale. Phys Rev Res 5:023186
21. Berberich J, Fink D, Pranjić D, Tutschku C, Holm C (2024) Training robust and generalizable quantum models. Phys Rev Res 6:043326
22. Abbas A, Sutter D, Zoufal C, Lucchi A, Figalli A, Woerner S (2021) The power of quantum neural networks. Nat Comput Sci 1:403–409
23. Banchi L, Pereira J, Pirandola S (2021) Generalization in quantum machine learning: a quantum information standpoint. PRX Quantum 2:040321
24. Caro MC, Gil-Fuster E, Meyer JJ, Eisert J, Sweke R (2021) Encoding-dependent generalization bounds for parametrized quantum circuits. Quantum 5:582
25. Huang H-Y, Bourghton M, Mohseni M, Babbush R, Boixo S, Neven H, McClean JR (2021) Power of data in quantum machine learning. Nat Commun 12(1):1–9
26. Caro MC, Huang H-Y, Cerezo M, Sharma K, Sornborger A, Cincio L, Coles PJ (2022) Generalization in quantum machine learning from few training data. Nat Commun 13:4919
27. Caro MC, Huang H-Y, Ezzell N, Gibbs J, Sornborger AT, Cincio L, Coles PJ, Holmes Z (2023) Out-of-distribution generalization for learning quantum dynamics. Nat Commun 14(1)
28. Peters E, Schuld M (2023) Generalization despite overfitting in quantum machine learning models. Quantum 7:1210
29. Jerbi S, Gyurik C, Marshall SC, Molteni R, Dunjko V (2024) Shadows of quantum machine learning. Nat Commun 15:5676
30. Gil-Fuster E, Eisert J, Bravo-Prieto C (2024) Understanding quantum machine learning also requires rethinking generalization. Nat Commun 15:2277
31. Benedetti M, Lloyd E, Sack S, Fiorentini M (2019) Parameterized quantum circuits as machine learning models. Quantum Sci Technol 4:043001
32. Havlíček V, Córcoles AD, Temme K, Harrow AW, Kandala A, Chow JM, Gambetta JM (2019) Supervised learning with quantum-enhanced feature spaces. Nature 567:209–212
33. Schuld M, Killoran N (2019) Quantum machine learning in feature Hilbert spaces. Phys Rev Lett 122:040504
34. Cerezo M, Arrasmith A, Babbush R, Benjamin SC, Endo S, Fujii K, McClean JR, Mitarai K, Yuan X, Cincio L, Coles PJ (2021) Variational quantum algorithms. Nat Rev Phys 3:625–644
35. Fazlyab M, Robey A, Hassani H, Morari M, Pappas GJ (2019) Efficient and accurate estimation of Lipschitz constants for deep neural networks. In: Advances in neural information processing systems, vol 32. https://proceedings.neurips.cc/paper_files/paper/2019/file/95e1533eb1b20a97777749fb94fdb944-Paper.pdf
36. Revay M, Wang R, Manchester IR (2021) A convex parameterization of robust recurrent neural networks. IEEE Control Syst Lett 5(4):1363–1368
37. Pauli P, Koch A, Berberich J, Kohler P, Allgöwer F (2022) Training robust neural networks using Lipschitz bounds. IEEE Control Syst Lett 6:121–126
38. Pérez-Salinas A, Cervera-Lierta A, Gil-Fuster E, Latorre JI (2020) Data re-uploading for a universal quantum classifier. Quantum 4:226
39. Jaderberg B, Gentile AA, Berrada YA, Shishenina E, Elfving VE (2024) Let quantum neural networks choose their own frequencies. Phys Rev A 109:042421
40. Sweke R, Wilde F, Meyer J, Schuld M, Fährmann PK, Meynard-Piganeau B, Eisert J (2020) Stochastic gradient descent for hybrid quantum-classical optimization. Quantum 4:314
41. Berberich J, Fink D, Holm C (2024) Robustness of quantum algorithms against coherent control errors. Phys Rev A 109:012417
42. Recio-Armengol E, Eisert J, Meyer JJ (2025) Single-shot quantum machine learning. Phys Rev A 111:042420
43. Kempkes M, Ijaz A, Gil-Fuster E, Bravo-Prieto C, Spiegelberg J, van Nieuwenburg E, Dunjko V (2025) Double descent in quantum machine learning. arXiv:2501.10077

44. Mohri M, Rostamizadeh A, Talwalkar A (2018) Foundations of machine learning, 2nd ed. Adaptive computation and machine learning. MIT Press, Cambridge, MA. ISBN: 9780262039406
45. Mitarai K, Negoro M, Kitagawa M, Fujii K (2018) Quantum circuit learning. Phys Rev A 98:032309
46. Schuld M, Bocharov A, Svore KM, Wiebe N (2020) Circuit-centric quantum classifiers. Phys Rev A 101:032308
47. Fellner T (2025) RobustnessGeneralizationQML, May 2025. https://github.com/tobias-fllnr/RobustnessGeneralizationQML
48. May RM (1976) Simple mathematical models with very complicated dynamics. Nature 261(5560):459–467 Jun
49. Devaney R (2018) An introduction to chaotic dynamical systems, 2nd edn. CRC Press, Boca Raton Mar.
50. Mihailović DT, Mimić G, Arsenić I (2014) Climate predictions: the chaos and complexity in climate models. Adv Meteorol 2014(1):878249
51. LeBaron B, Grenfell BT, May RM, Tong H (1994) Chaos and nonlinear forecastability in economics and finance. Philos Trans R Soc Lond Ser A Phys Eng Sci 348(1688):397–404
52. Fellner T, Kreplin D, Tovey S, Holm C (2025) Quantum vs. classical: a comprehensive benchmark study for predicting time series with variational quantum machine learning, Apr 2025
53. Bergholm V, Izaac J, Schuld M, Gogolin C, Ahmed S, Ajith V, Alam MS, Alonso-Linaje G, AkashNarayanan B, Asadi A, Arrazola JM, Azad U, Banning S, Blank C, Bromley TR, Cordier BA, Ceroni J, Delgado A, Matteo OD, Dusko A, Garg T, Guala D, Hayes A, Hill R, Ijaz A, Isacsson T, Ittah D, Jahangiri S, Jain P, Jiang E, Khandelwal A, Kottmann K, Lang RA, Lee C, Loke T, Lowe A, McKiernan K, Meyer JJ, Montañez-Barrera JA, Moyard R, Niu Z, O'Riordan LJ, Oud S, Panigrahi A, Park C-Y, Polatajko D, Quesada N, Roberts C, Sá N, Schoch I, Shi B, Shu S, Sim S, Singh A, Strandberg I, Soni J, Száva A, Thabet S, Vargas-Hernández RA, Vincent T, Vitucci N, Weber M, Wierichs D, Wiersema R, Willmann M, Wong V, Zhang S, Killoran N (2018) Pennylane: automatic differentiation of hybrid quantum-classical computations. arXiv:1811.04968
54. Kingma DP, Ba J (2014) Adam: a method for stochastic optimization. arXiv:1412.6980
55. Wendlinger M, Tscharke K, Debus P (2024) A comparative analysis of adversarial robustness for quantum and classical machine learning. In: Proceedings of IEEE international conference quantum computing and engineering (QCE), pp 1447–1457
56. Meyer N, Berberich J, Mutschler C, Scherer DD (2024) Robustness and generalization in quantum reinforcement learning via Lipschitz regularization. Quantum Mach Intell. https://arxiv.org/abs/2410.21117
57. Georgiou P, Jose ST, Simeone O (2024) Adversarial quantum machine learning: an information-theoretic generalization analysis. In: Proceedings of IEEE international symposium information theory (ISIT), pp 789–794
58. Georgiou P, Thomas AM, Jose ST (2025) On the generalization of adversarially trained quantum classifiers. arXiv:2504.17690

Chapter 10
Verifying Adversarial Robustness in Quantum Machine Learning: From Theory to Physical Validation Via a Software Tool

Ji Guan and Mingsheng Ying

Abstract As with classical neural networks, quantum machine learning (QML) models are vulnerable to small input perturbations that can significantly alter output predictions. Certifying the robustness of QML models, particularly on NISQ hardware, is therefore a fundamental step toward trustworthy quantum AI. This chapter reviews our recently developed comprehensive formal framework for *verifying adversarial robustness* in QML. The core of this framework is a fidelity-based robustness lower bound computable directly from the measurement outcome distribution, which enables both formal verification and empirical estimation on real quantum devices. Additionally, the optimal bound can be computed via semidefinite programming (SDP) with full knowledge of the quantum machine learning models. We incorporate these results into: (1) an efficient formal verification framework; (2) VeriQR, the first dedicated QML robustness verification tool; and (3) the first experimental benchmark of quantum adversarial robustness on a 20-qubit superconducting processor. Together, these systematic advances enable scalable, physically grounded robustness evaluation of QML models.

10.1 Introduction

Quantum machine learning (QML) has emerged as a promising paradigm at the intersection of quantum computing and artificial intelligence, offering the potential to accelerate data-driven tasks such as classification, pattern recognition, and quan-

J. Guan (✉)
Key Laboratory of System Software (Chinese Academy of Sciences), Institute of Software, Chinese Academy of Sciences, Beijing, China
e-mail: guanj@ios.ac.cn

M. Ying
Centre for Quantum Software and Information, University of Technology Sydney, Ultimo, NSW, Australia
e-mail: Mingsheng.Ying@uts.edu.au

M. Usman (ed.), *Quantum Robustness in Artificial Intelligence*, Quantum Science and Technology, https://doi.org/10.1007/978-3-032-11153-1_10

tum system identification [1, 2]. Enabled by rapid progress in quantum hardware platforms—particularly superconducting circuits—the implementation and empirical study of QML algorithms is advancing swiftly [3–10]. Industrial momentum is also building; for instance, Google has introduced *TensorFlow Quantum*, integrating quantum circuit training into the mainstream classical ML framework TensorFlow [11].

However, like their classical counterparts, QML models are susceptible to *adversarial perturbations*—subtle modifications to the input quantum state that can cause incorrect predictions or degrade performance [12, 13]. These perturbed inputs, known as *adversarial examples*, present a serious threat to the reliability and security of quantum AI, particularly in the noisy intermediate-scale quantum (NISQ) era where hardware noise is pervasive. Ensuring that QML models are robust against such perturbations is therefore a pressing concern [14].

While adversarial robustness has been widely studied in classical deep learning [15, 16], its quantum counterpart poses distinct challenges [17]. Quantum states are probabilistic, subject to decoherence and readout noise, and constrained by physical laws that restrict allowable perturbations to those preserving trace and positivity. This necessitates new verification frameworks grounded in quantum-specific geometry, such as fidelity-based distances.

In this chapter, we review our recently developed comprehensive framework for **verifying adversarial robustness in quantum machine learning**, grounded in rigorous theoretical analysis and experimental validation supported by software tool. Our approach spans three interlinked dimensions:

- **Theoretical Foundations**. We formalize robustness using fidelity-based metrics and introduce a sound, instance-specific *robustness lower bound* that can be computed directly from the measurement outcome distribution [18]. This bound enables certification even in black-box scenarios without model access, which is used in the physical validation. Additionally, we formulate the exact robustness radius as a semidefinite program (SDP), enabling provably optimal verification when the model is known. Based on these results, we design a set of verification algorithms that certify robustness under various information assumptions.
- **Algorithmic Realization**. We implement these ideas in VERIQR [19], the first dedicated tool for QML robustness verification. VERIQR supports both exact (sound and complete) and approximate robustness verification algorithms, and can identify quantum adversarial examples that may be used for adversarial training. The tool accepts models in OpenQASM 2.0 format and simulates realistic noise to reflect NISQ hardware characteristics, offering a unified benchmarking framework compatible with multiple quantum platforms.
- **Experimental Validation**. To assess practical viability, the first physical benchmark of QML robustness was conducted on a 20-qubit superconducting processor [20] based on our theoretical idea and using the VeriQR tool. It was demonstrated that the fidelity-based lower bound provides a tight and stable estimate of

robustness in the presence of real quantum noise and validates its correlation with robustness upper bounds obtained by adversarial attack methods and SDP-derived optimal bounds.

Together, the above components establish a scalable, hardware-compatible, and formally grounded approach to robustness in quantum machine learning. They show that rigorous certification is feasible on near-term devices and can be integrated seamlessly into the QML development lifecycle.

Organization. The remainder of this review is organized as follows. Section 10.2 introduces the necessary preliminaries, including quantum classifiers, fidelity-based distance measures, and the formal definition of the robustness verification problem. Section 10.3 develops a series of robustness bounds that form the theoretical foundation for solving the verification task. Section 10.4 presents verification algorithms derived from these bounds. Section 10.5 describes the design and functionality of the VERIQR tool, which implements these verification techniques. Section 10.6 details the experimental setup and results from benchmarking adversarial robustness on real superconducting quantum hardware using the robustness bounds. Finally, Sect. 10.7 concludes with a discussion of open problems and future research directions.

10.2 Preliminaries

This section presents the foundational concepts essential for verifying adversarial robustness in quantum machine learning. We begin by introducing the structure and semantics of quantum classifiers, followed by the definition of fidelity-based distance measures used to quantify adversarial perturbations. These perturbations can lead to adversarial examples, which are input states that induce misclassification. Building on these definitions, we then formalize the robustness verification problem, which forms the central focus of this chapter.

10.2.1 Quantum Classifiers

Let $\mathcal{H}$ be a 2^n-dimensional Hilbert space on an n-qubit quantum system. A *quantum state* $\rho \in \mathcal{D}(\mathcal{H})$ is a positive semidefinite operator ($\rho \succeq 0$) on $\mathcal{H}$ with trace one ($\mathrm{Tr}(\rho) = 1$). Here $\mathcal{D}(\mathcal{H})$ represents the set of quantum states on $\mathcal{H}$.

As illustrated in Fig. 10.1, a *quantum classifier* is a quantum algorithm that takes quantum input states and produces classical output labels, corresponding to predefined classes of interest. Formally, a quantum classifier over the Hilbert space $\mathcal{H}$ is defined as a pair:

$$\mathcal{A} = (\mathcal{E}, \{M_c\}_{c\in C}),$$

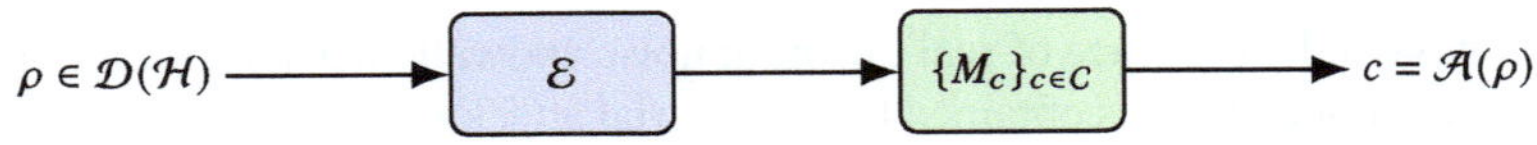

Fig. 10.1 **Quantum classifier pipeline**. The input quantum state ρ is processed by a quantum channel $\mathcal{E}$, followed by measurement via a POVM $\{M_c\}_{c\in C}$, to produce a classical class label $c = \mathcal{A}(\rho)$

where

- $\mathcal{E} : \mathcal{D}(\mathcal{H}) \to \mathcal{D}(\mathcal{H})$ is a *quantum channel*, a completely positive and trace-preserving (CPTP) map, that models the quantum evolution (including noise) of the algorithm;
- $\{M_c\}c \in C$ is a *positive operator-valued measure* (POVM), where each $M_c \succeq 0$ and $\sum_c M_c = I$, used to perform quantum measurement and extract classical outputs;
- C is a finite set of possible output labels representing the classification categories.

Given an input quantum state $\rho \in \mathcal{D}(\mathcal{H})$, the classifier outputs a label determined by the most probable measurement outcome:

$$\mathcal{A}(\rho) := \arg\max_{c\in C} \mathrm{Tr}[M_c\mathcal{E}(\rho)],$$

where $\mathrm{Tr}[M_c\mathcal{E}(\rho)]$ is the probability of obtaining outcome c upon measuring the output state $\mathcal{E}(\rho)$ of $\mathcal{E}$ with the POVM $\{M_c\}_{c\in C}$.

10.2.2 Distance Between Quantum States

A central component of robustness analysis is the choice of a metric to quantify the similarity between quantum states. In the context of quantum computing, the standard metric used to characterize adversarial perturbations is based on *quantum fidelity* [17].

Definition 1 (*Fidelity* [17]) Let $\rho, \sigma \in \mathcal{D}(\mathcal{H})$ be two quantum states. The fidelity between ρ and σ is defined as

$$F(\rho, \sigma) := \left(\mathrm{Tr}\sqrt{\sqrt{\rho}\sigma\sqrt{\rho}}\right)^2.$$

Fidelity measures the closeness of two quantum states. It ranges from 0 to 1, where $F(\rho, \sigma) = 1$ if and only if $\rho = \sigma$, and $F(\rho, \sigma) = 0$ when ρ and σ have orthogonal support.

Definition 2 (*Fidelity Distance*) The *fidelity distance* (also called *infidelity*) between two quantum states is defined as

$$D_F(\rho, \sigma) := 1 - F(\rho, \sigma).$$

Fidelity distance provides a quantitative measure of how distinguishable two quantum states are under optimal measurements [17]. In robustness verification, it serves as a natural metric for evaluating the magnitude of adversarial perturbations: smaller distances imply higher similarity, while larger distances indicate more significant deviations from the original state.

10.2.3 Adversarial Robustness

To assess the adversarial robustness of a quantum classifier at a given input state ρ, we require that all quantum states sufficiently close to ρ be classified identically to ρ. A violation of this condition indicates the presence of an *adversarial example*—a nearby state that is misclassified.

Definition 3 (*Adversarial Example*) Let $\mathcal{A}$ be a quantum classifier, $\rho \in \mathcal{D}(\mathcal{H})$ an input state, and $\varepsilon > 0$ a perturbation threshold. A quantum state σ is called an ε-*adversarial example* of ρ if

$$\mathcal{A}(\sigma) \neq \mathcal{A}(\rho) \quad \text{and} \quad D_F(\rho, \sigma) \leq \varepsilon.$$

If such a state σ exists, then ε is referred to as an *adversarial perturbation* of ρ.

Here, the condition $\mathcal{A}(\sigma) \neq \mathcal{A}(\rho)$ indicates that σ is classified differently from ρ, while the condition $D_F(\rho, \sigma) \leq \varepsilon$ ensures that σ is sufficiently close to ρ in fidelity distance. Thus, an ε-adversarial example is a small perturbation of ρ that causes the classifier to mispredict.

Note that this definition implicitly assumes that ρ is correctly classified. If $\mathcal{A}(\rho)$ is already incorrect, identifying σ such that $\mathcal{A}(\sigma) \neq \mathcal{A}(\rho)$ no longer signifies a robustness issue but a correctness one. Therefore, in the subsequent discussions, we restrict attention to correctly classified input states.

The absence of adversarial examples within a given neighborhood defines robustness.

Definition 4 (*Adversarial Robustness*) A quantum classifier $\mathcal{A}$ is said to be ε-*robust* at state ρ if there exists no ε-adversarial example of ρ.

Equivalently, $\mathcal{A}$ is ε-robust at ρ if

$$\forall \sigma \in \mathcal{D}(\mathcal{H}), \quad D_F(\rho, \sigma) \leq \varepsilon \Rightarrow \mathcal{A}(\sigma) = \mathcal{A}(\rho).$$

This notion allows us to define the exact robustness radius of a quantum state.

Definition 5 (*Robustness Radius*) Let $\mathcal{A}$ be a quantum classifier and ρ a correctly classified input state. The *robustness radius* of ρ, denoted $\varepsilon^*(\rho)$, is the maximum value ε such that $\mathcal{A}$ is ε-robust at ρ:

$$\varepsilon^*(\rho) := \sup_{\substack{\sigma \in \mathcal{D}(\mathcal{H}) \\ \mathcal{A}(\sigma)=\mathcal{A}(\rho)}} D_F(\rho, \sigma).$$

Intuitively, $\varepsilon^*(\rho)$ quantifies the largest allowable fidelity degradation that preserves the classifier's decision. For any $\varepsilon > \varepsilon^*(\rho)$, there exists an ε-adversarial example σ that is misclassified relative to ρ.

The robustness radius gives rise to the central verification task addressed in this chapter:

Proposition 1 (Robustness Verification Problem) *Given a quantum classifier $\mathcal{A}$, an input state $\rho \in \mathcal{D}(\mathcal{H})$, and a threshold $\varepsilon > 0$, determine whether*

$$\varepsilon \leq \varepsilon^*(\rho).$$

If so, $\mathcal{A}$ is ε-robust at ρ; otherwise, ε is an adversarial perturbation, and a violating state σ can be returned as an ε-adversarial example.

The objective is to determine whether adversarial examples exist within an ε-fidelity ball centered at ρ. The remainder of this chapter develops exact and approximate approaches to solving this verification problem.

Finally, robustness can be aggregated across a dataset to evaluate a classifier's overall robustness:

Definition 6 (*Robust Accuracy*) Let $\mathcal{A}$ be a quantum classifier. The *ε-robust accuracy* of $\mathcal{A}$ is the proportion of correctly classified input states in the dataset that are also ε-robust.

This definition extends the notion of robustness from individual quantum states to an entire dataset. Depending on the application, the dataset may consist of training samples, validation samples, or a combination of both, under the assumption that all included quantum states are correctly labeled.

10.3 Robustness Bounds

This section presents a series of robustness bounds that address the robustness verification problem formulated in Problem 1. These bounds characterize the robustness radius $\varepsilon^*(\rho)$ through exact, under-approximate, and over-approximate techniques. We derive these bounds using a variety of methods, including semidefinite programming, analysis of measurement outcome distributions, and adversarial attack construction.

10.3.1 Optimal Robustness Bound Via Semidefinite Programming

We first show that the robustness radius $\varepsilon^*(\rho)$ can be exactly computed using *semidefinite programming (SDP)*, assuming full knowledge of the quantum classifier.

Given the explicit description of the quantum channel $\mathcal{E}$ and the POVM $\{M_c\}_{c\in C}$, the robustness radius can be formulated as an SDP [18]:

Theorem 1 (Optimal Robustness Bound via SDP [18]) *Let $\mathcal{A} = (\mathcal{E}, \{M_c\}_{c\in C})$ be a quantum classifier. The exact robustness radius is given by*

$$\varepsilon^*(\rho) = \min_{\substack{c\in C \\ c\neq\mathcal{A}(\rho)}} \varepsilon_c^*(\rho),$$

where each $\varepsilon_c^*(\rho)$ is the solution to the following SDP:

$$\begin{aligned} \textit{minimize:}\quad & D_F(\rho, \sigma) \\ \textit{subject to:}\quad & \sigma \succeq 0, \\ & \mathrm{Tr}(\sigma) = 1, \\ & \mathrm{Tr}[(M_{\mathcal{A}(\rho)} - M_c)\mathcal{E}(\sigma)] \leq 0. \end{aligned}$$

If this SDP is infeasible for some c, then $\varepsilon_c^*(\rho) = \infty$, indicating that no adversarial example of ρ exists which is misclassified as class c.

This SDP formulation allows exact and efficient computation of the robustness radius using convex optimization. The tractability of this approach is rooted in a fundamental principle of quantum mechanics: the linearity of quantum operations. Both the quantum channel $\mathcal{E}$ and the measurement process are linear maps, which leads to a convex feasible set and objective in the optimization problem.

This stands in contrast to classical neural networks, where nonlinearities such as ReLU activations or pooling layers make the robustness verification problem non-convex. As a result, certifying adversarial robustness in classical settings is NP-complete [21], and typically requires approximations or linear encodings of nonlinear functions (e.g., NSVerify [22], MIPVerify [23], ILP [24], ImageStar [25]).

However, computing $\varepsilon^*(\rho)$ using Theorem 1 assumes full access to the internal structure of the classifier, including its noise-free quantum evolution. In practice, especially on near-term noisy intermediate-scale quantum (NISQ) devices, this assumption rarely holds due to unknown or device-specific noise effects.

To address this, we introduce lower and upper bounds on $\varepsilon^*(\rho)$ that can be estimated under partial or empirical knowledge of the model. While the SDP method offers a gold-standard verification backend, these bounds provide practical alternatives for applications where exact model information is unavailable. In the following subsections, we present such bounds and discuss how they partially solve the robustness verification problem (Problem 1).

10.3.2 Robustness Lower Bound Via Measurement Distribution

We now present a certified lower bound for the robustness radius $\varepsilon^*(\rho)$ based on the measurement outcome distribution of a quantum classifier.

Let $\mathcal{A} = (\mathcal{E}, \{M_c\}_{c\in C})$ be a quantum classifier. For a given input state $\rho \in \mathcal{D}(\mathcal{H})$, denote the probability of observing class c as

$$p_c^\rho := \mathrm{Tr}[M_c\mathcal{E}(\rho)].$$

Let $c^* := \mathcal{A}(\rho)$ be the predicted class label of ρ. Then, the measurement distribution $\{p_c^\rho\}_{c\in C}$ enables the following certified lower bound:

Theorem 2 (Robustness Lower Bound from Measurement Distribution [18]) *Let $\rho \in \mathcal{D}(\mathcal{H})$ and $c^* = \mathcal{A}(\rho)$. Then*

$$\varepsilon_{\mathrm{RLB}}(\rho) := \min_{c\neq c^*} \frac{1}{2}\left(\sqrt{p_{c^*}^\rho} - \sqrt{p_c^\rho}\right)^2$$

is a certified robustness lower bound: for all σ such that $D_F(\rho, \sigma) \leq \varepsilon_{\mathrm{RLB}}(\rho)$, it holds that $\mathcal{A}(\sigma) = \mathcal{A}(\rho)$.

This robustness lower bound has the following key features:

- **Efficient to Compute**. The bound depends only on the measurement probabilities $\{p_c^\rho\}_{c\in C}$, and can be evaluated directly from measurement outcomes of ρ without searching for adversarial perturbations. This efficiency allows for fast robustness certification and dataset-level evaluation of robust accuracy.
- **Model-agnostic**: Since it requires no access to the internal structure of $\mathcal{E}$, this bound is particularly suited for hardware-level evaluation. In real-device settings, one can estimate p_c^ρ by repeated execution of $\mathcal{E}$ on quantum hardware and compute $\varepsilon_{\mathrm{RLB}}(\rho)$ from the empirical outcome distribution.

10.3.3 Robustness Upper Bound Via Attack Generation

In contrast to certified lower bounds, robustness upper bounds estimate the minimum perturbation required to fool a quantum classifier. These bounds can be obtained through adversarial attack strategies and provide empirical insights into the model's worst-case vulnerability.

Definition 7 (*Empirical Robustness Upper Bound*) Let $\rho \in \mathcal{D}(\mathcal{H})$ be an input quantum state. An *adversarial attack method* constructs a perturbed state σ_{adv} such that

$$\mathcal{A}(\sigma_{\text{adv}}) \neq \mathcal{A}(\rho), \quad \text{and} \quad \varepsilon_{\text{RUB}}(\rho) := D_F(\rho, \sigma_{\text{adv}}),$$

where D_F is the fidelity distance. Then, $\varepsilon_{\text{RUB}}(\rho)$ serves as an *empirical robustness upper bound* for $\varepsilon^*(\rho)$.

Since any adversarial example must lie outside the certified robust region, the inequality

$$\varepsilon^*(\rho) < \varepsilon_{\text{RUB}}(\rho)$$

always holds. Thus, ε_{RUB} provides an over-approximation of the robustness radius. Although ε_{RUB} is not a certified bound, it provides practical evidence of vulnerability and complements robustness lower bounds. It is especially valuable in real-device settings where exact robustness certification via SDP is infeasible.

In classical adversarial learning, numerous algorithms have been developed to craft adversarial examples [26]. These methods are typically categorized as either white-box or black-box attacks, depending on the attacker's access to model parameters. A widely used white-box approach is the *Fast Gradient Sign Method (FGSM)* [27], which perturbs a legitimate input $\boldsymbol{x}$ as follows:

$$\boldsymbol{x}' = \boldsymbol{x} + \varepsilon \cdot \text{sgn}(\nabla_{\boldsymbol{x}} \mathcal{L}), \tag{10.1}$$

where ε denotes the perturbation magnitude, $\nabla_{\boldsymbol{x}} \mathcal{L}$ is the gradient of the loss function $\mathcal{L}$ with respect to the input and the function $\text{sgn}(\cdot)$ stands for the sign function to extract the direction of a gradient without its magnitude.

In the quantum setting, generating analogous adversarial examples requires gradient estimation using the parameter-shift rule [28]. This introduces significant computational overhead on quantum hardware: producing a single adversarial example typically demands $2N_s \cdot N_x$ circuit executions, where N_s is the number of shots per measurement and $N_x = \dim(\boldsymbol{x})$ is the input dimension. Consequently, applying FGSM to high-dimensional quantum data becomes impractical, particularly in batch settings. This overhead highlights a major limitation of naive gradient-based attacks in quantum machine learning, especially on current noisy intermediate-scale quantum (NISQ) devices [29, 30].

To mitigate the overhead of full-gradient-based attacks, a localized variant of the FGSM, termed *Mask FGSM*, has been proposed for efficient adversarial sample generation in quantum machine learning (QML) experiments [20]. This method restricts perturbations to a strategically selected sparse subset of input features, specified by a binary mask $\mathcal{M} = (m_1, m_2, \ldots, m_{\dim(\boldsymbol{x})})^T$. The mask identifies which components of the input are perturbed, reducing both computational cost and the influence of noisy gradient estimates in experimental settings.

Given a legitimate input $\boldsymbol{x}$, adversarial examples are generated by applying a perturbation vector $\boldsymbol{\delta}$ such that $\boldsymbol{x}' = \boldsymbol{x} + \boldsymbol{\delta}$, where each component δ_i is computed as

$$\delta_i = \begin{cases} \varepsilon \cdot \operatorname{sgn}\left(\frac{\partial \mathcal{L}}{\partial x_i}\right), & \text{if } m_i = 1, \\ 0, & \text{if } m_i = 0, \end{cases} \tag{10.2}$$

with ε denoting the perturbation strength and $\mathcal{L}$ the loss function. The mask $\mathcal{M}$ is constructed by ranking gradient magnitudes, allowing targeted perturbation of the most influential input features (see [20] for implementation details).

This sparse attack strategy achieves a favorable trade-off between efficiency and effectiveness, as confirmed in the quantum hardware experiments on EMNIST and LCEI classification tasks [20], which will be detailed in Sect. 10.6.

10.3.4 Visualizing the Bounds

Based on the certified lower bound and the empirical upper bound presented in the previous subsections, we can formalize their relationship through the following result:

Theorem 3 (Sandwich Robustness Bound) *Given a quantum input state ρ, a certified lower bound $\varepsilon_{\text{RLB}}(\rho)$ (Theorem 2), and an adversarially generated state σ_{adv}, we have*

$$\varepsilon_{\text{RLB}}(\rho) \leq \varepsilon^*(\rho) \leq \varepsilon_{\text{RUB}}(\rho), \tag{10.3}$$

where $\varepsilon_{\text{RUB}}(\rho) = D_F(\rho, \sigma_{adv})$.

This relation is illustrated in Fig. 10.2. Each quantity in inequality (10.3) has a distinct role in robustness analysis:

- $\varepsilon_{\text{RLB}}(\rho)$: a certified lower bound used for formal robustness guarantees;
- $\varepsilon^*(\rho)$: the exact robustness radius, computable via SDP;
- $\varepsilon_{\text{RUB}}(\rho)$: an empirical upper bound derived from adversarial attacks.

Tightness Assessment. The gap $\Delta := \varepsilon_{\text{RUB}}(\rho) - \varepsilon_{\text{RLB}}(\rho)$ quantifies the precision of the robustness estimation. In the superconducting hardware experiments presented later [20], this certified lower bound is compared against an empirical robustness upper bound $\varepsilon_{\text{RUB}}(\rho)$, obtained using gradient-based adversarial attacks (introduced in the next section). The observed gap between the two bounds is typically less than 3×10^{-3}, demonstrating that $\varepsilon_{\text{RLB}}(\rho)$ provides a tight and practically useful certificate of robustness.

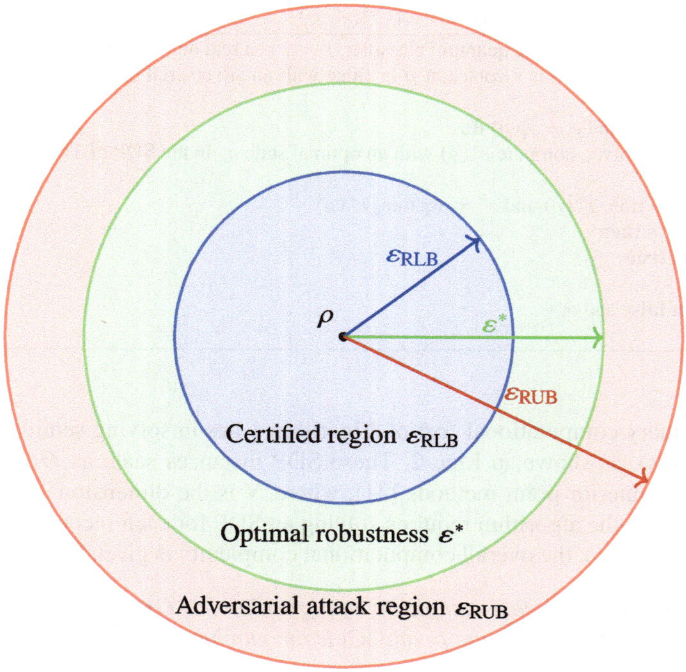

Fig. 10.2 Visualization of robustness bounds as nested fidelity-distance regions around the input state ρ. The certified lower bound $\varepsilon_{\mathrm{RLB}}$ defines a guaranteed-safe region, the upper bound $\varepsilon_{\mathrm{RUB}}$ corresponds to successful adversarial attacks, and the optimal robustness ε^* lies in between

10.4 Robustness Verification Algorithms

In this section, we present several algorithms for formally verifying the robustness of quantum classifiers, building upon the theoretical lower and optimal bounds introduced in the previous section.

10.4.1 State Robustness Verification

We begin by considering the robustness of a specific quantum input state ρ under a quantum classifier $\mathcal{A}$. The robustness verification problem (Problem 1) is to determine whether $\mathcal{A}$ is ε-robust at ρ for a given threshold $\varepsilon > 0$.

Note that once the exact robustness radius $\varepsilon^*(\rho)$ is computed, verifying ε-robustness reduces to a simple comparison: $\mathcal{A}$ is ε-robust at ρ if and only if $\varepsilon \leq \varepsilon^*(\rho)$. This observation, together with Theorem 1, leads to Algorithm 1, which simultaneously determines the robustness status of ρ and computes the exact robustness radius $\varepsilon^*(\rho)$.

Algorithm 1 StateRobustnessVerifier($\mathcal{A}$, ε, ρ)

Input: $\mathcal{A} = (\mathcal{E}, \{M_c\}_{c\in C})$ is a quantum classifier, $\varepsilon < 1$ is a real number, ρ is an input state of $\mathcal{A}$
Output: **true** indicates $\mathcal{A}$ is ε-robust at ρ or **false** with an adversarial example σ indicates $\mathcal{A}$ is not ε-robust at ρ

1: **for each** $c \in C$ and $c \neq \mathcal{A}(\rho)$ **do**
2:     By a SDP solver, compute $\varepsilon_c^*(\rho)$ with an optimal state σ_c in the SDP of Theorem 1
3: **end for**
4: Let $\varepsilon^*(\rho) = \min_c \varepsilon_c^*(\rho)$ and $c^* = \arg\min_c \varepsilon^*(\rho)$
5: **if** $\varepsilon^*(\rho) > \varepsilon$ **then**
6:     **return true**
7: **else**
8:     **return false** and σ_{c^*}
9: **end if**

The primary computational cost of Algorithm 1 lies in solving semidefinite programs (SDPs), as shown in Line 2. These SDP instances scale as $O(N^{6.5})$ when solved using interior-point methods [31], where N is the dimension of the Hilbert space $\mathcal{H}$. Since the algorithm requires solving an SDP for each incorrect class label (i.e., $|C| - 1$ times), the overall computational complexity is given by the following:

Theorem 4 *The worst-case complexity of Algorithm 1 is $O(|C| \cdot N^{6.5})$, where N is the dimension of the input state ρ, and $|C|$ is the number of class labels.*

Algorithm 2 RobustnessVerifier($\mathcal{A}$, ε, T)

Input: $\mathcal{A} = (\mathcal{E}, \{M_c\}_{c\in C})$ is a well-trained quantum classifier, $\varepsilon < 1$ is a real number, $T = \{(\rho_i, l_i)\}$ is a dataset indicating ρ_i be correctly classified into class l_i.
Output: The robust accuracy RA and a set $R = \{< \sigma_j, i_j >\}$, where for each j, σ_j is an ε-adversarial example of ρ_{i_j}; R can be an empty set if $\mathcal{A}$ is ε-robust at any state in T.

$R = \emptyset$ be an empty set. *// Recording adversarial examples and corresponding indexes of states in dataset T*
for each $(\rho_i, l_i) \in T$ **do**
    Let $p_c = \mathrm{Tr}[M_c\mathcal{E}(\rho_i)]$ be the measurement probability of observing outcome c for input ρ.
    Obtain the lower bound $\varepsilon_{\mathrm{RLB}}(\rho_i) = \min_{c\neq c^*} \frac{1}{2}\left(\sqrt{p_{c^*}} - \sqrt{p_c}\right)^2$ *// Applying the robust bound in Theorem 2*
    if $\varepsilon > \varepsilon_{\mathrm{RLB}}(\rho)$ **then**
        if StateRobustnessVerifier ($\mathcal{A}$, ε, ρ_i) == **false then**
            σ be the output state of StateRobustnessVerifier ($\mathcal{A}$, ε, ρ_i)
            $R = R \cup \{(\sigma, i)\}$
        end if
    end if
end for
return $RA = 1 - \frac{|R|}{|T|}$, R *// $|R| = 0$ if R is a empty set*

10.4.2 Classifier Robustness Verification

We now turn our attention to verifying the robustness of a quantum classifier $\mathcal{A}$ as a whole. Algorithm 2 is developed for this purpose by combining the exact robustness computation procedure (Algorithm 1) with the certified lower bound technique from Theorem 2.

One key advantage of formal robustness verification in classical machine learning is its ability to identify counterexamples, i.e., adversarial examples, for specific inputs (see, e.g., [25, 32–34]). This advantage is preserved in the quantum setting through Algorithm 2, which enables both verification and counterexample generation. In particular, this facilitates a natural extension of adversarial training [35] to quantum machine learning. Specifically, when a quantum state ρ fails the ε-robustness check, the algorithm automatically identifies an adversarial example σ. By augmenting the training dataset with the misclassified pair (σ, l), where l is the correct label, the classifier $\mathcal{A}$ can be retrained to enhance its robustness. This iterative refinement process allows quantum models to gradually become more resilient to adversarial perturbations.

To analyze the complexity of Algorithm 2, we begin by noting from Theorem 1 that evaluating the robustness of a quantum classifier $\mathcal{A}$—specifically, computing its robust accuracy and identifying adversarial examples—requires invoking Algorithm 1 on each quantum state in the given dataset. Since the cost of a single call to Algorithm 1 is $O(|C| \cdot N^{6.5})$, the overall complexity for a dataset T of size $|T|$ is

$$O(|T| \cdot |C| \cdot N^{6.5}),$$

where $|C|$ is the number of classes and $N = \dim(\mathcal{H})$ is the dimension of the input state space.

However, the certified robustness lower bound provided in Theorem 2 offers a way to significantly reduce this computational cost. This bound can be computed efficiently in $O(|C| \cdot N^5)$ time, which corresponds to performing $|C| \cdot N^2$ matrix multiplications for $N \times N$ density matrices by noting that the number of Kraus operator in a super-operator $\mathcal{E}$ can be limited to be less than N^2.

A practical strategy for efficient robustness verification is thus as follows: we first use the certified lower bound to pre-screen the training dataset T, identifying all potentially non-robust states and collecting them in a subset $T' \subseteq T$. We then apply the exact robustness-checking Algorithm 1 only to the states in T', using a separate set R to record the discovered adversarial examples and their indices.

This two-stage approach reduces the overall complexity to

$$O(|T'| \cdot |C| \cdot N^{6.5}),$$

where typically $|T'| \ll |T|$, as empirically confirmed by our simulation results in [18]. This demonstrates the practical effectiveness of using the robust lower bound for accelerating formal verification of robustness in quantum classifiers.

10.4.3 Approximate Robustness Verification

Finally, we introduce an efficient algorithm (Algorithm 3) for under-approximating the robust accuracy computed by Algorithm 2. Specifically, Algorithm 3 serves as a lightweight subroutine that avoids calling an SDP solver whenever a potential non-robust state can be directly detected using the robustness lower bound established in Theorem 2.

Algorithm 3 UnderRobustAccuracy($\mathcal{A}$, ε, T)

Input: $\mathcal{A} = (\mathcal{E}, \{M_c\}_{c\in C})$ is a well-trained quantum classifier, $\varepsilon < 1$ is a real number, $T = \{(\rho_i, l_i)\}$ is a dataset indicating ρ_i be correctly classified into class l_i.
Output: A under-approximation of robust accuracy URA

```
r = 0.                                   // Record the number of potential non-robust states
for each (ρ_i, l_i) ∈ T do
    Let p_c = Tr[M_c E(ρ_i)] be the measurement probability of observing outcome c for input ρ.
    Obtain the lower bound ε_RLB(ρ_i) = min_{c≠c*} 1/2 (√p_{c*} − √p_c)^2   // Applying the robust bound in Theorem 2
    if ε > ε_RLB(ρ_i) then
        r = r + 1
    end if
end for
return URA= 1 − r/|T|.
```

We have empirically compared Algorithms 2 and 3 in terms of verification time and output accuracy across various quantum classifiers simulated on classical hardware, as reported in [18]. The results demonstrate that Algorithm 3 exhibits favorable scalability and achieves significantly faster runtimes than Algorithm 2, while still providing accurate approximations of robust accuracy. These findings confirm the practical tightness of the proposed robustness lower bound.

10.4.4 Verification Algorithm Comparison

To conclude this section, we summarize the robustness verification algorithms based on different robustness bounds in Table 10.1. These algorithms vary in computational complexity and verification guarantees. In practice, the condition $|T'| \ll |T|$ often holds, confirming the utility of robustness lower bounds in efficiently identifying non-robust quantum states.

Thanks to the linearity of quantum learning models—an outcome of the foundational postulates of quantum mechanics—robustness verification for quantum classifiers can often be performed efficiently, with polynomial-time complexity in the dimension of the input state. This is in stark contrast to classical machine learn-

Table 10.1 Summary of robustness verification algorithms based on different bounds

Robustness verification algorithms			
	Robustness lower bound	Robustness optimal bound	Mixed strategy
Method	Matrix multiplication (MM)	Semidefinite programming (SDP)	MM & SDP
Complexity	$O(\|T\| \cdot \|C\| \cdot N^5)$	$O(\|T\| \cdot \|C\| \cdot N^{6.5})$	$O(\|T'\| \cdot \|C\| \cdot N^{6.5})$
Robust accuracy	Under-approximate	Exact	Exact

ing models, such as deep neural networks (DNNs), which are highly nonlinear and non-convex. In such classical settings, even verifying simple properties can be NP-complete [21].

However, this computational advantage in the quantum setting diminishes when verification is restricted to *pure states*. Unlike the convex set of mixed states, the set of pure states is non-convex. Consequently, the optimization problem underlying the computation of the robustness radius is no longer a semidefinite program (SDP) but becomes a *quadratically constrained quadratic program* (QCQP)—an optimization class where both the objective and constraints are quadratic [18]. Solving QCQPs is NP-hard in general.

To address this challenge, Algorithm 1 can be adapted to pure-state robustness verification by replacing the SDP solver in Line 2 with a QCQP solver. The resulting variant can be integrated into Algorithm 2 to compute robust accuracy and identify adversarial examples constrained to pure states. This adaptation has been evaluated in case studies such as MNIST classification, where input samples are encoded as pure quantum states [18].

10.5 VeriQR: A Tool for Robustness Verification

To translate the robustness verification algorithms (Algorithms 2 and 3) developed in the previous section into practical workflows, we introduce VERIQR [19], the first dedicated software tool for certifying adversarial robustness in quantum machine learning. VERIQR bridges formal verification, adversarial evaluation, and robustness enhancement, providing a flexible and efficient platform for real-world quantum applications.

VERIQR is available at https://github.com/Veri-Q/VeriQR.

As illustrated in Fig. 10.3, the core capabilities of VERIQR include

1. *Tool Integration*: A built-in parser compatible with OpenQASM, supporting frontends such as Qiskit, Cirq, and MindSpore Quantum;
2. *Noisy Simulation*: Injection of various noise models, including random noise, depolarizing, bit-flip, phase-flip, and user-defined noise;

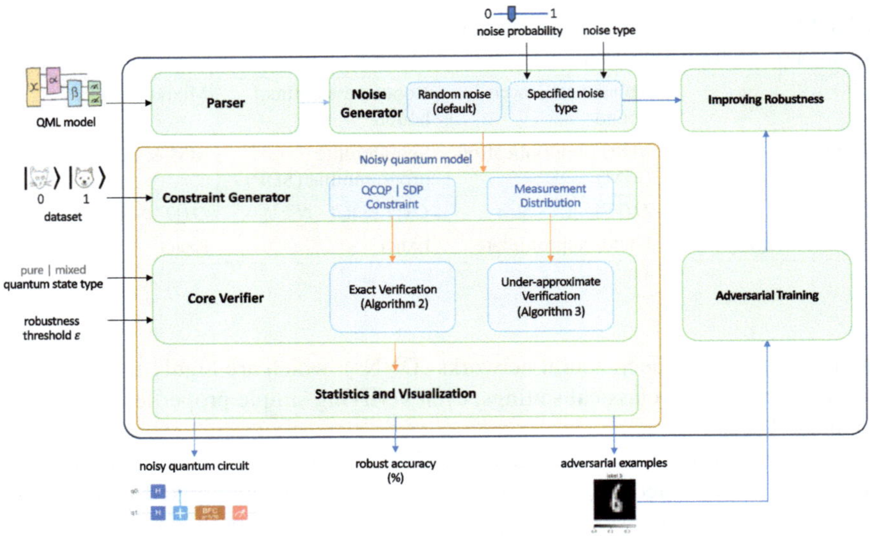

Fig. 10.3 System architecture of VeriQR

3. *Robustness Verification*: Formal verification of adversarial robustness for quantum classifiers based on certified bounds;
4. *GUI and Visualization*: Interactive interface for diagnostics and visualizing robustness landscapes of quantum learning models;
5. *Robustness Improvement*: Enhancement of classifier robustness via adversarial training using counterexamples identified through formal verification or adding specified quantum noise.

In the following, we detail these features.

10.5.1 Setups and Inputs

VeriQR is a graphical user interface (GUI) tool implemented in C++, leveraging the widely adopted Qt framework for GUI development [36].

As shown in the upper left corner of Fig. 10.3, users begin by importing a QML model along with a dataset of quantum states and corresponding ground truth labels, drawn from either the training or testing phase. VeriQR supports models in the following formats, each representing a quantum circuit with a final measurement stage:

1. *NumPy data file (.npz format)*: This format packages the quantum circuit, measurement operators, and dataset into a single file. It is particularly suitable for users familiar with classical machine learning and formal methods but less expe-

rienced in quantum computing. VERIQR includes several example .npz models, enabling new users to quickly begin robustness verification.
2. *OpenQASM 2.0 file (.qasm format)*: OpenQASM is a widely accepted standard for describing quantum circuits, originally introduced by IBM [37]. QML models trained using various quantum platforms can be converted into this format for unified robustness analysis within VERIQR.

Once a model is loaded, users can configure verification parameters via the control panel, as depicted in the left and top portions of Fig. 10.3. The configurable options include:

1. **Noise Model**. Select among several quantum noise types, including *depolarizing*, *phase-flip*, and *bit-flip* noise [38]. Random noise is enabled by default. Additionally, users may define custom noise models or apply combinations of different types.
2. **Quantum State Type**. Specify whether the input quantum states are *mixed* or *pure*. By default, VERIQR assumes mixed states, as they better represent realistic NISQ system states.
3. **Robustness Threshold**. Set the perturbation tolerance parameter ε for robustness verification. This determines the minimum fidelity distance required to guarantee robustness certification.

10.5.2 Verifying Robustness

As shown in the center of Fig. 10.3, the robustness verification workflow in VERIQR consists of five key modules:

(1) **Parser**. This module reads the quantum classifier file and constructs the corresponding quantum circuit object for downstream analysis.
(2) **Noise Generator**. Taking a quantum circuit as input, this module supports two noise injection modes:

   - In the default mode, random noise is applied to each qubit at random positions within the circuit, with randomly sampled noise probabilities, simulating the effects of real-world quantum hardware noise.
   - Alternatively, users can specify noise types (e.g., depolarizing, bit-flip, phase-flip, or custom Kraus operators) and inject them at the end of the circuit, a commonly adopted assumption. This functionality also enables robustness enhancement, as discussed in [39].

(3) **Constraint Generator**. This module generates the necessary mathematical constraints from the (possibly noisy) quantum circuit and dataset, preparing them for the verification engine.
(4) **Core Verifier**. Given the generated constraints, a perturbation threshold ε, and the type of quantum state (mixed or pure), this module selects an appropriate

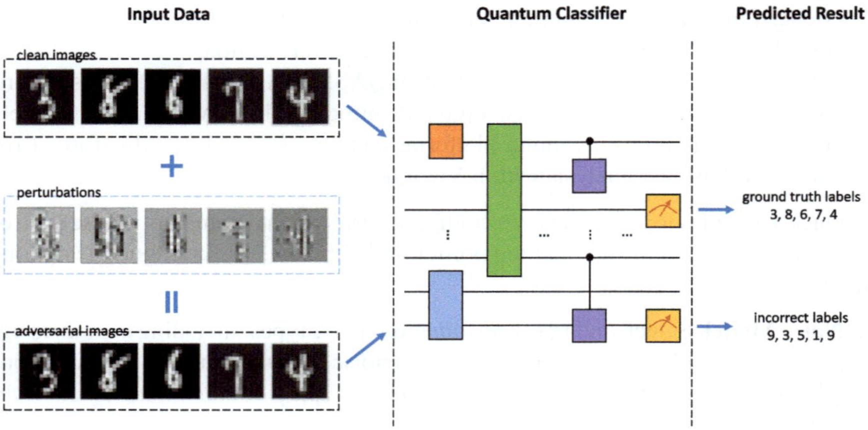

Fig. 10.4 The adversarial examples and the corresponding adversarial perturbations found by VERIQR in MINIST handwritten digit classification

solver: an SDP solver for mixed states or a QCQP solver for pure states [18]. It executes both exact and under-approximate verification procedures via Algorithms 2 and 3, respectively, to certify ε-robustness.

(5) **Statistics and Visualization**. This module visualizes verification results in the GUI. It reports the robust accuracy of the classifier and stores detected adversarial examples in a NumPy file for further analysis or adversarial training. Additionally, it displays the quantum circuit diagrams before and after noise injection. For MNIST digit classification tasks, VERIQR also presents images of adversarial examples corresponding to digits selected by the user, as illustrated in Fig. 10.4.

10.5.3 Improving Robustness

VERIQR supports two complementary strategies to enhance the adversarial robustness of QML models: adversarial training and the deliberate injection of specific quantum noise.

Adversarial Training. VERIQR extends classical adversarial training techniques to the quantum domain. When the ε-robustness of a state ρ with true label l fails, the embedded verification algorithms automatically generate an adversarial example σ. The pair (σ, l) can then be added to the training dataset, allowing the QML model to be retrained with adversarial samples. This iterative procedure effectively improves the model's robustness against future adversarial perturbations.

Injection of Specific Noise. Prior studies [40–42] have shown that carefully introducing quantum noise at selected points within a circuit can enhance robustness, potentially outperforming random noise. VERIQR provides users with the ability to inject either standard or user-defined noise models—using Kraus operators—at specific positions in the quantum circuit. This functionality enables systematic exploration and optimization of noise-assisted robustness strategies for QML models.

10.6 Experimental Benchmark on Superconducting Hardware

To evaluate the practical hardware effectiveness of our robustness verification framework, the first experimental benchmark of adversarial robustness in QML was conducted on real superconducting quantum hardware [20]. This section outlines the experimental setup, summarizes the benchmark results, and highlights insights obtained from physical validation.

The formal framework was validated by executing QML classifiers on a 20-qubit superconducting processor [20]. As shown in Fig. 10.5a, the processor comprises 72 qubits and 126 couplers arranged in a 2D lattice; 20 qubits with high fidelity, highlighted in green, were selected for the experiments.

Two types of classification tasks were considered—classical and quantum—both implemented using trained quantum neural network (QNN) classifiers. As depicted in Fig. 10.5b, each QNN architecture consists of three components: a state preparation circuit that encodes data into a quantum input state, a variational circuit that performs the learning task, and a basis transformation followed by quantum measurement to extract output predictions.

1. **Classical Task**: The classification of handwritten characters "Q" and "T" was performed using the *EMNIST dataset* [43], as illustrated in Fig. 10.5c.
2. **Quantum Task**: The task of *Linear Cluster State Excitation Identification (LCEI)* [11, 18] involved distinguishing between excited and non-excited 20-qubit cluster states. These states were generated using a synthetic dataset, with excitation determined by the rotation angle α of an $R_x(\alpha)$ operation applied at the end of the cluster state preparation circuit (Fig. 10.5d).

For both tasks, classification was based on the expectation value of the output qubit with respect to the Pauli-Z operator, σ_z [17]. The resulting expectation value was converted into a probability via the transformation $p = (\langle\sigma_z\rangle + 1)/2$, with a threshold of $p = 0.5$ used to define the decision boundary.

For each classification task, adversarial robustness benchmarking is conducted on a well-trained binary quantum classifier using 10 randomly selected input quantum states (5 from each class). The experimental procedure involves the following steps:

1. **Measurement**: For each selected input quantum state ρ, execute the quantum classifier circuit on the quantum hardware to obtain the binary measurement outcome probabilities $p_1, p_2 = 1 - p_1$. Without loss of generality, it is assumed that $p_1 \geq p_2$ to facilitate subsequent analysis.

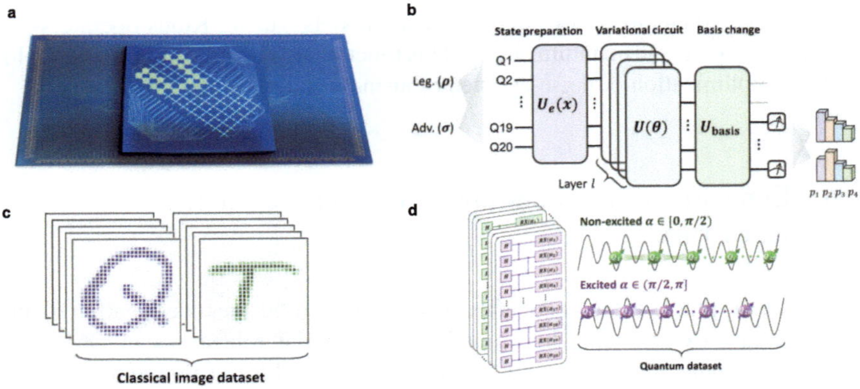

Fig. 10.5 Experimental schematic for QNN robustness evaluation. **a** Schematic diagram of the superconducting quantum processor, comprising 72 qubits and 126 couplers arranged in a 2D lattice. The 20 qubits selected for the experiment are highlighted in green. **b** Architecture of the quantum neural network (QNN) classifier, including the state preparation circuit, an l-layer variational circuit, and pre-measurement basis transformation gates. **c** Sample visualization of handwritten letters "Q" and "T" from the EMNIST dataset, used for the classical image classification task. **d** Quantum circuit used to generate the LCEI dataset, showing the application of an $R_x(\alpha)$ rotation following the linear cluster state. States are labeled as "excited" or "non-excited" based on the rotation angle α

2. **Robustness Lower Bound**: Compute the robustness lower bound $\varepsilon_{\mathrm{RLB}}(\rho)$ based on the measured probabilities using Theorem 2.
3. **Robustness Upper Bound**: Apply the Mask FGSM attack (as described in Sect. 10.3.3) on the quantum hardware to estimate the robustness upper bound $\varepsilon_{\mathrm{RUB}}(\rho)$ and identify the corresponding adversarial example σ_{adv}.
4. **Adversarial Training**: Retrain the quantum neural network classifier by incorporating the adversarial examples of all 10 quantum states into the training set, using their correct labels, in order to enhance the model's robustness.
5. **Adversarial Analysis**: Compare the robustness bounds before and after adversarial training to evaluate the tightness and stability of the certification framework.

As shown in Fig. 10.6, the experimental results validate two key aspects of the robustness verification framework:

1. **Tightness of Robustness Bounds**: Fig. 10.6a and c compare the experimentally measured robustness upper bounds $\varepsilon_{\mathrm{RUB}}$ (denoted as R_{UB}) against the certified robustness lower bounds $\varepsilon_{\mathrm{RLB}}$ (denoted as R_{LB}), computed from five randomly selected samples per class. In both EMNIST and LCEI tasks, the upper bounds consistently exceed the corresponding lower bounds, with average gaps of 2.96×10^{-3} and 3.25×10^{-3}, respectively. These results demonstrate the near-optimality of the Mask FGSM attack strategy and affirm the tightness of the certified lower bounds derived in Theorem 2.

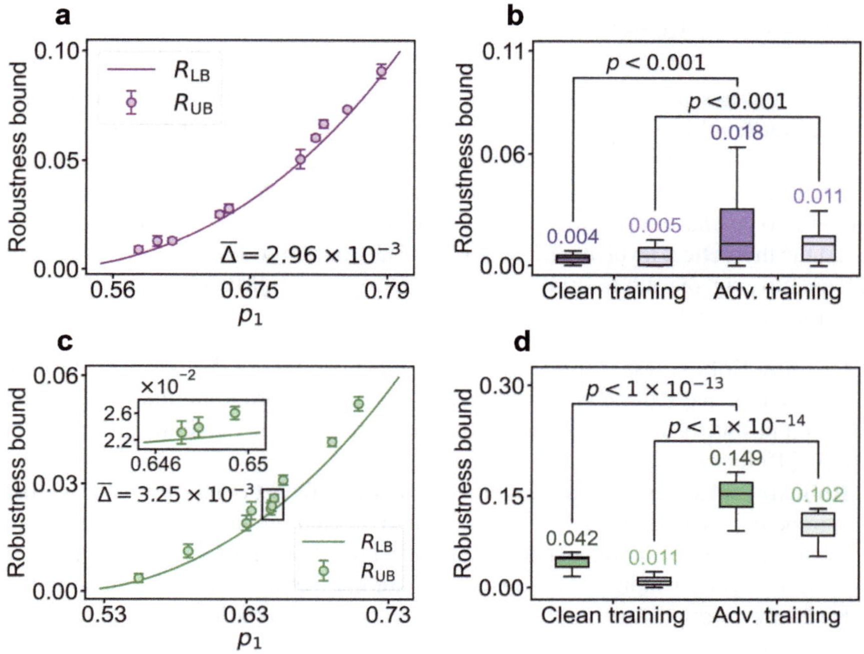

Fig. 10.6 Robustness bound verification experiments. **a**, **c**, Comparison of the experimental upper bound R_{UB} (from 10 randomly selected samples, 5 per class) versus theoretical R_{LB}. Error bars indicate the root mean square error from fitting $D(\rho, \sigma)$. $\overline{\Delta}$ denotes the average gap between R_{UB} and R_{LB} of the 10 samples. **b**, **d**, The robustness bounds for critical samples under clean and adversarial training. Adversarial training significantly increased the average robustness lower bound, with dark and light colors denoting distinct classes. p-value ≤ 0.001, indicating statistically significant differences. Panels **a** and **b** correspond to EMNIST dataset, while **c** and **d** to LCEI

2. **Improvement through Adversarial Training**: Critical samples, which are defined as the 20% of instances with the lowest robustness under clean training, were used to evaluate the effect of adversarial training. As depicted in Fig. 10.6b and d, adversarial training significantly increased the mean certified robustness lower bound for these critical samples, by a factor of 4.22 in EMNIST and 4.74 in LCEI. This improvement underscores the utility of adversarial training in enhancing the robustness of QMLs and mitigating misclassification under adversarial perturbations.

10.7 Summary and Outlook

This review has presented a comprehensive overview of our recently developed formal verification framework for certifying the adversarial robustness of quantum machine learning (QML) classifiers. The framework unifies rigorous theoretical foundations, algorithmic advances, toolchain development, and experimental validation into an end-to-end pipeline for trustworthy QML on quantum devices.

At the theoretical level, the framework introduces two certified robustness bounds for quantifying the adversarial robustness radius of a QML classifier at a given quantum input state:

- **Optimal Robustness Bound.** This bound exactly quantify the adversarial robustness radius and can be computable via semidefinite programming (SDP) when the input state is mixed, and via quadratically constrained quadratic programming (QCQP) for pure states.
- **Robustness Lower Bound.** This efficiently computable bound, derived from measurement statistics, certifies a region around a given input state where the classifier's output remains unchanged. It is especially valuable for identifying potentially non-robust inputs with low computational overhead and is well suited for hardware-level evaluation, as it does not require access to the internal evolution of the quantum classifier.

These certified bounds enable both exact and under-approximate verification. Furthermore, they are connected through the **robustness upper bound**, an empirical robustness estimate derived from adversarial attacks (e.g., the Mask FGSM method). This empirical bound serves as a benchmark for evaluating the tightness of the certified lower bound in practical scenarios.

To operationalize the theoretical results, the software tool VeriQR has been developed, serving as the first dedicated platform for robustness verification of QML models. VeriQR supports exact and approximate verification, simulates various noise models, enables adversarial training, and integrates with major quantum programming frameworks (e.g., Qiskit, Cirq). A graphical user interface (GUI) supports intuitive robustness diagnostics and visualization.

Notably, this framework has been experimentally validated on a 20-qubit superconducting quantum processor. These experiments constitute the first hardware-based benchmark of adversarial robustness in QML. The results show that:

- The certified robustness lower bounds closely match empirical upper bounds across classification tasks.
- Adversarial training substantially improves the robustness guarantees of quantum classifiers, particularly for inputs most vulnerable to perturbation.

With these results, our framework marks a promising step toward trustworthy QML on noisy intermediate-scale quantum (NISQ) hardware and opens promising directions for scalable and hardware-integrated robustness certification. In the following, we outline several key avenues for future research and development.

1. Hardware-integrated robustness verification. While the current implementation of VERIQR operates within classical simulation environments, a critical next step is to extend the framework for execution directly on quantum hardware. This would enable robustness certification under real-world conditions, accounting for hardware-specific noise characteristics, calibration drift, and temporal fluctuations. Such integration would bridge the practical gap between formal robustness guarantees and physical device behavior, enabling more faithful certification of QML models in practical deployment scenarios.

2. Certification of dynamically trained quantum classifiers. Most existing robustness verification approaches, including those presented in this review, assume a fixed and pre-trained quantum classifier. However, many real-world applications require models that learn from data in real time or adapt to adversarial perturbations. A promising direction is to develop verification methods that support dynamic or online QML training. In this setting, the quantum model evolves continuously, and robustness certification must be integrated into the training process to ensure that robustness guarantees are preserved or improved over time. This requires the design of adaptive certification techniques that can operate in tandem with quantum training loops, like their classical counterparts [44].

3. Beyond robustness: certifying fairness, interpretability, and privacy. While this framework focuses on adversarial robustness, many other aspects of trustworthiness are equally important in quantum AI systems. Future research may extend the verification methodology to properties such as fairness across quantum finance, interpretability of quantum decision-making, and differential privacy under quantum data access. Initial efforts in these directions have already begun [41, 45, 46], leveraging the robustness-based verification framework. A unified formal foundation for verifying such multifaceted trust properties would significantly strengthen the overall reliability of QML.

4. Cross-platform robustness benchmarking. The experimental results in this work are based on superconducting quantum hardware. Expanding the robustness benchmarking to other quantum platforms, such as trapped ions, neutral atoms, or photonic systems, would offer deeper insights into the relationship between hardware noise characteristics and model robustness. Establishing standard robustness benchmarks across different architectures will be essential for evaluating and comparing the trustworthiness of QML implementations on heterogeneous quantum systems.

These directions outline a broader agenda for building verifiable, trustworthy, and scalable quantum AI systems. As quantum computing advances toward practical and widespread use, formal verification tools like those presented in this framework may play an important role. By embedding trust guarantees into the lifecycle of quantum machine learning from model design and training to deployment and adaptation, future research can ensure that quantum AI not only performs well, but also behaves reliably and predictably in complex and uncertain environments. It is hoped that this review will foster broader engagement from the formal methods and quantum AI communities toward this critical goal.

References

1. Biamonte J, Wittek P, Pancotti N, Rebentrost P, Wiebe N, Lloyd S (2017) Quantum machine learning. Nature 549(7671):195–202
2. Cerezo M, Verdon G, Huang H-Y, Cincio L, Coles PJ (2022) Challenges and opportunities in quantum machine learning. Nat Comput Sci 2(9):567–576
3. Havlíček V, Córcoles AD, Temme K, Harrow AW, Kandala A, Chow JM, Gambetta JM (2019) Supervised learning with quantum-enhanced feature spaces. Nature 567:209–212
4. Herrmann J, Llima SM, Remm A, Zapletal P, McMahon NA, Scarato C, Swiadek F, Andersen CK, Hellings C, Krinner S, Lacroix N, Lazar S, Kerschbaum M, Zanuz DC, Norris GJ, Hartmann MJ, Wallraff A, Eichler C (2022) Realizing quantum convolutional neural networks on a superconducting quantum processor to recognize quantum phases. Nat Commun 13:12
5. Gong M, Huang HL, Wang S, Guo C, Li S, Wu Y, Zhu Q, Zhao Y, Guo S, Qian H, Ye Y, Zha C, Chen F, Ying C, Yu J, Fan D, Wu D, Su H, Deng H, Rong H, Zhang K, Cao S, Lin J, Xu Y, Sun L, Guo C, Li N, Liang F, Sakurai A, Nemoto K, Munro WJ, Huo YH, Lu CY, Peng CZ, Zhu X, Pan JW (2023) Quantum neuronal sensing of quantum many-body states on a 61-qubit programmable superconducting processor. Sci Bullet 68:906–912
6. Tacchino F, Macchiavello C, Gerace D, Bajoni D (2019) An artificial neuron implemented on an actual quantum processor. npj Quantum Inf 5:12
7. Huang HL, Du Y, Gong M, Zhao Y, Wu Y, Wang C, Li S, Liang F, Lin J, Xu Y, Yang R, Liu T, Hsieh MH, Deng H, Rong H, Peng CZ, Lu CY, Chen YA, Tao D, Zhu X, Pan JW (2021) Experimental quantum generative adversarial networks for image generation. Phys Rev Appl 16:8
8. Huang K, Wang Z-A, Song C, Xu K, Li H, Wang Z, Guo Q, Song Z, Liu Z-B, Zheng D, et al (2021) Quantum generative adversarial networks with multiple superconducting qubits. npj Quantum Inf 7(1):165
9. Zhang C, Lu Z, Zhao L, Xu S, Li W, Wang K, Chen J, Wu Y, Jin F, Zhu X, et al (2024) Quantum continual learning on a programmable superconducting processor. arXiv:2409.09729
10. Chen J, Wu Y, Yang Z, Xu S, Ye X, Li D, Wang K, Zhang C, Jin F, Zhu X et al (2025) Quantum ensemble learning with a programmable superconducting processor. arXiv:2503.11047
11. Broughton M, Verdon G, McCourt T, Martinez AJ, Yoo JH, Isakov SV, Massey P, Halavati R, Niu MY, Zlokapa A et al (2020) Tensorflow quantum: a software framework for quantum machine learning. arXiv:2003.02989
12. Lu S, Duan LM, Deng DL (2020) Quantum adversarial machine learning. Phys Rev Res 2:8
13. Liu N, Liu N, Liu N, Wittek P, Wittek P, Wittek P, Wittek P (2020) Vulnerability of quantum classification to adversarial perturbations. Phys Rev A 101:6
14. Franco N, Sakhnenko A, Stolpmann L, Thuerck D, Petsch F, Rüll A, Lorenz JM (2024) Predominant aspects on security for quantum machine learning: Literature review. In: 2024 IEEE international conference on quantum computing and engineering (QCE), vol 1. IEEE, pp 1467–1477
15. Huang L, Joseph AD, Nelson B, Rubinstein BIP, Tygar JD (2011) Adversarial machine learning. In: Proceedings of the 4th ACM workshop on security and artificial intelligence, pp 43–58
16. Goodfellow IJ, Shlens J, Szegedy C (2015) Explaining and harnessing adversarial examples. In: Bengio Y, LeCun Y (eds) 3rd international conference on learning representations, ICLR 2015, San Diego, CA, USA, 7–9 May 2015, conference track proceedings
17. Nielsen MA, Chuang IL (2010) Quantum computation and quantum information. Cambridge University Press
18. Guan J, Fang W, Ying M (2021) Robustness verification of quantum classifiers. In: International conference on computer aided verification. Springer, pp 151–174
19. Lin Y, Guan J, Fang W, Ying M, Su Z (2024) A robustness verification tool for quantum machine learning models. In: International symposium on formal methods. Springer, pp 403–421
20. Zhang H-F, Chen Z-Y, Wang P, Guo L-L, Wang T-L, Yang X-Y, Zhao R-Z, Zhao Z-A, Zhang S, Du L et al (2025) Experimental robustness benchmark of quantum neural network on a superconducting quantum processor. arXiv:2505.16714

21. Katz G, Barrett C, Dill DL, Julian K, Kochenderfer MJ (2017). Reluplex: an efficient smt solver for verifying deep neural networks. In: International conference on computer aided verification. Springer, pp 97–117
22. Lomuscio A, Maganti L (2017) An approach to reachability analysis for feed-forward relu neural networks. arXiv:1706.07351
23. Tjeng V, Xiao KY, Tedrake R (2019) Evaluating robustness of neural networks with mixed integer programming. In: 7th international conference on learning representations, ICLR 2019, New Orleans, LA, USA, 6–9 May 2019. OpenReview.net
24. Bastani O, Ioannou Y, Lampropoulos L, Vytiniotis D, Nori AV, Criminisi A (2016) Measuring neural net robustness with constraints. In: Lee DD, Sugiyama M, von Luxburg U, Guyon I, Garnett R (eds) Advances in neural information processing systems 29: annual conference on neural information processing systems 2016, 5–10 Dec 2016, Barcelona, Spain, pp 2613–2621
25. Tran H-D, Bak S, Xiang W, Johnson TT (2020) Verification of deep convolutional neural networks using imagestars. In: International conference on computer aided verification. Springer, pp 507–526
26. Ren K, Zheng T, Qin Z, Liu X (2020) Adversarial attacks and defenses in deep learning. Engineering 6(3):346–360
27. Goodfellow IJ, Shlens J, Szegedy C (2014) Explaining and harnessing adversarial examples. arXiv:1412.6572
28. Wierichs D, Izaac J, Wang C, Lin CY-Y (2022) General parameter-shift rules for quantum gradients. Quantum 6:677
29. Mitarai K, Negoro M, Kitagawa M, Fujii K (2018) Quantum circuit learning. Phys Rev A 98:9
30. Schuld M, Bergholm V, Gogolin C, Izaac J, Killoran N (2019) Evaluating analytic gradients on quantum hardware. Phys Rev A 99(3):032331
31. Zhang RY, Lavaei J (2018) Sparse semidefinite programs with near-linear time complexity. In: 2018 IEEE conference on decision and control (CDC). IEEE, pp 1624–1631
32. Elboher YY, Gottschlich J, Katz G (2020) An abstraction-based framework for neural network verification. In: International conference on computer aided verification. Springer, pp 43–65
33. Fremont DJ, Chiu J, Margineantu DD, Osipychev D, Seshia SA (2020) Formal analysis and redesign of a neural network-based aircraft taxiing system with verifai. In: International conference on computer aided verification. Springer, pp 122–134
34. Kwiatkowska MZ (2019) Safety verification for deep neural networks with provable guarantees (invited paper). In: Fokkink WJ, van Glabbeek R (eds) 30th international conference on concurrency theory, CONCUR 2019, 27–30 Aug 2019, Amsterdam, The Netherlands, vol 140. LIPIcs. Schloss Dagstuhl-Leibniz-Zentrum für Informatik, pp 1:1–1:5
35. Madry A, Makelov A, Schmidt L, Tsipras D, Vladu A (2018) Towards deep learning models resistant to adversarial attacks. In: 6th international conference on learning representations, ICLR 2018, Vancouver, BC, Canada, April 30–May 3, 2018, conference track proceedings. OpenReview.net
36. Blanchette J, Summerfield M (2006) C++ GUI programming with Qt 4. Prentice Hall Professional
37. Cross AW, Bishop LS, Smolin JA, Gambetta JM (2017) Open quantum assembly language. arXiv:1707.03429
38. Nielsen MA, Chuang IL (2001) Quantum computation and quantum information. Phys Today 54(2):60
39. Lin Y, Guan J, Fang W, Ying M, Su Z (2024) Veriqr: a robustness verification tool for quantum machine learning models. arXiv:2407.13533
40. Yuxuan D, Hsieh M-H, Liu T, Tao D, Liu N (2021) Quantum noise protects quantum classifiers against adversaries. Phys Rev Res 3(2):023153
41. Guan J, Fang W, Ying M (2022) Verifying fairness in quantum machine learning. In: Computer aided verification: 34th international conference, CAV 2022, Haifa, Israel, Aug 7–10, 2022, proceedings, part II. Springer, pp 408–429
42. Huang J-C, Tsai Y-L, Yang C-HH, Su C-F, Yu C-M, Chen P-Y, Kuo S-Y (2023) Certified robustness of quantum classifiers against adversarial examples through quantum noise. In:

ICASSP 2023-2023 IEEE international conference on acoustics, speech and signal processing (ICASSP). IEEE, pp 1–5
43. Cohen G, Afshar S, Tapson J, Van Schaik A (2017) Emnist: extending mnist to handwritten letters. In: 2017 international joint conference on neural networks (IJCNN). IEEE, pp 2921–2926
44. Huasong Meng M, Bai G, Teo SG, Hou Z, Xiao Y, Lin Y, Dong JS (2022) Adversarial robustness of deep neural networks: a survey from a formal verification perspective. IEEE Trans Dependable Secure Comput
45. Guan J, Fang W, Huang M, Ying M (2023) Detecting violations of differential privacy for quantum algorithms. In: Proceedings of the 2023 ACM SIGSAC conference on computer and communications security, pp 2277–2291
46. Guan J (2025) Optimal mechanisms for quantum local differential privacy. In: Proceedings of the 2025 ACM SIGSAC Conference on Computer and Communications Security, pp 3737–3749

Chapter 11
Adversarial Robustness in Distributed Quantum Machine Learning

Pouya Kananian and Hans-Arno Jacobsen

Abstract Studying adversarial robustness of quantum machine learning (QML) models is essential in order to understand their potential advantages over classical models and build trustworthy systems. Distributing QML models allows leveraging multiple quantum processors to overcome the limitations of individual devices and build scalable systems. However, this distribution can affect their adversarial robustness, potentially making them more vulnerable to new attacks. Key paradigms in distributed QML include federated learning, which, similar to classical models, involves training a shared model on local data and sending only the model updates, as well as circuit distribution methods inherent to quantum computing, such as circuit cutting and teleportation-based techniques. These quantum-specific methods enable the distributed execution of quantum circuits across multiple devices. This chapter reviews the differences between these distribution methods, summarizes existing approaches on the adversarial robustness of QML models when distributed using each paradigm, and discusses open questions in this area.

11.1 Introduction

Quantum machine learning (QML) is a rapidly developing area of research [3, 33, 51, 165, 185]. When evaluating quantum versus classical classifiers and studying the potential advantages of quantum models, adversarial robustness emerges as an important consideration [298]. As a result, adversarial robustness in quantum machine learning has recently attracted significant attention [11, 75, 77, 108, 151, 179, 183, 189, 295]. By distributing QML models across multiple quantum processors, we can overcome the limitations of individual devices and enable scalable quantum systems. However, this distribution introduces unique vulnerabilities, which adversaries can

P. Kananian (✉) · H.-A. Jacobsen
Department of Electrical and Computer Engineering, University of Toronto, Toronto, Canada
e-mail: pouya.kananian@mail.utoronto.ca

H.-A. Jacobsen
e-mail: jacobsen@eecg.toronto.edu

M. Usman (ed.), *Quantum Robustness in Artificial Intelligence*, Quantum Science and Technology, https://doi.org/10.1007/978-3-032-11153-1_11

exploit to launch more sophisticated and scalable attacks. Grasping and addressing these challenges is critical to ensuring the security and reliability of distributed quantum machine learning systems, as well as to understanding their potential benefits compared to classical models.

Federated learning [158, 158, 202] is one of the key paradigms in distributed machine learning. It allows multiple data owners to collaboratively train a shared model without sharing their local private data. Quantum computing can be incorporated into federated learning via quantum data [53, 304], quantum machine learning models [135, 161], or quantum communication [171, 317, 339]. A key benefit of adding quantum capabilities to federated learning is the ability to encode model parameters as quantum states, enabling secure and efficient communication through quantum channels [148, 171, 221, 247, 272, 275, 316, 339]. Furthermore, quantum models such as overparameterized variational classifiers may possess intrinsic robustness against adversarial attacks [109, 123, 161, 220]. However, one of the reasons classical federated learning has attracted significant attention is the widespread availability of hardware resources—such as IoT devices, smartphones, and edge servers—that can be leveraged for training machine learning models in a decentralized manner. In contrast, quantum computing lacks such flexibility due to the scarcity and high cost of quantum hardware. Therefore, one important paradigm in distributed quantum machine learning is how to distribute the execution of a single quantum model across multiple quantum processors, using either classical communication [209, 226] or quantum entanglement and teleportation protocols [24, 36, 37]. This research direction is especially important given the current limitations of NISQ-era [235] quantum devices, such as limited qubit counts, short coherence times, and noisy operations, which constrain the scalability of quantum machine learning algorithms on individual devices.

While the robustness of quantum federated learning systems to adversarial and privacy-leaking attacks has been extensively studied [59, 63, 161, 167, 171, 198, 220, 307, 339], adversarial robustness in quantum machine learning models when their circuits are distributed across multiple processors has only recently begun to receive attention [151]. This chapter provides an overview of quantum federated learning and circuit distribution methods in quantum computing, outlining existing work on the adversarial robustness of quantum models distributed under each paradigm. Although alternative distribution paradigms exist in quantum machine learning [58, 67, 142, 155, 222, 223, 263, 283, 302, 333], we exclude them from the scope of this study. These paradigms have received comparatively less attention in the context of adversarial robustness. Section 11.2 provides background on classical and quantum federated learning, as well as circuit distribution methods in quantum computing using classical or quantum communication, i.e., circuit cutting [209, 210, 226] and teleportation-based methods. In Sect. 11.3, we review adversarial attacks in quantum federated learning and potential defenses, while Sect. 11.4 focuses on the adversarial robustness of quantum models with partitioned circuits.

11.2 Background

11.2.1 Federated Learning

Federated learning (FL) [158, 202] is a machine learning approach that enables multiple data owners to collaboratively train a shared model while keeping their local data private. Moreover, the shared model should achieve accuracy close to what would have been obtained if all the data had been centrally aggregated and used to train a single model [321]. Instead of sending data to a central server, clients can train models locally and only communicate model parameters. Federated learning systems often employ diverse hardware and use datasets that are typically non-IID and imbalanced in terms of size, diversity, and quality [172].

Federated learning is typically classified into three primary types: vertical, horizontal, and a variant known as federated transfer learning [252, 297, 340]. In horizontal (or homogeneous) federated learning, clients share the same types of features in their data, though the actual data samples differ among them [117]. Clients keep their private data local, exchanging only global and local model parameters with the server during communication. The server trains the global model by collecting and aggregating model parameters or gradients from the clients. In contrast, vertical (or heterogeneous) federated learning refers to situations where clients share the same data samples but have different sets of features [184, 329]. Typically, one client is assumed to possess all the labeled data and is termed the guest or active client, whereas the remaining clients, which do not have labels, are called host or passive clients [4, 62]. In contrast to horizontal federated learning, which produces a unified global model, the vertical setup results in distinct local models for each party, requiring collaboration among clients to perform inference [186]. In federated transfer learning, both the features and samples vary between datasets, albeit with some overlap [186].

To build machine learning systems that are both scalable and capable of leveraging rich, diverse data sources, it is essential to move beyond centralized training and embrace distributed approaches like federated learning. In many real-world scenarios, data is naturally distributed across a wide range of devices or organizations, and collecting it centrally is impractical due to privacy concerns, bandwidth limitations, or regulatory constraints. However, like other distributed systems, federated learning introduces additional attack surfaces and potential vulnerabilities related to security, privacy, and fairness [296, 340]. Although private data is not shared in federated learning, the exchanged models and gradients can still expose sensitive information, and FL systems remain vulnerable to inference and poisoning attacks [6, 149, 236, 313, 336, 340]. As a result, there is a growing need to develop trustworthy federated learning systems that ensure robust protections while maintaining the benefits of decentralized learning [340]. Trustworthy federated learning systems should ensure privacy by safeguarding sensitive data from exposure, maintain robustness even under adversarial conditions, uphold fairness, and support explainability—both through transparent, interpretable system design and through external mechanisms that help elucidate the model's decisions [340]. Therefore, to build such systems, it is imperative to study adversarial robustness in federated learning, along with the closely related challenge of privacy preservation.

11.2.2 Quantum Federated Learning

Quantum computing can be integrated into federated learning through the use of quantum data [53, 304], quantum machine learning models [53, 135, 161, 304], or quantum communication [153, 171, 216, 317, 339]. Numerous frameworks for quantum federated learning have been proposed [53, 60, 114, 116, 135, 143, 192, 238, 243, 244, 272, 304, 334], as well as quantum-inspired approaches [271, 319]. For instance, Chen et al. [60] pioneered a quantum federated learning framework integrating hybrid quantum-classical networks, where classical neural networks extract features that are subsequently processed by quantum circuits. Xia et al. [304] propose a framework in which multiple quantum nodes train a quantum neural network [66, 77, 150, 230] using their local quantum data. Additionally, there exist frameworks that allow quantum model training across classical clients [270]. In the work proposed by Song et al. [270], shadow tomography [132, 133] is used by the server to generate a classical approximation of the quantum model, allowing clients to calculate local gradients on their own data. To learn more about developments in this area, readers can refer to several available surveys [54, 115, 200, 239, 246, 252].

One advantage of incorporating quantum capabilities into federated learning is the ability to encode classical data using a logarithmic number of qubits. This allows for inference and training of certain machine learning models via gradient descent with exponentially lower communication costs compared to frameworks relying on classical communication [105]. Moreover, encoding model parameters into quantum states facilitates secure and efficient communication over quantum channels [148, 171, 221, 247, 272, 275, 316, 339], employing methods such as quantum key distribution [22, 23, 204, 267], quantum secret sharing [126], and blind quantum computing [43, 234]. Leveraging quantum mechanical principles like the no-cloning theorem [19, 74, 208, 301], these approaches offer a secure alternative that lessens reliance on computationally intensive encryption [263, 339]. Furthermore, certain quantum machine learning models, such as overparameterized variational classifiers, may offer inherent resilience to adversarial attacks [109, 123, 161, 220]. The potential advantages of quantum federated learning frameworks in terms of privacy preservation and resilience to adversarial attacks are explored further in Sect. 11.3.3.

11.2.3 Circuit Cutting

Noisy Intermediate-Scale Quantum (NISQ) [235] devices represent the current generation of quantum computers, characterized by having tens to a few hundred qubits without full error correction capabilities. These devices mark a significant step toward practical quantum computing but are still limited by short coherence times, gate errors, and noise that degrade computational accuracy. Due to the limited qubit capacity of NISQ-era devices, a major challenge is that some quantum circuits exceed the size that current quantum processors can handle. To address this limitation, various methods [40, 85, 98, 209, 210, 226, 299, 332] have been proposed that leverage classical processing to enable execution on these constrained devices. A significant category of these approaches is circuit cutting [187]—also known as circuit knitting [232], circuit decomposition, or circuit fragmentation [285]— with most methods in this category relying on quasiprobability simulation, a core method also widely applied in quantum error mitigation [90, 233, 279] and classical simulation of quantum systems [131, 224, 258, 259]. Circuit cutting entails dividing quantum circuits into smaller sub-circuits. After executing these sub-circuits, their outcomes could be combined using classical post-processing to simulate quantum circuits that need more qubits than are available on a specific quantum processor. Circuit cutting techniques generally fall into two categories. In wire cutting [41, 120, 121, 173, 187, 225, 226, 284], the quantum identity channel is decomposed into a linear combination of measure-and-prepare channels, while gate cutting [121, 209, 210, 232, 255, 285, 286] involves breaking down a non-local channel using a sum of tensor products of local channels.

11.2.3.1 Quasiprobability Decomposition

Quasiprobability simulation [90, 209, 224, 232, 279] and circuit cutting [209, 210, 226] have mostly been explored in the context of circuits where the output is the expectation value of an observable O. The goal of these circuits is to estimate the expected value $\langle O \rangle = \mathrm{Tr}(O\mathcal{E}(\sigma))$, where $\mathcal{E}$ represents the quantum channel realized by the circuit, and σ is the input quantum state. Quasiprobability simulation involves replacing a quantum channel $\mathcal{V}$ with a linear combination of implementable channels $\{\mathcal{E}_i\}$, according to the following decomposition:

$$\mathcal{V}(\rho) = \sum_i c_i \mathcal{E}_i(\rho), \tag{11.1}$$

where $c_i \in \mathbb{R}$ and ρ is a quantum state. The term *quasiprobability* comes from the fact that the coefficients can be negative; therefore, they are not true probabilities. Rewriting the decomposition (11.1) as follows allows us to obtain the expectation value of the circuit using Monte Carlo sampling.

$$\mathcal{V}(\rho) = \sum_i p_i \mathcal{E}_i(\rho).\text{sign}(c_i)\left(\sum_i |c_i|\right),$$

where $p_i := |c_i|/(\sum_i |c_i|)$. Using Monte Carlo sampling, each shot of the circuit samples an index i randomly according to the probability distribution $\{p_i\}$, and replaces the channel $\mathcal{V}$ with the corresponding channel $\mathcal{E}_i$. The measurement outcome from this shot is then multiplied by the weight $\text{sign}(c_i)(\sum_i |c_i|)$, where c_i is the quasiprobability coefficient associated with $\mathcal{E}_i$. This process is repeated over many shots, and the final estimate of the observable's expectation value is obtained by averaging the weighted measurement outcomes. In this approach, measurement outcomes are reweighted by a factor proportional to $\kappa = \sum_i |c_i|$, which introduces a sampling overhead. Specifically, using Hoeffding's inequality, to estimate the expectation value of an observable to within an additive error ϵ with probability at least $1 - \delta$, the number of required circuit executions (shots) scales as

$$N = a.\left(\sum_i |c_i|\right)^2 \epsilon^{-2} \ln \frac{1}{2\delta} = O(\kappa^2),$$

for some constant a.

11.2.3.2 Gate Cutting

Similar to Sect. 11.2.3.1, for gate cutting and wire cutting, consider a circuit whose goal is to estimate the expectation value of an observable. Suppose the qubits are divided into two partitions, α and β. Consider a multi-qubit V that acts across the two partitions, and let $\mathcal{V}(.) = V(.)V^\dagger$ denote its corresponding unitary channel. The objective of gate cutting is to express this unitary channel as a decomposition:

$$\mathcal{V}(\rho) = \sum_i c_i \left(\mathcal{V}_i^\alpha(\rho^\alpha) \otimes \mathcal{V}_i^\beta(\rho^\beta)\right), \tag{11.2}$$

where $\mathcal{V}_i^\alpha$ and $\mathcal{V}_i^\beta$ are local channels acting on partitions α and β, respectively, and $\rho = \rho^\alpha \otimes \rho^\beta$, with ρ^α and ρ^β denoting the marginal states of ρ corresponding to these two partitions. Figure 11.1 illustrates the application of gate cutting to a simple quantum circuit. Let $\mathcal{U}$ and $\mathcal{W}$ denote the unitary channels corresponding to $U^\alpha \otimes U^\beta$ and $W^\alpha \otimes W^\beta$, respectively. The expectation value of the observable O is given by

$$\langle O \rangle = \text{Tr}(O\mathcal{W} \circ \mathcal{V} \circ \mathcal{U}(\sigma)),$$

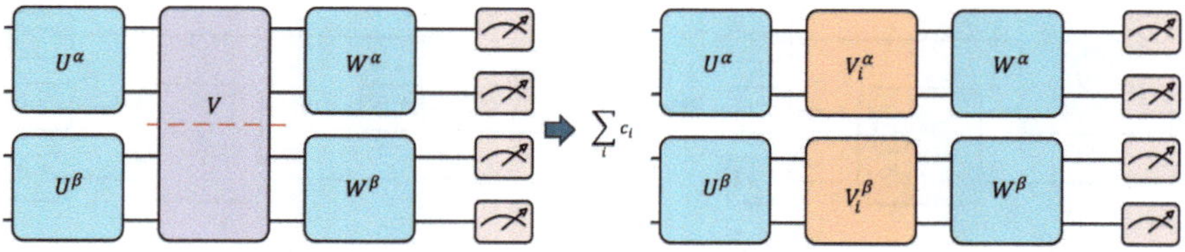

Fig. 11.1 By applying gate cutting to the quantum circuit on the left, we can quasiprobabilistically decompose gate V using Eq. (11.2), replacing it with a linear combination of local gates. To simulate the original circuit on the left, one can execute the corresponding sub-circuits on the right and reconstruct the result using classical post-processing

where $\mathcal{U}(.) = (U^\alpha \otimes U^\beta)(.)(U^\alpha \otimes U^\beta)^\dagger$, $\mathcal{W}(.) = (W^\alpha \otimes W^\beta)(.)(W^\alpha \otimes W^\beta)^\dagger$, and σ denotes the input quantum state. After applying gate cutting, the expectation value can be expressed as follows:

$$\begin{aligned}\langle O\rangle &= \sum_i c_i \mathrm{Tr}((O^\alpha \otimes O^\beta)\left((\mathcal{W}^\alpha \circ \mathcal{V}_i^\alpha \circ \mathcal{U}^\alpha)(\sigma^\alpha) \otimes (\mathcal{W}^\beta \circ \mathcal{V}_i^\beta \circ \mathcal{U}^\beta)(\sigma^\beta)\right)\\ &= \sum_i c_i \mathrm{Tr}\left(O^\alpha(\mathcal{W}^\alpha \circ \mathcal{V}_i^\alpha \circ \mathcal{U}^\alpha)(\sigma^\alpha)\right)\mathrm{Tr}\left(O^\beta(\mathcal{W}^\beta \circ \mathcal{V}_i^\beta \circ \mathcal{U}^\beta)(\sigma^\beta)\right)\\ &= \sum_i c_i \langle O^\alpha\rangle_i \langle O^\beta\rangle_i,\end{aligned}$$

where $O = O^\alpha \otimes O^\beta$, with O^α and O^β acting on partitions α and β, respectively. Here, $\sigma = \sigma^\alpha \otimes \sigma^\beta, \mathcal{U} = \mathcal{U}^\alpha \otimes \mathcal{U}^\beta$, and $\mathcal{W} = \mathcal{W}^\alpha \otimes \mathcal{W}^\beta$, where $\mathcal{U}^\alpha(.) = U^\alpha(.)U^{\alpha\dagger}(.)$, $\mathcal{U}^\beta(.) = U^\beta(.)U^{\beta\dagger}(.)$, $\mathcal{W}^\alpha(.) = W^\alpha(.)W^{\alpha\dagger}(.)$, and $\mathcal{W}^\beta(.) = W^\beta(.)W^{\beta\dagger}(.)$. We employ $\langle O^\alpha\rangle_i$ and $\langle O^\beta\rangle_i$ to denote the expectation values of O^α and O^β, respectively, when $\mathcal{V}$ is replaced by $\mathcal{V}_i^\alpha$ and $\mathcal{V}_i^\beta$.

As discussed in Sect. 11.2.3.1, Monte Carlo sampling can be used to estimate the expectation value. The sampling overhead associated with gate cutting scales as $O(\kappa^2)$, where $\kappa = \sum_i |c_i|$ [224, 233, 285]. If m gates are cut in the circuit, the sampling overhead scales exponentially with m. When these gates are cut separately by applying decomposition (11.2) to each one, the sampling overhead becomes $O(\kappa^{2m})$. However, more efficient methods for jointly cutting multiple gates have been proposed [121, 232, 255, 285, 286], which reduce the sampling overhead—though it still scales exponentially with m.

11.2.3.3 Wire Cutting

Wire cutting was originally introduced by Peng et al. [226]. The objective of wire cutting is to replace a wire—i.e., an identity channel—with a linear combination of

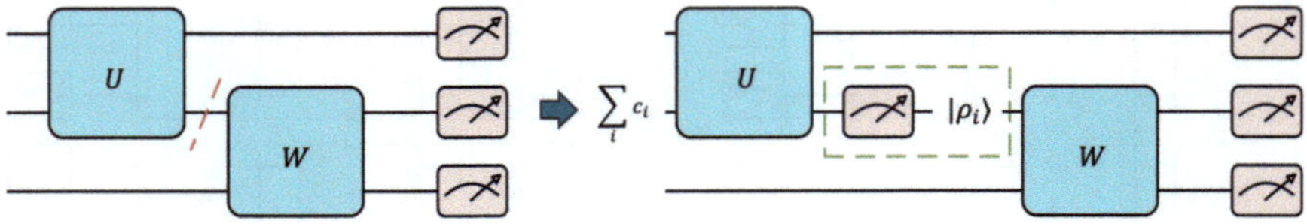

Fig. 11.2 This figure depicts a simple quantum circuit that has been partitioned using wire cutting by applying Eq. (11.3) to the wire connecting gates U and W. As with gate cutting, the original circuit can be simulated by executing the sub-circuits on the right and combining their outcomes via classical post-processing

measurement and state preparation operations. The single-qubit identity channel can be decomposed as follows:

$$\mathcal{I}(\rho) = \sum_i c_i \rho_i \mathrm{Tr}(O_i \rho), \tag{11.3}$$

where $c_i \in \mathbb{R}$, ρ and ρ_i are density matrices, and O_i denotes an observable corresponding to a measurement. After applying wire cutting to the simple circuit shown in Fig. 11.2, the expectation value $\langle O \rangle$ can be expressed as follows:

$$\begin{aligned}
\langle O \rangle &= \sum_i c_i \mathrm{Tr}\left((O_i \otimes O)(\mathcal{U}(\sigma^\alpha) \otimes \mathcal{W}(\rho_i \otimes \sigma^\beta))\right) \\
&= \sum_i c_i \mathrm{Tr}\left((O_i \otimes (O^\alpha \otimes O^\beta))(\mathcal{U}(\sigma^\alpha) \otimes \mathcal{W}(\rho_i \otimes \sigma^\beta))\right) \\
&= \sum_i c_i \mathrm{Tr}\Big((O_i \otimes O^\alpha)\mathcal{U}(\sigma^\alpha)\Big) \mathrm{Tr}\left(O^\beta \mathcal{W}(\rho_i \otimes \sigma^\beta))\right) \\
&= \sum_i c_i \langle O_i \otimes O^\alpha \rangle \langle O^\beta \rangle_i ,
\end{aligned}$$

where $\mathcal{U}(.) = U(.)U^\dagger$ and $\mathcal{W}(.) = W(.)W^\dagger$ represent unitary channels, and $\sigma = \sigma^\alpha \otimes \sigma^\beta$, with σ^α and σ^β denoting the marginal states of σ associated with the top and bottom sub-circuits, respectively. Similarly, the observable is factorized as $O = O^\alpha \otimes O^\beta$, where O^α and O^β correspond to these two sub-circuits. Here, $\langle O_i \otimes O^\alpha \rangle$ and $\langle O^\beta \rangle_i$ represent the expectation values of $O_i \otimes O^\alpha$ and O^β, when the wire is replaced by the measure-and-prepare channel corresponding to O_i and ρ_i.

The following is an example of a set of observables, quantum states, and real coefficients $\{O_i, \rho_i, c_i\}_{i=0}^8$ that satisfies Eq. (11.3) [226]:

$$\begin{aligned}
\{O_1 = I, \quad & \rho_1 = |0\rangle\langle 0|, & c_1 = +\frac{1}{2}, \\
O_2 = I, \quad & \rho_2 = |1\rangle\langle 1|, & c_2 = +\frac{1}{2}, \\
O_3 = X, \quad & \rho_3 = |+\rangle\langle +|, & c_3 = +\frac{1}{2}, \\
O_4 = X, \quad & \rho_4 = |-\rangle\langle -|, & c_4 = -\frac{1}{2}, \\
O_5 = Y, \quad & \rho_5 = |+i\rangle\langle +i|, & c_5 = +\frac{1}{2}, \\
O_6 = Y, \quad & \rho_6 = |-i\rangle\langle -i|, & c_6 = -\frac{1}{2}, \\
O_7 = Z, \quad & \rho_7 = |0\rangle\langle 0|, & c_7 = +\frac{1}{2}, \\
O_8 = Z, \quad & \rho_8 = |1\rangle\langle 1|, & c_8 = -\frac{1}{2}\},
\end{aligned} \tag{11.4}$$

where I, X, Y, and Z denote single-qubit Pauli matrices, with $|\pm\rangle$ and $|\pm i\rangle$ representing $(|0\rangle \pm |1\rangle)/\sqrt{2}$ and $(|0\rangle \pm i|1\rangle)/\sqrt{2}$, respectively. This set can be obtained because any 2×2 density operator ρ can be expanded using the normalized Pauli matrices, which form an orthonormal basis:

$$\rho = \sum_{\tilde{B} \in \{I,X,Y,Z\}/\sqrt{2}} \mathrm{Tr}(\tilde{B}\rho)\tilde{B} = \frac{1}{2} \sum_{B \in \{I,X,Y,Z\}} \mathrm{Tr}(B\rho)B. \tag{11.5}$$

Expanding each Pauli matrix in its eigenbasis in Eq. (11.5) reveals how the set (11.4) satisfies Eq. (11.3) [226]. Decomposition (11.3) can be generalized to account for cutting m parallel wires [120]:

$$\mathcal{I}^{\otimes m}(\rho) = \frac{1}{2^m} \sum_{P \in \{I,X,Y,Z\}^{\otimes m}} \mathrm{Tr}(P\rho)P, \tag{11.6}$$

where P represents an $m-$qubit Pauli string. Since each Pauli matrix in the Pauli strings can be expanded in its eigenbasis, each term $\mathrm{Tr}(P(.))P$ reflects a measurement-preparation process in which the expectation value of P is measured, followed by the preparation of its eigenstates as input to the following sub-circuit.

Similar to gate cutting, the sampling overhead associated with wire cutting scales as $O(\kappa^{2m})$ when m wires are independently cut using decomposition (11.3), where $\kappa = \sum_i |c_i|$. When the values from set (11.4) are substituted into decomposition (11.3), we obtain $\kappa = \sum_{i=1}^{8} 1/2 = 4$, resulting in the sampling overhead of $O(\kappa^{2m}) = O(16^m)$. Using decomposition (11.6) to cut m parallel wires results in a similar sampling overhead as decomposition (11.3). However, like gate cutting, more efficient methods have been proposed for the joint cutting of m

wires [41, 120, 121, 187, 225]. For example, Harada et al. [120] propose a decomposition that not only achieves optimal sampling overhead for cutting m parallel wires but also minimizes the number of quantum channels required for this task. More efficient approaches exist for cutting non-parallel wires; however, they require assistance from ancilla bits [41]. The decomposition proposed by Harada et al. [120] achieves a sampling overhead of $O((2^{m+1} - 1)^2)$ for cutting m parallel wires and $O(9^m)$ for arbitrarily located wires. In contrast, the decomposition introduced by Brenner et al. [41] achieves a sampling overhead of $O((2^{m+1} - 1)^2)$ for cutting non-parallel wires, which has been shown to be optimal when wire cutting can utilize arbitrary local operations and classical communication (LOCC) [41].

11.2.3.4 Circuit Cutting and Quantum Machine Learning

Circuit cutting has attracted growing interest in recent years [16, 20, 101, 215, 228, 229, 282], with research focusing on a wide range of areas. Notable examples include approximate circuit reconstruction [55–57, 178], intelligent qubit assignment across processors [38], finding optimal cut locations [268, 277, 278], distributed scheduling of circuit partitions [32, 260], and the intersection of circuit cutting and quantum error mitigation [154, 169, 182, 196]. Circuit cutting can be used to distribute the execution of a quantum machine learning circuit across multiple quantum processors [231, 260]. Moreover, it can be integrated with federated learning systems to distribute the training of local models across multiple participants, enhancing the suitability of such systems for noisy, resource-constrained quantum processors [253]. However, implementing circuit cutting in quantum machine learning presents notable challenges [112, 151, 199].

The exponential sampling overhead associated with circuit cutting becomes especially problematic when applied to strongly entangled ansätze (see Fig. 11.3), which are commonly used in quantum machine learning [151]. All qubits are interconnected in this ansätze; therefore, for an n-qubit circuit, for instance, in wire cutting, at least n wires need to be cut to obtain two separate sub-circuits. For these ansätze, we can turn to approximation techniques to reduce the cost of circuit reconstruction at the expense of some accuracy [199].

Ansätze based on tree tensor networks [264, 274] (see Fig. 11.4) are more compatible with integration with circuit cutting [112]. When cutting them, each tensor block could correspond to a sub-circuit. This allows them to be executed on a processor with fewer qubits, while the number of circuit evaluations needed to estimate the expectation value of the original circuit increases polynomially with the number of tensor blocks [112]. Quantum convolutional neural networks (QCNNs) [66] are among the variational quantum algorithms that utilize tensor-network-inspired and hierarchical architectures [118, 119, 141]. While such structures often avoid barren plateaus, recent research suggests that architectures which provably do not exhibit barren plateaus can result in loss landscapes that are classically simulable using polynomial-time algorithms [27, 52]. This implies that although a quantum computer might be necessary for initial data collection and producing shadows of

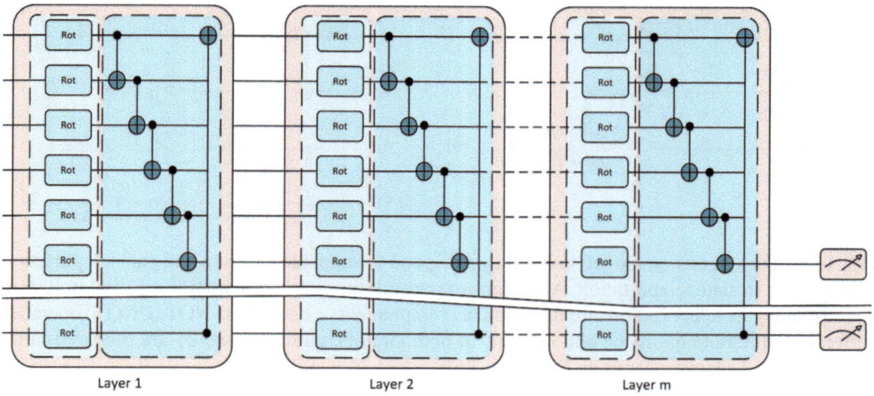

Fig. 11.3 A strongly entangled quantum circuit with m layers. In this circuit, each layer comprises a series of rotation gates followed by an entangling layer that fully entangles all the qubits

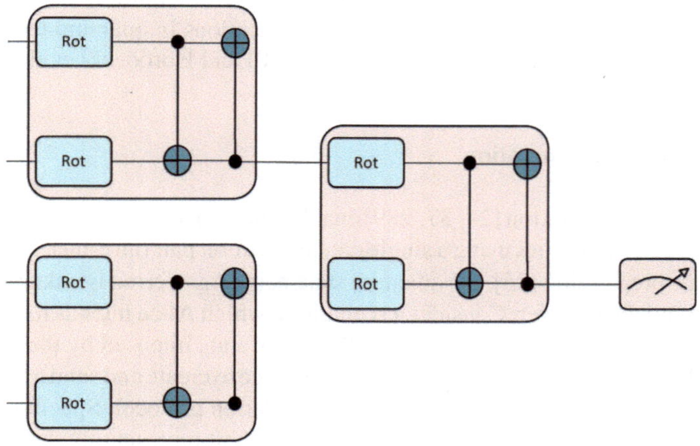

Fig. 11.4 A simple tree tensor quantum circuit

the input data, a hybrid classical-quantum optimization loop is not required, and the parameterized quantum circuit need not be implemented on a quantum processor. Further research is needed to determine whether non-classically simulable ansätze exist that are both practically useful for quantum machine learning and compatible with circuit cutting.

11.2.4 Teleportation-Based Methods

Notable variants of quantum teleportation include state teleportation, entanglement swapping, and gate teleportation, which enable one-way state transfer, bi-directional

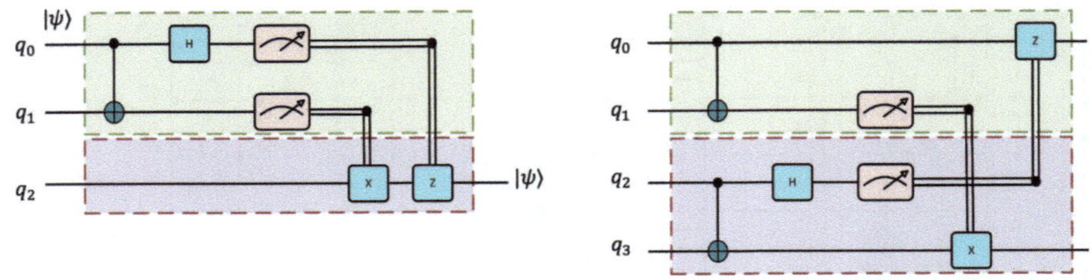

Fig. 11.5 The circuits on the left and right correspond to state teleportation and gate teleportation, respectively. In state teleportation, Alice aims to transmit the quantum state $\rho = |\psi\rangle\langle\psi|$ to Bob. In contrast, the gate teleportation circuit shown here implements a controlled-NOT (CNOT) operation between two remote qubits q_0 and q_3 [69]. In both circuits, qubits q_1 and q_2 are prepared in the Bell state $|\Phi^+\rangle$, also known as an EPR pair [86]

entanglement distribution, and remote gate application, respectively [20]. Here, we only review state and gate teleportation (Fig. 11.5). To learn more about other variants of quantum teleportation and their important applications in quantum technologies, readers can refer to the reviews by Barral et al. [20] and Horodecki et al. [129].

11.2.4.1 State Teleportation

Quantum state teleportation [24, 37, 289] enables the transfer of an arbitrary quantum state ρ between two parties using a single entangled qubit pair (an e-bit). In contrast to remote state preparation [25], the quantum state being transferred is unknown to both the sender and the receiver. Consider a scenario in which Alice intends to transmit an arbitrary quantum state ρ to Bob. Due to the constraints imposed by the no-cloning theorem [19, 74, 208, 301], it is not possible for Alice to create and send a duplicate of the state. Instead, she employs a quantum teleportation protocol. Specifically, Alice conducts a Bell-state measurement (BSM) on the qubit representing the state ρ and one half of an entangled pair that she shares with Bob. This measurement projects her two qubits randomly into one of the four maximally entangled Bell states [217]—$|\Phi^\pm\rangle$ or $|\Psi^\pm\rangle$—each with equal probability, where

$$|\Phi^\pm\rangle = \frac{1}{\sqrt{2}}(|00\rangle \pm |11\rangle), \, |\Psi^\pm\rangle = \frac{1}{\sqrt{2}}(|01\rangle \pm |10\rangle).$$

Bob's qubit, which is entangled with Alice's, collapses into the state $B^\dagger \rho B$, where $B \in \{I, Z, X, ZX\}$ corresponds to the outcome of the BSM. Alice then classically communicates the result of her measurement to Bob via two classical bits. After getting the message, Bob applies the correct operation to his qubit to recover the original state ρ.

11.2.4.2 Gate Teleportation

Gate teleportation [87, 138, 146, 207] allows the implementation of a two-qubit controlled unitary operation on unknown control and target states using a single entangled qubit pair, without physically transferring either qubit between two quantum processors. Similar to state teleportation, gate teleportation also requires classical communication between the two parties to enable the application of appropriate corrections.

Unlike quantum circuit cutting, quantum teleportation does not introduce an exponential sampling overhead. However, teleportation consumes entangled pairs of qubits shared between parties. Minimizing the e-bit cost as well as optimizing partitioning strategies for distributed quantum circuits has been an active area of research [18, 44, 70, 94, 95, 99, 125, 130, 218, 350].

11.3 Adversarial Robustness in Quantum Federated Learning

In this section, we begin by reviewing adversarial attacks that target classical federated learning architectures, encompassing both privacy breaches and attempts to degrade or disrupt system performance. This review, presented in Sect. 11.3.1, is essential, as quantum federated learning systems may also be susceptible to similar types of attacks. While Fig. 11.7 presents an overview of defense mechanisms applicable to classical federated learning systems, we do not investigate them in depth. Rather, our primary focus in Sects. 11.3.3 and 11.3.4 is on the defense strategies proposed for quantum federated learning (Fig. 11.6). Some of these defense methods are extensions of classical federated learning techniques adapted to the quantum realm, including differential privacy [59, 344], homomorphic encryption [63, 96, 249, 317], and secure multi-party computation [47, 167, 323] for privacy preservation, as well as adversarial training [198] to enhance system robustness. Other methods, however, are intrinsic to quantum computing itself, such as the inherent resilience of overparameterized variational quantum classifiers to adversarial attacks [123, 161, 220], and secure quantum protocols that rely on quantum communication, including quantum key distribution [22], quantum secret sharing [126], and blind quantum computing [43, 171, 234].

11.3.1 Overview of Adversarial Attacks in Classical Federated Learning

Adversarial attacks in federated learning can be broadly categorized into privacy-leakage attacks and integrity-oriented adversarial attacks, with threats potentially

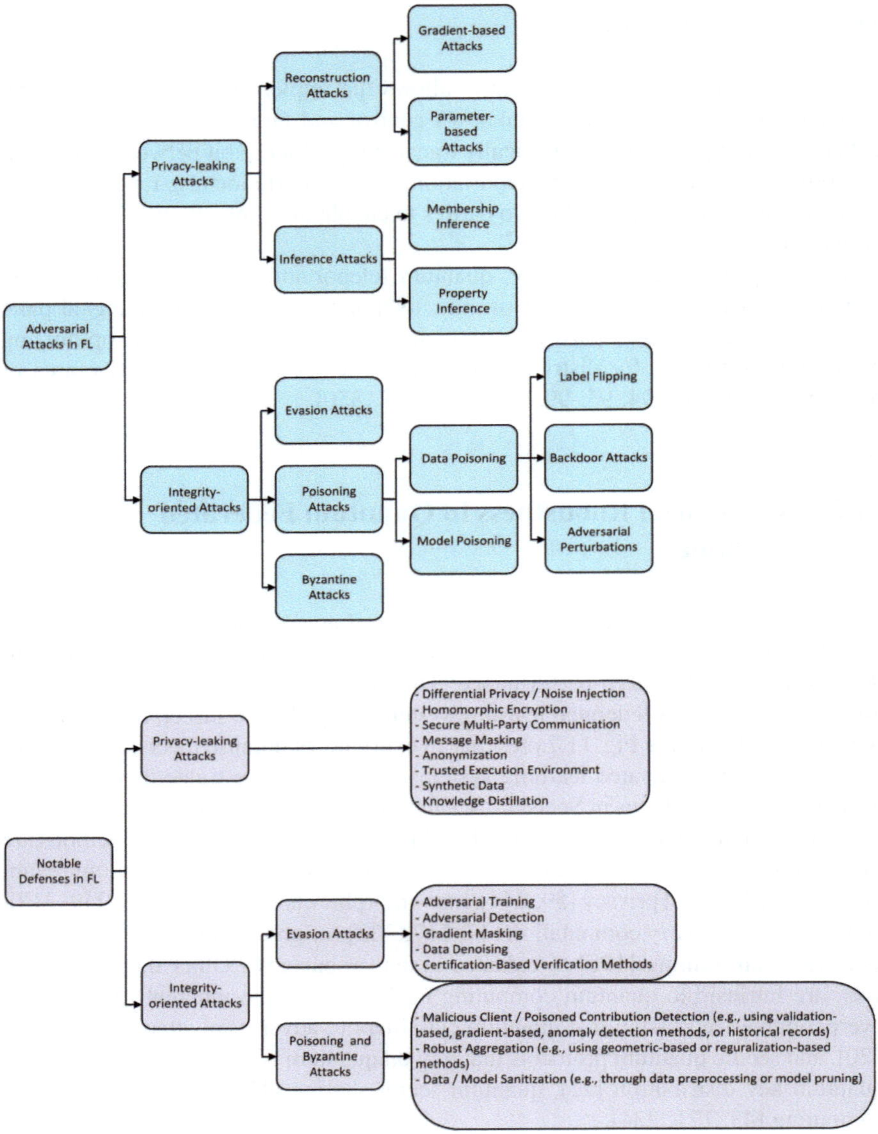

Fig. 11.6 An overview of the defense methods discussed in Sects. 11.3.3 and 11.3.4. There is relatively less research on the adversarial robustness of quantum federated learning systems against integrity-based attacks compared to privacy-leaking attacks, and the defense methods proposed to date are typically built upon classical methods

originating from both the server and client sides [296, 340, 346, 347]. Privacy-leakage attacks include reconstruction and inference attacks [340]. Reconstruction attacks aim to retrieve clients' datasets and include both gradient-based and parameter-based attacks [340]. Gradient-based attacks exploit shared gradients to extract original data samples [342, 349], while parameter-based attacks apply to scenarios where clients share model parameters with the server rather than gradients [128, 202, 293, 318, 331]. In contrast to reconstruction attacks, inference attacks—which include membership inference and property inference—aim to uncover properties of the data rather than reconstructing it [340]. Membership inference [97, 177, 188, 266] aims to identify whether a specific sample was included in the training dataset, while property inference [191, 206, 211, 291, 315, 337] focuses on extracting key attributes from data that clients have not explicitly disclosed.

Evasion, poisoning, and Byzantine attacks are among the primary categories of integrity-oriented adversarial attacks in federated learning [160, 296]. Evasion attacks involve generating adversarial examples by introducing small perturbations to input samples—typically preserving perceptual similarity for human observers—to deceive trained models into making incorrect predictions [273]. Poisoning attacks consist of data and model poisoning: the former involves tampering with training data to disrupt global model training, while the latter corrupts the model by altering the training process instead of the data itself [296]. Model poisoning could involve altering local model updates before sharing them with the server to manipulate the global model's outcomes [28, 92, 144]. Data poisoning encompasses both denial-of-service attacks that halt the learning process and more subtle strategies that target specific learning objectives while leaving most of the model's training unaffected [296], often implemented via label-flipping [281], backdoor insertion [17, 310], or adversarial perturbations [93, 157, 214, 262, 348]. Backdoor attacks [17, 111, 156, 175] subtly modify a small portion of the training data, usually by embedding a specific pattern known as a trigger into a chosen class. This allows attackers to manipulate the model's predictions later by including the trigger in test-time inputs. In a federated learning setup, backdoor attacks can be carried out by corrupting the contributions of multiple clients, allowing the attacker to insert hidden behaviors into the shared model [17, 205, 290, 310]. Poisoning attacks based on adversarial perturbations involve tweaking training samples through gradient-based methods [214]. A Byzantine failure [35, 163] in federated learning refers to a subset of nodes behaving arbitrarily or maliciously. If the server aggregates these corrupted updates, the federated learning process could be disrupted [340]. Data and model poisoning attacks are sometimes regarded as a type of Byzantine failure, while other Byzantine failures in federated learning could be a result of unreliable communication and noisy data samples or models [7, 92, 251, 281, 340]. Figure 11.7 provides an overview of the different categories of adversarial attacks and key defense techniques in federated learning. For a more detailed examination of adversarial attacks in classical federated learning and potential defense strategies, readers can refer to several comprehensive surveys in this field [6, 136, 160, 296, 313, 340].

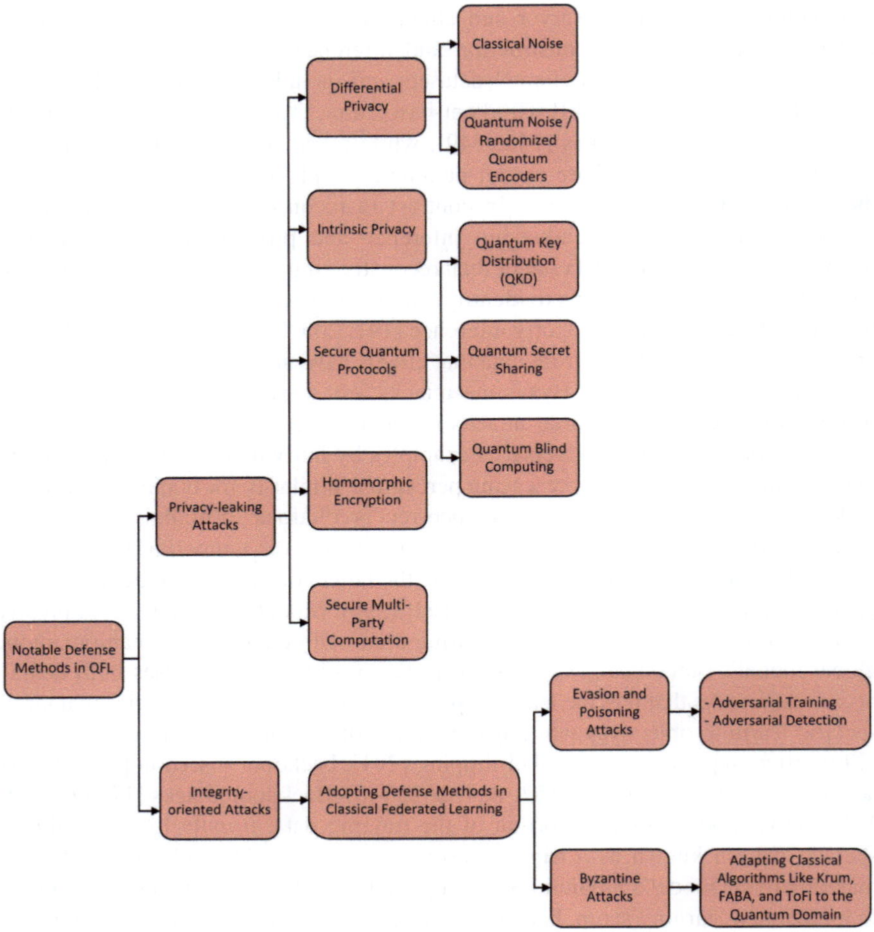

Fig. 11.7 This figure summarizes various categories of adversarial attacks in (classical) federated learning and notable defense methods. For further details on these attacks and the corresponding defense strategies, refer to the relevant surveys on the subject [160, 296, 340]

11.3.2 Adversarial Attacks in Quantum Federated Learning

Adversarial attacks in quantum federated learning can be categorized similar to those in classical federated learning, although the methods for implementing these attacks may differ. For instance, consider evasion attacks on quantum machine learning models in federated learning systems. These models may be trained on either native quantum data or classical data encoded into quantum states. In classical machine learning, adversarial examples are typically created by making small perturbations to the inputs of classifiers. In quantum machine learning, similar adversarial modifications can be introduced by either applying unitary perturbation operators to quantum input states or by perturbing classical inputs before they are encoded into quantum states [11, 108, 151, 189]. To ensure that the perturbations remain small, such unitary perturbation operators are often constrained to be close to the identity operator.

In the context of privacy-leaking attacks, one way they can be implemented in federated learning systems using quantum communication is through eavesdropping techniques, such as the intercept-resend or Trojan horse attacks [339]. In quantum communication, an intercept-resend attack occurs when an attacker intercepts the qubits in transit and then prepares and sends new qubits to the receiver [46, 71, 73, 181, 197, 320]. Trojan horse attacks [73, 106, 174, 190, 288] involve sending covert, unauthorized optical pulses into a legitimate quantum communication device. A portion of these Trojan signals becomes modulated with the legitimate information and is subsequently reflected back into the communication channel. By examining the modulated reflections, the attacker can eavesdrop on the communication [190]. For a more comprehensive overview of various attacks in quantum communication, readers may refer to the survey by Kumar et al. [159].

11.3.3 Resilience to Privacy-Leakage Attacks

11.3.3.1 Differential Privacy

Differential privacy (DP) [83] is a mathematical framework for sharing aggregate statistics about a dataset while limiting the amount of information leaked about specific individuals. Informally, an algorithm satisfies differential privacy if, upon observing its output, one cannot determine whether any individual's data was included in the computation. Consequently, the behavior of a differentially private algorithm remains nearly unchanged when a single individual is added to or removed from the dataset. This guarantee applies to every individual, providing a formal guarantee that limits information leakage. The core idea of differential privacy is to add carefully calibrated noise to statistical outputs in a way that preserves the privacy of individuals. Due to its relatively straightforward implementation and competitive computational and communication costs, differential privacy is one of the leading privacy-preservation mechanisms in machine learning [219, 296], especially in settings like federated learning. It also has applications in other areas, such as distributed optimization [139, 140] and mechanism design in game theory [91, 152, 203].

Definition 1 (Differential Privacy [83]) A randomized mechanism $\mathcal{M} : \mathcal{D} \rightarrow \mathcal{R}$ satisfies $(\epsilon, \delta)-$differentially privacy if for any pair of adjacent input sets $X, X' \subseteq \mathcal{D}$ and every subset of outputs $O \in \mathcal{R}$, we have

$$\mathbb{P}[\mathcal{M}(X) \in O] \leq e^{\epsilon}\mathbb{P}[\mathcal{M}(X') \in O] + \delta.$$

Here, $\mathcal{D}$ and $\mathcal{R}$ denote the domain and range of the mechanism, respectively, while ϵ and δ control the intended privacy guarantees. Generally, smaller values yield stronger privacy guarantees, though at the cost of introducing more noise, which can negatively impact output quality.

Differential privacy has been extended into the quantum realm [344], with various definitions of quantum differential privacy (QDP) emerging based on the choice of distance metrics used to define neighboring quantum states [1, 10, 109, 127, 344]. Here, we present the definition based on the trace distance [344]. For a discussion of alternative definitions of quantum differential privacy and the distance metrics they employ, see the survey by Zhao et al. [343].

Definition 2 (Quantum Differential Privacy [344]) A quantum operation $\mathcal{E}$ satisfies (ϵ, δ)−differentially privacy if for any pair of adjacent inputs ρ, σ such that $\tau(\rho, \sigma) \leq d$, and for every POVM $M = \{M_m\}$ and all $O \subseteq \Omega(M)$, we have

$$\mathbb{P}[\mathcal{E}(\rho) \in_M O] \leq e^{\epsilon}\mathbb{P}[\mathcal{E}(\sigma) \in_M O] + \delta,$$

where $\mathcal{E}$ and τ denote a completely positive and trace-preserving (CPTP) map and the trace distance, respectively. Furthermore, ρ and σ are density operators, $d \in (0, 1]$, POVM stands for Positive Operator-Valued Measure, and $\Omega(M) = \{m\}$ represents the set of all possible outcomes of M.

Differential privacy has been shown to provide certified robustness against adversarial attacks in classical classifiers [64]. This connection between differential privacy and certified robustness has also been explored in the context of quantum machine learning [77, 109, 134, 295, 300, 303]. The depolarization noise present in NISQ-era quantum classifiers can make them inherently quantum differentially private and naturally resilient to adversaries [77]. Differential privacy can be achieved in quantum algorithms through internal or external randomization mechanisms. These mechanisms may influence the state preparation phase, quantum circuits, or the measurement process [343].

Encoding classical information into quantum states can inherently yield (ϵ, δ)-differential privacy (Definition 1) [9]. Furthermore, in quantum algorithms that use classical data encoded as quantum states, noise can be added to the classical data prior to encoding [261]. Since the classical input to the algorithm satisfies ϵ-differential privacy in this scenario, the quantum algorithm also satisfies ϵ-differential privacy (Definition 1), by the post-processing property of differential privacy [84, 344]. Quantum mechanisms for achieving differential privacy for classical inputs have also been studied by Yoshida and Hayashi under the framework of classical-quantum differential privacy [327]. They show that the performance of classical differential privacy mechanisms can match that of classical-quantum differential privacy mechanisms when the cardinality of the private data domain is 2. However, Yoshida [326] shows that this equivalence no longer holds once the domain size reaches 3 or more. To achieve quantum differential privacy (Definition 2), randomized encoding can be applied to quantum states [78, 109, 134]. QDP and certified robustness against adversarial attacks can be attained by introducing quantum noise through random rotation gates applied to the input states of quantum classifiers [134], or by randomly encoding the inputs using unitary transformations or quantum error correction encoders [109]. In variational quantum classifiers [50], where classical optimizers are used to tune the parameters of a parameterized quantum circuit, one approach to achieving differential privacy is to introduce noise into the classical optimization process [250, 294]. Conversely, shot noise inherent in quantum measurements can naturally induce QDP in quantum algorithms [176]. Differentially private quantum algorithms can also be realized through the deliberate manipulation of quantum measurements [8].

Numerous studies have explored the role of differential privacy in quantum federated learning [29, 59, 171, 213, 250, 287, 317]. For instance, Rofougaran et al. [250] propose an approach that incorporates differential privacy into the training of each local client by adding noise to the classical optimization process used to train the parameters of variational quantum classifiers. Bhatia et al. [29] also achieve local differential privacy by clipping random samples from clients' data and adding noise to the clipped gradients. In contrast, Chen et al. [59] leverage quantum noise to mitigate privacy leakage and enhance robustness against adversarial attacks in quantum federated learning.

11.3.3.2 Intrinsic Privacy

In federated learning, sharing gradients with a honest-but-curious server could potentially compromise clients' private information [137, 161, 342]. An honest-but-curious server carries out gradient aggregation but may also attempt a local gradient inversion attack to infer information about client data. Kumar et al. [161] show that in a federated learning environment where variational quantum circuits [21, 50] are used in place of classical neural networks, highly expressive and overparameterized circuits provide a form of intrinsic protection from gradient inversion attacks. Here, high expressiveness refers to the presence of a large number of distinct, non-degenerate Fourier frequencies when the circuit's output is represented in the Fourier domain [104, 124, 164, 257, 265]. Specifically, Kumar et al. [161] explore overparameterized, hardware efficient ansätze [212], where the number of trainable parameters grows exponentially with the qubit count, alongside expressive encoding schemes that induce an exponential growth in the number of frequencies per input dimension when expressing the output and cost function gradients. They demonstrate that performing gradient inversion to recover the data leads to solving systems of high-degree multivariate Chebyshev polynomial equations, where the polynomial degree grows exponentially with the number of qubits. The number of such equations depends on the amount of shared gradient information, which is, in turn, tied to the number of trainable parameters. Furthermore, they show that both the time and memory required to solve these equations—whether exactly or approximately—increase exponentially with the number of qubits. Another approach to recovering client data involves machine learning-based methods that generate dummy gradients and train an attack model to replicate the client's gradients [88, 137, 342]. These attacks typically minimize the absolute distance between the dummy and real gradients by optimizing a dummy input vector designed to approximate the client's original input [89, 100, 349]—though various adaptations of this approach exist [100, 325]. As shown by Kumar et al. [161], these methods also fail when the attack model is underparameterized and highly expressive. This is primarily due to the attack model's loss landscape, which contains an exponential number of isolated local minima, making the model effectively untrainable.

One drawback of the work of Kumar et al. [161] is that overparameterization of the ansatz may result in barren plateau problems [201] and hinder trainability [123]. The relationship between barren plateaus and adversarial robustness in variational quantum classifiers has also been explored by Gong et al. [109], who show that adding randomized encoders to quantum circuits can lead to barren plateaus, obscuring gradient information from potential adversaries and impeding gradient-based adversarial algorithms in generating adversarial perturbations. A barren plateau refers to a region in the loss landscape where the loss values become exponentially concentrated as the problem size increases, with the loss gradients vanishing with high probability for randomly selected parameter values [165, 201]. Along with poor local minima [12, 34] and limited expressivity [280], barren plateaus represent one of the three major obstacles to the trainability of variational quantum algorithms, and a significant body of work has been dedicated to studying them [15, 27, 49, 52, 165, 237, 292]. In a recent study, Heredge et al. [123] theoretically investigate the trade-off between privacy protection and trainability in variational quantum classifiers, establishing a connection between privacy vulnerabilities in these models and the dimension of the Lie algebra of the generators of their circuits.

Despite the potential resilience of overparameterized ansätze against gradient inversion attacks [161], which may, however, lead to barren plateau problems and hinder trainability [109, 123], recent work by Papadopoulos et al. [220] introduces an inversion attack capable of recovering private training data from variational quantum classifiers in a federated learning setting. This approach integrates adaptive low-pass filters into the Finite Difference Method (FDM) for numerically computing gradients, helping the optimization process find the global minimum when minimizing the absolute distance between dummy and real gradients, despite the presence of many local minima. This is achieved by tuning the filter's window to suppress frequencies associated with local minima.

11.3.3.3 Secure Quantum Protocols

Integrating quantum capabilities into federated learning enables secure and efficient communication via quantum channels, leveraging methods like quantum key distribution [23, 204] and quantum secret sharing [126] to reduce reliance on resource-intensive encryption [263, 339]. Quantum Key Distribution (QKD) [22, 23, 204, 267] is a cryptographic method that leverages the principles of quantum mechanics—such as the no-cloning theorem [19, 74, 208, 301]— to enable two parties to generate and share a secret key with information-theoretic security. It ensures that any attempt at eavesdropping introduces detectable disturbances in the quantum states, allowing the communicating parties to identify potential security breaches. Numerous studies [148, 221, 247, 275, 316] have explored the application of quantum communication and QKD in federated learning, including those that focus on the allocation of quantum communication resources (such as key generation rates and QKD links) and the routing of data or key material across the network [148, 247, 316]. Building upon the quantum secret sharing protocol [126], Zhang et al. [339] present a secure aggregation framework based on GHZ states [110] that is applicable to both classical and federated learning with conventional models such as neural networks and quantum federated learning employing variational quantum circuits. This framework ensures security against both external eavesdroppers seeking to infer private information and internal semi-honest participants—those who follow the protocol correctly but attempt to covertly extract sensitive data. Assuming malicious participants do not collude, the physical properties of quantum communication enable the detection of both external eavesdropping and internal attacks [23, 267].

An additional method that can be incorporated into federated learning via quantum communication is blind quantum computing (BQC) [2, 234]. This technique allows clients to offload quantum computations to an untrusted server while keeping their data and algorithms confidential [54, 242]. Li et al. [171] propose a method for federated and private distributed learning based on the universal blind quantum computation protocol (UBQC) introduced by Broadbent et al. [43]. This blind quantum computation protocol offers unconditional security without relying on computational assumptions, enabling a client to offload a quantum computation to a server without disclosing any information about the computation, including its inputs or outputs. The client requires neither quantum memory nor significant computational power—only the ability to prepare qubits randomly selected from a finite set. The server performs the computation by receiving these qubits and following measurement instructions sent by the client via classical communication. In the protocol proposed by Li et al. [171], clients employ the UBQC protocol to outsource their gradient computations to the server. To ensure differential privacy and mitigating the risk of gradient inversion attacks by potential eavesdroppers during model training, noise is added to the computed gradients before uploading them. This work achieves differential privacy through classical, not quantum, noise.

11.3.3.4 Homomorphic Encryption

Homomorphic Encryption (HE) [96, 102, 249] is a privacy-preserving technique that allows computations to be performed directly on ciphertexts without decrypting it. HE produces an encrypted result that, when decrypted, matches the result of operations performed on the plaintext. The applications of HE in classical federated learning have been extensively studied, especially in the context of reducing the encryption and communication overhead it introduces [13, 147, 241, 314, 335]. More recently, HE has also been explored in the context of quantum federated learning [63, 81, 82, 170, 317]. For instance, Xu et al. [317] apply local differential privacy and homomorphic encryption in a federated learning framework for autonomous vehicular networks that incorporates quantum communication. In their simulations, the system employs classical convolutional neural networks as the underlying machine learning model. Quantum Homomorphic Encryption (QHE) [42, 79, 195] is the quantum counterpart to classical homomorphic encryption, allowing quantum circuits to be evaluated on encrypted quantum data. Chu et al. [63] integrate QHE into a federated learning framework that leverages quantum communication for training quantum neural networks (QNNs) [3, 66, 80, 145, 227, 256]. In their approach, gradients are encoded into quantum states and homomorphically aggregated using quantum adders [162, 254].

On the other hand, some encryption methods are vulnerable to attacks by quantum computers, as a malicious attacker equipped with quantum capabilities could potentially access plaintext data. Consequently, there has been increasing interest in developing secure federated learning systems that are resilient to quantum threats, including the advancement of homomorphic encryption techniques designed to safeguard against these attacks [14, 65, 103, 113, 240, 322, 328, 338, 341, 351].

11.3.3.5 Secure Multi-Party Computation

Secure Multi-Party Computation (SMPC) [47, 68, 107, 122, 323] allows multiple participants to collaboratively compute a function over their private data without revealing that data to each other or to a central server. No party learns anything beyond their own input and the final result they are meant to receive. SMPC can be applied in federated learning to protect interactions between parties without disclosing private data [245]. For example, it can be used to securely aggregate gradients during model training [330].

A key benefit of integrating quantum capabilities into federated learning is the ability to efficiently encode classical data using a logarithmic number of qubits. By encoding model contributions into quantum states, Li et al. [167] propose two frameworks for secure model aggregation in federated learning systems that leverage quantum communication: one utilizing private inner product estimation and the other implementing incremental learning. Incremental learning refers to the process where a model learns continuously, incorporating new data or updates over time without retraining from scratch. In the first protocol, secure model aggregation is reformulated as a correlation estimation task, allowing the blind quantum bipartite correlator (BQBC) algorithm [168] to be adapted for use in a multi-party setting. Built on quantum counting [39, 276], the BQBC algorithm facilitates blind quantum machine learning [345] by enabling inner product estimation, a fundamental operation in many widely used machine learning methods. In the second protocol, clients engage in multi-party computation without server participation until the final stage, where the server retrieves the aggregated gradient information. By using quantum communication, these privacy-preserving mechanisms achieve reduced communication costs compared to classical approaches.

11.3.4 Robustness to Integrity-Oriented Attacks

11.3.4.1 Byzantine Attacks

Byzantine faults [48, 76, 163] refer to failures in a distributed system where components may act arbitrarily or maliciously, including lying or sending conflicting information to different parts of the system. Several Byzantine-tolerant algorithms have been proposed for applications in classical distributed machine learning [5, 35, 61, 72, 305, 306, 309, 311, 312, 324]. Xia et al. [307] analyze the differences in Byzantine problems between classical distributed learning and quantum federated learning, and adapt several Byzantine-resilient algorithms [35, 305, 308] designed for classical distributed learning to fit the quantum federated learning framework proposed in their earlier work [304]. These Byzantine-robust algorithms include Krum [35], FABA [305], and ToFi [308]. Krum and FABA are geometric-based approaches [340] that assume malicious updates lie far from benign ones in terms of distance. ToFi operates based on a reference dataset. It evaluates the loss on this dataset for each weight update uploaded by participants. Updates that lead to high losses are assumed to originate from Byzantine nodes and are removed.

11.3.4.2 Evasion and Poisoning Attacks

Similar to their classical counterparts, quantum federated learning systems are vulnerable to evasion attacks, such as adversarial input perturbations [198], as well as poisoning attacks, such as label-flipping [30, 31]. Bhatia et al. [30, 31] investigate the robustness of a quantum federated learning system, using variational quantum classifiers for lithography hotspot detection, against label-flipping attacks, which may result from either malicious intent or human error. Hotspots refer to specific areas on the wafer in semiconductor manufacturing that signal potential irregularities or defects during production, often arising from challenges associated with the printability of particular layout patterns [180]. To counter label-flipping attacks, Bhatia et al. [30] propose a defense strategy that detects such attacks by analyzing pairwise Euclidean distances between clients, where large distances indicate that a specific client's update is adversarial and detrimental to training the global model. In a related work, Lee et al. [166] propose an auction-based approach to filter out untrustworthy clients in federated learning systems utilizing quantum neural networks. However, their work primarily focuses on mitigating the non-IID data distribution issues in federated learning, rather than addressing poisoning or Byzantine attacks. Client selection and outlier exclusion in quantum federated learning have also been explored by Son and Park [269], who propose an approach based on assessing class imbalance in local models using entropy, as well as measuring quantum state dissimilarity between the global target model and individual local models. Ma et al. [193] propose a decentralized framework for quantum kernel learning that employs a clipping-based aggregation mechanism to mitigate the impact of corrupted updates from faulty or adversarial nodes. In this approach, client data is clipped prior to aggregation.

To enhance resilience against adversarial input perturbations, Maouaki et al. [198] present a framework where the clients' variational quantum classifiers are adversarially trained within a quantum federated learning system. For adversarial training, they leverage adversarial examples generated using PGD-based methods [194]. Their results demonstrate that adversarial training, even when applied to only a subset of clients, enhances robustness against adversarial attacks. In fact, varying the number of adversarially trained clients, as well as the strength of the perturbation, reveals a trade-off between accuracy on clean data and resilience to adversarial attacks.

11.4 Circuit Cutting and Adversarial Robustness

The adversarial robustness of quantum classifiers subjected to circuit cutting has recently been studied [151]. When quantum communication is unavailable, circuit cutting can be used to distribute the execution of a quantum circuit across multiple quantum processors. However, as with other distribution methods, partitioning these circuits may increase their susceptibility to adversarial attacks. The sub-circuits produced by circuit cutting can be executed in a distributed manner across multiple devices. If an adversary gains access to any of these sub-circuits, they could attempt to infer private information or launch various attacks to disrupt the system's functionality. When the outputs of these sub-circuits are combined to reconstruct the original circuit's outcome, any manipulation of one or more sub-circuits by an adversary can cause the reconstructed result to differ from the intended one.

One possible method of attacking the sub-circuits is through evasion attacks and adversarial perturbations. Adversarial perturbations typically refer to slight modifications made to the input data of classifiers to trick the models into producing incorrect predictions [273]. In quantum classifiers, adversarial perturbations can be introduced either by applying perturbation gates to the quantum input states or by altering the classical inputs prior to their encoding into quantum states (see Sect. 11.3.2). When a circuit is partitioned by wire cutting, the input states of the resulting sub-circuits are either inherited from the original circuit's inputs or prepared specifically as part of the wire cutting process. As shown in Fig. 11.8, when an adversary adds adversarial perturbation gates to the input states generated through wire cutting, this modification leads to the implementation of an adversarial gate within intermediate layers of the reconstructed quantum circuit. In a recent work [151], Kananian and Jacobsen theoretically and experimentally study the implications of such an attack.

The attacks illustrated in Fig. 11.8 for wire cutting could be extended to a scenario where the quantum circuit is partitioned through state teleportation instead of wire cutting. An adversary with physical access to a sub-circuit at the receiving end of the quantum teleportation protocol can add perturbations to its received input state, resulting in the introduction of an adversarial gate within the original circuit's layers. Alternatively, a malicious node executing a sender sub-circuit can adversarially perturb a state prior to teleporting it to another sub-circuit. Beyond evasion

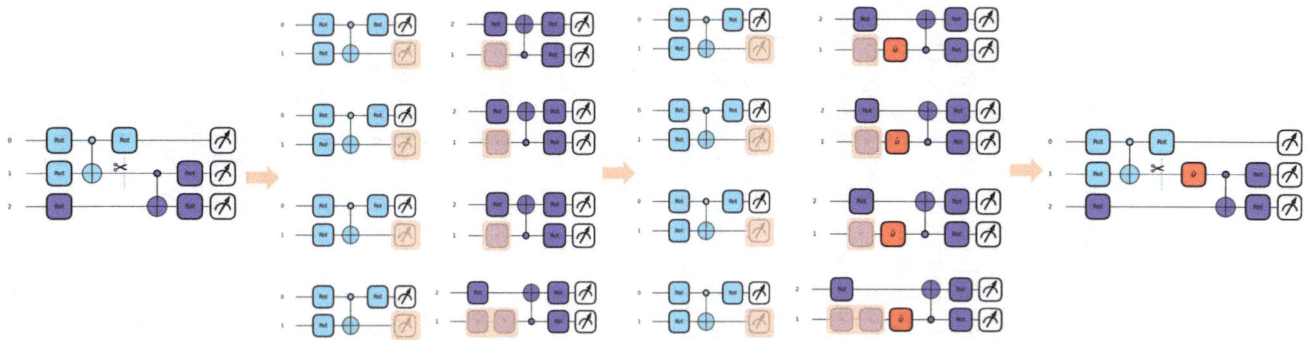

Fig. 11.8 This figure, drawn with the aid of Pennylane [26], illustrates a simple quantum circuit undergoing wire cutting, resulting in multiple sub-circuits. The sub-circuits shown in purple are then attacked by applying an adversarial perturbation gate $\hat{U}$ to their input states. After combining the results of these sub-circuits to reconstruct the original circuit, this adversarial attack results in implementing an adversarial gate $\hat{U}$ within intermediate layers of the reconstructed circuit. Wire cutting involves replacing identity channels (i.e., wires) with a linear combination of measurement and state preparation channels. These measurement and state preparation operators are highlighted in the sub-circuits using cream-colored boxes

attacks, future research should investigate other potential attack scenarios targeting partitioned quantum classifiers and possible methods for defending against them.

11.5 Conclusions

This chapter has reviewed the current literature on adversarial robustness in quantum federated learning and partitioned quantum classifiers. Both paradigms are important in distributed quantum machine learning. Studying the adversarial robustness of quantum models when deployed using these paradigms provides valuable insights into the potential advantages of QML models over classical ones, and contributes to the design of systems that are resilient and trustworthy.

There remains significant opportunity to explore a broader range of attack and defense scenarios in quantum federated learning, particularly regarding integrity-oriented attacks, which have been less studied compared to privacy-leaking attacks. While applying circuit distribution methods to quantum machine learning is increasingly important, especially in light of current limitations in quantum hardware, the adversarial robustness of partitioned quantum models has only recently begun to receive attention in the literature [151]. It is therefore essential to conduct a deeper investigation into the robustness of these systems and to develop effective defense mechanisms tailored to their unique characteristics.

References

1. Aaronson S, Rothblum GN (2019) Gentle measurement of quantum states and differential privacy. In: Proceedings of the 51st annual ACM SIGACT symposium on theory of computing (STOC), pp 322–333
2. Aaronson S, Cojocaru A, Gheorghiu A, Kashefi E (2017) Complexity-theoretic limitations on blind delegated quantum computation. arXiv:1704.08482
3. Abbas A, Sutter D, Zoufal C, Lucchi A, Figalli A, Woerner S (2021) The power of quantum neural networks. Nat Comput Sci 1(6):403–409
4. Aledhari M, Razzak R, Parizi RM, Saeed F (2020) Federated learning: a survey on enabling technologies, protocols, and applications. IEEE Access 8:140699–140725
5. Alistarh D, Allen-Zhu Z, Li J (2018) Byzantine stochastic gradient descent. In: Proceedings of the 32nd conference on neural information processing systems (NeurIPS)
6. Almutairi S, Barnawi A (2023) Federated learning vulnerabilities, threats and defenses: a systematic review and future directions. Internet Things 24:100947
7. Ang F, Chen L, Zhao N, Chen Y, Wang W, Yu FR (2020) Robust federated learning with noisy communication. IEEE Trans Commun 68(6):3452–3464
8. Angrisani A, Kashefi E (2022) Quantum local differential privacy and quantum statistical query model. arXiv:2203.03591
9. Angrisani A, Doosti M, Kashefi E (2022) Differential privacy amplification in quantum and quantum-inspired algorithms. arXiv:2203.03604
10. Angrisani A, Doosti M, Kashefi E (2023) A unifying framework for differentially private quantum algorithms. arXiv:2307.04733

11. Anil G, Vinod V, Narayan A (2024) Generating universal adversarial perturbations for quantum classifiers. In: Proceedings of the 38th AAAI conference on artificial intelligence, vol 38, pp 10891–10899
12. Anschuetz ER (2021) Critical points in quantum generative models. arXiv:2109.06957
13. Aono Y, Hayashi T, Wang L, Moriai S et al (2017) Privacy-preserving deep learning via additively homomorphic encryption. IEEE Trans Inf Forensics Secur 13(5):1333–1345
14. Appiah B, Osei I, Frimpong BK, Commey D, Owusu-Agymang K, Assamah G (2025) Enhanced federated learning for secure medical data collaboration. J Anal Sci Technol 16(1):13
15. Arrasmith A, Holmes Z, Cerezo M, Coles PJ (2022) Equivalence of quantum barren plateaus to cost concentration and narrow gorges. Quantum Sci Technol 7(4):045015
16. Ayanzadeh R, Alavisamani N, Das P, Qureshi M (2023) Frozenqubits: boosting fidelity of qaoa by skipping hotspot nodes. In: Proceedings of the 28th ACM international conference on architectural support for programming languages and operating systems, vol 2 (ASPLOS), pp 311–324
17. Bagdasaryan E, Veit A, Hua Y, Estrin D, Shmatikov V (2020) How to backdoor federated learning. In: Proceedings of the 23rd international conference on artificial intelligence and statistics (AISTATS, pp 2938–2948
18. Baker JM, Duckering C, Hoover A, Chong FT (2020) Time-sliced quantum circuit partitioning for modular architectures. In: Proceedings of the 17th ACM international conference on computing frontiers, pp 98–107
19. Barnum H, Caves CM, Fuchs CA, Jozsa R, Schumacher B (1996) Noncommuting mixed states cannot be broadcast. Phys Rev Lett 76(15):2818
20. Barral D, Cardama FJ, Díaz G, Faílde D, Llovo IF, Juane MM, Vázquez-Pérez J, Villasuso J, Piñeiro C, Costas N et al (2024) Review of distributed quantum computing. from single qpu to high performance quantum computing. arXiv:2404.01265
21. Benedetti M, Lloyd E, Sack S, Fiorentini M (2019) Parameterized quantum circuits as machine learning models. Quantum Sci Technol 4(4):043001
22. Bennett CH, Brassard G (1984) Quantum cryptography: Public key distribution and coin tossing. In: Proceedings of the 1984 IEEE international conference on computers, systems and signal processing (ICCSP), pp 175–179
23. Bennett CH, Brassard G (2014) Quantum cryptography: public key distribution and coin tossing. Theoret Comput Sci 560:7–11
24. Bennett CH, Brassard G, Crépeau C, Jozsa R, Peres A, Wootters WK (1993) Teleporting an unknown quantum state via dual classical and einstein-podolsky-rosen channels. Phys Rev Lett 70(13):1895
25. Bennett CH, DiVincenzo DP, Shor PW, Smolin JA, Terhal BM, Wootters WK (2001) Remote state preparation. Phys Rev Lett 87(7):077902
26. Bergholm V, Izaac J, Schuld M, Gogolin C, Ahmed S, Ajith V, Alam MS, Alonso-Linaje G, AkashNarayanan B, Asadi A et al (2018) Pennylane: Automatic differentiation of hybrid quantum-classical computations. arXiv:1811.04968
27. Bermejo P, Braccia P, Rudolph MS, Holmes Z, Cincio L, Cerezo M (2024) Quantum convolutional neural networks are (effectively) classically simulable. arXiv:2408.12739
28. Bhagoji AN, Chakraborty S, Mittal P, Calo S (2019) Analyzing federated learning through an adversarial lens. In: Proceedings of the 36th international conference on machine learning (ICML), pp 634–643
29. Bhatia AS, Neira DEB (2024) Federated hierarchical tensor networks: a collaborative learning quantum ai-driven framework for healthcare. arXiv:2405.07735
30. Bhatia AS, Kais S, Alam MA (2024a) On the robustness of variational quantum classifier against "label flipping attacks" in federated learning for semiconductor manufacturing. In: Proceedings of the 2024 IEEE international conference on quantum computing and engineering (QCE), vol 1, pp 161–168
31. Bhatia AS, Kais S, Alam MA (2024b) Robustness of quantum federated learning (qfl) against "label flipping attacks" for lithography hotspot detection in semiconductor manufacturing.

In: Proceedings of the 2024 IEEE international reliability physics symposium (IRPS). IEEE, pp 1–4
32. Bhoumik D, Majumdar R, Saha A, Sur-Kolay S (2023) Distributed scheduling of quantum circuits with noise and time optimization. arXiv:2309.06005
33. Biamonte J, Wittek P, Pancotti N, Rebentrost P, Wiebe N, Lloyd S (2017) Quantum machine learning. Nature 549(7671):195–202
34. Bittel L, Kliesch M (2021) Training variational quantum algorithms is np-hard. Phys Rev Lett 127(12):120502
35. Blanchard P, El Mhamdi EM, Guerraoui R, Stainer J (2017) Machine learning with adversaries: byzantine tolerant gradient descent. In: Proceedings of the 31st conference on neural information processing systems (NeurIPS)
36. Boschi D, Branca S, De Martini F, Hardy L, Popescu S (1998) Experimental realization of teleporting an unknown pure quantum state via dual classical and Einstein-Podolsky-Rosen channels. Phys Rev Lett 80(6):1121
37. Bouwmeester D, Pan JW, Mattle K, Eibl M, Weinfurter H, Zeilinger A (1997) Experimental quantum teleportation. Nature 390(6660):575–579
38. Brandhofer S, Polian I, Krsulich K (2023) Optimal partitioning of quantum circuits using gate cuts and wire cuts. IEEE Trans Quantum Eng
39. Brassard G, Høyer P, Tapp A (1998) Quantum counting. In: Proceedings of the 25th international colloquium on automata, languages and programming (ICALP), pp 820–831
40. Bravyi S, Smith G, Smolin JA (2016) Trading classical and quantum computational resources. Phys Rev X 6(2):021043
41. Brenner L, Piveteau C, Sutter D (2023) Optimal wire cutting with classical communication. arXiv:2302.03366
42. Broadbent A, Jeffery S (2015) Quantum homomorphic encryption for circuits of low t-gate complexity. In: Annual cryptology conference, pp 609–629
43. Broadbent A, Fitzsimons J, Kashefi E (2009) Universal blind quantum computation. In: Proceedings of the 50th annual IEEE symposium on foundations of computer science (FOCS), pp 517–526
44. Burt F, Chen KC, Leung KK (2024) Generalised circuit partitioning for distributed quantum computing. In: Proceedings of the 2024 IEEE international conference on quantum computing and engineering (QCE), vol 2, pp 173–178
45. Cacciapuoti AS, Caleffi M, Van Meter R, Hanzo L (2020) When entanglement meets classical communications: quantum teleportation for the quantum internet. IEEE Trans Commun 68(6):3808–3833
46. Cai QY (2006) Eavesdropping on the two-way quantum communication protocols with invisible photons. Phys Lett A 351(1–2):23–25
47. Canetti R (2000) Security and composition of multiparty cryptographic protocols. J Cryptol 13:143–202
48. Castro M, Liskov B, et al. (1999) Practical byzantine fault tolerance. In: Proceedings of the 3rd USENIX symposium on operating systems design and implementation (OSDI), vol 99, pp 173–186
49. Cerezo M, Coles PJ (2021) Higher order derivatives of quantum neural networks with barren plateaus. Quantum Sci Technol 6(3):035006
50. Cerezo M, Arrasmith A, Babbush R, Benjamin SC, Endo S, Fujii K, McClean JR, Mitarai K, Yuan X, Cincio L et al (2021) Variational quantum algorithms. Nat Rev Phys 3(9):625–644
51. Cerezo M, Verdon G, Huang HY, Cincio L, Coles PJ (2022) Challenges and opportunities in quantum machine learning. Nat Comput Sci 2(9):567–576
52. Cerezo M, Larocca M, García-Martín D, Diaz NL, Braccia P, Fontana E, Rudolph MS, Bermejo P, Ijaz A, Thanasilp S et al (2023) Does provable absence of barren plateaus imply classical simulability? or, why we need to rethink variational quantum computing. arXiv:2312.09121
53. Chehimi M, Saad W (2022) Quantum federated learning with quantum data. In: Proceedings of the 47th IEEE international conference on acoustics, speech and signal processing (ICASSP), pp 8617–8621

54. Chehimi M, Chen SYC, Saad W, Towsley D, Debbah M (2023) Foundations of quantum federated learning over classical and quantum networks. IEEE Netw 38(1):124–130
55. Chen D, Baheri B, Chaudhary V, Guan Q, Xie N, Xu S (2022) Approximate quantum circuit reconstruction. In: Proceedings of the 3rd IEEE international conference on quantum computing and engineering (QCE), pp 509–515
56. Chen DT, Hansen EH, Li X, Kulkarni V, Chaudhary V, Ren B, Guan Q, Kuppannagari S, Liu J, Xu S (2023a) Efficient quantum circuit cutting by neglecting basis elements. In: Proceedings of the 37th IEEE international parallel and distributed processing symposium workshops (IPDPSW), pp 517–523
57. Chen DT, Hansen EH, Li X, Orenstein A, Kulkarni V, Chaudhary V, Guan Q, Liu J, Zhang Y, Xu S (2023b) Online detection of golden circuit cutting points. In: Proceedings of the 4th IEEE international conference on quantum computing and engineering (QCE), vol 1, pp 26–31
58. Chen KC, Liu CY, Shang Y, Burt F, Leung KK (2025) Distributed quantum neural networks on distributed photonic quantum computing. arXiv:2505.08474
59. Chen L, Yan L, Zhang S (2024) Robust quantum federated learning with noise. Phys Scr 99(7):076003
60. Chen SYC, Yoo S (2021) Federated quantum machine learning. Entropy 23(4):460
61. Chen Y, Su L, Xu J (2017) Distributed statistical machine learning in adversarial settings Byzantine gradient descent 2:1–25
62. Cheng K, Fan T, Jin Y, Liu Y, Chen T, Papadopoulos D, Yang Q (2021) Secureboost: a lossless federated learning framework. IEEE Intell Syst 36(6):87–98
63. Chu C, Jiang L, Chen F (2023) Cryptoqfl: quantum federated learning on encrypted data. In: Proceedings of the 2023 IEEE international conference on quantum computing and engineering (QCE), vol 1, pp 1231–1237
64. Cohen J, Rosenfeld E, Kolter Z (2019) Certified adversarial robustness via randomized smoothing. In: Proceedings of the 36th international conference on machine learning (ICML), pp 1310–1320
65. Commey D, Crosby GV (2025) Pqs-bfl: a post-quantum secure blockchain-based federated learning framework. arXiv:2505.01866
66. Cong I, Choi S, Lukin MD (2019) Quantum convolutional neural networks. Nat Phys 15(12):1273–1278
67. Cowlessur H, Thapa C, Alpcan T, Camtepe S (2024) A hybrid quantum neural network for split learning. arXiv:2409.16593
68. Cramer R, Damgård IB, Nielsen JB (2015) Secure multiparty computation and secret sharing. Cambridge University Press
69. Cuomo D, Caleffi M, Cacciapuoti AS (2020) Towards a distributed quantum computing ecosystem. IET Quantum Commun 1(1):3–8
70. Cuomo D, Caleffi M, Krsulich K, Tramonto F, Agliardi G, Prati E, Cacciapuoti AS (2023) Optimized compiler for distributed quantum computing. ACM Trans Quantum Comput 4(2):1–29
71. Curty M, Lütkenhaus N (2005) Intercept-resend attacks in the Bennett-Brassard 1984 quantum-key-distribution protocol with weak coherent pulses. Phys Rev A-Atomic Mol Opt Phys 71(6):062301
72. Damaskinos G, Guerraoui R, Patra R, Taziki M et al (2018) Asynchronous byzantine machine learning (the case of sgd). In: Proceedings of the 35th international conference on machine learning (ICML), pp 1145–1154
73. Deng FG, Li XH, Zhou HY, Zj Z (2005) Improving the security of multiparty quantum secret sharing against trojan horse attack. Phys Rev A 72(4):044302
74. Dieks D (1982) Communication by epr devices. Phys Lett A 92(6):271–272
75. Dowling N, West MT, Southwell A, Nakhl AC, Sevior M, Usman M, Modi K (2024) Adversarial robustness guarantees for quantum classifiers. arXiv:2405.10360
76. Driscoll K, Hall B, Paulitsch M, Zumsteg P, Sivencrona H (2004) The real byzantine generals. In: Proceedings of the 23rd digital avionics systems conference (DASC), vol 2, pp 6–D

77. Du Y, Hsieh MH, Liu T, Tao D, Liu N (2021) Quantum noise protects quantum classifiers against adversaries. Phys Rev Res 3(2):023153
78. Du Y, Hsieh MH, Liu T, You S, Tao D (2022) Quantum differentially private sparse regression learning. IEEE Trans Inf Theory 68(8):5217–5233
79. Dulek Y, Schaffner C, Speelman F (2016) Quantum homomorphic encryption for polynomial-sized circuits. In: Proceedings of the 36th annual international cryptology conference, advances in cryptology (CRYPTO), pp 3–32
80. Dunjko V, Briegel HJ (2018) Machine learning and artificial intelligence in the quantum domain: a review of recent progress. Rep Prog Phys 81(7):074001
81. Dutta S, Innan N, Yahia SB, Shafique M, Neira DEB (2024a) Mqfl-fhe: multimodal quantum federated learning framework with fully homomorphic encryption. arXiv:2412.01858
82. Dutta S, Karanth PP, Xavier PM, de Freitas IL, Innan N, Yahia SB, Shafique M, Neira DEB (2024b) Federated learning with quantum computing and fully homomorphic encryption: a novel computing paradigm shift in privacy-preserving ml. arXiv:2409.11430
83. Dwork C, McSherry F, Nissim K, Smith A (2006) Calibrating noise to sensitivity in private data analysis. In: Proceedings of the 3rd theory of cryptography conference (TCC), pp 265–284
84. Dwork C, Roth A, et al. (2014) The algorithmic foundations of differential privacy. Found Trends® Theoret Comput Sci 9(3–4):211–407
85. Eddins A, Motta M, Gujarati TP, Bravyi S, Mezzacapo A, Hadfield C, Sheldon S (2022) Doubling the size of quantum simulators by entanglement forging. PRX Quantum 3(1):010309
86. Einstein A, Podolsky B, Rosen N (1935) Can quantum-mechanical description of physical reality be considered complete? Phys Rev 47(10):777
87. Eisert J, Jacobs K, Papadopoulos P, Plenio MB (2000) Optimal local implementation of nonlocal quantum gates. Phys Rev A 62(5):052317
88. Eloul S, Silavong F, Kamthe S, Georgiadis A, Moran SJ (2022) Enhancing privacy against inversion attacks in federated learning by using mixing gradients strategies. arXiv:2204.12495
89. Eloul S, Silavong F, Kamthe S, Georgiadis A, Moran SJ (2024) Mixing gradients in neural networks as a strategy to enhance privacy in federated learning. In: Proceedings of the 2024 IEEE/CVF winter conference on applications of computer vision (WACV), pp 3956–3965
90. Endo S, Benjamin SC, Li Y (2018) Practical quantum error mitigation for near-future applications. Phys Rev X 8(3):031027
91. Fain B, Goel A, Munagala K (2016) The core of the participatory budgeting problem. In: Proceedings of the 12th international conference on web and internet economics (WINE), pp 384–399
92. Fang M, Cao X, Jia J, Gong N (2020) Local model poisoning attacks to {Byzantine-Robust} federated learning. In: Proceedings of the 29th USENIX security symposium, pp 1605–1622
93. Feng J, Cai QZ, Zhou ZH (2019) Learning to confuse: Generating training time adversarial data with auto-encoder. In: Proceedings of the 33rd conference on neural information processing systems (NeurIPS), vol 32
94. Ferrari D, Cacciapuoti AS, Amoretti M, Caleffi M (2021) Compiler design for distributed quantum computing. IEEE Trans Quantum Eng 2:1–20
95. Ferrari D, Carretta S, Amoretti M (2023) A modular quantum compilation framework for distributed quantum computing. IEEE Trans Quantum Eng 4:1–13
96. Fontaine C, Galand F (2007) A survey of homomorphic encryption for nonspecialists. EURASIP J Inf Secur 2007:1–10
97. Fredrikson M, Jha S, Ristenpart T (2015) Model inversion attacks that exploit confidence information and basic countermeasures. In: Proceedings of the 22nd ACM SIGSAC conference on computer and communications security (CCS), pp 1322–1333
98. Fujii K, Mizuta K, Ueda H, Mitarai K, Mizukami W, Nakagawa YO (2022) Deep variational quantum eigensolver: a divide-and-conquer method for solving a larger problem with smaller size quantum computers. PRX Quantum 3(1):010346
99. G Sundaram R, Gupta H, Ramakrishnan C (2021) Efficient distribution of quantum circuits. In: 35th International symposium on distributed computing (DISC 2021), pp 41–1

100. Geiping J, Bauermeister H, Dröge H, Moeller M (2020) Inverting gradients-how easy is it to break privacy in federated learning? In: Proceedings of the 34th conference on neural information processing systems (NeurIPS), pp 16937–16947
101. Gentinetta G, Metz F, Carleo G (2024) Overhead-constrained circuit knitting for variational quantum dynamics. Quantum 8:1296
102. Gentry C (2009) A fully homomorphic encryption scheme. Stanford university
103. Gharavi H, Granjal J, Monteiro E (2025) Pqbfl: a post-quantum blockchain-based protocol for federated learning. arXiv:2502.14464
104. Gil Vidal FJ, Theis DO (2020) Input redundancy for parameterized quantum circuits. Front Phys 8:297
105. Gilboa D, Michaeli H, Soudry D, McClean J (2024) Exponential quantum communication advantage in distributed inference and learning. In: Proceedings of the 38th conference on neural information processing systems (NeurIPS), vol 37, pp 30425–30473
106. Gisin N, Fasel S, Kraus B, Zbinden H, Ribordy G (2006) Trojan-horse attacks on quantum-key-distribution systems. Phys Rev A 73(2):022320
107. Goldreich O, Micali S, Wigderson A (2019) How to play any mental game, or a completeness theorem for protocols with honest majority. In: Providing sound foundations for cryptography: on the work of Shafi Goldwasser and Silvio Micali, pp 307–328
108. Gong W, Deng DL (2022) Universal adversarial examples and perturbations for quantum classifiers. Nat Sci Rev 9(6):nwab130
109. Gong W, Yuan D, Li W, Deng DL (2024) Enhancing quantum adversarial robustness by randomized encodings. Phys Rev Res 6(2):023020
110. Greenberger DM, Horne MA, Zeilinger A (1989) Going beyond bell's theorem. In: Bell's theorem, quantum theory and conceptions of the universe, pp 69–72
111. Gu T, Dolan-Gavitt B, Garg S (2017) Badnets: Identifying vulnerabilities in the machine learning model supply chain. arXiv:1708.06733
112. Guala D, Zhang S, Cruz E, Riofrío CA, Klepsch J, Arrazola JM (2023) Practical overview of image classification with tensor-network quantum circuits. Sci Rep 13(1):4427
113. Gurung D, Pokhrel S, Li G (2023a) Secure communication model for quantum federated learning: a proof of concept. In: The 1st tiny papers track at the 11th international conference on learning representations (ICLR)
114. Gurung D, Pokhrel SR, Li G (2023b) Decentralized quantum federated learning for metaverse: analysis, design and implementation. arXiv:2306.11297
115. Gurung D, Pokhrel SR, Li G (2023c) Quantum federated learning: analysis, design and implementation challenges. arXiv:2306.15708
116. Gurung D, Pokhrel SR, Li G (2025) Quantum federated learning for metaverse: analysis, design and implementation. IEEE Trans Netw Serv Manag
117. Haddadpour F, Mahdavi M (2019) On the convergence of local descent methods in federated learning. arXiv:1910.14425
118. Haghshenas R (2021) Optimization schemes for unitary tensor-network circuit. Phys Rev Res 3(2):023148
119. Haghshenas R, Gray J, Potter AC, Chan GKL (2022) Variational power of quantum circuit tensor networks. Phys Rev X 12(1):011047
120. Harada H, Wada K, Yamamoto N (2024) Doubly optimal parallel wire cutting without ancilla qubits. PRX Quantum 5(4):040308
121. Harrow AW, Lowe A (2024) Optimal quantum circuit cuts with application to clustered hamiltonian simulation. arXiv:2403.01018
122. Hayashi M, Koshiba T (2019) Verifiable quantum secure modulo summation. arXiv:1910.05976
123. Heredge J, Kumar N, Herman D, Chakrabarti S, Yalovetzky R, Sureshbabu SH, Li C, Pistoia M (2024) Prospects of privacy advantage in quantum machine learning. arXiv:2405.08801
124. Herman D, Raymond R, Li M, Robles N, Mezzacapo A, Pistoia M (2023) Expressivity of variational quantum machine learning on the boolean cube. IEEE Trans Quantum Eng 4:1–18

125. Heunen C, Martinez PA (2019) Automated distribution of quantum circuits. Phys Rev A 100:032308
126. Hillery M, Bužek V, Berthiaume A (1999) Quantum secret sharing. Phys Rev A 59(3):1829
127. Hirche C, Rouzé C, França DS (2023) Quantum differential privacy: an information theory perspective. IEEE Trans Inf Theory 69(9):5771–5787
128. Hitaj B, Ateniese G, Perez-Cruz F (2017) Deep models under the gan: information leakage from collaborative deep learning. In: Proceedings of the 2017 ACM SIGSAC conference on computer and communications security (CCS), pp 603–618
129. Horodecki R, Horodecki P, Horodecki M, Horodecki K (2009) Quantum Entanglem Rev Mod Phys 81(2):865–942
130. Houshmand M, Mohammadi Z, Zomorodi-Moghadam M, Houshmand M (2020) An evolutionary approach to optimizing teleportation cost in distributed quantum computation. Int J Theor Phys 59(4):1315–1329
131. Howard M, Campbell E (2017) Application of a resource theory for magic states to fault-tolerant quantum computing. Phys Rev Lett 118(9):090501
132. Huang HY, Kueng R, Preskill J (2020) Predicting many properties of a quantum system from very few measurements. Nat Phys 16(10):1050–1057
133. Huang HY, Kueng R, Preskill J (2021) Efficient estimation of Pauli observables by derandomization. Phys Rev Lett 127(3):030503
134. Huang JC, Tsai YL, Yang CHH, Su CF, Yu CM, Chen PY, Kuo SY (2023) Certified robustness of quantum classifiers against adversarial examples through quantum noise. In: Proceedings of the 48th IEEE international conference on acoustics, speech and signal processing (ICASSP), pp 1–5
135. Huang R, Tan X, Xu Q (2022) Quantum federated learning with decentralized data. IEEE J Sel Top Quantum Electron 28(4: Mach Learn Photon Commun Meas Syst):1–10
136. Huang W, Ye M, Shi Z, Wan G, Li H, Du B, Yang Q (2024) Federated learning for generalization, robustness, fairness: a survey and benchmark. IEEE Trans Pattern Anal Mach Intell
137. Huang Y, Gupta S, Song Z, Li K, Arora S (2021b) Evaluating gradient inversion attacks and defenses in federated learning. In: Proceedings of the 35th conference on neural information processing systems (NeurIPS), pp 7232–7241
138. Huang YF, Ren XF, Zhang YS, Duan LM, Guo GC (2004) Experimental teleportation of a quantum controlled-not gate. Phys Rev Lett 93(24):240501
139. Huang Z, Mitra S, Vaidya N (2015) Differentially private distributed optimization. In: Proceedings of the 16th international conference on distributed computing and networking, pp 1–10
140. Huang Z, Hu R, Guo Y, Chan-Tin E, Gong Y (2019) Dp-admm: Admm-based distributed learning with differential privacy. IEEE Trans Inf Forensics Secur 15:1002–1012
141. Huggins W, Patil P, Mitchell B, Whaley KB, Stoudenmire EM (2019) Towards quantum machine learning with tensor networks. Quantum Sci Technol 4(2):024001
142. Hwang K, Lim HT, Kim YS, Park DK, Kim Y (2024) Distributed quantum machine learning via classical communication. Quantum Sci Technol 10(1):015059
143. Innan N, Khan MAZ, Marchisio A, Shafique M, Bennai M (2024) Fedqnn: federated learning using quantum neural networks. In: Proceedings of the 2024 international joint conference on neural networks (IJCNN), pp 1–9
144. Jagielski M, Oprea A, Biggio B, Liu C, Nita-Rotaru C, Li B (2018) Manipulating machine learning: Poisoning attacks and countermeasures for regression learning. In: Proceedings of the 39th IEEE symposium on security and privacy (SP), pp 19–35
145. Jeswal S, Chakraverty S (2019) Recent developments and applications in quantum neural network: a review. Arch Comput Methods Eng 26(4):793–807
146. Jiang L, Taylor JM, Sørensen AS, Lukin MD (2007) Distributed quantum computation based on small quantum registers. Phys Rev A 76(6):062323
147. Jin W, Yao Y, Han S, Gu J, Joe-Wong C, Ravi S, Avestimehr S, He C (2023) Fedml-he: an efficient homomorphic-encryption-based privacy-preserving federated learning system. arXiv:2303.10837

148. Kaewpuang R, Xu M, Niyato D, Yu H, Xiong Z, Shen XS (2023) Adaptive resource allocation in quantum key distribution (qkd) for federated learning. In: Proceedings of the 2023 international conference on computing, networking and communications (ICNC), pp 71–76
149. Kairouz P, McMahan HB, Avent B, Bellet A, Bennis M, Bhagoji AN, Bonawitz K, Charles Z, Cormode G, Cummings R et al (2021) Advances and open problems in federated learning. Found Trends® Mach Learn 14(1–2):1–210
150. Kak SC (1995) Quantum neural computing. Adv Imaging Electron Phys 94:259–313
151. Kananian P, Jacobsen HA (2025) Adversarial robustness of partitioned quantum classifiers. arXiv:2502.20403
152. Kananian P, Zahedi SM (2024) Asymptotically fair and truthful allocation of public goods. arXiv:2404.15996
153. Kannan E, MJ CMB, Ravikumar S, Kannan S, Vijay K, et al. (2024) Quantum-safe federated learning: enhancing data privacy and security. In: Proceedings of the 2024 international conference on emerging research in computational science (ICERCS), pp 1–6
154. Karim A, Zhang S, Usman M (2024) Low depth virtual distillation of quantum circuits by deterministic circuit decomposition. Phys Rev Res 6(3):033223
155. Kawase Y (2024) Distributed quantum neural networks via partitioned features encoding. Quantum Mach Intell 6(1):15
156. Khaddaj A, Leclerc G, Makelov A, Georgiev K, Salman H, Ilyas A, Madry A (2023) Rethinking backdoor attacks. In: Proceedings of the 40th international conference on machine learning (ICML), pp 16216–16236
157. Koh PW, Steinhardt J, Liang P (2022) Stronger data poisoning attacks break data sanitization defenses. Mach Learn 1–47
158. Konečný J, McMahan HB, Ramage D, Richtárik P (2016) Federated optimization: distributed machine learning for on-device intelligence. arXiv:1610.02527
159. Kumar A, Garhwal S (2021) State-of-the-art survey of quantum cryptography. Arch Comput Methods Eng 28:3831–3868
160. Kumar KN, Mohan CK, Cenkeramaddi LR (2023) The impact of adversarial attacks on federated learning: a survey. IEEE Trans Pattern Anal Mach Intell 46(5):2672–2691
161. Kumar N, Heredge J, Li C, Eloul S, Sureshbabu SH, Pistoia M (2023b) Expressive variational quantum circuits provide inherent privacy in federated learning. arXiv:2309.13002
162. Kumar PK, Rao PP, Kishore KH (2017) Optimal design of reversible parity preserving new full adder/full subtractor. In: Proceedings of the 11th international conference on intelligent systems and control (ISCO), pp 368–373
163. Lamport L, Shostak R, Pease M (2019) The byzantine generals problem. In: Concurrency: the works of leslie lamport, pp 203–226
164. Landman J, Thabet S, Dalyac C, Mhiri H, Kashefi E (2022) Classically approximating variational quantum machine learning with random fourier features. arXiv:2210.13200
165. Larocca M, Thanasilp S, Wang S, Sharma K, Biamonte J, Coles PJ, Cincio L, McClean JR, Holmes Z, Cerezo M (2024) A review of barren plateaus in variational quantum computing. arXiv:2405.00781
166. Lee H, Son SB, Chen SYC, Park S (2025) Auction-based trustworthy and resilient quantum distributed learning. IEEE Int Things J
167. Li C, Kumar N, Song Z, Chakrabarti S, Pistoia M (2024) Privacy-preserving quantum federated learning via gradient hiding. Quantum Sci Technol 9(3):035028
168. Li C, Li B, Amer O, Shaydulin R, Chakrabarti S, Wang G, Xu H, Tang H, Schoch I, Kumar N et al (2024) Blind quantum machine learning with quantum bipartite correlator. Phys Rev Lett 133(12):120602
169. Li P, Liu J, Patil HP, Hovland P, Zhou H (2023) Enhancing virtual distillation with circuit cutting for quantum error mitigation. In: Proceedings of the 41st IEEE international conference on computer design (ICCD), pp 94–101
170. Li W, Deng DL (2024) Quantum delegated and federated learning via quantum homomorphic encryption. Quantum technologies, research directions, pp 1–6

171. Li W, Lu S, Deng DL (2021) Quantum federated learning through blind quantum computing. Sci China Phys Mech Astron 64(10):100312
172. Li X, Gu Y, Dvornek N, Staib LH, Ventola P, Duncan JS (2020) Multi-site fmri analysis using privacy-preserving federated learning and domain adaptation: Abide results. Med Image Anal 65:101765
173. Li X, Kulkarni V, Chen DT, Guan Q, Jiang W, Xie N, Xu S, Chaudhary V (2024c) Efficient circuit wire cutting based on commuting groups. In: Proceedings of the 5th IEEE international conference on quantum computing and engineering (QCE), vol 1, pp 117–123
174. Li XH, Deng FG, Zhou HY (2006) Improving the security of secure direct communication based on the secret transmitting order of particles. Phys Rev A 74(5):054302
175. Li Y, Zhai T, Wu B, Jiang Y, Li Z, Xia S (2020b) Rethinking the trigger of backdoor attack. arXiv:2004.04692
176. Li Y, Zhao Y, Zhang X, Zhong H, Pan M, Zhang C (2024d) Differential privacy preserving quantum computing via projection operator measurements. In: Proceedings of the 2024 international conference on quantum communications, networking, and computing (QCNC), pp 236–243
177. Li Z, Zhang Y (2021) Membership leakage in label-only exposures. In: Proceedings of the 2021 ACM SIGSAC conference on computer and communications security (CCS), pp 880–895
178. Lian H, Xu J, Zhu Y, Fan Z, Liu Y, Shan Z (2023) Fast reconstruction algorithm based on hmc sampling. Sci Rep 13(1):17773
179. Liao H, Convy I, Huggins WJ, Whaley KB (2021) Robust in practice: adversarial attacks on quantum machine learning. Phys Rev A 103(4):042427
180. Liebmann L, Mansfield S, Han G, Culp J, Hibbeler J, Tsai R (2006) Reducing dfm to practice: the lithography manufacturability assessor. In: Proceedings of the 4th design and process integration for microelectronic manufacturing, vol 6156, pp 178–189
181. Lin J, Yang CW, Tsai CW, Hwang T (2013) Intercept-resend attacks on semi-quantum secret sharing and the improvements. Int J Theor Phys 52:156–162
182. Liu J, Gonzales A, Saleem ZH (2022) Classical simulators as quantum error mitigators via circuit cutting. arXiv:2212.07335
183. Liu N, Wittek P (2020) Vulnerability of quantum classification to adversarial perturbations. Phys Rev A 101(6):062331
184. Liu Y, Kang Y, Xing C, Chen T, Yang Q (2020) A secure federated transfer learning framework. IEEE Intell Syst 35(4):70–82
185. Liu Y, Arunachalam S, Temme K (2021) A rigorous and robust quantum speed-up in supervised machine learning. Nat Phys 17(9):1013–1017
186. Liu Y, Kang Y, Zou T, Pu Y, He Y, Ye X, Ouyang Y, Zhang YQ, Yang Q (2024) Vertical federated learning: concepts, advances, and challenges. IEEE Trans Knowl Data Eng 36(7):3615–3634
187. Lowe A, Medvidović M, Hayes A, O'Riordan LJ, Bromley TR, Arrazola JM, Killoran N (2023) Fast quantum circuit cutting with randomized measurements. Quantum 7:934
188. Lu H, Liu C, He T, Wang S, Chan KS (2020a) Sharing models or coresets: a study based on membership inference attack. arXiv:2007.02977
189. Lu S, Duan LM, Deng DL (2020) Quantum adversarial machine learning. Phys Rev Res 2(3):033212
190. Luo T, Liu Q, Sun X, Huang C, Chen Y, Zhang Z, Wei K (2024) Security analysis against the trojan horse attack on practical polarization-encoding quantum key distribution systems. Phys Rev A 109(4):042608
191. Luo X, Wu Y, Xiao X, Ooi BC (2021) Feature inference attack on model predictions in vertical federated learning. In: Proceedings of the 37th IEEE international conference on data engineering (ICDE), pp 181–192
192. Lusnig L, Sagingalieva A, Surmach M, Protasevich T, Michiu O, McLoughlin J, Mansell C, de'Petris G, Bonazza D, Zanconati F, et al (2024) Hybrid quantum image classification and federated learning for hepatic steatosis diagnosis. Diagnostics 14(5):558

193. Ma W, Chen KC, Yu S, Liu M, Deng R (2025) Robust decentralized quantum kernel learning for noisy and adversarial environment. arXiv:2504.13782
194. Madry A, Makelov A, Schmidt L, Tsipras D, Vladu A (2017) Towards deep learning models resistant to adversarial attacks. arXiv:1706.06083
195. Mahadev U (2020) Classical homomorphic encryption for quantum circuits. SIAM J Comput 52(6):FOCS18–189
196. Majumdar R, Wood CJ (2022) Error mitigated quantum circuit cutting. arXiv:2211.13431
197. Makarov* V, Hjelme DR, (2005) Faked states attack on quantum cryptosystems. J Mod Opt 52(5):691–705
198. Maouaki WE, Innan N, Marchisio A, Said T, Bennai M, Shafique M (2025) Qfal: quantum federated adversarial learning. arXiv:2502.21171
199. Marshall SC, Gyurik C, Dunjko V (2023) High dimensional quantum machine learning with small quantum computers. Quantum 7:1078
200. Mathur A, Gupta A, Das SK (2025) When federated learning meets quantum computing: survey and research opportunities. arXiv:2504.08814
201. McClean JR, Boixo S, Smelyanskiy VN, Babbush R, Neven H (2018) Barren plateaus in quantum neural network training landscapes. Nat Commun 9(1):4812
202. McMahan B, Moore E, Ramage D, Hampson S, Yarcas BA (2017) Communication-efficient learning of deep networks from decentralized data. In: Proceedings of the 20th international conference on artificial intelligence and statistics (AISTATS 2017), pp 1273–1282
203. McSherry F, Talwar K (2007) Mechanism design via differential privacy. In: Proceedings of the 48th annual IEEE symposium on foundations of computer science (FOCS), pp 94–103
204. Mehic M, Niemiec M, Rass S, Ma J, Peev M, Aguado A, Martin V, Schauer S, Poppe A, Pacher C et al (2020) Quantum key distribution: a networking perspective. ACM Comput Surv (CSUR) 53(5):1–41
205. Mei H, Li G, Wu J, Zheng L (2023) Privacy inference-empowered stealthy backdoor attack on federated learning under non-iid scenarios. In: Proceedings of the 2023 international joint conference on neural networks (IJCNN), pp 1–10
206. Melis L, Song C, De Cristofaro E, Shmatikov V (2019) Exploiting unintended feature leakage in collaborative learning. In: Proceedings of the 2019 IEEE symposium on security and privacy (SP), pp 691–706
207. Meter III RDV (2006) Architecture of a quantum multicomputer optimized for shor's factoring algorithm. arXiv preprint quant-ph/0607065
208. Milonni PW, Hardies M (1982) Photons cannot always be replicated. Phys Lett A 92(7)
209. Mitarai K, Fujii K (2021) Constructing a virtual two-qubit gate by sampling single-qubit operations. New J Phys 23(2):023021
210. Mitarai K, Fujii K (2021) Overhead for simulating a non-local channel with local channels by quasiprobability sampling. Quantum 5:388
211. Mo F, Borovykh A, Malekzadeh M, Haddadi H, Demetriou S (2020) Layer-wise characterization of latent information leakage in federated learning. arXiv:2010.08762
212. Moll N, Barkoutsos P, Bishop LS, Chow JM, Cross A, Egger DJ, Filipp S, Fuhrer A, Gambetta JM, Ganzhorn M et al (2018) Quantum optimization using variational algorithms on near-term quantum devices. Quantum Sci Technol 3(3):030503
213. Moore E, Rezapour S, Amini MH (2025) Quantum encryption for secure federated learning against generative adversarial network attacks. Authorea Preprints
214. Muñoz-González L, Biggio B, Demontis A, Paudice A, Wongrassamee V, Lupu EC, Roli F (2017) Towards poisoning of deep learning algorithms with back-gradient optimization. In: Proceedings of the 10th ACM workshop on artificial intelligence and security (AISEC), pp 27–38
215. Nagai R, Kanno S, Sato Y, Yamamoto N (2023) Quantum channel decomposition with preselection and postselection. Phys Rev A 108(2):022615
216. Narottama B, Shin SY (2023) Federated quantum neural network with quantum teleportation for resource optimization in future wireless communication. IEEE Trans Veh Technol 72(11):14717–14733

217. Nielsen MA, Chuang IL (2010) Quantum computation and quantum information. Cambridge University Press
218. Nikahd E, Mohammadzadeh N, Sedighi M, Zamani MS (2021) Automated window-based partitioning of quantum circuits. Phys Scr 96(3):035102
219. Pan K, Ong YS, Gong M, Li H, Qin AK, Gao Y (2024) Differential privacy in deep learning: a literature survey. Neurocomputing, p 127663
220. Papadopoulos G, Eloul S, Satsangi Y, Heredge J, Kumar N, Chen CF, Pistoia M (2025) A numerical gradient inversion attack in variational quantum neural-networks. arXiv:2504.12806
221. Park H, Lee J (2023) Hqk-fl: hybrid-quantum-key-based secure federated learning for distributed multi-center clinical studies. Hum-Centric Comput Inf Sci 13
222. Park JJ, Cha J, Chen SYC, Tseng HH, Yoo S (2025) Addressing the current challenges of quantum machine learning through multi-chip ensembles. arXiv:2505.08782
223. Park S, Baek H, Kim J (2023) Quantum split learning for privacy-preserving information management. In: Proceedings of the 32nd ACM international conference on information and knowledge management, pp 4239–4243
224. Pashayan H, Wallman JJ, Bartlett SD (2015) Estimating outcome probabilities of quantum circuits using quasiprobabilities. Phys Rev Lett 115(7):070501
225. Pednault E (2023) An alternative approach to optimal wire cutting without ancilla qubits. arXiv:2303.08287
226. Peng T, Harrow AW, Ozols M, Wu X (2020) Simulating large quantum circuits on a small quantum computer. Phys Rev Lett 125(15):150504
227. Pérez-Salinas A, Cervera-Lierta A, Gil-Fuster E, Latorre JI (2020) Data re-uploading for a universal quantum classifier. Quantum 4:226
228. Pérez-Salinas A, Draškić R, Tura J, Dunjko V (2023) Shallow quantum circuits for deeper problems. Phys Rev A 108(6):062423
229. Perlin MA, Saleem ZH, Suchara M, Osborn JC (2021) Quantum circuit cutting with maximum-likelihood tomography. NPJ Quantum Inf 7(1):64
230. Pesah A, Cerezo M, Wang S, Volkoff T, Sornborger AT, Coles PJ (2021) Absence of barren plateaus in quantum convolutional neural networks. Phys Rev X 11(4):041011
231. Pira L, Ferrie C (2023) An invitation to distributed quantum neural networks. Quantum Mach Intell 5(2):1–24
232. Piveteau C, Sutter D (2023) Circuit knitting with classical communication. IEEE Trans Inf Theory
233. Piveteau C, Sutter D, Woerner S (2022) Quasiprobability decompositions with reduced sampling overhead. NPJ Quantum Inf 8(1):12
234. Polacchi B, Leichtle D, Limongi L, Carvacho G, Milani G, Spagnolo N, Kaplan M, Sciarrino F, Kashefi E (2023) Multi-client distributed blind quantum computation with the qline architecture. Nat Commun 14(1):7743
235. Preskill J (2018) Quantum computing in the nisq era and beyond. Quantum 2:79
236. Qammar A, Ding J, Ning H (2022) Federated learning attack surface: taxonomy, cyber defences, challenges, and future directions. Artif Intell Rev 55(5):3569–3606
237. Qi H, Wang L, Zhu H, Gani A, Gong C (2023) The barren plateaus of quantum neural networks: review, taxonomy and trends. Quantum Inf Process 22(12):435
238. Qi J, Zhang XL, Tejedor J (2023b) Optimizing quantum federated learning based on federated quantum natural gradient descent. In: Proceedings of the 2023 IEEE international conference on acoustics, speech and signal processing (ICASSP), pp 1–5
239. Qiao C, Li M, Liu Y, Tian Z (2024) Transitioning from federated learning to quantum federated learning in internet of things: a comprehensive survey. IEEE Commun Surv Tutor
240. Qin X, Xu R (2025) Efficient post-quantum cross-silo federated learning based on key homomorphic pseudo-random function. Mathematics 13(9):1404
241. Qiu P, Zhang X, Ji S, Fu C, Yang X, Wang T (2024) Hashvfl: defending against data reconstruction attacks in vertical federated learning. IEEE Trans Inf Forensics Secur 19:3435–3450

242. Qu GJ, Wang MM (2021) Secure multi-party quantum computation based on blind quantum computation. Int J Theor Phys 60(8):3003–3012
243. Qu Z, Li Y, Liu B, Gupta D, Tiwari P (2023) Dtqfl: a digital twin-assisted quantum federated learning algorithm for intelligent diagnosis in 5g mobile network. IEEE J Biomed Health Inf
244. Qu Z, Zhang L, Tiwari P (2024) Quantum fuzzy federated learning for privacy protection in intelligent information processing. IEEE Trans Fuzzy Syst
245. Ravikumar S, Chandralekha E, Vijay K, Antony Kumar K, Pretty Diana Cyril C (2023) Quantum-secured collaborative machine learning: facilitating privacy-protecting quantum federated learning. In: Proceedings of the 3rd international conference on computing and communication networks, pp 537–550
246. Ren C, Yan R, Zhu H, Yu H, Xu M, Shen Y, Xu Y, Xiao M, Dong ZY, Skoglund M et al (2023) Towards quantum federated learning. arXiv:2306.09912
247. Ren C, Xu M, Yu H, Xiong Z, Zhang Z, Niyato D (2024) Variational quantum circuit and quantum key distribution-based quantum federated learning: A case of smart grid dynamic security assessment. In: Proceedings of the 2024 IEEE international conference on communications (ICC), pp 1115–1120
248. Ren JG, Xu P, Yong HL, Zhang L, Liao SK, Yin J, Liu WY, Cai WQ, Yang M, Li L et al (2017) Ground-to-satellite quantum teleportation. Nature 549(7670):70–73
249. Rivest RL, Adleman L, Dertouzos ML et al (1978) On data banks and privacy homomorphisms. Found Secure Comput 4(11):169–180
250. Rofougaran R, Yoo S, Tseng HH, Chen SYC (2024) Federated quantum machine learning with differential privacy. In: Proceedings of the 49th IEEE international conference on acoustics, speech and signal processing (ICASSP), pp 9811–9815
251. Rong D, He Q, Chen J (2022) Poisoning deep learning based recommender model in federated learning scenarios. arXiv:2204.13594
252. Saha S, Hota A, Chattopadhyay AK, Nag A, Nandi S (2024) A multifaceted survey on privacy preservation of federated learning: progress, challenges, and opportunities. Artif Intell Rev 57(7):184
253. Sahu H, Gupta HP (2024) Nac-qfl: noise aware clustered quantum federated learning. arXiv:2406.14236
254. Sarma R, Jain R (2018) Quantum gate implementation of a novel reversible half adder and subtractor circuit. In: Proceedings of the 2018 international conference on intelligent circuits and systems (ICICS), pp 72–76
255. Schmitt L, Piveteau C, Sutter D (2023) Cutting circuits with multiple two-qubit unitaries. arXiv:2312.11638
256. Schuld M, Sinayskiy I, Petruccione F (2014) The quest for a quantum neural network. Quantum Inf Process 13:2567–2586
257. Schuld M, Sweke R, Meyer JJ (2021) Effect of data encoding on the expressive power of variational quantum-machine-learning models. Phys Rev A 103(3):032430
258. Seddon JR, Campbell ET (2019) Quantifying magic for multi-qubit operations. Proc R Soc A 475(2227):20190251
259. Seddon JR, Regula B, Pashayan H, Ouyang Y, Campbell ET (2021) Quantifying quantum speedups: improved classical simulation from tighter magic monotones. PRX Quantum 2(1):010345
260. Seitz P, Geiger M, Mendl CB (2024) Multithreaded parallelism for heterogeneous clusters of qpus. In: Proceedings of the 39th international supercomputing conference (ISC) high performance, pp 1–8
261. Senekane M, Mafu M, Taele BM (2017) Privacy-preserving quantum machine learning using differential privacy. In: Proceedings of the 2017 IEEE AFRICON conference, pp 1432–1435
262. Shafahi A, Huang WR, Najibi M, Suciu O, Studer C, Dumitras T, Goldstein T (2018) Poison frogs! targeted clean-label poisoning attacks on neural networks. In: Proceedings of the 32nd conference on neural information processing systems (NeurIPS), p 31
263. Sheng YB, Zhou L (2017) Distributed secure quantum machine learning. Sci Bull 62(14):1025–1029

264. Shi YY, Duan LM, Vidal G (2006) Classical simulation of quantum many-body systems with a tree tensor network. Phys Rev A-Atomic Mol Opt Phys 74(2):022320
265. Shin S, Teo YS, Jeong H (2023) Exponential data encoding for quantum supervised learning. Phys Rev A 107(1):012422
266. Shokri R, Stronati M, Song C, Shmatikov V (2017) Membership inference attacks against machine learning models. In: Proceedings of the 2017 IEEE symposium on security and privacy (SP), pp 3–18
267. Shor PW, Preskill J (2000) Simple proof of security of the bb84 quantum key distribution protocol. Phys Rev Lett 85(2):441
268. Smith KN, Perlin MA, Gokhale P, Frederick P, Owusu-Antwi D, Rines R, Omole V, Chong F (2023) Clifford-based circuit cutting for quantum simulation. In: Proceedings of the 50th annual international symposium on computer architecture (ISCA), pp 1–13
269. Son SB, Park S (2024) Toward uniform quantum federated aggregation: heterogeneity exclusion using entropy and fidelity. IEEE Int Things J
270. Song Y, Wu Y, Wu S, Li D, Wen Q, Qin S, Gao F (2024) A quantum federated learning framework for classical clients. Sci China Phys Mech Astron 67(5):250311
271. Subramanian G, Chinnadurai M (2024) Hybrid quantum enhanced federated learning for cyber attack detection. Sci Rep 14(1):32038
272. Swaminathan A, Akgün M (2024) Distributed and secure kernel-based quantum machine learning. arXiv:2408.10265
273. Szegedy C, Zaremba W, Sutskever I, Bruna J, Erhan D, Goodfellow I, Fergus R (2013) Intriguing properties of neural networks. arXiv:1312.6199
274. Tagliacozzo L, Evenbly G, Vidal G (2009) Simulation of two-dimensional quantum systems using a tree tensor network that exploits the entropic area law. Phys Rev B-Condensed Matter Mater Phys 80(23):235127
275. Tanbhir G, Shahriyar MF (2025) Quantum-inspired privacy-preserving federated learning framework for secure dementia classification. arXiv:2503.03267
276. Tang H, Li B, Wang G, Xu H, Li C, Barr A, Cappellaro P, Li J (2023) Communication-efficient quantum algorithm for distributed machine learning. Phys Rev Lett 130(15):150602
277. Tang W, Martonosi M (2022) Scaleqc: a scalable framework for hybrid computation on quantum and classical processors. arXiv:2207.00933
278. Tang W, Tomesh T, Suchara M, Larson J, Martonosi M (2021) Cutqc: using small quantum computers for large quantum circuit evaluations. In: Proceedings of the 26th ACM international conference on architectural support for programming languages and operating systems (ASPLOS), pp 473–486
279. Temme K, Bravyi S, Gambetta JM (2017) Error mitigation for short-depth quantum circuits. Phys Rev Lett 119(18):180509
280. Tikku A, Kim IH (2022) Circuit depth versus energy in topologically ordered systems. arXiv:2210.06796
281. Tolpegin V, Truex S, Gursoy ME, Liu L (2020) Data poisoning attacks against federated learning systems. In: Proceedings of the 25th European symposium on research in computer security (ESORICS), pp 480–501
282. Tomesh T, Saleem ZH, Perlin MA, Gokhale P, Suchara M, Martonosi M (2023) Divide and conquer for combinatorial optimization and distributed quantum computation. In: Proceedings of the 4th IEEE international conference on quantum computing and engineering (QCE), vol 1, pp 1–12
283. Tüysüz C, Clemente G, Crippa A, Hartung T, Kühn S, Jansen K (2023) Classical splitting of parametrized quantum circuits. Quantum Mach Intell 5(2):34
284. Uchehara G, Aamodt TM, Di Matteo O (2022) Rotation-inspired circuit cut optimization. In: Proceedings of the 3rd IEEE/ACM international workshop on quantum computing software (QCS), pp 50–56
285. Ufrecht C, Periyasamy M, Rietsch S, Scherer DD, Plinge A, Mutschler C (2023) Cutting multi-control quantum gates with zx calculus. Quantum 7:1147

286. Ufrecht C, Herzog LS, Scherer DD, Periyasamy M, Rietsch S, Plinge A, Mutschler C (2024) Optimal joint cutting of two-qubit rotation gates. Phys Rev A 109(5):052440
287. Ullah S, Shah MH, Anjum A (2024) Quantum enhanced federated learning with differential privacy. In: Proceedings of the 2024 international conference on frontiers of information technology (FIT), pp 1–6
288. Vakhitov A, Makarov V, Hjelme DR (2001) Large pulse attack as a method of conventional optical eavesdropping in quantum cryptography. J Mod Opt 48(13):2023–2038
289. Van Meter R, Nemoto K, Munro W, Itoh KM (2006) Distributed arithmetic on a quantum multicomputer. ACM SIGARCH computer architecture news 34(2):354–365
290. Wang H, Sreenivasan K, Rajput S, Vishwakarma H, Agarwal S, Sohn Jy, Lee K, Papailiopoulos D (2020) Attack of the tails: yes, you really can backdoor federated learning. In: Proceedings of the 34th conference on neural information processing systems (NeurIPS), vol 33, pp 16070–16084
291. Wang L, Xu S, Wang X, Zhu Q (2019a) Eavesdrop the composition proportion of training labels in federated learning. arXiv:1910.06044
292. Wang S, Fontana E, Cerezo M, Sharma K, Sone A, Cincio L, Coles PJ (2021) Noise-induced barren plateaus in variational quantum algorithms. Nat Commun 12(1):6961
293. Wang Z, Song M, Zhang Z, Song Y, Wang Q, Qi H (2019b) Beyond inferring class representatives: user-level privacy leakage from federated learning. In: Proceedings of the 38th IEEE INFOCOM conference on computer communications (INFOCOM), pp 2512–2520
294. Watkins WM, Chen SYC, Yoo S (2023) Quantum machine learning with differential privacy. Sci Rep 13(1):2453
295. Weber M, Liu N, Li B, Zhang C, Zhao Z (2021) Optimal provable robustness of quantum classification via quantum hypothesis testing. NPJ Quantum Inf 7(1):76
296. Wei W, Liu L (2025) Trustworthy distributed ai systems: robustness, privacy, and governance. ACM Comput Surv 57(6):1–42
297. Wen J, Zhang Z, Lan Y, Cui Z, Cai J, Zhang W (2023) A survey on federated learning: challenges and applications. Int J Mach Learn Cybern 14(2):513–535
298. West MT, Tsang SL, Low JS, Hill CD, Leckie C, Hollenberg LC, Erfani SM, Usman M (2023) Towards quantum enhanced adversarial robustness in machine learning. Nat Mach Intell 1–9
299. Wiersema R, Guerini L, Carrasquilla JF, Aolita L (2022) Circuit connectivity boosts by quantum-classical-quantum interfaces. Phys Rev Res 4(4):043221
300. Winderl D, Franco N, Lorenz JM (2023) Quantum neural networks under depolarization noise: exploring white-box attacks and defenses. arXiv:2311.17458
301. Wootters WK, Zurek WH (1982) A single quantum cannot be cloned. Nature 299(5886):802–803
302. Wu J, Hu T, Li Q (2024) Distributed quantum machine learning: federated and model-parallel approaches. IEEE Internet Comput 28(2):65–72
303. Wu Y, Adermann E, Thapa C, Camtepe S, Suzuki H, Usman M (2023) Radio signal classification by adversarially robust quantum machine learning. arXiv:2312.07821
304. Xia Q, Li Q (2021) Quantumfed: a federated learning framework for collaborative quantum training. In: Proceedings of the 2021 IEEE global communications conference (GLOBECOM), pp 1–6
305. Xia Q, Tao Z, Hao Z, Li Q (2019) Faba: an algorithm for fast aggregation against byzantine attacks in distributed neural networks. In: Proceedings of the 28th international joint conference on artificial intelligence (IJCAI)
306. Xia Q, Tao Z, Li Q (2020) Defenses against byzantine attacks in distributed deep neural networks. IEEE Trans Netw Sci Eng 8(3):2025–2035
307. Xia Q, Tao Z, Li Q (2021a) Defending against byzantine attacks in quantum federated learning. In: Proceedings of the 17th international conference on mobility, sensing and networking (MSN), pp 145–152
308. Xia Q, Tao Z, Li Q (2021b) Tofi: An algorithm to defend against byzantine attacks in federated learning. In: Proceedings of the 17th EAI international conference on security and privacy in communication networks (SecureComm), pp 229–248

309. Xie C, Koyejo O, Gupta I (2018) Generalized byzantine-tolerant sgd. arXiv:1802.10116
310. Xie C, Huang K, Chen PY, Li B (2019a) Dba: distributed backdoor attacks against federated learning. In: Proceedings of the 7th international conference on learning representations (ICLR)
311. Xie C, Koyejo S, Gupta I (2019b) Zeno: distributed stochastic gradient descent with suspicion-based fault-tolerance. In: Proceedings of the 36th international conference on machine learning (ICML), PMLR, pp 6893–6901
312. Xie C, Koyejo S, Gupta I (2020) Zeno++: robust fully asynchronous sgd. In: Proceedings of the 37th international conference on machine learning (ICML), pp 10495–10503
313. Xie X, Hu C, Ren H, Deng J (2024) A survey on vulnerability of federated learning: a learning algorithm perspective. Neurocomputing 573:127225
314. Xu G, Li H, Liu S, Yang K, Lin X (2019) Verifynet: secure and verifiable federated learning. IEEE Trans Inf Forensics Secur 15:911–926
315. Xu M, Li X (2020) Subject property inference attack in collaborative learning. In: Proceedings of the 12th international conference on intelligent human-machine systems and cybernetics (IHMSC), vol 1, pp 227–231
316. Xu M, Niyato D, Yang Z, Xiong Z, Kang J, Kim DI, Shen X (2022) Privacy-preserving intelligent resource allocation for federated edge learning in quantum internet. IEEE J Sel Top Signal Proc 17(1):142–157
317. Xu Q, Zhao L, Su Z, Fang D, Li R (2023) Secure federated learning in quantum autonomous vehicular networks. IEEE Netw 37(6):240–247
318. Xu X, Wu J, Yang M, Luo T, Duan X, Li W, Wu Y, Wu B (2020) Information leakage by model weights on federated learning. In: Proceedings of the 2020 workshop on privacy-preserving machine learning (PPMLP), pp 31–36
319. Yamany W, Moustafa N, Turnbull B (2021) Oqfl: an optimized quantum-based federated learning framework for defending against adversarial attacks in intelligent transportation systems. IEEE Trans Intell Transp Syst 24(1):893–903
320. Yang CW, Hwang T, Luo YP (2013) Enhancement on "quantum blind signature based on two-state vector formalism". Quantum Inf Process 12:109–117
321. Yang Q, Liu Y, Chen T, Tong Y (2019) Federated machine learning: concept and applications. ACM Trans Intell Syst Technol (TIST) 10(2):1–19
322. Yang S, Chen Y, Tu S, Yang Z (2022) A post-quantum secure aggregation for federated learning. In: Proceedings of the 12th international conference on communication and network security (ICCNS), pp 117–124
323. Yao ACC (1986) How to generate and exchange secrets. In: Proceedings of the 27th annual symposium on foundations of computer science (SFCS), pp 162–167
324. Yin D, Chen Y, Kannan R, Bartlett P (2018) Byzantine-robust distributed learning: towards optimal statistical rates. In: Proceedings of the 35th international conference on machine learning (ICML), pp 5650–5659
325. Yin H, Mallya A, Vahdat A, Alvarez JM, Kautz J, Molchanov P (2021) See through gradients: image batch recovery via gradinversion. In: Proceedings of the 34th IEEE/CVF conference on computer vision and pattern recognition (CVPR), pp 16337–16346
326. Yoshida Y (2024) Mathematical comparison of classical and quantum mechanisms in optimization under local differential privacy. J Phys A: Math Theor 58(3):035301
327. Yoshida Y, Hayashi M (2020) Classical mechanism is optimal in classical-quantum differentially private mechanisms. In: Proceedings of the 2020 IEEE international symposium on information theory (ISIT), pp 1973–1977
328. Yu B, Shen H, Xu Q, He W, Mao W, Zhang Q, Zhang F (2024) Hqsfl: A novel training strategy for constructing high-performance and quantum-safe federated learning. In: Proceedings of the 19th ACM Asia conference on computer and communications security (ASIACCS), pp 512–521
329. Yu F, Zhang W, Qin Z, Xu Z, Wang D, Liu C, Tian Z, Chen X (2020) Heterogeneous federated learning. arXiv:2008.06767

330. Yu K, Gao F, Lin S (2022) Quantum federated learning for distributed quantum networks. arXiv:2212.12913
331. Yuan X, Ma X, Zhang L, Fang Y, Wu D (2021) Beyond class-level privacy leakage: breaking record-level privacy in federated learning. IEEE Int Things J 9(4):2555–2565
332. Yuan X, Sun J, Liu J, Zhao Q, Zhou Y (2021) Quantum simulation with hybrid tensor networks. Phys Rev Lett 127(4):040501
333. Yun WJ, Baek H, Kim J (2022a) Quantum split neural network learning using cross-channel pooling. arXiv:2211.06524
334. Yun WJ, Kim JP, Jung S, Park J, Bennis M, Kim J (2022b) Slimmable quantum federated learning. arXiv:2207.10221
335. Zhang C, Li S, Xia J, Wang W, Yan F, Liu Y (2020) {BatchCrypt}: efficient homomorphic encryption for {Cross-Silo} federated learning. In: Proceedings of the 2020 USENIX annual technical conference (USENIX ATC), pp 493–506
336. Zhang R, Guo S, Wang J, Xie X, Tao D (2022a) A survey on gradient inversion: attacks, defenses and future directions. arXiv:2206.07284
337. Zhang W, Tople S, Ohrimenko O (2021) Leakage of dataset properties in {Multi-Party} machine learning. In: Proceedings of the 30th USENIX security symposium, pp 2687–2704
338. Zhang X, Deng H, Wu R, Ren J, Ren Y (2024) Pqsf: post-quantum secure privacy-preserving federated learning. Sci Rep 14(1):23553
339. Zhang Y, Zhang C, Zhang C, Fan L, Zeng B, Yang Q (2022b) Federated learning with quantum secure aggregation. arXiv:2207.07444
340. Zhang Y, Zeng D, Luo J, Fu X, Chen G, Xu Z, King I (2024) A survey of trustworthy federated learning: issues, solutions, and challenges. ACM Trans Intell Syst Technol 15(6):1–47
341. Zhang Y, Behnia R, Yavuz AA, Ebrahimi R, Bertino E (2025) Efficient full-stack private federated deep learning with post-quantum security. IEEE Trans Depend Secure Comput
342. Zhao B, Mopuri KR, Bilen H (2020) idlg: improved deep leakage from gradients. arXiv:2001.02610
343. Zhao Y, Zhong H, Zhang X, Li Y, Zhang C, Pan M (2024) Bridging quantum computing and differential privacy: insights into quantum computing privacy. In: Proceedings of the 2024 IEEE international conference on quantum computing and engineering (QCE), vol 1, pp 13–24
344. Zhou L, Ying M (2017) Differential privacy in quantum computation. In: Proceedings of the 30th IEEE computer security foundations symposium (CSF), pp 249–262
345. Zhou X, Qiu D (2021) Blind quantum machine learning based on quantum circuit model. Quantum Inf Process 20(11):363
346. Zhou Z, Xu C, Wang M, Ma T, Yu S (2021) Augmented dual-shuffle-based moving target defense to ensure cia-triad in federated learning. In: Proceedings of the 2021 IEEE global communications conference (GLOBECOM), pp 01–06
347. Zhou Z, Xu C, Wang M, Kuang X, Zhuang Y, Yu S (2022) A multi-shuffler framework to establish mutual confidence for secure federated learning. IEEE Trans Depend Secure Comput 20(5):4230–4244
348. Zhu C, Huang WR, Li H, Taylor G, Studer C, Goldstein T (2019a) Transferable clean-label poisoning attacks on deep neural nets. In: Proceedings of the 36th international conference on machine learning (ICML), pp 7614–7623
349. Zhu L, Liu Z, Han S (2019b) Deep leakage from gradients. In: Proceedings of the 33rd conference on neural information processing systems (NeurIPS)
350. Zomorodi-Moghadam M, Houshmand M, Houshmand M (2018) Optimizing teleportation cost in distributed quantum circuits. Int J Theor Phys 57:848–861
351. Zuo R, Tian H, Zhang F (2023) Post-quantum dropout-resilient aggregation for federated learning via lattice-based prf. In: Proceedings of the 2023 international conference on artificial intelligence security and privacy (AIS&P), pp 382–399

Chapter 12
Adversarially Robust Quantum Transfer Learning

Amena Khatun and Muhammad Usman

Abstract Quantum machine learning (QML) has emerged as a promising area of research for enhancing the performance of classical machine learning systems by leveraging quantum computational principles. However, practical deployment of QML remains limited due to current hardware constraints such as limited number of qubits and quantum noise. This chapter introduces a hybrid quantum-classical architecture that combines the advantages of quantum computing with transfer learning techniques to address high-resolution image classification. Specifically, we propose a Quantum Transfer Learning (QTL) model that integrates classical convolutional feature extraction with quantum variational circuits. Through extensive simulations on diverse datasets including Ants and Bees, CIFAR-10, and Road Sign Detection, we demonstrate that QTL achieves superior classification performance compared to both conventional and quantum models trained without transfer learning. Additionally, we also investigate the model's vulnerability to adversarial attacks and demonstrate that incorporating adversarial training significantly boosts the robustness of QTL, enhancing its potential for deployment in security-sensitive applications.

Keywords Quantum machine learning · Quantum transfer learning · Adversarial attack · Adversarial robustness

A. Khatun (✉) · M. Usman
Quantum Systems, Data61, CSIRO, Clayton 3168, Australia
e-mail: amena.khatun@data61.csiro.au

M. Usman
School of Physics, The University of Melbourne, Parkville, Victoria 3010, Australia
e-mail: muhammad.usman@data61.csiro.au

M. Usman (ed.), *Quantum Robustness in Artificial Intelligence*, Quantum Science and Technology, https://doi.org/10.1007/978-3-032-11153-1_12

12.1 Introduction

Machine learning (ML) has witnessed rapid adoption across numerous domains due to its capability to autonomously learn patterns from data and make predictive decisions. Its transformative impact spans critical areas such as autonomous driving [1], intelligent surveillance systems [2], facial recognition technologies [3], object detection [4], person re-identification [5–9], behavior analysis [10], and anomaly detection [11]. However, the increasing complexity and size of datasets, combined with the computational intensity of training deep neural networks, have motivated the exploration of Quantum Machine Learning (QML).

QML employs the principles of quantum mechanics such as superposition, entanglement, and interference to develop fundamentally new learning algorithms [13–23]. These algorithms offer the potential for exponential speedups and improved accuracy in solving complex learning problems. A particularly notable development in this space has been the development of hybrid quantum-classical models, including quantum convolutional neural networks (QCNNs), which have demonstrated promising performance in classification tasks while maintaining relatively low model complexity [24, 25]. These models show effectiveness particularly in low-parameter environments [26, 27], yet their scalability to high-dimensional, real-world datasets remains constrained. To overcome these scalability issues and to extend the applicability of QML to more complex datasets, researchers have proposed Quantum Transfer Learning (QTL) as a hybrid strategy that integrates classical deep learning with quantum neural networks (QNNs) [12, 28–32]. In a QTL framework, a classical model often pre-trained on large datasets is used for feature extraction, while a quantum variational circuit handles classification. This effectively reduces the quantum resource burden while still leveraging the advantages of quantum computation.

Transfer Learning (TL) itself has become a widely used method in deep learning for accelerating training and improving performance, especially when labeled data is limited [33]. It enables the reuse of parameters (such as weights and biases) from a model trained on one task (e.g., Task A using Dataset X) for a related task (e.g., Task B using Dataset Y). By freezing the early layers of a deep network that capture generic visual features like edges, textures, and patterns, and fine-tuning the later layers, TL achieves efficient adaptation to new domains with minimal computational overhead [34, 35]. This strategy significantly reduces training time while improving model generalization. Figure 12.1 illustrates the overview of TL approach.

Despite the notable performance improvements offered by QML models, handling high-dimensional data remains a key challenge. Firstly, the number of qubits required to represent high-dimensional feature spaces scales with data complexity, but current quantum devices are limited in qubit count. Secondly, increasing the number of input features typically necessitates deeper quantum circuits, which are more susceptible to noise and decoherence. Thirdly, encoding high-dimensional classical data into quantum states efficiently is still an unresolved issue, with ongoing research into effective quantum data-encoding strategies [36–39]. These challenges motivate the design of hybrid QTL models that balance quantum computational efficiency with

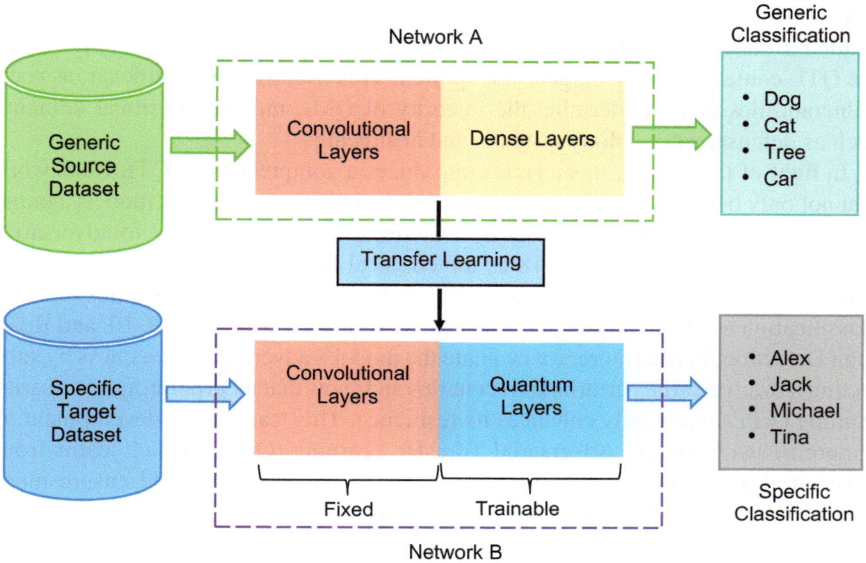

Fig. 12.1 Overview of Quantum Transfer Learning (QTL). Network A is trained on a source dataset which is typically large and diverse. The purpose of training Network A is to learn general features, patterns, and representations that are applicable to the broader domain. Here, Network A will learn to recognize basic shapes, textures, and patterns commonly found in images of various animals and objects. Network B is the target model that is used for a different but related task. Instead of starting the training of Network B from scratch, the weights and parameters are initialized and learned by Network A. QTL model performs well on a smaller or completely different dataset compared to the source dataset as it learns features and representations from a source task to initialize a model for the target task. Even if the source and target domains are different, many lower-level features (edges, colors, and textures) and patterns are common to both domains. After initialization, Network B is further trained on the target dataset. During this fine-tuning process, the model adapts its representations to the specific characteristics and requirements of the target task. This fine-tuning allows the model to specialize while retaining the general knowledge from Network A. This figure is reproduced under terms of the CC-BY license [12]. Copyright 2024, Amena Khatun and Muhammad Usman. Advanced Quantum Technologies published by Wiley-VCH GmbH

classical preprocessing capabilities. Several QTL approaches have recently been developed with various applications such as waste detection [40], image-based crack identification [32], tuberculosis diagnosis [30], and breast cancer detection from mammograms [31]. Reference [41] presented a QTL model for speech command recognition. While these QTL approaches show promising results, these studies lack a comprehensive comparison between QTL and classical TL methods. Without such comparisons, it is difficult to understand if the observed performance gains in QTL are outperforming the classical approaches.

Another notable limitation of the current QTL approaches is the lack of research on their vulnerability to adversarial attacks. Adversarial attacks introduce subtle, often imperceptible perturbations to input data to trick the model to make mistakes, which is a serious threat to the reliability of ML and QML systems. Although extensively studied in classical settings [42–47], their impact on QML models, especially within the QTL context, remains largely unexplored. This oversight is significant, as such vulnerabilities could undermine the integrity of QML models in critical domains such as defense, autonomous systems, and healthcare.

In light of these gaps, this chapter introduces a comprehensive QTL framework that not only benchmarks the performance of quantum-enhanced TL models against classical baselines but also rigorously examines their susceptibility to adversarial attacks. By employing ResNet-18 as the classical feature extractor and a quantum variational circuit as the classifier, we build a hybrid model that demonstrates superior classification accuracy across datasets such as Ants and Bees, CIFAR-10, and Road Sign Detection. Furthermore, we evaluate the model's adversarial robustness by subjecting it to adversarial perturbation scenarios and show that incorporating adversarial training (AT) significantly enhances its resilience. This leads to the development of a more robust Quantum Adversarial Transfer Learning (QATL) which learns from adversarial examples during training to mitigate vulnerabilities and ensure more reliable predictions in real-world environments.

12.2 Literature Review

This section provides an overview of related research in hybrid quantum-classical convolutional neural networks, quantum transfer learning, adversarial vulnerability in quantum models, and methods to enhance their robustness through AT.

12.2.1 Quantum-Classical Hybrid Convolutional Neural Networks (QCHCNNs)

The performance of classical ML algorithms is limited by the computational demands of processing large-scale data. Quantum computing, with its potential to perform certain tasks faster than classical systems, offers a promising direction for developing QML models capable of addressing this growing complexity. Hence, the field of QML has attracted substantial attention, with several recent works proposing quantum algorithms aimed at outperforming their classical counterparts [48–51]. Motivated by these developments, researchers have introduced various quantum convolutional neural network (QCNN) architectures developed to near-term quantum devices. Reference [24] proposed a QCNN that requires only $O(\log(N))$ variational parameters for N qubits, making it scalable for practical implementations. Their architecture integrates concepts from quantum error correction and multiscale entanglement renormalization. Similarly, Ref. [52] developed a quantum clustering algorithm inspired by classical k-means, which can achieve exponential speedups under certain data conditions. Expanding on conventional QCNN architectures, Ref. [53] introduced the branching QCNN, which enhances feature extraction by leveraging global information from all the qubits that are measured at the pooling layers. This design enables the model to capture long-range correlations between qubits as this approach is not only focusing on the local information. In high-energy physics applications, Ref. [54] proposed a hybrid QCNN for event classification and demonstrated its superior learning efficiency and test accuracy compared to classical CNNs, despite using a similar number of parameters.

Due to the wide range of applications of image classification across domains, such as autonomous driving [55, 56], defense, healthcare [57, 58], and security [59], QCNNs have also been developed for image classification tasks. Quanvolutional Neural Network (QNN) is introduced in Ref. [60], which incorporates a quantum transformation layer using randomized quantum circuits. This method is evaluated on MNIST dataset, and demonstrated performance improvements in test accuracy and training speed over classical baselines. Further innovations in quantum feature extraction have been proposed by Ref. [26] that integrates Multi-scale Entanglement Renormalization Ansatz and fractal feature extraction via box-counting within a QCNN architecture for binary classification. Their model outperformed classical CNNs on a breast cancer dataset. Reference [27] developed a fully parameterized QCNN using shallow-depth circuits and two-qubit interactions, which is well suited for noisy intermediate-scale quantum (NISQ) hardware. Their model achieved strong results on MNIST and Fashion-MNIST datasets, despite having a small number of parameters.

12.2.2 Quantum Transfer Learning

While hybrid QCNNs have shown promising capabilities in small-scale image classification tasks, their extension to more complex, real-world datasets is challenging due to current hardware limitations, specifically the small number of available qubits and relatively high noise levels. To address these challenges, in recent years, researchers have explored QTL to combine classical ML scalability with QNNs.

Reference [28] introduced the concept of QTL and proposed a quantum variational circuit (QVC) to a classical pre-trained model, such as ResNet, to leverage classical feature extraction while performing final classification in the quantum domain. This method proved particularly effective for high-dimensional input data, as the classical component handles heavy preprocessing, thereby minimizing the quantum hardware load. Building on this foundation, Ref. [29] extended QTL to image classification with small, high-dimensional datasets like Eurosat and UC Merced. They used a hybrid approach involving classical CNNs and multi-qubit QML networks to classify $256 \times 256 \times 3$ images, demonstrating improved performance through strong entanglement and amplitude encoding in the quantum layer. Reference [30] further validated QTL's potentiality by applying it to diverse tasks such as chest X-ray tuberculosis detection, concrete crack recognition, and trash material classification. Their findings underscored that pre-trained classical extractors like DenseNet, AlexNet, and VGG19 provide advantages when combined with a quantum classifier, especially with limited quantum resources. Reference [31] adapted this methodology for breast cancer detection in full-image mammograms, further proving that quantum-enhanced TL can enhance the performance of purely classical models. Similarly, Ref. [32] applied a hybrid QTL framework to grayscale crack detection in images and reported improved accuracy, replicating Ref. [28] architecture and showing its generalizability across domains.

Going beyond the visual domains, Ref. [41] proposed a QTL architecture for spoken command recognition using a classical 1D CNN for feature extraction and a QVC-based quantum classifier. This work illustrated that QTL's benefits are not limited to image classification but extend to other modalities such as audio processing. Recent advancements have further deepened the understanding of QTL's capabilities. Reference [61] proposed a QTL framework using QVC and demonstrated that expressibility linked to circuit depth, structure, and qubit count has a strong correlation with classification accuracy. This study highlights the importance of circuit design in enhancing QTL efficiency and found that annular QVCs combined with a ResNet-18 backbone significantly reduced training time while maintaining high accuracy. Reference [62] approached QTL from a theoretical perspective, studying how pre-trained QVCs can be fine-tuned across domains. Reference [63] extended QTL into natural language processing by using embedding vectors from large language models to perform sentence acceptability judgments. The QTL model not only outperformed classical performance but also showed better handling of complex sentence structures, pointing toward a quantum advantage in future NLP tasks.

The existing QTL studies have made notable progress in addressing key challenges in image classification, particularly limited qubits and high noise levels. Several QTL approaches have been proposed to leverage the advantages of TL in the quantum field. However, these studies generally do not explore the performance of QTL models in the absence of transfer learning, that is, without the use of pre-trained classical networks. Furthermore, direct comparisons between QTL and their classical counterparts remain limited, hindering a comprehensive understanding of QTL's performance over classical TL. While current QTL methods demonstrate promising potential, more extensive evaluations are needed to assess their practical benefits and limitations in real-world applications. Notably, most existing QTL studies have focused on low-resolution datasets such as EuroSAT ($64 \times 64 \times 3$), UC Merced Land Use ($256 \times 256 \times 3$), and MNIST ($28 \times 28 \times 3$). In contrast, our proposed method evaluates QTL on high-resolution image datasets including Ants and Bees ($768 \times 512 \times 3$) and Road Sign Detection ($1024 \times 1024 \times 3$), thereby addressing a significant gap in the literature.

12.2.3 Adversarial Vulnerabilities in QML

It is well established that even the most advanced classical ML models are vulnerable to small, deliberately crafted perturbations applied to even a few pixels of the input data. These subtle perturbations are referred to as adversarial examples which can lead to significant misclassification, raising serious concerns about the security and robustness of ML applications [42–47]. This vulnerability raises similar concerns for QML systems, particularly as their adoption grows in sensitive domains. Recent studies have demonstrated that quantum learning systems much like their classical counterparts are susceptible to adversarial manipulation, regardless of whether the input data is classical or quantum in nature [64–71]. Quantum classifiers, including those that achieve near state-of-the-art performance, can be misled by inputs containing imperceptible perturbations, resulting in incorrect predictions. Reference [66] systematically investigated the vulnerability of QML models under both white-box and black-box attack scenarios. Their experiments included handwritten digit classification on MNIST, phase recognition in physical systems, and quantum data classification. In the white-box setting, where the attacker has full access to the model architecture and parameters, even minor perturbations led to substantial drops in quantum classifier accuracy. In the black-box setting, adversarial examples generated using classical models were successfully transferred to quantum models, highlighting the transferability of adversarial examples and the inherent susceptibility of quantum classifiers to such attacks. In contrast, Ref. [72] presented a more optimistic view of QML robustness. By comparing the adversarial performance of QVCs with classical models like CNNs and ResNet-18, their results demonstrated that quantum models exhibit greater resilience to attacks crafted for classical classifiers. This difference was attributed to the distinct feature representations learned by QVCs, which may be less aligned with those exploited by traditional adversarial attacks. These findings suggest that while QML models are not immune to adversarial threats, their internal feature representations may offer some degree of inherent robustness, especially against cross-model attacks.

Despite these advancements, defending QML models from adversarial attacks remains a challenging and open research problem. While theoretical frameworks and empirical results are emerging, practical implementations of adversarial defenses for quantum models are still limited. As QML moves toward real-world applications, ensuring the robustness and security will be critical for their deployment in security-sensitive applications.

12.2.4 Quantum Adversarial Training

Enhancing the robustness of quantum classifiers is essential to ensure the security and reliability of QML applications. While AT has been widely recognized as an effective defense mechanism in classical ML, its adoption in the quantum domain remains limited. Only a few of studies have explored its applicability and impact within QML frameworks [15, 66, 73]. Reference [66] demonstrated the vulnerability of quantum classifiers to adversarial perturbations, comparable to their classical counterparts. The study first introduced AT in QML and reported a notable improvement in robustness of quantum classifier against adversarial attacks generated using the basic iterative method [74]. However, the evaluation was limited to binary classification tasks on the downscaled MNIST dataset and did not explore the model's applicability to more complex, multi-class, and high-dimensional image datasets. Expanding on this foundation, Ref. [73] presented a comprehensive framework to assess and benchmark the adversarial robustness of both quantum and classical ML models. This study evaluated QVCs, CNNs, and ResNet-18 architectures under both classical and quantum adversarial attacks, including Fast Gradient Sign Method (FGSM) [64], Projected Gradient Descent [75], and AutoAttack [76]. The results showed that while AT significantly enhanced the robustness of classical models against classical attacks, the improvements observed in QVCs were relatively modest. The authors concluded that QML models, particularly QVCs, exhibit inherent robustness against certain types of adversarial perturbations, and the additional benefits provided by AT are limited in such cases. These findings suggest different interpretations, while Ref. [66] reported substantial gains from AT, especially under BIM attacks in a binary setting, Ref. [73] found that quantum models are already resilient to many classical attacks and therefore do not benefit substantially from AT in more diverse and realistic settings. Notably, the methodological scope of these two works differs significantly in terms of dataset complexity, number of classes, and variety of attack methods. Although early research supports the use of AT as a defense mechanism in QML, its overall effectiveness across architectures and datasets remains an open research question. A systematic investigation involving diverse QML architectures, multi-class datasets, and a range of adversarial attack models is necessary to fully understand the impact and limitations of AT in the quantum setting.

12.3 Methodology

12.3.1 Knowledge Transfer from Classical CNNs to QCHCNNs

The proposed QTL framework incorporates TL principles to train QCHCNNs, effectively leveraging the representational power of large datasets to improve classification on specific target datasets such as Ants and Bees, Road Sign Detection, and CIFAR-10. In this architecture, a widely used deep residual network, ResNet-18 [77], is employed as a pre-trained model to transfer knowledge from a source domain to the target domain. Residual connections in ResNet significantly improve gradient flow through the network, ensuring stable learning and faster convergence. These connections allow information to bypass certain layers, minimizing training time and offer optimal parameter tuning. While deeper networks typically suffer from degradation in performance as their depth increases, ResNet mitigates this through the use of residual mappings, enabling the network to model complex relationships without performance saturation [77]. The ResNet-18 model is pre-trained on the extensive ImageNet dataset [78], which contains over 1.2 million training images across 1,000 classes, and serves as a highly capable feature extractor [34, 35, 79]. The initial processing of input data involves a convolutional layer with a 7×7 filter and stride of 2, followed by a max-pooling layer with a 3×3 filter and stride of 2, as shown in Fig. 12.2a. Each subsequent layer contains two residual blocks, each with two weighted layers connected through skip connections and ReLU activations. When input and output dimensions match, identity shortcuts are used. In the case when

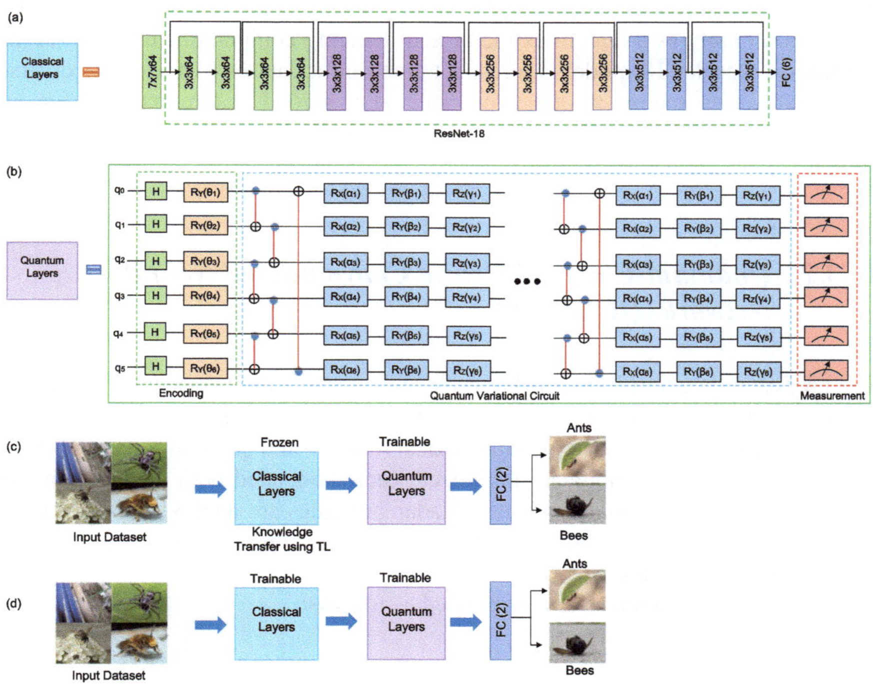

Fig. 12.2 The architecture of the proposed QTL framework: **a** illustrates the classical layers utilizing ResNet-18, which comprises 18 layers organized in residual blocks with skip connections. **b** depicts the trainable quantum components, including data encoding, a Quantum Variational Circuit (QVC), and measurement layers. The QVC integrates CNOT gates and single-qubit rotation gates (R_X, R_Y, R_Z). **c** shows the complete QTL workflow, where input data passes through frozen ResNet-18 layers before reaching trainable quantum layers for final classification. **d** demonstrates the architecture of the hybrid quantum-classical model without transfer learning, where all classical and quantum layers are updated during training. The Ants and Bees dataset used is publicly available at https://www.kaggle.com/datasets/gauravduttakiit/ants-bees. This figure is reproduced under terms of the CC-BY license [12]. Copyright 2024, Amena Khatun and Muhammad Usman. Advanced Quantum Technologies published by Wiley-VCH GmbH

dimensionality is increased, a projection shortcut technique is employed by 1×1 convolutions to match the dimension. Downsampling is performed via convolutional layers with a stride of 2, and batch normalization follows each convolutional layer. The core residual block operation is mathematically represented by

$$y = \sum F(x, W_i) + x, \tag{12.1}$$

where x denotes the input, y the output, and $F(x, W_i)$ is the residual function parameterized by weights W_i. The final output from ResNet-18 is a 512-dimensional feature map, which is flattened and reduced to a six-dimensional feature vector using a fully connected layer. These six features are then passed into the quantum variational circuit for the classification task, as illustrated in Fig. 12.2c. During training, all ResNet-18 layers remain frozen except for the last dense layer, enabling the network to function purely as a feature extractor and reducing computational costs. As the pre-trained ResNet-18 model is already trained on a dataset with a lot of diverse image categories, the pre-train network will act as a good feature extractor for our data (Ants and Bees, Road Sign Detection, CIFAR-10), even if the new data are from completely different classes.

12.3.2 Quantum-Classical Hybrid Convolutional Neural Network (QCHCNN)

The proposed quantum framework consists of three standard stages: classical data encoding into quantum states, quantum variational circuit, and quantum measurement. Angle encoding is used to map classical features into quantum states. With six extracted features, six qubits are initialized, each associated with a corresponding feature. To initiate the quantum states, Hadamard gates are applied to bring qubits into superposition states, followed by R_Y rotation gates to encode the classical feature values as rotation angles.

The encoded quantum state is expressed as follows:

$$\begin{aligned} |\theta\rangle &= |\theta_1\rangle \otimes |\theta_2\rangle \otimes |\theta_3\rangle \otimes |\theta_4\rangle \otimes |\theta_5\rangle \otimes |\theta_6\rangle \\ = \begin{bmatrix} cos(\theta_1) \\ sin(\theta_1) \end{bmatrix} \otimes \begin{bmatrix} cos(\theta_2) \\ sin(\theta_2) \end{bmatrix} &\otimes \begin{bmatrix} cos(\theta_3) \\ sin(\theta_3) \end{bmatrix} \otimes \begin{bmatrix} cos(\theta_4) \\ sin(\theta_4) \end{bmatrix} \otimes \begin{bmatrix} cos(\theta_5) \\ sin(\theta_5) \end{bmatrix} \otimes \begin{bmatrix} cos(\theta_6) \\ sin(\theta_6) \end{bmatrix} \\ &= (\otimes_{i=1}^{6} R_y(2x_i))|\theta_1\rangle^{\otimes^6}. \end{aligned} \tag{12.2}$$

Thus, the final encoded quantum state becomes

$$(\otimes_{i=1}^{6} R_y(2x_i))|\theta_1\rangle^{\otimes 6} = \begin{bmatrix} cos(\pi\theta_1) \\ sin(\pi\theta_1) \end{bmatrix} \otimes \begin{bmatrix} cos(\pi\theta_2) \\ sin(\pi\theta_2) \end{bmatrix} \otimes \begin{bmatrix} cos(\pi\theta_3) \\ sin(\pi\theta_3) \end{bmatrix} \otimes \begin{bmatrix} cos(\pi\theta_4) \\ sin(\pi\theta_4) \end{bmatrix} \otimes \begin{bmatrix} cos(\pi\theta_5) \\ sin(\pi\theta_5) \end{bmatrix} \otimes \begin{bmatrix} cos(\pi\theta_6) \\ sin(\pi\theta_6) \end{bmatrix}. \tag{12.3}$$

A QVC formulates the optimization problem directly within the quantum computational framework by embedding tunable gate parameters into the circuit design. These parameters are subsequently refined through classical optimization techniques to identify optimal solutions for the targeted task. The QVC employs a hybrid structure of two types of quantum operations: entangling gates and single-qubit rotation gates. Specifically, the Controlled-NOT (CNOT) gate creates entanglement between pairs of qubits, enabling quantum correlations that surpass the capabilities of classical systems. Here, the single-qubit rotation gates are R_X, R_Y, and R_Z which apply rotations around the X-, Y-, and Z-axes, respectively. The rotation angles, α_i, β_i, and γ_i, control these operations and are treated as trainable parameters throughout the learning process. By fine-tuning these angles, the state of each qubit can be manipulated to effectively encode complex quantum information.

The output quantum states from the QVC circuit are evaluated by computing the expectation values of six observables, denoted as $z = [z_1, z_2, z_3, z_4 z_5, z_6]$. These observables are selected to capture distinct characteristics of the quantum system. The objective is to estimate the average values of these observables, representing the expected outcomes obtained from repeated measurements of the same quantum state.

$$M : |\theta\rangle \rightarrow z = \langle\theta|z|\theta\rangle, \tag{12.4}$$

where M represents the measurement operation. The QNN model is optimized using stochastic gradient descent (SGD), updating parameters according to

$$\theta(t+1) = \theta(t) - \eta \cdot \nabla L(\theta(t)), \tag{12.5}$$

where $\theta(t)$ is the vector of model parameters at iteration t; $\theta(t+1)$ is the updated vector of model parameters at iteration $(t+1)$; η is the learning rate, a hyperparameter that determines the step size for the update; and $\nabla L(\theta(t))$ is the gradient of the loss function L with respect to the model parameters. For a small value of ϵ, the finite-difference method is used to approximate the partial derivative as

$$\frac{\partial L(\theta)}{\partial \theta} \approx \frac{L(\theta + \varepsilon) - L(\theta)}{\varepsilon}. \tag{12.6}$$

Finally, the cross-entropy loss, a commonly used objective function in classical ML, is employed to quantify the discrepancy between the predicted probability distribution and the actual distribution of the target classes. For binary classification, the loss function is defined as follows:

$$L_{classification} = -[y \cdot \log(p) + (1 - y) \cdot \log(1 - p)]. \tag{12.7}$$

Here, y represents the true class label, p denotes the predicted probability of the sample belonging to the positive class (i.e., class 1), and $1 - p$ corresponds to the predicted probability of the sample belonging to the negative class (i.e., class 0). When $y = 1$, the loss depends solely on $\log(p)$, encouraging the model to predict p close to 1, indicating a correct positive classification. Conversely, when $y = 0$, the loss is determined by $-\log(1 - p)$, pushing the model to predict p near 0, which reflects a correct negative classification. For multi-class scenarios, the loss can be represented as

$$L_{classification} = -\sum_{i=1}^{n} p_i \log \bar{p}_i = -\log \bar{p}_t, \tag{12.8}$$

where p_i is the true distribution, $\bar{p}_i$ is the predicted probability for class i, and t is the target class.

12.3.3 QCHCNN Without Transfer Learning

To conduct a comprehensive comparison with the proposed QTL framework, we also design a hybrid classical-quantum architecture that operates without incorporating transfer learning. In this configuration, although ResNet-18 serves as the initial feature extractor, the entire model, including all layers of ResNet-18, is fine-tuned from scratch rather than leveraging pre-trained weights. As illustrated in Fig. 12.2(d), all network parameters, both classical and quantum, are updated throughout the training process. The classical component of this hybrid architecture performs the initial preprocessing of raw input data and encodes it into a representation suitable for quantum processing. While ResNet-18 generates high-dimensional feature maps that capture complex hierarchical patterns, directly working with such high-dimensional data can be computationally prohibitive and inefficient. To address this, the extracted features are compressed into a lower-dimensional latent space using a fully connected (Dense) layer, which learns meaningful and compact representations. These compressed features are subsequently passed to the quantum module, where quantum encoding techniques prepare the data for quantum state manipulation. After processing within the quantum circuit, measurement operations are performed to extract classical outputs from the resulting quantum states. These outputs are then forwarded to a final classical fully connected layer, connected through a non-trainable matrix, which integrates quantum-derived features and produces the final predictions. The use of a non-trainable matrix ensures that quantum information is seamlessly incorporated without introducing additional learnable parameters at this stage. Training this hybrid classical-quantum architecture involves jointly optimizing both the classical layers and the quantum circuit parameters. The learning process is driven by a loss function that determines the difference between predicted outputs and ground truth labels. Standard gradient-based optimization algorithms are employed to iteratively adjust the network parameters, minimizing the loss and improving predictive accuracy.

However, due to the complexity of quantum state manipulation and the high-dimensional optimization landscape, training quantum circuits presents significant computational challenges. Techniques such as the parameter-shift rule and classical gradient descent methods are utilized to fine-tune the parameters of the quantum circuit. Unlike the QTL framework, which benefits from faster convergence and improved performance through TL, this architecture demands greater computational resources and more extensive training iterations as it learns classical and quantum representations entirely from the ground up.

12.3.4 Classical Transfer Learning

The classical transfer learning architecture is illustrated in Fig. 12.3. This approach utilizes ResNet-18 as the backbone of the model, pre-trained on the extensive and diverse ImageNet dataset, which contains millions of labeled images spanning a wide array of object categories. In this framework, the pre-acquired knowledge from the source domain is effectively leveraged to improve performance on a new, yet related, target domain task. The detailed design and functionality of ResNet-18 have already been discussed comprehensively in Sect. 12.3.1. During the fine-tuning process, all convolutional layers of ResNet-18 remain frozen, with only the final dense layers left trainable. These fully connected layers act as the new classifier, specifically adapted

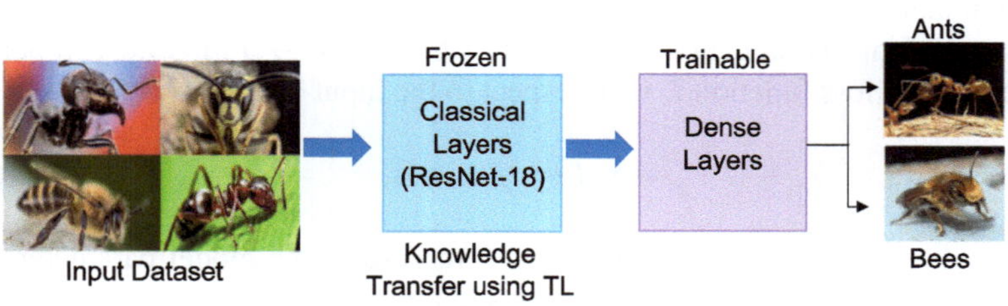

Fig. 12.3 The architecture of classical transfer learning. ResNet-18 consists of 18 layers as illustrated in Fig. 12.2a and pre-trained on a larger and diverse ImageNet dataset. The knowledge of the pre-trained network is transferred from source domain to the target domain. For classical TL, all the layers of ResNet-18 are frozen except the last dense layer. Only the dense layers are trainable which are fully connected layers for the final classification task. When there is no transfer learning involved, all the classical layers and dense layers are trainable. This figure is reproduced under terms of the CC-BY license [12]. Copyright 2024, Amena Khatun and Muhammad Usman. Advanced Quantum Technologies published by Wiley-VCH GmbH

for the target domain task. In contrast, when TL is not applied, the entire ResNet-18 model, including all convolutional and dense layers, is trained end to end from scratch.

12.3.5 Adversarial Attacks on Quantum Models

Adversarial attacks are intentionally crafted to deceive ML models, particularly the classifiers, by introducing subtle perturbations to input data that lead to incorrect predictions. Consider a classifier defined as $f : X \to Y$, where X is a subset of R^d, representing the input space, and Y denotes the set of possible class labels. The goal of an adversarial attack is to construct a perturbed input $x^\wedge$ that remains close to a valid input sample x under a specified distance metric, while ensuring that the classifier's output for the perturbed input differs from its original prediction, i.e., $f(x^\wedge) \neq f(x)$. If the attacker's sole objective is to induce any incorrect classification without targeting a specific label, it is classified as a non-targeted attack. Conversely, in a targeted attack, the adversary aims to manipulate the classifier so that the perturbed input $x^\wedge$ is misclassified as a predetermined target label y, which can be represented as $f(x^\wedge) = y$. Regardless of the attack type, the perturbation introduced to the original input should remain minimal to avoid detection. This perturbation is commonly measured using the L_p norm, where typical values of p include 0, 1, 2, or ∞. These norms characterize the perturbation in different ways: the L_1 norm measures the sum of absolute differences, the L_2 norm corresponds to the Euclidean distance, and the L_∞ norm measures the maximum absolute difference across all dimensions. An illustration of adversarial attack process is provided in Fig. 12.4. In our proposed QTL framework, we utilize the FGSM attack to generate adversarial examples. FGSM introduces perturbations by modifying the input data in the direction that most rapidly increases the loss of the model, thereby maximizing the likelihood of misclassification. Given a classifier $f : X \to Y$, FGSM computes the gradient of the loss function L with respect to the input sample x as follows:

$$\Delta_x L(y, f(x; \theta)), \tag{12.9}$$

where y denotes the true class label and θ represents the model parameters. This gradient identifies the direction in which small changes to the input will most significantly affect the loss.The adversarial example $x^\wedge$ is then generated by applying a perturbation in the direction of the gradient sign:

$$x^\wedge = x + \epsilon \cdot \text{sign}(\nabla_x L(y, f(x; \theta))), \tag{12.10}$$

where ϵ is a small positive constant that controls the magnitude of the perturbation. The $sign$ function extracts the direction of the gradient for each input feature, indicating whether the perturbation should increase or decrease the corresponding input

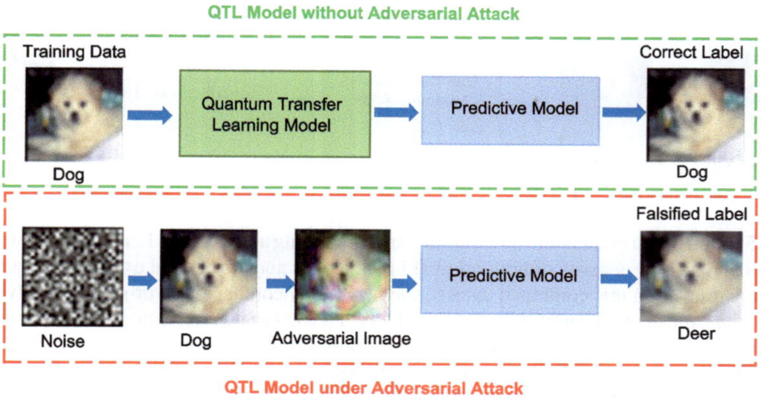

Fig. 12.4 Overview of quantum adversarial attack. If we consider the input image of a dog, the QTL model correctly classifies it as a dog when the model is not under adversarial attack. However, a carefully crafted perturbation is applied to the input image in a way that exploits the vulnerabilities of the proposed QTL model. When the perturbed images are fed to the model for prediction, the QTL model incorrectly classified the image as a deer rather than a dog. This is an indication that the QTL model is vulnerable to small changes in input data, leading to significant implications for the reliability and security of the model. The original image of dog in this figure is from the publicly available dataset, CIFAR-10 [80]. This figure is reproduced under terms of the CC-BY license. [12]. Copyright 2024, Amena Khatun and Muhammad Usman. Advanced Quantum Technologies published by Wiley-VCH GmbH

value. This process effectively leads the model to make incorrect predictions while maintaining the perturbed input visually close to the original data.

12.3.6 Adversarial Training for Quantum Robustness

Adversarial training is a widely adopted defense mechanism in ML aimed at improving the resilience of neural networks against adversarial perturbations. These perturbations introduce subtle, carefully crafted modifications to input data with the intent of misleading a model's predictions. The conceptual framework of adversarial training is illustrated in Fig. 12.5, where the proposed hybrid QTL model is trained using a combination of both original and adversarially perturbed images. Here, we consider a scenario in which the QTL network is tasked with classifying animal images. An adversarial attack might introduce imperceptible changes to an image of a dog, causing the model to incorrectly classify it as a deer. The purpose of adversarial training is to improve the model's robustness against such deceptive manipulations, ensuring reliable predictions even in the presence of adversarially modified inputs. This is achieved by exposing the model to adversarial examples during training, allowing it to learn representations that are less sensitive to small input perturbations.

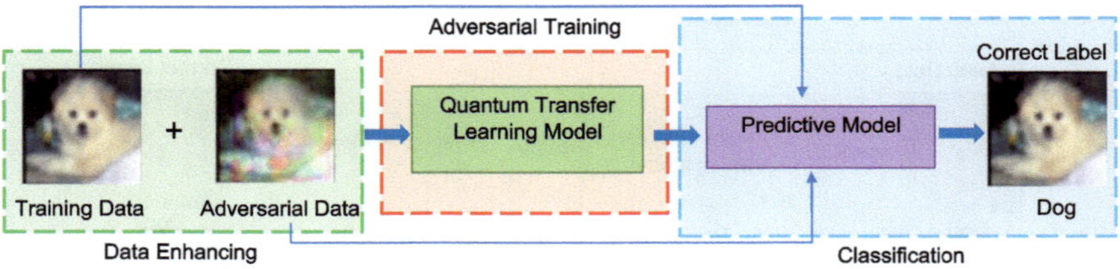

Fig. 12.5 Illustration of quantum adversarial transfer learning architecture. For adversarial training, the training data now consists of both original input images and adversarial examples. As the QTL network is trained on this combined dataset, the network encounters both original and adversarial images. Thus, the network learns to resist the effects of adversarial perturbations and make accurate predictions even when presented with adversarial examples. As a result of adversarial training, the QTL network becomes more robust against adversarial attacks. When presented with a perturbed image that was crafted to deceive the network, it is now less likely to be misclassified. In the figure, where a dog image was misclassified as a deer, the adversarial training would help the network correctly classify it as a dog. The original image of dog in this figure is taken from the publicly available dataset, CIFAR-10 [80]. This figure is reproduced under terms of the CC-BY license [12]. Copyright 2024, Amena Khatun and Muhammad Usman. Advanced Quantum Technologies published by Wiley-VCH GmbH

Mathematically, adversarial training formulates the learning objective as a min-max optimization problem:

$$\min_{\theta} \max_{\delta \in \Delta} \mathcal{L}(f_{\theta}(x + \delta)), y, \tag{12.11}$$

where θ represents the model parameters, x is the original input sample, y is the corresponding ground truth label, and $f_{\theta}(x)$ denotes the model's prediction. The loss function $\mathcal{L}$ calculates the error between the model's predictions and the true labels. The term δ represents the perturbation applied to the input, and Δ defines the set of allowable perturbations under the threat model. In this context, the threat model is specified as $\Delta = \delta : |\delta|\infty \leq \varepsilon$, constraining the perturbations to have a maximum magnitude bounded by ϵ under the $\mathcal{L}\infty$ norm. Directly solving the inner maximization over the perturbation space Δ is computationally intensive. Therefore, FGSM is used as an efficient approximation to generate adversarial perturbations. The optimal perturbation is computed as follows:

$$\delta^* = \varepsilon \cdot \text{sign}(\nabla_x \mathcal{L}(f(x), y)), \tag{12.12}$$

where ∇_x denotes the gradient of the loss function with respect to the input x. This gradient identifies the direction in which the input should be perturbed to maximize the loss. During training, the model parameters θ are updated using gradient descent, taking into account the adversarially perturbed inputs generated by δ^*. After each update, the perturbation δ may push the modified input outside the permissible bounds defined by Δ. To ensure the perturbation remains within the threat model, it is projected back into the allowed set by clipping each element of δ to the range $[-\varepsilon, \varepsilon]$. By iteratively applying this process, adversarial training strengthens the model's defenses against adversarial attacks, enabling it to maintain reliable classification performance even when exposed to adversarially manipulated inputs.

12.4 Network Setup and Training

Given the current limitations of quantum hardware such as the limited number of qubits, high error rates, and decoherence, all experimental simulations were performed using PennyLane [81], an open-source QML framework developed by Xanadu. The models were trained for a total of 50 epochs, requiring approximately 12 minutes and 27 s to complete. For the classical components of the framework, we used PyTorch [82].The Adam optimizer was employed across all experiments, using a batch size of 16 and an initial learning rate of 0.0004. This learning rate was kept constant during the first 25 epochs and then gradually decayed linearly to zero over the remaining 25 epochs to ensure smooth convergence. In the quantum circuit configuration, we initialized the model with 6 qubits and a circuit depth of 6 variational layers. Consequently, the total number of trainable parameters for the QVC is 108, calculated as $6 \times 6 \times 3$. In comparison, the classical TL baseline included 1,026 trainable parameters. While increasing the depth of the quantum circuit could potentially improve performance, it also increases the risk of encountering the well-known barren plateau problem, where the optimization landscape becomes flat, leading to ineffective training. However, since the QVC models demonstrated strong performance during both training and evaluation, we believe further increasing the circuit depth would likely yield only marginal gains. To further support efficient convergence, a learning rate scheduling strategy was applied, reducing the learning rate by a factor of 0.1 every 10 epochs. The quantum model weights were initialized with a small random spread, set to 0.01, to ensure stable optimization at the start of training.

12.5 Experimental Results

12.5.1 Datasets

To evaluate the effectiveness and adversarial robustness of the proposed QTL framework, we employed three widely used high-resolution image datasets: Ants and Bees, CIFAR-10, and Road Sign Detection as illustrated in Fig. 12.6. Unlike many existing QML/QAML studies that primarily rely on low-resolution datasets such as MNIST ($28 \times 28 \times 1$) or CIFAR-10 ($32 \times 32 \times 3$), our evaluation is conducted on higher-dimensional image datasets to demonstrate scalability and real-world applicability.

Fig. 12.6 Overview of datasets used to evaluate the proposed QTL method. In the figure, **a** represents Ants and Bees dataset, taken from kaggle (https://www.kaggle.com/datasets/gauravduttakiit/ants-bees), **b** illustrates Road sign detection dataset, taken from kaggle (https://www.kaggle.com/datasets/andrewmvd/road-sign-detection), and **c** represents CIFAR-10 dataset [80]. Each dataset encapsulates distinct challenges and contexts, collectively contributing to a comprehensive assessment of the QTL method's performance. All the datasets are publicly available. This figure is reproduced under terms of the CC-BY license [12]. Copyright 2024, Amena Khatun and Muhammad Usman. Advanced Quantum Technologies published by Wiley-VCH GmbH

Ants and Bees[1] is relatively compact dataset and consists of only 120 labeled images categorized into ants and bees. Additionally, there are 75 validation images for each class. All images have a resolution of 768×512 pixels. Due to the limited size of the dataset, training a deep model from scratch would be inefficient and prone to overfitting. However, our QTL framework capitalizes on knowledge transferred from a pre-trained model, allowing it to adapt efficiently and perform well despite the small size datasets.

CIFAR-10 (Canadian Institute for Advanced Research) [80] is a benchmark dataset extensively used in computer vision for evaluating classification algorithms. It contains a total of 60,000 color images across 10 distinct object categories. The training subset comprises 50,000 samples, evenly distributed among the classes, while the test subset consists of 10,000 images. Each image is of dimension 32×32 pixels and includes three color channels (RGB). Although the images are low in resolution, the dataset remains challenging due to intra-class variability and visual complexity across categories.

Road Sign Detection Dataset[2] is a high-resolution dataset which contains 877 images, each with a resolution of 1024×1024 pixels. It is curated to support the detection and classification of road signage. The dataset encompasses four primary categories: traffic lights, stop signs, speed limit signs, and pedestrian crossings. We employed this dataset to train and validate our QTL model for traffic sign classification tasks, a critical component in intelligent transportation systems and autonomous vehicle applications. For experimental consistency, the dataset was divided into an 80% training set and a 20% testing set.

[1] https://www.kaggle.com/datasets/gauravduttakiit/ants-bees.

[2] https://www.kaggle.com/datasets/andrewmvd/road-sign-detection.

12.5.2 *Performance Comparison Between QTL and Classical TL Approaches*

The results of our QTL framework against classical TL baselines across three datasets are reported in Table 12.1. For the Ants and Bees dataset, the QTL model achieves a classification accuracy of 96.1%, surpassing the classical TL model which attained 94.7%. Similarly, on the CIFAR-10 and Road Sign Detection datasets, QTL attained accuracies of 95.8% and 94.62%, outperforming classical TL models that achieved 92.1% and 92.47%, respectively. These results demonstrate consistent improvements of 1.4, 3.7, and 2.2% in classification accuracy across the datasets when using the QTL framework. To further validate the importance of TL in both classical and quantum contexts, we also evaluated performance when training the models from scratch, i.e., without TL. The architectural setups used for these evaluations followed the designs outlined in Fig. 12.2 and Sect. 12.3.3. Without TL, the quantum model achieved 65.36% accuracy on Ants and Bees, representing a 30.74% drop compared to the QTL counterpart. A similar decline is evident in the CIFAR-10 and Road Sign Detection datasets, where quantum models trained without TL reached 51.7% and 48.39% accuracies, respectively, reflecting performance drops of 44.1% and 46.23% compared to QTL. This notable degradation underscores the importance of TL in enhancing model performance, particularly in quantum neural networks, by enabling effective feature reuse and faster convergence. These empirical findings reinforce the effectiveness of the proposed QTL framework compared to classical and non-transfer learning quantum models.

Table 12.1 Comparison of classification accuracy across Classical TL, Classical without TL, Quantum TL, and Quantum without TL methods on Ants and Bees [82], CIFAR-10 [80], and Road Sign Detection [83] datasets. This table is reproduced under terms of the CC-BY license [12]. Copyright 2024, Amena Khatun and Muhammad Usman. Advanced Quantum Technologies published by Wiley-VCH GmbH

Method	Dataset	Accuracy (%)
Classical TL	Ants and Bees	94.70
Classical without TL	Ants and Bees	62.75
Quantum TL	Ants and Bees	96.10
Quantum without TL	Ants and Bees	65.36
Classical TL	CIFAR-10	92.10
Classical without TL	CIFAR-10	72.9
Quantum TL	CIFAR-10	95.80
Quantum without TL	CIFAR-10	51.70
Classical TL	Road Sign Detection	92.47
Classical without TL	Road Sign Detection	60.22
Quantum TL	Road Sign Detection	94.62
Quantum without TL	Road Sign Detection	48.39

12.5.3 Adversarial Attack Evaluation and Robustness Through Adversarial Training

To evaluate the robustness of our proposed QTL framework against adversarial attack, we subjected the trained model to adversarial perturbations generated using the FGSM method, as described in Sect. 12.3.5. The results for Ants and Bees dataset are reported in Table 12.2. We investigated the impact of adversarial noise at different perturbation magnitudes (ϵ), specifically at $\epsilon = 0.1, 0.2, \text{and} 0.3$, across four settings: Classical TL, Classical without TL, Quantum TL, and Quantum without TL. For the QTL model, we observe a sharp drop in performance from a clean accuracy of 96.1 to 53.85% when subjected to an adversarial strength of $\epsilon = 0.1$, corresponding to a 42.25% decrease. Increasing the attack strength to $\epsilon = 0.2$ causes a further degradation of 2.95%. The deterioration in classification accuracy is a clear indication that the quantum TL model is vulnerable against adversarial attack as the classical TL method.

Table 12.2 Simulation results of the proposed QTL approach on Ants and Bees dataset under adversarial attack with a different attack strength are compared to the accuracy of classical TL under attack. The experiments are also performed on AT to evaluate the robustness of the proposed method. This table is reproduced under terms of the CC-BY license [12]. Copyright 2024, Amena Khatun and Muhammad Usman. Advanced Quantum Technologies published by Wiley-VCH GmbH

Method	Attack strength	Clean accuracy (%)	Accuracy under attack (%)	Clean Accuracy with adversarial data (%)	Adversarial training accuracy (%)
Classical TL	0.1	94.70	50.64	92.38	64.10
Classical TL	0.2	94.70	47.70	89.17	61.10
Classical TL	0.3	94.70	45.79	86.93	59.10
Classical without TL	0.1	62.75	44.87	59.48	47.64
Classical without TL	0.2	62.75	44.50	57.54	47.01
Classical without TL	0.3	62.75	44.23	56.44	46.79
Quantum TL	0.1	96.10	53.85	94.76	71.87
Quantum TL	0.2	96.10	50.90	94.53	69.30
Quantum TL	0.3	96.10	47.45	94.12	67.82
Quantum without TL	0.1	65.36	45.68	49.23	48.12
Quantum without TL	0.2	65.36	45.12	48.74	48.0
Quantum without TL	0.3	65.36	45.12	48.06	47.98

To address this vulnerability, we introduced adversarial training (detailed in Sect. 12.3.6) into the QTL pipeline. When the QTL model is adversarially trained at perturbation levels of $\epsilon = 0.1$, 0.2, and 0.3, the model achieves significantly higher accuracies of 71.87%, 69.30%, and 67.82%, respectively. These results represent notable gains of 18.02, 18.4, and 20.35% compared to their non-robust QTL counterparts under attack. We also report results for QNN without TL, reinforcing the insight that transfer learning greatly enhances both clean and adversarial robustness performance. Furthermore, we observed that classical models without TL exhibit substantial performance degradation under adversarial influence, underscoring the critical role of TL in building robust representations. Our findings are aligned with prior literature such as [84], which also demonstrated significant susceptibility of classical CNNs under adversarial attack.

The effectiveness of AT was similarly confirmed on the Road Sign Detection and CIFAR-10 datasets, with detailed results reported in Tables 12.3 and 12.4. For Road Sign Detection, applying an adversarial attack of strength $\epsilon = 0.1$ caused a drop in performance from 95.8 to 89.6%. Interestingly, the performance only declined by an additional 2.1% when ϵ increased to 0.3. Hence, we increase the attack strength to 1.0 and notice a significant reduction of model performance to 60.4%. On the CIFAR-10 dataset, the QTL model's accuracy decreased marginally from 50.9 to 50.0% under adversarial attacks of increasing intensity ($\epsilon = 0.1, 0.3$, and 1.0). This suggests a relative robustness on CIFAR-10 dataset, although performance still benefits from AT. Across both datasets, applying AT substantially mitigated performance loss, with accuracy levels approaching the clean baseline for the Road Sign Detection dataset in particular. Figure 12.7 visualizes model accuracy as a function of adversarial strength ϵ for multiple settings: QTL, QTL-AT (QTL with adversarial training), quantum without TL, Classical TL (CTL), CTL-AT (CTL with adversarial training), and classical without TL. The QTL-AT curve consistently outperforms

Table 12.3 Simulation results of the proposed QTL approach on Road Sign Detection dataset under adversarial attack with a different attack strength are compared to the accuracy of classicalTL under attack. The experiments are also performed on AT to evaluate the robustness of the proposed method. This table is reproduced under terms of the CC-BY license [12]. Copyright 2024, Amena Khatun and Muhammad Usman. Advanced Quantum Technologies published by Wiley-VCH GmbH

Method	Attack strength	Clean accuracy (%)	Accuracy under attack (%)	Adversarial training accuracy (%)
Classical TL	0.1	93.7	85.4	93.7
Classical TL	0.3	93.7	83.3	92.4
Classical TL	1.0	93.7	58.7	89.5
Quantum TL	0.1	95.8	89.6	95.8
Quantum TL	0.3	95.8	87.5	93.7
Quantum TL	1.0	95.8	60.4	92.1

Table 12.4 Simulation results of the proposed QTL approach on CIFAR-10 dataset under adversarial attack with a different attack strength are compared to the accuracy of classical TL under attack. The experiments are also performed on AT to evaluate the robustness of the proposed method. This table is reproduced under terms of the CC-BY license [12]. Copyright 2024, Amena Khatun and Muhammad Usman. Advanced Quantum Technologies published by Wiley-VCH GmbH

Method	Attack strength	Clean accuracy (%)	Accuracy under attack (%)	Adversarial training accuracy (%)
Classical TL	0.1	92.1	50.1	85.8
Classical TL	0.3	92.1	50.2	84.9
Classical TL	1.0	92.1	49.2	80.8
Quantum TL	0.1	95.8	50.9	86.6
Quantum TL	0.3	95.8	50.0	86.2
Quantum TL	1.0	95.8	50.0	85.0

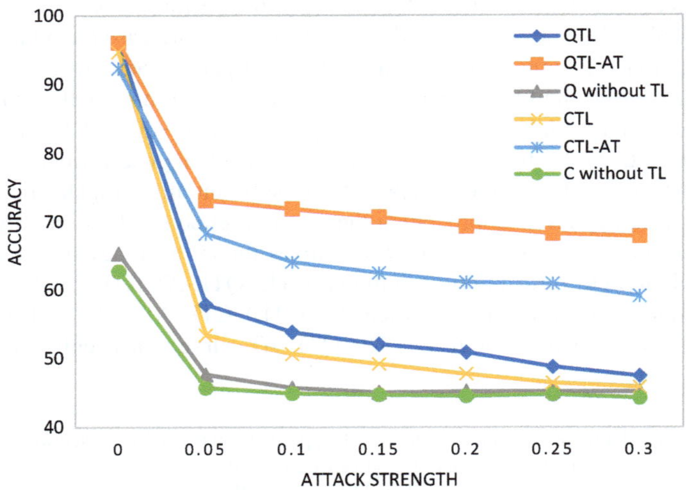

Fig. 12.7 The plot of accuracy versus attack strength on Ants and Bees for QTL (quantum transfer learning), QTL-AT (quantum transfer learning with adversarial training), Q without TL (quantum without transfer learning), CTL (classical transfer learning), CTL-AT (classical transfer learning with adversarial training), and C without TL (classical without transfer learning). This figure is reproduced under terms of the CC-BY license [12]. Copyright 2024, Amena Khatun and Muhammad Usman. Advanced Quantum Technologies published by Wiley-VCH GmbH

all other approaches across perturbation levels, clearly demonstrating the superior adversarial resilience of our proposed method.

12.6 Analysis and Interpretation of Results

The results presented in this work demonstrate the effectiveness of the proposed QTL framework, which combines classical and quantum ML techniques for image classification. Across the three benchmark datasets, Ants and Bees, CIFAR-10, and Road Sign Detection, the QTL model consistently outperformed both classical TL and QNNs trained without TL. One of the main advantages of the QTL approach is its ability to address the limitations of current quantum resources, such as limited number of qubits, while still handling high-dimensional input data. This is achieved by using a pre-trained classical CNN (ResNet-18) as a feature extractor. The ResNet-18 model extracts relevant and compact representations from raw images, which are then passed to a QVC for further processing. This knowledge transfer from a large-scale classical CNN to a small-scale QVC-based QNN leads the proposed QTL to achieve gain in performance. The results clearly highlight the benefits of this hybrid approach. On the Ants and Bees dataset, the QTL model achieved 96.1% accuracy, which is 1.4% higher than the classical TL model and 30.74% higher than QNN without TL. Similar performance improvements were observed for CIFAR-10 and Road Sign Detection datasets, confirming that the QTL framework can generalize well across different tasks. In addition to evaluating performance under clean data conditions, we also tested the robustness of the QTL model against adversarial attacks. As described in Sect. 12.5.3, adversarial attacks led to a significant drop in model accuracy. However, when AT was incorporated within the proposed framework, the performance of the QTL model improved notably, demonstrating that AT is effective in increasing the model's resistance to perturbations. This improvement was observed across all datasets, highlighting the importance of integrating robustness techniques into QTL framework.

Quantum computing systems are known to produce computational patterns that are counter-intuitive and believed to be beyond the efficient reach of classical hardware [49]. It has been suggested that by exploiting the quantum superposition and entanglement, quantum models are capable of processing and representing information in fundamentally different ways than their classical counterparts. Recent work has also shown that QML models, including those explored in this study, exhibit a higher degree of robustness against adversarial perturbations [73], a property that naturally arises from the inherent characteristics of quantum mechanics. We believe that combining quantum and classical models via TL captures the advantages of both domains by leveraging the representational power of quantum systems while utilizing the learning efficiency of classical models. The performance gains observed in our experiments suggest that hybrid quantum-classical models such as QTL are capable of outperforming purely classical approaches, particularly in terms of both accuracy and robustness. Since classical systems lack access to quantum phenomena, it is not possible to train classical systems to match the performance of quantum models. Our results indicate that QTL can act as a bridge between today's limited quantum hardware resources and the demands of high-dimensional learning tasks, even when limited training data is available. In future work, we aim to deploy and benchmark our QTL models on real quantum processors to test their practical performance in real-world applications, particularly when fault-tolerant quantum hardware becomes accessible in the coming years.

12.7 Conclusion

In this work, we introduced a comprehensive QTL framework that integrates the capabilities of quantum computing and classical ML into a unified architecture. Extensive simulations conducted on diverse datasets, including Ants and Bees, CIFAR-10, and Road Sign Detection, demonstrated the superiority of the proposed QTL approach over traditional classical TL methods. The results consistently showed that QTL achieved higher classification accuracy across all datasets, highlighting its effectiveness even in scenarios with limited training data. Additionally, we explored the susceptibility of QTL model to adversarial attacks. By applying adversarial perturbations to the input data, we exposed potential security vulnerabilities in quantum-enhanced models. To address this, we incorporated AT into the QTL framework, which significantly improved the resilience and overall classification performance under attack scenarios. The results from our comparative analysis underscore the potential of QTL in advancing QML applications, particularly in improving accuracy and robustness. Our simulation results and extensive comparison of QTL with its classical counterparts open up avenues for future research in quantum-enhanced ML.

Acknowledgements The authors acknowledge the use of CSIRO HPC (High-Performance Computing) and NCI's Gadi supercomputer for conducting the experiments. A.K. and M.U. also acknowledge CSIRO's Quantum Technologies Future Science Platform for providing the opportunity to work on quantum machine learning.

References

1. Grigorescu S, Trasnea B, Cocias T, Macesanu G (2020) A survey of deep learning techniques for autonomous driving. J Field Robot 37(3):362–386
2. Sreenu G, Durai S (2019) Intelligent video surveillance: a review through deep learning techniques for crowd analysis. J Big Data 6(1):1–27
3. Schroff F, Kalenichenko D, Philbin J (2015) Facenet: a unified embedding for face recognition and clustering. In: Proceedings of the IEEE conference on computer vision and pattern recognition, pp 815–823
4. Ren S, He K, Girshick R, Sun J (2015) Faster r-cnn: towards real-time object detection with region proposal networks. Advances in neural information processing systems, vol 28

5. Khatun A, Denman S, Sridharan S, Fookes C (2018) A deep four-stream siamese convolutional neural network with joint verification and identification loss for person re-detection. In: 2018 IEEE winter conference on applications of computer vision (WACV). IEEE, pp 1292–1301
6. Khatun A, Denman S, Sridharan S, Fookes C (2020) Semantic consistency and identity mapping multi-component generative adversarial network for person re-identification. In: Proceedings of the IEEE/CVF winter conference on applications of computer vision, pp 2267–2276
7. Khatun A, Denman S, Sridharan S, Fookes C (2020) Joint identification-verification for person re-identification: a four stream deep learning approach with improved quartet loss function. Comput Vis Image Underst 197:102989
8. Khatun A, Denman S, Sridharan S, Fookes C (2021) End-to-end domain adaptive attention network for cross-domain person re-identification. IEEE Trans Inf Forensics Secur 16:3803–3813
9. Khatun A, Denman S, Sridharan S, Fookes C (2023) Pose-driven attention-guided image generation for person re-identification. Pattern Recogn 137:109246
10. Sturman O, von Ziegler L, Schläppi C, Akyol F, Privitera M, Slominski D, Grimm C, Thieren L, Zerbi V, Grewe B et al (2020) Deep learning-based behavioral analysis reaches human accuracy and is capable of outperforming commercial solutions. Neuropsychopharmacology 45(11):1942–1952
11. Pang G, Shen C, Cao L, Van Den Hengel A (2021) Deep learning for anomaly detection: a review. ACM Comput Surv (CSUR) 54(2):1–38
12. Khatun A, Usman M (2025) Quantum transfer learning with adversarial robustness for classification of high-resolution image datasets. Adv Quantum Technol 8(1):2400268
13. Cerezo M, Verdon G, Huang H-Y, Cincio L, Coles PJ (2022) Challenges and opportunities in quantum machine learning. Nat Comput Sci 2(9):567–576
14. West MT, Sevior M, Usman M (2023) Reflection equivariant quantum neural networks for enhanced image classification. Mach Learn: Sci Technol 4(3):035027
15. Ren W, Li W, Shibo X, Wang K, Jiang W, Jin F, Zhu X, Chen J, Song Z, Zhang P et al (2022) Experimental quantum adversarial learning with programmable superconducting qubits. Nat Comput Sci 2(11):711–717
16. West MT, Heredge J, Sevior M, Usman M (2023) Provably trainable rotationally equivariant quantum machine learning. arXiv:2311.05873
17. Huang H-Y, Broughton M, Cotler J, Chen S, Li J, Mohseni M, Neven H, Babbush R, Kueng R, Preskill J et al (2022) Quantum advantage in learning from experiments. Science 376(6598):1182–1186
18. West MT, Sevior M, Usman M (2023) Boosted ensembles of qubit and continuous variable quantum support vector machines for b meson flavour tagging. arXiv:2305.02729
19. Tsang SL, West MT, Erfani SM, Usman M (2023) Hybrid quantum-classical generative adversarial network for high resolution image generation. IEEE transactions on quantum engineering
20. Khatun A, Yeter-Aydeniz K, Weinstein YS, Usman M (2024) Quantum generative learning for high-resolution medical image generation. Science and technology. Machine learning
21. Wang Z, Wang F, Li L, Wang Z, van der Laan T, Leon RCC, Huang J-K, Usman M, Quantum kernel learning for small dataset modeling in semiconductor fabrication: application to ohmic contact. Advanced science, p e06213
22. Gill SS, Kumar A, Singh H, Singh M, Kaur K, Usman M, Buyya R (2022) Quantum computing: a taxonomy, systematic review and future directions. Soft: Pract Exp 52(1):66–114
23. Wang Z, van der Laan T, Usman M (2025) Self-adaptive quantum kernel principal component analysis for compact readout of chemiresistive sensor arrays. Adv Sci 12(15):2411573
24. Cong I, Choi S, Lukin MD (2019) Quantum convolutional neural networks. Nat Phys 15(12):1273–1278
25. Beer K, Bondarenko D, Farrelly T, Osborne TJ, Salzmann R, Scheiermann D, Wolf R (2020) Training deep quantum neural networks. Nat Commun 11(1):808
26. Chen G, Chen Q, Long S, Zhu W, Yuan Z, Yilin W (2023) Quantum convolutional neural network for image classification. Pattern Anal Appl 26(2):655–667

27. Hur T, Kim L, Park DK (2022) Quantum convolutional neural network for classical data classification. Quantum Mach Intell 4(1):3
28. Mari A, Bromley TR, Izaac J, Schuld M, Killoran N (2020) Transfer learning in hybrid classical-quantum neural networks. Quantum 4:340
29. Otgonbaatar S, Schwarz G, Datcu M, Kranzlmueller D (2022) Quantum transfer learning for real-world, small, and large-scale datasets. arXiv:2209.07799
30. Mogalapalli H, Abburi M, Nithya B, Bandreddi SKV (2022) Classical–quantum transfer learning for image classification. SN Comput Sci 3(1):20
31. Azevedo V, Silva C, Dutra I (2022) Quantum transfer learning for breast cancer detection. Quantum Mach Intell 4(1):5
32. Geng A, Moghiseh A, Redenbach C, Schladitz K (2023) Hybrid quantum transfer learning for crack image classification on nisq hardware
33. Bengio Y (2012) Deep learning of representations for unsupervised and transfer learning. In: Proceedings of ICML workshop on unsupervised and transfer learning. JMLR Workshop and Conference Proceedings, pp 17–36
34. Neyshabur B, Sedghi H, Zhang C (2020) What is being transferred in transfer learning? Adv Neural Inf Process Syst 33:512–523
35. Kornblith S, Shlens J, Le QV (2019) Do better imagenet models transfer better? In: Proceedings of the IEEE/CVF conference on computer vision and pattern recognition, pp 2661–2671
36. Nakaji K, Uno S, Suzuki Y, Raymond R, Onodera T, Tanaka T, Tezuka H, Mitsuda N, Yamamoto N (2022) Approximate amplitude encoding in shallow parameterized quantum circuits and its application to financial market indicators. Phys Rev Res 4(2):023136
37. LaRose R, Coyle B (2020) Robust data encodings for quantum classifiers. Phys Rev A 102(3):032420
38. Wu V, Adermann E, Thapa C, Camtepe S, Suzuki H, Usman M (2023) Radio signal classification by adversarially robust quantum machine learning. arXiv:2312.07821
39. West MT, Nakhl AC, Heredge J, Creevey FM, Hollenberg LCL, Sevior M, Usman M (2023) Drastic circuit depth reductions with preserved adversarial robustness by approximate encoding for quantum machine learning. arXiv:2309.09424
40. Mogalapalli H, Abburi M, Nithya B, Bandreddi SKV (2022) Trash classification using quantum transfer learning. In: AIP conference proceedings, vol 2424. AIP Publishing
41. Qi J, Tejedor J (2022) Classical-to-quantum transfer learning for spoken command recognition based on quantum neural networks. In: ICASSP 2022-2022 IEEE international conference on acoustics, speech and signal processing (ICASSP). IEEE, pp 8627–8631
42. Szegedy C, Zaremba W, Sutskever I, Bruna J, Erhan D, Goodfellow I, Fergus R (2013) Intriguing properties of neural networks. arXiv:1312.6199
43. Huang L, Joseph AD, Nelson B, Rubinstein BIP, Tygar JD (2011) Adversarial machine learning. In: Proceedings of the 4th ACM workshop on Security and artificial intelligence, pp 43–58
44. Goodfellow IJ, Shlens J, Szegedy C (2015) Explaining and harnessing adversarial examples. In: 3rd international conference on learning representations, ICLR 2015, San Diego, CA, USA, 7–9 May 2015. Conference Track Proceedings
45. Kurakin A, Goodfellow IJ, Bengio S (2017) Adversarial examples in the physical world. In: 5th international conference on learning representations, ICLR 2017, Toulon, France, 24–26 Apr 2017. Workshop Track Proceedings
46. Ilyas A, Engstrom L, Athalye A, Lin J (2018) Black-box adversarial attacks with limited queries and information. In: International conference on machine learning. PMLR, pp 2137–2146
47. Tjeng V, Xiao KY, Tedrake R (2019) Evaluating robustness of neural networks with mixed integer programming. In: 7th international conference on learning representations, ICLR 2019, New Orleans, LA, USA, 6–9 May 2019. OpenReview.net
48. Rebentrost P, Mohseni M, Lloyd S (2014) Quantum support vector machine for big data classification. Phys Rev Lett 113(13):130503
49. Biamonte J, Wittek P, Pancotti N, Rebentrost P, Wiebe N, Lloyd S (2017) Quantum machine learning. Nature 549(7671):195–202

50. Lloyd S, Mohseni M, Rebentrost P (2014) Quantum principal component analysis. Nat Phys 10(9):631–633
51. Lloyd S, Mohseni M, Rebentrost P (2013) Quantum algorithms for supervised and unsupervised machine learning. arXiv:1307.0411
52. Kerenidis I, Landman J, Luongo A, Prakash A (2019) q-means: a quantum algorithm for unsupervised machine learning. Advances in neural information processing systems, p 32
53. MacCormack I, Delaney C, Galda A, Aggarwal N, Narang P (2022) Branching quantum convolutional neural networks. Phys Rev Res 4(1):013117
54. Yen-Chi Chen S, Wei T-C, Zhang C, Yu H, Yoo S (2022) Quantum convolutional neural networks for high energy physics data analysis. Phys Rev Res 4(1):013231
55. Chen X, Ma H, Wan J, Li B, Xia T (2017) Multi-view 3d object detection network for autonomous driving. In: Proceedings of the IEEE conference on computer vision and pattern recognition, pp 1907–1915
56. Maturana D, Scherer S (2015) Voxnet: a 3d convolutional neural network for real-time object recognition. In: 2015 IEEE/RSJ international conference on intelligent robots and systems (IROS). IEEE, pp 922–928
57. McKinney SM, Sieniek M, Godbole V, Godwin J, Antropova N, Ashrafian H, Back T, Chesus M, Corrado GS, Darzi A et al (2020) International evaluation of an ai system for breast cancer screening. Nature 577(7788):89–94
58. Esteva A, Kuprel B, Novoa RA, Ko J, Swetter SM, Blau HM, Thrun S (2017) Dermatologist-level classification of skin cancer with deep neural networks. Nature 542(7639):115–118
59. Taigman Y, Yang M, Ranzato MA, Wolf L (2014) Deepface: closing the gap to human-level performance in face verification. In: Proceedings of the IEEE conference on computer vision and pattern recognition, pp 1701–1708
60. Henderson M, Shakya S, Pradhan S, Cook T (2020) Quanvolutional neural networks: powering image recognition with quantum circuits. Quantum Mach Intell 2(1):2
61. Zhang S, Wang A (2024) Quantum transfer learning algorithm based on parameterized quantum circuit. In: 2024 5th international seminar on artificial intelligence, networking and information technology (AINIT). IEEE, pp 1883–1887
62. Tseng H-H, Lin H-Y, Yen-Chi Chen S, Yoo S (2025) Transfer learning analysis of variational quantum circuits. arXiv:2501.01507
63. Buonaiuto G, Guarasci R, Minutolo A, De Pietro G, Esposito M (2024) Quantum transfer learning for acceptability judgements. Quantum Mach Intell 6(1):13
64. Goodfellow I, Shlens J, Szegedy C (2015) Explaining and harnessing adversarial examples. In: International conference on learning representations
65. Liao H, Convy I, Huggins WJ, Whaley KB (2021) Robust in practice: adversarial attacks on quantum machine learning. Phys Rev A 103(4):042427
66. Sirui L, Duan L-M, Deng D-L (2020) Quantum adversarial machine learning. Phys Rev Res 2(3):033212
67. Anand A, Romero J, Degroote M, Aspuru-Guzik A (2021) Noise robustness and experimental demonstration of a quantum generative adversarial network for continuous distributions. Adv Quantum Technol 4(5):2000069
68. Guan J, Fang W, Ying M (2021) Robustness verification of quantum classifiers. In: Computer aided verification: 33rd international conference, CAV 2021, virtual event, 20–23 July 2021, Proceedings, Part I 33, pp 151–174. Springer
69. Gong W, Yuan D, Li W, Deng D-L (2022) Enhancing quantum adversarial robustness by randomized encodings. arXiv:2212.02531
70. Geng A, Moghiseh A, Redenbach C, Schladitz K (2023) Hybrid quantum transfer learning for crack image classification on nisq hardware. arXiv:2307.16723
71. Khatun A, Usman M (2025) Classical autoencoder distillation of quantum adversarial manipulations. arXiv:2504.09216
72. West MT, Tsang S-L, Low JS, Hill CD, Leckie C, Hollenberg LCL, Erfani SM, Usman M (2023) Towards quantum enhanced adversarial robustness in machine learning. Nat Mach Intell 1–9

73. West MT, Erfani SM, Leckie C, Sevior M, Hollenberg LCL, Usman M (2023) Benchmarking adversarially robust quantum machine learning at scale. Phys Rev Res 5(2):023186
74. Feinman R, Curtin RR, Shintre S, Gardner AB (2017) Detecting adversarial samples from artifacts. arXiv:1703.00410
75. Madry A, Makelov A, Schmidt L, Tsipras D, Vladu A (2017) Towards deep learning models resistant to adversarial attacks. arXiv:1706.06083
76. Croce F, Hein M (2020) Reliable evaluation of adversarial robustness with an ensemble of diverse parameter-free attacks. In: International conference on machine learning. PMLR, pp 2206–2216
77. He K, Zhang X, Ren S, Sun J (2016) Deep residual learning for image recognition. In: Proceedings of the IEEE conference on computer vision and pattern recognition, pp 770–778
78. Russakovsky O, Deng J, Hao S, Krause J, Satheesh S, Ma S, Huang Z, Karpathy A, Khosla A, Bernstein M et al (2015) Imagenet large scale visual recognition challenge. Int J Comput Vis 115:211–252
79. Huh M, Agrawal P, Efros AA (2016) What makes imagenet good for transfer learning? arXiv:1608.08614
80. Krizhevsky A, Hinton G et al (2009) Learning multiple layers of features from tiny images
81. Bergholm V, Izaac J, Schuld M, Gogolin C, Ahmed S, Ajith V, Sohaib Alam M, Alonso-Linaje G, AkashNarayanan B, Asadi A et al (2018) Pennylane: Automatic differentiation of hybrid quantum-classical computations. arXiv:1811.04968
82. Paszke A, Gross S, Massa F, Lerer A, Bradbury J, Chanan G, Killeen T, Lin Z, Gimelshein N, Antiga L et al (2019) Pytorch: an imperative style, high-performance deep learning library. Adv Neural Inf Proc Syst 32 <error l="308" c="Invalid command: paragraph not started." />
83. Mogelmose A, Trivedi MM, Moeslund TB (2012) Vision-based traffic sign detection and analysis for intelligent driver assistance systems: perspectives and survey. IEEE trans Intell Transp Syst 13(4):1484–1497
84. Bai T, Luo J, Zhao J, Wen B, Wang Q (2021) Recent advances in adversarial training for adversarial robustness. In: Zhou Z-H (ed) Proceedings of the thirtieth international joint conference on artificial intelligence, IJCAI-21, Aug 2021, pp 4312–4321

Chapter 13
Efficient State Preparation for Quantum Machine Learning

Azar C. Nakhl, Maxwell West, and Muhammad Usman

Abstract One of the key considerations in the development of Quantum Machine Learning (QML) protocols is the encoding of classical data onto a quantum device. In this chapter, we introduce the matrix product state representation of quantum systems and show how it may be used to construct circuits which encode a desired state. Putting this in the context of QML we show how this process may be modified to give a low-depth approximate encoding and crucially that this encoding does not hinder classification accuracy and indeed exhibits an increased robustness against classical adversarial attacks. This is illustrated by demonstrations of adversarially robust variational quantum classifiers for the MNIST and FMNIST dataset, as well as a small-scale experimental demonstration on a superconducting quantum device.

13.1 Introduction

Crucial to the implementation of quantum machine learning algorithms on quantum devices is the loading of classical data onto a quantum processor. To do so a number of different encodings exist, such as basis encoding for the encoding of binary data, or angle encoding in which each value in a vector is assigned a qubit which is rotated proportionally to its value. However, a more intuitive encoding exists in *amplitude encoding* wherein some input vector v_i is encoded onto the quantum processor as $\sum_i v_i |i\rangle$. The inherent advantage of this is that it requires exponentially fewer qubits than those which require one qubit for each feature. The primary set back, however, is that the quantum circuit to prepare such an arbitrary state is, in general, costly with a depth scaling as $\mathcal{O}(4^n)$ [19], where n is the number of qubits. Wishing to decrease the depth of the quantum circuit without necessitating an increase in the number of qubits, one may turn to heuristic methods such as variational encoding [1] or genetic

A. C. Nakhl (✉) · M. West · M. Usman
School of Physics, The University of Melbourne, Victoria, Australia
e-mail: chris.nakhl@unimelb.edu.au

M. Usman
Data61, CSIRO, Victoria, Australia

M. Usman (ed.), *Quantum Robustness in Artificial Intelligence*, Quantum Science and Technology, https://doi.org/10.1007/978-3-032-11153-1_13

algorithms [2] which find some approximation to the desired state. Alternatively, a more direct method of constructing these states with possibly low depth may be found by considering the Matrix Product State (MPS) representation of the state which can iteratively find circuits that better approximates the state without relying on heuristics.

In the following chapter, we will focus on the MPS-assisted state preparation method by introducing MPS, their construction given an input vector v_i and the algorithm used to construct a circuit that prepares $|v\rangle$ from the MPS representation. We will then briefly summarize the paradigmatic QML algorithm, the Quantum Variational Classifier (QVC) and provide a demonstration of how QVCs are resilient toward noise in the input state and how one can use this to their advantage when preparing circuits for the input state. Finally, we show the intriguing result whereby QVCs trained on noisy input states appear to exhibit an increased resilience against classical adversarial attacks.

13.2 Introduction to Matrix Product States

MPSs are a one-dimensional tensor network representation of many-body quantum systems that have found significant application in the simulation of quantum computing [4–8], and many-body physical systems [9–11]. States represented as an MPS have the form:

$$|\psi\rangle = \sum_{i_1, i_2, \ldots, i_N} A^{(1)}_{i_1} A^{(2)}_{i_2} \ldots A^{(N)}_{i_N} |i_1 i_2 \ldots i_N\rangle , \tag{13.1}$$

where N is the total number of qubits, $i_s \in \{0, 1\}$ and $A^{(s)}_{i_s}$ are matrices of size $\chi \times \chi$ where χ is the bond dimension of the state[1]. The matrices in the MPS in some sense encode the entanglement of the state, which can be more directly probed by considering the singular values of the matrix.

MPS, and tensor networks in general have a convenient graphical notation which one would be remiss not to bring up. The building blocks of *Penrose Notation* are solid shapes denoting tensors, with lines protruding from them indicating each of the indices of the tensor. For example, a matrix is a tensor with two indices and hence may be represented using a solid shape with two protruding lines.

$A = a_{ij} =$ 

Connecting the lines of two tensors in this notation represents summing over an index. For example, in matrix multiplication

[1] Note that the matrices at the end of the matrix product are of size $1 \times \chi$ and $\chi \times 1$, respectively.

$$AB = \sum_j a_{ij} b_{jk} = a_{ij} b_{jk} =$$ 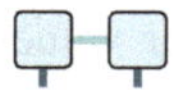

For those familiar with the*Einstein summation convention* this representation should look pretty familiar. In this graphical notation then, an MPS may be represented as

$$|\psi\rangle =$$ 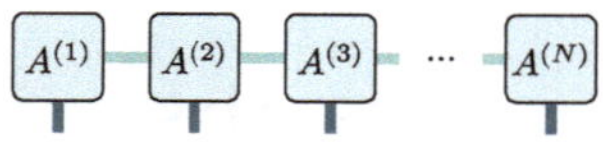

(13.2)

where each tensor $A^{(s)} = \sum_{i=0,1} |i\rangle \otimes A_i^{(s)}$.

13.2.1 Creating the MPS

Given that we are ostensibly starting off with classical data which we wish to encode, we must set up the protocol which takes some classical input vector v_i and forms an MPS which encodes $\sum_i v_i |i\rangle$. Luckily for us, the tools to do this are already commonplace and hopefully familiar to those who have encountered linear algebra before. The two core operations needed to construct an MPS from a state vector are matrix decomposition and reshaping.

There are a number of ways to decompose some matrix into a product of different matrices, but the one that is perhaps most of interest to us is the Singular Value Decomposition (SVD). This decomposition takes some matrix A and returns $A = USV^\dagger$ where U and V are unitary matrices, and S is a diagonal matrix of singular values. Of importance is that if some singular values are sufficiently small then they may be discarded and hence one may be able to reduce the number of columns in U and V. This results in a more memory-efficient representation of the state, and will have positive consequences when it comes to determining a circuit representation of the state.

In order to be able to utilize SVD (or indeed any matrix decomposition) one must of course have a matrix to begin with, this is where reshaping comes in. Fundamentally this is a rather simple operation where one redistributes data over a different set of indices but such that the tensor is still of the same size overall. Computers will use either a row-major or column-major order when reshaping data along a given set of indices, but fundamentally this ordering does not matter to us just that one remains consistent when reading, writing, and reshaping data. To start building our MPS we must first reshape the input vector of size 2^N to a matrix of size $2 \times 2^{N-1}$. Following this one may then use SVD to result in three matrices of size $2 \times s_0$, $s_0 \times s_0$, and $s_0 \times 2^{N-1}$. The $2 \times s_0$ matrix is $U = A^{(0)}$ as per Eq. (13.2).

To build the rest of the MPS we reshape the $V^\dagger$ matrix to $2s_{i-1} \times 2^{N-i}$ where i counts how many SVDs have been performed thus far and then perform another SVD. The resulting U of size $2s_{i-1} \times s_i$ may be reshaped into $2 \times s_{i-1} \times s_i$ and stored as $A^{(i)}$. The s_i are the number of singular values resulting from each SVD and in this context are referred to as the bond dimension at site i. Note that the matrices which store the singular values that result from the SVD may either be absorbed into the $A^{(i)}$ or $A^{(i+1)}$, or simply just left as is and inserted along the internal bond connecting $A^{(i)}$ and $A^{(i+1)}$.

With the MPS now in hand, we may now begin to answer the question of how do we build a low-depth circuit that encodes v_i.

13.3 State Preparation Using MPS

Now that we've gone to all that trouble to build the MPS, we must now destroy it. This is done by determining a series of unitary operators that sequentially disentangle each qubit from the rest of the system. Strictly speaking, your state does not need to be represented as an MPS to perform this procedure, but it does make the implementation more efficient.

The procedure begins by finding the reduced density matrix $\rho_{0...k}$ of the state over the first k sites. For an MPS in canonical form[2], finding the reduced density matrix over some linear bipartition is computationally efficient. We can then find the eigendecomposition of the reduced density matrix as

$$\rho_{0...k} = \sum_{i=0}^{2^k} \lambda_i \left|i\right\rangle \left\langle i\right| . \tag{13.3}$$

If $k = \lceil \log(\chi) \rceil + 1$ where χ is the largest bond dimension of the MPS. We note that for this selection of k, the rank of the reduced density matrix is 2^{k-1} and therefore there are only 2^{k-1} non-zero eigenvalues. Given this decomposition one can construct the unitary operator

$$U_0 = \sum_{i=0}^{2^k} \left|i\right\rangle \left\langle \phi_i\right| , \tag{13.4}$$

where the eigenstates are ordered in order of decreasing eigenvalue. When applied to the reduced density matrix $\rho_{0...k}$ this gives

$$U_o^\dagger \rho U_0 = \sum_{i=0}^{2^{k-1}} \lambda_i \left|i\right\rangle \left\langle i\right| . \tag{13.5}$$

It is apparent that this operation completely disentangles the first qubit from the rest of the system. This process can then be repeated for the subsystems $1 \ldots k+1, 2 \ldots k+2, \ldots, N-k \ldots N$. The U_i can then be used to reconstruct the original state:

---

[2] An MPS is in a *canonical form* [12] when it is left orthogonal at all sites to the left of some site i, and right orthogonal to the right of i. Site i is referred to as the orthogonality center.

$$|\psi\rangle = \sum_i v_i |i\rangle = \prod_{i=N-k+1}^{0} U_i^\dagger |0\rangle . \tag{13.6}$$

This process works well for MPS with low maximum bond dimension, but it will not produce a useful set of unitaries for a highly entangled MPS. The way around this is to select a smaller value of k. In this case, Eq. (13.5) does not hold, however, provided that the eigenvalues λ_i are not all equal at any given step j then it is evident that the probability of measuring $|0\rangle$ at site j will increase. Hence, at the end of the procedure, we have not completely disentangled the state ρ but have increased the probability of measuring $|0\rangle^{\otimes N}$. One can then repeat the process from the beginning with the new ρ to further destroy the state. After many repetitions one should finally arrive at $\rho = (|0\rangle\langle 0|)^{\otimes N}$ and a series of unitaries U_i^j which can be used to reconstruct the original ρ. In general, for any small fixed k the number of repetitions required to achieve this perfect disentangling scales is $O(2^N)$, however, one is also free to terminate the algorithm before this point and has a set of unitaries which only approximately prepare the state. To keep track of the fidelity that would be achieved using the unitaries from j iterations, one can keep track of $\langle 0| \rho_j |0\rangle$ where ρ_j here is the state after j iterations of the state preparation algorithm.

Typically, one will select $k = 2$ as the resulting two-qubit unitary gates can be decomposed into no more than three CNOT gates and six single-qubit rotation gates. The resulting circuit as per Fig. 13.1a features only nearest-neighbor operations which may be somewhat beneficial depending on the quantum hardware architecture

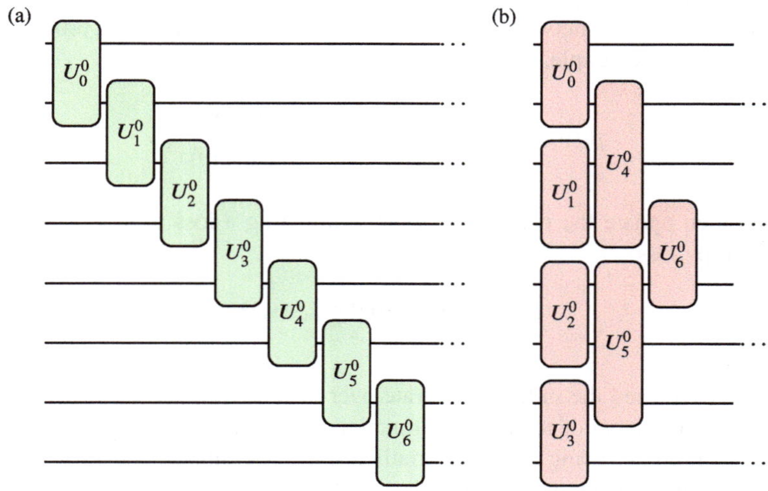

Fig. 13.1 The state preparation circuits resulting from the procedure outlined in Sect. 13.3 **a** uses the conventional procedure as described with $k = 2$ whereas **b** is seeking to minimize the depth of the each individual layer. Note that in **(b)** the "disentangled" qubit is the top qubit for unitaries acting on the first four qubit and the bottom qubit elsewhere and that U_4^0 and U_5^0 are two-qubit unitaries acting on the top and bottom qubits only

available. Indeed, if qubit connectivity is a restraint that needs to be accounted for, the procedure above can be altered to account for that by instead disentangling a different set of subsystems [3]. In this case, however, an MPS representation may not be as useful in forming the reduced density matrices.

Lastly, one efficiency that can be made if we are to relax the requirement of exact disentangling is that several disjoint subsystems may be processed at once, this can result in shallower circuits as per Fig. 13.1b. Note, however, that this circuit results in a higher depth circuit over multiple layers.

13.4 Quantum Variational Classifiers

The paradigmatic QML algorithm is the Quantum Variational Classifier (QVC)[3]. The concept behind this algorithm is rather simple, one has some unitary operation $U(\boldsymbol{\theta})$ parameterized by some angles $\boldsymbol{\theta}$ whereby classification is determined by measuring some observable set of observables $\{O_i\}_{i=0}^{m}$ where m is the number of classes. Predictably the classification corresponds to the observable with the largest expectation values, i.e.:

$$\hat{y}(v) = \arg\max_j \langle v| U(\boldsymbol{\theta})^\dagger O_j U(\boldsymbol{\theta}) |v\rangle \,. \tag{13.7}$$

Typically the parameterized unitary is composed of many repeating layers of local one- and two-qubit gates generally chosen such that some notion of locality is maintained in the two-qubit gates, e.g., [16] or [17].

Following classical networks, training is performed by assigning a probability of classification, for example, using Softmax normalization:

$$p_i(v) = \frac{\exp\left(\langle v| U(\boldsymbol{\theta})^\dagger O_i U(\boldsymbol{\theta}) |v\rangle\right)}{\sum_j \exp\left(\langle v| U(\boldsymbol{\theta})^\dagger O_j U(\boldsymbol{\theta}) |v\rangle\right)}. \tag{13.8}$$

One can then update the parameters $\boldsymbol{\theta}$ by minimizing a loss function such as the cross-entropy loss:

$$\mathcal{L} = \frac{1}{N}\sum_{i=1}^{N} - \log(p_{y(v_i)}(v_i)), \tag{13.9}$$

where we have used the index i to iterate over different inputs v_i and where $y(v_i)$ refers to the true label of v_i. The loss function, although determined on a quantum device, is minimized using a classical gradient-based optimizer such as ADAM [18].

The primary advantage of QVCs is that they are easy to implement and are rather versatile, being able to handle generic classification tasks. However, they are

[3] You will sometimes find these referred to as Variational Quantum Classifiers (VQCs) or Quantum Neural Networks (QNNs) in the literature.

susceptible to barren plateaus [13] which significantly hinders the training of deep variational quantum circuits. Luckily though, one may adapt the variational circuit to attempt to mitigate this effect [14, 15].

13.5 Approximate Encoding in QVCs

As it turns out, one does not necessarily need highly accurate input states to perform classification tasks. Intriguingly, as reported in [1] and reproduced in Fig. 13.2, it appears that classification accuracy appears to be somewhat resilient to random perturbations on the input. On the face of it, this suggests that perhaps an approximate encoding may be sufficient for classification tasks. An important consideration, however, is that the approximate encoding of the kind that is described above is not akin to random perturbations. For example, if we consider the approximate preparation of images in Fashion-MNIST dataset using the above procedure as reported in [1], we see in Fig. 13.3 this results in images that are blocky with alternating vertical stripes. These visual artifacts appear to be a result of the "compressed" encoding of the images:

$$v_j \mapsto \frac{1}{||v||} \sum_{i=0}^{2^{N-1}-1} (v_{2i} + \mathrm{i} v_{2i+1}) |i\rangle \,, \tag{13.10}$$

where one can reduce the number of required qubits by one by encoding adjacent qubits (in this case, the horizontal qubits) in the real and imaginary components of the basis state, respectively. If one were to use a more conventional amplitude encoding scheme without any compression, one would observe that the approximate encoding is "blocky" in nature (see Supplemental Fig. 13.2 of [1]). Nevertheless, we empirically observe that the classification accuracy when using these approximate

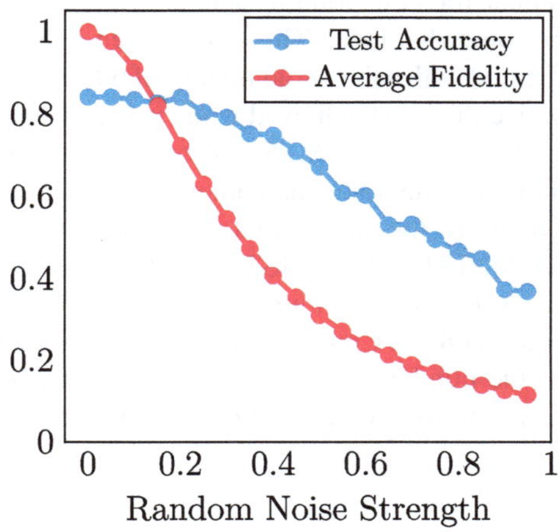

Fig. 13.2 Test accuracy for 500 images of the MNIST dataset [21] in the presence of random perturbations to the input amplitudes on a trained QVC with many layers. Note that the test accuracy only dips after the input fidelity drops below 60%. Reprint from [1]

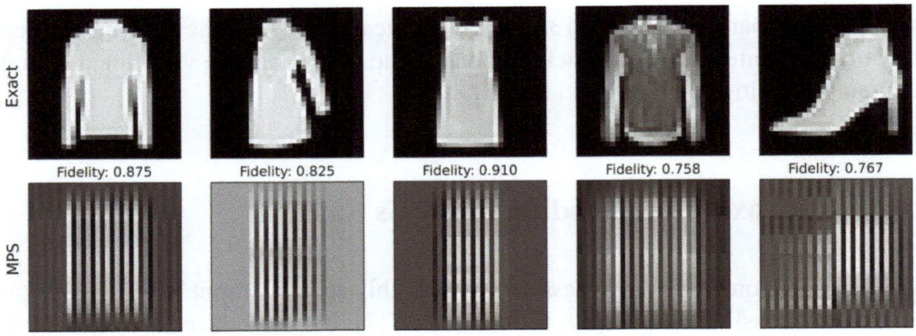

Fig. 13.3 Example images from the Fashion-MNIST dataset upon approximate encoding using the MPS-assisted protocol. Note that the vertical stripes that result from the encoding as per Eq. (13.10). Reprint from [1]

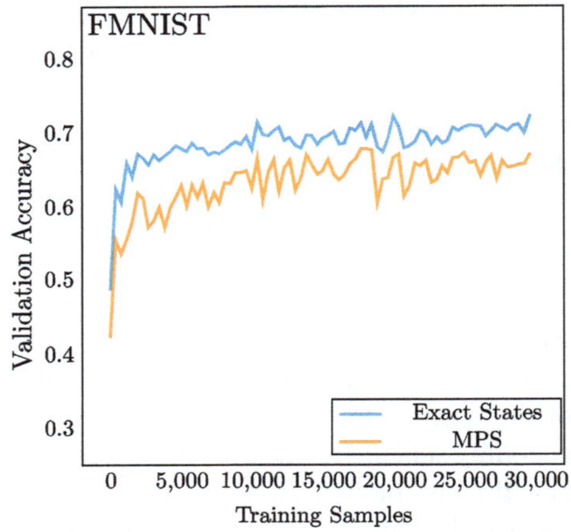

Fig. 13.4 The training accuracy for a deep nine qubit QVC using both an exact encoding and MPS-assisted encoding with a threshold accuracy of 60%. Figure adapted from Supplemental Material of [1]

inputs for both training and testing remains high as these artifacts do appear to be somewhat local. This can be seen in Fig. 13.4 where the MPS-assisted encoding method with a threshold fidelity of 60% results in a training accuracy only marginally worse than that achieved when using an exact encoding.

Noting that the MPS-assisted encoding runs over a discrete number of iterations and is less heuristic than other approximate encoding schemes, one finds that the true fidelity of the approximately encoded inputs can be rather variable and indeed for the MNIST and FMNIST dataset, it is frequently greater than the target fidelity. This is particularly pronounced for the FMNIST dataset where the fidelity distribution peaks at 83%. Consequently, the number of gates required, which is a direct function of the number of iterations, is less fixed but is orders of magnitude less than that resulting from conventional state preparation algorithms as per Fig. 13.5.

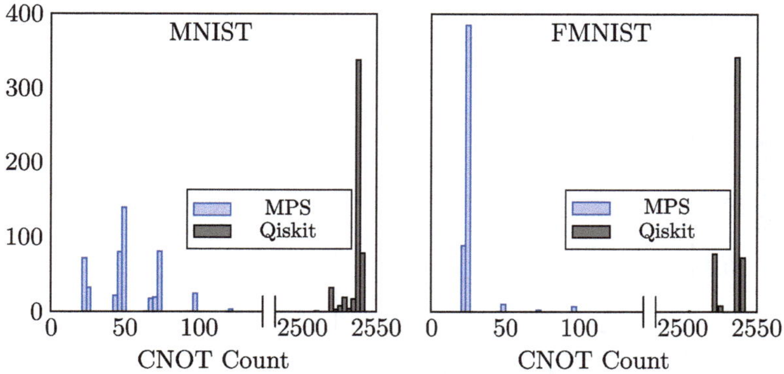

Fig. 13.5 CNOT gate count for 500 MNIST and FMNIST input states using the MPS method with a target fidelity of 60%, as well as the Qiskit [20] exact state preparation function based off [19]. Note that the MPS method results in a circuit with significantly fewer CNOT gates, suggesting that at most only seven iterations of the algorithm are required to reach the target fidelity. Figure adapted from [1]

13.5.1 Robustness Against Classical Adversarial Attacks

One of the interesting developments resulting from the advent of QML is the possibility that QML classifiers are robust toward classical adversaries [22, 23]. *A priori* this is not obviously true as adversaries trained on a given classical learner are seemingly transferable to other completely independent learners [24, 25]. Intriguingly, there is also evidence toward quantum adversaries being capable of successfully attacking classical networks [22]. As a result, there may be a niche in which QML may present an advantage wherein security is of utmost importance and quantum devices are not so readily available as to allow an attacker to train a quantum adversary.

The approximate encoding of input states also appears to present an increased robustness toward classical adversarial attacks [1]. In some sense, this is not surprising as random perturbations have been shown to increase the robustness of classical networks [26, 27] as the noisy input classifiers "wash-out" the adversarial perturbations that result from training on the exact input. Importantly though, the approximate state preparation discussed here is not stochastic in nature and so in principle one could train an adversary on the approximate inputs. However, given that the approximate input preparation can occur at runtime, adversarial perturbations may again be overcome.

The adversarial examples reported in [1] (and reproduced in Fig. 13.6) for the MNIST and FMNIST dataset show that QVCs trained and tested on approximate inputs show an increased test accuracy compared to QVCs with exact inputs. This improvement, although notable is in stark contrast to the accuracy of the classical convolutional neural network which drops off considerably under adversarial attack.

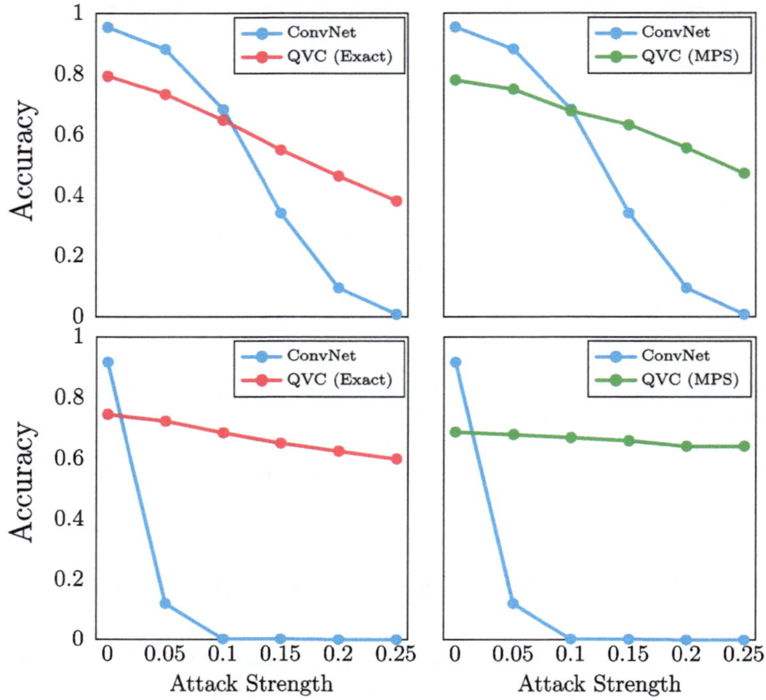

Fig. 13.6 (Top Row) Test accuracies for a trained Convolutional Neural Network (ConvNet) and QVC on adversarially perturbed images from the MNIST dataset. The adversary is trained on the ConvNet classifier with the attack strength corresponding to the step size in the Projective Gradient Decent (PGD) [28]. Further details on the adversary training can be found in [1]. (Bottom Row) Similar results using images from the FMNIST dataset. For both MNIST and FMNIST, it is found that while both the QVC with exact inputs and approximate inputs are robust against the adversarial perturbations, the approximate inputs show an increased robustness toward the perturbations. Figure adapted from [1]

13.5.2 *Experimental Demonstration on a Superconducting Quantum Computer*

Although quantum devices are incapable of running moderately deep circuits with high fidelity, one can experimentally demonstrate successful classification on a quantum device by considering simple datasets on few qubits. For example, the toy "Shapes" dataset considered in [1] consists of 8×8 images, requiring five qubits to encode, and only has two classes. This dataset was evaluated on the `ibm_algeirs` device with the Qiskit [20] exact state preparation method adapted from [19] achieving an average test accuracy of $45.8\% \pm 6.7$ compared to the MPS-assisted method which achieved $95\% \pm 2.3$. One also continues to see a robustness toward adversarial perturbations, albeit with a diminished test accuracy of $66.8\% \pm 0.3$. More results are presented in Fig. 13.3 of [1].

13.6 Conclusion

As more sophisticated QML algorithm continues to be developed, the input data and its encoding will remain a constant presence. In the context of specific classification tasks, one can always develop a bespoke encoding that relies on the underlying structure and symmetry of the task at hand [29, 30]. For general classification tasks, however, one will not have the ability to rely on this and in that context it remains crucial to be able to encode data efficiently in such a way that does not negate the possible exponential advantages that arise from QML algorithms. To that end, one may be able to utilize the inherent noise resilience of QML algorithms to encode states in a lossy fashion. This comes with the added advantage that such encodings lend themselves to being resilient toward classical adversarial attacks that are commonly trained using an exact representation of the input. An open question remains whether such an advantage exists when adversaries also have access to quantum devices or indeed have prior knowledge as to the encoding algorithm.

Acknowledgements We thank our collaborators Jamie Heredge, Floyd Creevey, Martin Sevior, and Lloyd Hollenberg with whom the results presented this chapter were originally published in [1]. C.N. and M.W. acknowledge the support of Australian Government Research Training Program Scholarship.

References

1. West MT, Nakhl AC, Heredge J, Creevey FM, Hollenberg LCL, Sevior M, Usman M (2024) Drastic circuit depth reductions with preserved adversarial robustness by approximate encoding for quantum machine learning. Intell Comput 3:0100
2. Creevey FM, Hill CD, Hollenberg LCL (2023) GASP: a genetic algorithm for state preparation on quantum computers. Sci Rep 13:11956
3. Nakhl AC (2021) Simulating noisy quantum algorithms and low depth quantum state preparation using matrix product states. University of Melbourne, Melbourne
4. Dang A, Hill CD, Hollenberg LC (2019) Optimising matrix product state simulations of shor's algorithm. Quantum 3:116
5. Nakhl AC, Quella T, Usman M (2024) Calibrating the role of entanglement in variational quantum circuits. Phys Rev A 109(3):032413
6. Nakhl AC, Harper B, West M, Dowling N, Sevior M, Quella T, Usman M (2025) Stabilizer tensor networks with magic state injection. Phys Rev Lett 134(19):190602
7. Banuls MC, Orús R, Latorre JI, Pérez A, Ruiz-Femenia P (2006) Simulation of many-qubit quantum computation with matrix product states. Phys Rev A-Atomic Mol Opt Phys 73(2):022344
8. Ballarin M, Mangini S, Montangero S, Macchiavello C, Mengoni R (2023) Entanglement entropy production in quantum neural networks. Quantum 7:1023
9. Amico L, Fazio R, Osterloh A, Vedral V (2008) Entanglement in many-body systems. Rev Mod Phys 80(2):517–576
10. Verstraete F, Cirac JI (2004) Renormalization algorithms for quantum-many body systems in two and higher dimensions. arXiv preprint cond-mat/0407066
11. Lanyon BP, Maier C, Holzäpfel M, Baumgratz T, Hempel C, Jurcevic P et al (2017) Efficient tomography of a quantum many-body system. Nat Phys 13(12):1158–1162

12. Orús R (2014) A practical introduction to tensor networks: matrix product states and projected entangled pair states. Ann Phys 349:117–158
13. McClean JR, Boixo S, Smelyanskiy VN, Babbush R, Neven H (2018) Barren plateaus in quantum neural network training landscapes. Nat Commun 9(1):4812
14. West MT, Heredge J, Sevior M, Usman M (2024) Provably trainable rotationally equivariant quantum machine learning. PRX Quantum 5(3):030320
15. Sannia A, Tacchino F, Tavernelli I et al (2024) Engineered dissipation to mitigate barren plateaus. NPJ Quantum Inf 10:81
16. Nakhl AC, Quella T, Usman M (2024) Calibrating the role of entanglement in variational quantum circuits. Phys Rev A 109(3):032413
17. Kandala A, Mezzacapo A, Temme K, Takita M, Brink M, Chow JM, Gambetta JM (2017) Hardware-efficient variational quantum eigensolver for small molecules and quantum magnets. Nature 549(7671): 242–246
18. Kingma DP, Ba J (2014) Adam: a methDod for stochastic optimization. arXiv:1412.6980
19. Shende VV, Bullock SS, Markov IL (2006) Synthesis of quantum-logic circuits. IEEE Trans Comput Aided Des Integr Circuits Syst 25(6):1000–1010
20. Javadi-Abhari A, Treinish M, Krsulich K, Wood CJ, Lishman J, Gacon J et al (2024) Quantum computing with Qiskit. arXiv:2405.08810
21. Deng L (2012) The mnist database of handwritten digit images for machine learning research. IEEE Signal Process Mag 29(6):141–142
22. West MT, Erfani SM, Leckie C, Sevior M, Hollenberg LC, Usman M (2023) Benchmarking adversarially robust quantum machine learning at scale. Phys Rev Res 5(2):023186
23. Dowling N, West MT, Southwell A, Nakhl AC, Sevior M, Usman M, Modi K (2026) Adversarial robustness guarantees in quantum classifiers. npj Quantum Information 12:16
24. Goodfellow IJ, Shlens J, Szegedy C (2024) Explaining and harnessing adversarial examples. arXiv:1412.6572
25. Szegedy C, Zaremba W, Sutskever I, Bruna J, Erhan D, Goodfellow I, Fergus R (2013) Intriguing properties of neural networks. arXiv:1312.6199
26. Cohen J, Rosenfeld E, Kolter Z (2019) Certified adversarial robustness via randomized smoothing. In international conference on machine learning. PMLR, pp 1310–1320
27. Li B, Chen C, Wang W, Carin L (2019) Certified adversarial robustness with additive noise. Adv Neural Inf Proces syst 32
28. Madry A, Makelov A, Schmidt L, Tsipras D, Vladu A (2017) Towards deep learning models resistant to adversarial attacks. arXiv:1706.06083
29. West MT, Heredge J, Sevior M, Usman M (2024) Provably trainable rotationally equivariant quantum machine learning. PRX Quantum 5(3):030320
30. Heredge J, Hill C, Hollenberg L, Sevior M (2024) Permutation invariant encodings for quantum machine learning with point cloud data. Quantum Mach Intell 6(1):25

Chapter 14
Geometric Quantum Machine Learning

Maxwell West, Jamie Heredge, Martin Sevior, and Muhammad Usman

Abstract The prospects of QML remain dampened by considerable technical challenges. A particularly significant issue is that generic QML models suffer from so-called barren plateaus in their training landscapes: large regions where cost function gradients vanish exponentially in the number of qubits employed, rendering large models effectively untrainable. A leading strategy for combating this effect, *geometric quantum machine learning*, is to build problem-specific models which take into account (for example) the symmetries of their data in order to focus on a smaller, relevant subset of Hilbert space.

14.1 Introduction

Symmetries are a fundamental aspect of physics, entering into the laws of nature at the deepest known levels [1]. Their presence allows for the vast simplification of innumerable problems; their suspected presence allows for the prediction of such novelties as never-before-detected fundamental particles [2]. At a much higher level, symmetries are pervasive in data. From the perspective of constructing machine learning models, they constitute a redundancy of description that can be "quotiented out" without compromising the capacity of the model to characterize the data. Explicitly building in a respect for symmetry at the architectural level—obviating the need for the model to independently discover the redundancy during the training process—is the endeavor of *geometric machine learning* [3]. The intuitive inference that this would ease the learning process is explicitly captured in the QML case through the relation between geometric quantum machine learning (GQML) and *barren plateaus*.

Indeed, barren plateaus have come to be understood to be linked to the expressibility of the ansatz being optimized [4–7], with generic, highly expressible models

M. West (✉) · J. Heredge · M. Sevior · M. Usman
School of Physics, The University of Melbourne, Parkville, VIC, Australia
e-mail: westm2@student.unimelb.edu.au

M. Usman
Data61, CSIRO, Clayton, VIC, Australia

M. Usman (ed.), *Quantum Robustness in Artificial Intelligence*, Quantum Science and Technology, https://doi.org/10.1007/978-3-032-11153-1_14

rendered effectively untrainable as the number of qubits increases. This suggests that the architectures of variational QML models should be tweaked on a per-problem basis, with domain-specific knowledge employed to construct ansatze with inductive biases compatible with the optimal solution. GQML is a natural example of exactly this, and has become a topic of some popularity in recently years [8–19]. For example, GQML has been used to classify images with spatial symmetries [16, 17, 19], and more general permutationally-invariant data [11, 12, 20]. By constraining the search space of the optimization procedure, such symmetry-informed models have in a few instances been proved to be free of barren plateaus [11, 21, 22].

14.2 Incorporating Symmetries Into QML

Let us now describe the mathematical formalism of GQML; further details may be found in the literature, e.g., Refs. [9, 10]. We will consider the standard supervised learning setting, wherein one is given datapoints $\boldsymbol{x}$ drawn from a set $\mathcal{X}$, and attempts to determine associated labels $y(\boldsymbol{x}) \in \mathcal{Y} = \{1, 2, \ldots, n_{\text{classes}}\}$ assigned by an unknown function $f : \mathcal{X} \to \mathcal{Y}$. The quantum models for this task we will describe in this chapter are comprised of three main components—an initial data encoding layer $\mathcal{E} : \mathcal{X} \to \mathbb{U}(\mathcal{H})$ (with $\mathbb{U}(\mathcal{H})$ the unitary group on the Hilbert space $\mathcal{H}$ of the quantum computer), a trainable variational section $U_\theta \in \mathbb{U}(\mathcal{H})$, and a final set of measurements. In the final stage n_{classes} self-adjoint operators $\{M_j\}_{j=1}^{n_{\text{classes}}} \subset \text{End}\,\mathcal{H}$ are measured, with the prediction $\hat{y}_\theta$ of the model on the input $\boldsymbol{x}$ defined to be the class corresponding to the operator with the highest expectation value, i.e.,

$$\hat{y}_{\boldsymbol{\theta}}(\boldsymbol{x}) = \underset{j}{\text{argmax}} \langle 0| \, \mathcal{E}(\boldsymbol{x})^\dagger U_{\boldsymbol{\theta}}^\dagger M_j U_{\boldsymbol{\theta}} \mathcal{E}(\boldsymbol{x}) \, |0\rangle \, . \tag{14.1}$$

During training, a loss function ℓ_θ is minimized in an attempt to enforce $\hat{y}_{\boldsymbol{\theta}}(\boldsymbol{x}) = y(\boldsymbol{x})$ (for $\boldsymbol{x}$ in the training set anyway, with the hope being that such agreement then generalizes to unseen datapoints). One says that an n-qubit model exhibits an (aforementioned) barren plateau if, for some $b > 1$,

$$\text{Var}_{\boldsymbol{\theta}}[\ell_\theta] = \mathcal{O}(b^{-n}). \tag{14.2}$$

Now, if $\mathcal{G} : \mathcal{X} \to \mathcal{X}$ is a symmetry of the data at the level of the classification labels, i.e., $y(g(\boldsymbol{x})) = y(\boldsymbol{x}) \; \forall \boldsymbol{x} \in \mathcal{X}, \; g \in \mathcal{G}$ (but *not* necessarily $g(\boldsymbol{x}) = \boldsymbol{x}$), then we wish to build models with inductive biases which automatically enforce respect for this symmetry, i.e., $\hat{y}_{\boldsymbol{\theta}}(\boldsymbol{x}) = \hat{y}_{\boldsymbol{\theta}}(g(\boldsymbol{x})) \; \forall \boldsymbol{x}, \boldsymbol{\theta}$. The encoding map $\mathcal{E} : \mathcal{X} \to \mathcal{H}$ induces a unitary representation[1] $R : G \to \mathbb{U}(\mathcal{H})$ of the symmetry group G on $\mathcal{H}$ making the following diagram commute:

[1] The notion of a representation is recalled and discussed in the next subsection.

$$\begin{array}{ccc} \mathcal{X} & \xrightarrow{g} & \mathcal{X} \\ \downarrow \mathcal{E} & & \downarrow \mathcal{E} \\ \mathcal{H} & \xrightarrow{R(g)} & \mathcal{H} \end{array}$$

The prediction of the model on a data point acted upon by $g \in \mathcal{G}$ is

$$\hat{y}_{\boldsymbol{\theta}}(g(\boldsymbol{x})) = \underset{j}{\operatorname{argmax}} \langle \psi(g(\boldsymbol{x}))| \, U_{\boldsymbol{\theta}}^{\dagger} M_j U_{\boldsymbol{\theta}} \, |\psi(g(\boldsymbol{x}))\rangle \tag{14.3}$$

$$= \underset{j}{\operatorname{argmax}} \langle \psi(\boldsymbol{x})| \, R_g^{\dagger} U_{\boldsymbol{\theta}}^{\dagger} M_j U_{\boldsymbol{\theta}} R_g \, |\psi(\boldsymbol{x})\rangle \tag{14.4}$$

$$= \underset{j}{\operatorname{argmax}} \langle 0| \, \mathcal{E}(\boldsymbol{x})^{\dagger} R_g^{\dagger} U_{\boldsymbol{\theta}}^{\dagger} M_j U_{\boldsymbol{\theta}} R_g \mathcal{E}(\boldsymbol{x}) \, |0\rangle \, . \tag{14.5}$$

with $|\psi(\boldsymbol{x})\rangle = \mathcal{E}(\boldsymbol{x})\,|0\rangle$ the encoded state of the data. For the predictions of our model to be invariant with respect to the action of the symmetry, then, equating Eqs. 14.1 and 14.5 we find a condition for equivariance,

$$\left[R(g), U_{\boldsymbol{\theta}}^{\dagger} M_j U_{\boldsymbol{\theta}}\right] = 0 \qquad \forall g \in G, \ \boldsymbol{\theta} \in \Theta, \ j \in \{1, \ldots, n_{\text{classes}}\} \tag{14.6}$$

i.e.,

$$U_{\boldsymbol{\theta}}^{\dagger} M_j U_{\boldsymbol{\theta}} \in \mathtt{comm}(R) = \{A \in \operatorname{End} \mathcal{H} : [A, R_g] = 0 \; \forall g \in G\}, \tag{14.7}$$

and the representation of the symmetry group enforces constraints on the architecture of the model. We stress that the states $|\psi(\boldsymbol{x})\rangle$ are not necessarily invariant under the action of R, merely their associated labels. The subspace $\mathtt{comm}(R)$—which is evidently critical to the whole endeavor—will depend on the nature of both the group G the encoding map $\mathcal{E}$. We will begin by exploring what can be said in general, before investigating some specific examples.

14.2.1 Representations and Commutants

Mathematically, symmetries are described by *groups*[2]; we recall that a group G is a set equipped with an associative operation $\cdot : G \times G \to G$ with respect to which there exist inverses and an identity element. In practice, groups often occur as objects which act on vector spaces via *representations*[3]:

[2] Typically, anyway; more generally one can encounter for example non-invertible symmetries [23].

[3] It is customary to use the word "representation" fairly loosely, overloading it to refer variously to the map from the group to the vector space, the vector space itself, or the operators on it that represent group elements.

Definition 1 A representation R of a group G on a vector space V is a structure-preserving map $R : G \to \mathbb{GL}(V)$. In this context, structure-preserving means that, for all $g, h \in G$, $R(g \cdot h) = R(g)R(h)$. A representation is said to be unitary if $R(G) \subseteq \mathbb{U}(V)$.

Indeed, in our case, the symmetry of the data manifests as a unitary representation of the corresponding group on the Hilbert space of the quantum computer, and our analysis will depend explicitly only on this representation—rather than the group *per se*. To see this, consider the trivial example $\mathcal{E}(\boldsymbol{x}) = \mathbb{1}_{\mathcal{H}}\ \forall \boldsymbol{x} \in \mathcal{X}$. Then, no matter what symmetries may or may not exist in the data, the induced representation is the trivial one, $R(g) = \mathbb{1}_{\mathcal{H}}\ \forall g \in G$, and *any* choice of U_θ and M_j satisfies Eq. (14.6). So, evidently, it is the representation that is paramount.

In order to obtain our desired characterization of the commutant of Eq. (14.7), we will need a few elementary notions and results from representation theory. We begin with some important definitions:

Definition 2 Let $R : G \to \mathbb{GL}(V)$ be a representation. A subspace $W \subseteq V$ is said to be G-invariant if, for all $w \in W$ and $g \in G$, $R(g)w \in W$.

Note that under the hypothesis of this definition, W also furnishes a representation of G. We continue on with a few more definitions:

Definition 3 Let $R : G \to \mathbb{GL}(V)$ be a representation. If the only G-invariant subspaces of V are the trivial subspaces $\{0\}$ and V itself, then the representation R is said to be irreducible. Otherwise, it is reducible. V is said to be completely reducible if there exists a decomposition

$$V = \bigoplus_i V_i \tag{14.8}$$

where each V_i is an irreducible representation of G.

Definition 4 Let $R : G \to \mathbb{GL}(V)$, $S : G \to \mathbb{GL}(W)$ be representations. A linear map $\varphi : V \to W$ is said to be G-equivariant (or "G-linear") if, for all $g \in G$, the following diagram commutes:

$$\begin{array}{ccc} V & \xrightarrow{R(g)} & V \\ \big\downarrow{\scriptstyle \varphi} & & \big\downarrow{\scriptstyle \varphi} \\ W & \xrightarrow{S(g)} & W \end{array}$$

i.e., $S(g) \circ \varphi = \varphi \circ R(g)\ \forall g \in G$.

We now have the vocabulary to state some important results:

Fact 1 *Any finite dimensional unitary representation is completely reducible.*

Fact 2 (Schur's lemma) *If φ is a G-linear map between irreducible G-representations V and W then either $\varphi = 0$ or V and W are isomorphic, with $\varphi = \lambda \mathbb{1}$ for some $\lambda \in \mathbb{C}$.*

This later fact allows for a precise characterization of the commutant of Eq. (14.7); indeed, we have [9]

Fact 3 *Let*

$$V = \bigoplus_i \mathbb{1}_{m_i} \otimes V_i \tag{14.9}$$

be a representation possessing a decomposition into inequivalent irreps V_i (which occur with multiplicity m_i, and have dimension (say) d_i). Then, the commutant is given exactly by

$$\mathtt{comm}(V) = \bigoplus_i O_i \otimes \mathbb{1}_{d_i}, \tag{14.10}$$

where the O_i are arbitrary operators on the multiplicity spaces.

We certainly see immediately that the right-hand side of Eq. (14.10) is contained within the left-hand side; the reverse inclusion relies on Schur's lemma, which forces elements of the commutant to act trivially on each irrep.

14.3 Examples

We now turn to two examples of the above theory: permutation and rotation equivariant QML models. The discussion of the former will be based on Ref. [11], and the latter on Ref. [19]. As we shall see, the examples are quite complementary, with the two approaches taken being somewhat different.

14.3.1 Permutation Symmetry

In our first example, we consider the permutation equivariant model of Ref. [11]. Certainly permutation symmetry is widespread, occurring for example in molecular systems [24–26] and learning tasks defined over sets with no particular ordering [27–29]. Mathematically, permutations (on n letters, say) are elements of the *symmetric group S_n*; the proceeding analysis, in which we follow Ref. [11] as they consider the application of the general framework espoused in Sect. 14.2 to S_n, is greatly facilitated by the depth with which its representation theory is understood [30].

Indeed, consider the natural representation $P : S_n \to \mathbb{GL}((\mathbb{C}^2)^{\otimes n})$ of the symmetric group on n qubits, given by

$$P(\pi)\bigotimes_{i=1}^{n}|\psi_i\rangle = \bigotimes_{i=1}^{n}\left|\psi_{\pi^{-1}(i)}\right\rangle . \tag{14.11}$$

As in Fact 3 the action of the group will induce a decomposition of the Hilbert space into irreps; in the S_n case specifically we have[4]

$$R(\pi \in S_n) = \bigoplus_{\substack{\lambda \vdash n \\ \mathrm{len}(\lambda)\leq 2}} \mathbb{1}_{m_\lambda} \otimes r_\lambda(\pi) , \tag{14.12}$$

where λ iterates over the partitions of n of length at most two; i.e., $\lambda \equiv [\lambda_1, \lambda_2]$, with $\lambda_1, \lambda_2 \in \mathbb{Z}_{\geq 0}$, $\lambda_1 \geq \lambda_2$, and $\lambda_1 + \lambda_2 = n$. Equation (14.12) is a manifestation of the famous *Schur-Weyl duality*, one of the most celebrated results in representation theory [30].

Dutifully following the recipe of Sect. 14.2, the next step is to pick a circuit architecture with generators and final measurement operators in the commutant (Eq. (14.10)) corresponding to Eq. (14.12). Happily, due to the simplicity of the action of S_n, one can do this by inspection. Indeed, in order to satisfy Eq. (14.6) it is clearly sufficient to take both the final measurement operator and the generators of the circuit to themselves be permutation symmetric; the specific choice of Ref. [11] is to take generators in the set

$$\mathcal{G} = \left\{ \frac{1}{n}\sum_{j=1}^{n} X_j, \frac{1}{n}\sum_{j=1}^{n} Y_j, \frac{2}{n(n-1)}\sum_{k<j} Z_j Z_k \right\} \tag{14.13}$$

and measurement operators in

$$M = \left\{ \frac{1}{n}\sum_{j=1}^{n} \chi_j, \frac{2}{n(n-1)}\sum_{k<j;j=1}^{n} \chi_j\chi_k, \prod_{j=1}^{n}\chi_j \right\} , \tag{14.14}$$

where χ is any Pauli matrix. Evidently, these choices were not forced; one could've picked any number of other sets of permutation-symmetric operators; nonetheless, as we shall see the choices of Eqs. (14.13) and (14.14) suffice to illustrate the important conceptual points. More generally, from Eq. (14.10) any unitary (and measurement observable) of the form

[4] Note that we here adopt a slight different basis to that chosen in the analogous equation in Ref. [11], where the action diagonalised is that of $\mathbb{SU}(2)$ Schur-Weyl-dual to our action of S_n; of course, it ultimately matters not.

$$U = \bigoplus_{\substack{\lambda \vdash n \\ \mathrm{len}(\lambda) \leq 2}} u_\lambda \otimes \mathbb{1}_{d_\lambda} \tag{14.15}$$

will lead to an equivariant model. Evidently one can quantify the extent to which respecting the symmetry has constrained the expressibility of the model by calculating $\sum_\lambda m_\lambda^2$; in this case one can show [11]

$$\sum_{\substack{\lambda \vdash n \\ \mathrm{len}(\lambda) \leq 2}} m_\lambda^2 = \binom{n+3}{3} \in \mathcal{O}(n^3), \tag{14.16}$$

i.e., the unitaries belong to a submanifold of *polynomial* dimension.

So, now armed with an equivariant model, have we avoided barren plateaus? The full result, derived in Ref [11], is that[5]

$$\mathrm{Var}_\theta[\partial_\mu \ell(\theta)] = \sum_\lambda \frac{2m_\lambda}{(m_\lambda^2 - 1)^2} \Delta(H_{\mu,\lambda})\Delta(O_\lambda)\Delta\left(\sum_{\nu=1}^{d_\lambda} \sigma_\lambda^\nu\right), \tag{14.17}$$

where the trainable parameter θ_μ with respect to which we are differentiating appears in the circuit as a multiplicative factor $e^{-i\theta_\mu H_\mu}$, σ is the weighted average of the input states, a subscript λ denote the restriction of an operator to that isotypic, and $\Delta(A) = \mathrm{tr}[A^2] - \mathrm{tr}[A]^2/(\dim B)$. Having verified that both m_λ and Δ of the terms in the model (Eqs. (14.13) and (14.14)) are $\Omega(1/\mathrm{poly}(n))$ [11], one concludes that, if too $\Delta\left(\sum_{\nu=1}^{d_\lambda} \sigma_\lambda^\nu\right) \in \Omega(1/\mathrm{poly}(n))$, then

$$\mathrm{Var}_\theta[\partial_\mu \ell(\theta)] \in \Omega(1/\mathrm{poly}(n)), \tag{14.18}$$

i.e., that, under the detailed assumptions, S_n *equivariant QNNs do not have barren plateaus*. As a final remark, we note that (despite the lengths gone in both cases to construct a clever ansatz) an assumption on the input data will appear also in our next example, and indeed more generally [6]. Upon reflection, this is not so surprising; indeed, consider the symmetry between the input state ρ and the measurement observable O in the relation $\mathrm{tr}[(U_\theta \rho U_\theta^\dagger)O] = \mathrm{tr}[\rho(U_\theta^\dagger O U_\theta)]$.

14.3.2 Rotation Symmetry

In our second example, we describe a family of rotationally equivariant QML models introduced in Ref. [19] for classifying two-dimensional data with labels that are invariant to rotations by $2\pi/2^k$ for $k \in \mathbb{N}$. In particular, these models are applicable to image data, the consideration of the symmetries of which in the GQML context

[5] Recalling again the mismatch $m_\lambda \leftrightarrow d_\lambda$ between our notation and theirs.

has previously been limited to reflections [16] and 90° rotations [17]. The architecture allows for the fraction of resources allocated to the processing of the radial and angular degrees of freedom to be controlled, enabling for an interpolation between a model with no explicit notion of rotational symmetry ($k = 0$) to models which respect continuous rotations ($k \to \infty$). This is facilitated by choosing an encoding strategy which ensures that the representation of the group action on the encoded quantum states is that of the regular representation of $\mathbb{Z}_{2^k}$, which can be diagonalised via a (quantum) Fourier transform, following which it becomes simple to write down an equivariant model, without needing to resort to explicit symmetrization techniques such as twirling [9]. In particular, we will remark upon the trainability of the models, finding that the scaling of the variance of the cost function with respect to the trainable parameters of the model differs qualitatively depending on the fraction of the qubits used to encode the radial and angular degree of freedom. Specifically, recently developed techniques [6, 7] which involve studying the dynamical Lie algebras of quantum circuits reveal that the models are free from barren plateaus when the number of qubits devoted to encoding the radial information grows at most logarithmically with the total number of qubits.

In addition to the analytical results, we will present the results of numerical tests of the performance of the models—and the performance of generic classifiers—on a 10-class classification task involving the analysis of scanning tunneling microscope (STM) images of phosphorus impurities implanted in silicon. It was found in Ref. [19] that the rotationally equivariant models achieve drastically better results and scale much more readily to deeper model sizes than the generic models, which suffer from significant trainability issues. This is achieved despite the generic models being in principle capable of finding any solution available to the equivariant models.

The encoding strategy $\tilde{\mathcal{E}}$ is a slight modification of the standard amplitude encoding method. For an image this would typically involve flattening the image into a one dimensional vector $\boldsymbol{x}$ and then mapping the elements of that vector to the amplitudes of a set of basis states, i.e., $\boldsymbol{x} \mapsto 1/\|\boldsymbol{x}\| \sum_i x_i |i\rangle$. This, however, would lead to complicated and non-local elements $R(g)$ of the representation of the symmetry group. In principle this is not a fundamental issue, and one could construct an equivariant model via (for example) "twirling" (i.e., explicitly projecting onto the invariant subspaces Eq. (14.10) of the representation) [9], but Ref. [19] instead opts to implement an encoding method more amenable to rotational symmetry, which allows one to easily write down an equivariant model in terms of familiar quantum gates. Expending some effort modifying the initial data encoding stage in order to simplify the equivariance constraints has previously been successfully employed in the context of reflection symmetry [16]. The strategy is parameterised by two hyper-parameters, n_{rad} and n_{orb} (respectively the number of qubits assigned to encoding radial and orbital degrees of freedom, with $n = n_{\text{rad}} + n_{\text{orb}}$ the total number of qubits), and involves constructing $2^{n_{\text{rad}}}$ regular $2^{n_{\text{orb}}}$-gons (see Fig. 14.1c (reproduced with permission from Ref. [19]) for examples with $n_{\text{orb}} = 3$ and $4 \leq n_{\text{rad}} \leq 11$). The image is sampled from at each vertex of each polygon, and, indexing the polygons by $r \in \{0, 1, \ldots, 2^{n_{\text{rad}}} - 1\}$, and the angular coordinate describing the position on the

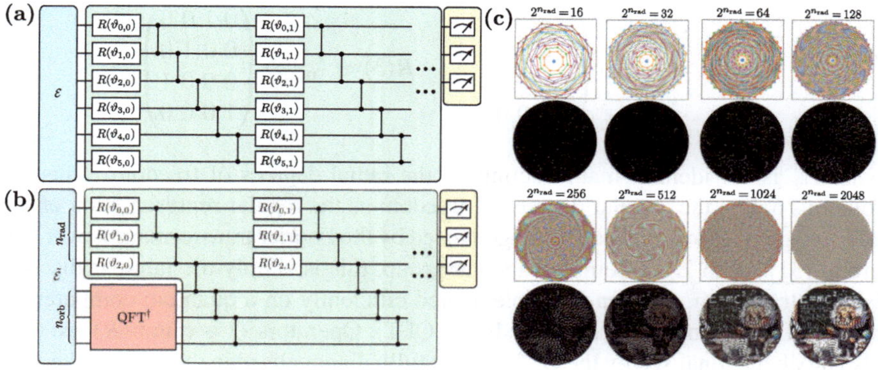

Fig. 14.1 **Rotationally Equivariant Quantum Machine Learning.** **a** A generic fully expressible QML model. Models such as these are known to suffer from *barren plateaus*, regions in their training landscape where the gradients of the cost function vanish exponentially, rendering the models untrainable. **b** A prototypical example of the architecture of the rotation equivariant models, which are less than fully expressible, having been restricted to the subspace of the full Hilbert space which respects rotation symmetry. The combination of the encoding choice $\tilde{\mathcal{E}}$ (see Eq. 14.19) and the quantum Fourier transform ensures that the induced representation of the rotation symmetry group will commute with any operation on the first n_{rad} qubits, and operations on the last n_{orb} qubits which are diagonal in the computational basis. Thus by employing an arbitrary circuit on the first n_{rad} qubits and (for example) CZ gates only on the remainder, we construct an equivariant model. **c** This choice of encoding leads to pixels being sampled from images at the vertices of $2^{n_{\mathrm{rad}}}$ regular $2^{n_{\mathrm{orb}}}$-gons (top row), with examples of the resulting reconstructed images for a 128×128 pixel image with $2^{n_{\mathrm{orb}}} = 8$ (i.e., octagons) and $16 \leq 2^{n_{\mathrm{rad}}} \leq 2048$ (bottom row). Reproduced with permission from Ref. [19]

polygon by $\phi = 2\pi k/2^{n_{\mathrm{orb}}}$, $k \in \{0, 1, \ldots, 2^{n_{\mathrm{orb}}} - 1\}$, the corresponding pixel values $x_{r,\phi}$ are used to construct the state

$$|\psi(\boldsymbol{x})\rangle = \frac{1}{\|\boldsymbol{x}\|} \sum_{r,\phi} x_{r,\phi} \, |r, \phi\rangle \, . \tag{14.19}$$

With this choice of data encoding the representation $R(g)$ of the operation of rotating by $2\pi g/2^{n_{\mathrm{orb}}}$ simply acts as

$$R(g) \, |\psi(\boldsymbol{x})\rangle = \sum_{r,\phi} x_{r,\phi} \left| r, \phi + 2\pi g/2^{n_{\mathrm{orb}}} \ (\mathrm{mod}\ 2\pi) \right\rangle . \tag{14.20}$$

Explicitly, with respect to this basis we have (in the case $n_{\mathrm{orb}} = 2$, for example)

$$R(0) = \mathbb{I}_{\mathrm{rad}} \otimes \begin{pmatrix} 1 & 0 & 0 & 0 \\ 0 & 1 & 0 & 0 \\ 0 & 0 & 1 & 0 \\ 0 & 0 & 0 & 1 \end{pmatrix}, \qquad R(1) = \mathbb{I}_{\mathrm{rad}} \otimes \begin{pmatrix} 0 & 0 & 0 & 1 \\ 1 & 0 & 0 & 0 \\ 0 & 1 & 0 & 0 \\ 0 & 0 & 1 & 0 \end{pmatrix}$$

$$R(2) = \mathbb{I}_{\text{rad}} \otimes \begin{pmatrix} 0 & 0 & 1 & 0 \\ 0 & 0 & 0 & 1 \\ 1 & 0 & 0 & 0 \\ 0 & 1 & 0 & 0 \end{pmatrix}, \qquad R(3) = \mathbb{I}_{\text{rad}} \otimes \begin{pmatrix} 0 & 1 & 0 & 0 \\ 0 & 0 & 1 & 0 \\ 0 & 0 & 0 & 1 \\ 1 & 0 & 0 & 0 \end{pmatrix}$$

where $\mathbb{I}_r$ is the identity matrix acting on the radial degrees of freedom. Thus the rotations act on the angular degrees of freedom as the regular representation of the additive group $\mathbb{Z}_{2^{n_{\text{orb}}}}$, and can be diagonalised by the Fourier transformation over that group [31]. With $\mathbb{Z}_{2^{n_{\text{orb}}}}$ being an abelian group, this is simply the familiar discrete Fourier transform, and can be implemented efficiently on a quantum computer by means of the quantum Fourier transform (QFT). Operators $O \in \mathtt{comm}(R)$ are then also block diagonalised by the QFT (in fact fully diagonalised in this case as the irreducible representations of abelian groups are one-dimensional), and so following the initial operations $\left(\mathbb{I}_{\text{rad}} \otimes \text{QFT}^{\dagger}_{\text{orb}}\right) \circ \tilde{\mathcal{E}}(\boldsymbol{x})$ one can construct equivariant models from components which are diagonal on the angular degrees of freedom and unconstrained on the radial degrees of freedom. In Ref. [19] the diagonal component is for simplicity variously taken to be either chains of nearest-neighbor CZ gates (see Fig. 14.1b), or a chain of controlled-Z rotations, with in either case a fully expressible ansatz on the radial component. More generally one could allow gates generated by Pauli strings that contain a Z at exactly k locations (all in the angular register), and I everywhere else, increasing the dimension of the DLA by $\binom{n_{\text{orb}}}{k}$, i.e., a polynomial of degree k in n_{orb}. All such gates are diagonal in the computational basis, thus preserving the symmetry properties of the model. Allowing parameterised Z rotations on the angular register, for example, would only linearly increase dim $\mathfrak{g}$, while simultaneously allowing strings with all values of $0 \leq k \leq n_{\text{orb}}$ would constitute an increase of $2^{n_{\text{orb}}}$, allowing for full expressibility while remaining symmetry preserving. Alternately, in the other direction, one could decrease dim $\mathfrak{g}$ by introducing additional constraints on the radial register. A potentially significant drawback of this encoding scheme, requiring as it does (in general) the preparation of highly entangled states, is that the time required to do so scales exponentially with the number of qubits. Moreover, as amplitude encoding creates a state of the same dimension as the input data vector, which we can presumably store and handle classically, simulating the entire quantum circuit is at most within a polynomial factor of what can be done with classical resources. With respect to the time taken necessary for state preparation, however, the situation may in practice be considerably more favorable than the worst case, as the data which one is interested in classifying can be highly structured (indeed, by assumption in the case considered here, containing non-trivial symmetries), potentially significantly lessening the difficulty of constructing the corresponding states. State preparation techniques can, for example, prepare approximations to amplitude encoded images with sufficiently high fidelity to allow for machine learning using drastically shallower circuits than a direct implementation [32]. Nonetheless, it would be interesting to attempt to generalize the techniques presented here to include encoding strategies which can be carried out more efficiently.

We now turn to the trainability of the rotationally equivariant models. The main analytical result is that, under some conditions, the models are free from barren plateaus. These results utilize the recent insights of Refs. [6, 7], which analyze the mean and variance of the cost function of QML models by studying the Lie closure of the generators of the circuit. In brief, given a layered quantum circuit of the form

$$\mathcal{U}(\boldsymbol{\theta}) = \prod_{i=1}^{L} U(\boldsymbol{\theta}_i) = \prod_{i=1}^{L}\prod_{j=1}^{J} e^{-i\theta_{ij}H_j} \tag{14.21}$$

one can relate the cost function mean and variance to the dimension of the *dynamical Lie algebra* $\mathfrak{g} = \text{span}\,\langle iH_0, iH_1, \ldots, iH_J\rangle_{\text{Lie}}$, where $\langle \mathcal{G}\rangle_{\text{Lie}}$ is the set formed from repeated nested commutators of the elements of $\mathcal{G}$. In particular, if $\dim\mathfrak{g}$ grows exponentially with the number of qubits then the model necessarily exhibits barren plateaus [6, 7]. Intuitively, the relevance of the nested commutators of the generators can be seen by applying the Baker-Campbell-Hausdorff formula to Eq. 14.21. In this case, one can show that the dimension of the dynamical Lie algebra of the models is exponential in n_{rad}, but can be linear in n_{orb}. Specifically [19]:

Proposition 1 *The dynamical Lie algebra $\mathfrak{g}$ of the rotationally equivariant model given by arbitrary single-qubit rotations on the radial register combined with a chain of controlled-Z rotations has a decomposition*

$$\mathfrak{g} \cong \mathfrak{su}(2^{n_{\text{rad}}}) \oplus \mathfrak{su}(2^{n_{\text{rad}}}) \oplus \mathbb{R}^{n_{\text{orb}}+1}$$

and thus dimension

$$\dim\mathfrak{g} = 2\cdot 4^{n_{\text{rad}}} + n_{\text{orb}} - 1$$

where n_{rad} (n_{orb}) is the number of qubits used to encode the radial (angular) degrees of freedom.

The exponential dependence on n_{rad} is a result of employing a fully expressible ansatz on the radial register; one could replace it with a variational structure possessing a DLA growing only polynomially in n_{rad}, but this constraint is not forced by rotational equivariance. Combined with the results of Ref. [6], the main result follows:

Corollary 1 *The rotationally equivariant models given by arbitrary single-qubit rotations on the radial register combined with a chain of controlled-Z rotations are, in the absence of noise, free from barren plateaus if and only if $4^{-n_{\text{rad}}}\mathcal{P}_{\mathfrak{g}_s}(\rho) \notin \mathcal{O}(1/b^n)$ for any $b > 2$, where $\rho = |\psi(\boldsymbol{x})\rangle\langle\psi(\boldsymbol{x})|$ is the initial state and $\mathcal{P}_{\mathfrak{g}_s}(\rho)$ its purity with respect to the semisimple component of the dynamical Lie algebra $\mathfrak{g}$.*

Here $\rho_{\mathfrak{g}}$ denotes the orthogonal projection of ρ onto $\mathfrak{g}$ (with respect to the Hilbert-Schmidt inner product) with purity $\mathcal{P}_{\mathfrak{g}}(\rho) := \text{tr}[(\rho_{\mathfrak{g}})^2]$. Proofs of Proposition 1 and Corollary 1 may be found in Ref. [19]. Corollary 1 gives the constraint within which these rotationally equivariant models must operate in order to not suffer from barren plateaus, and has several immediate consequences. Firstly, regardless of the choice of n_{rad}, the purity $\mathcal{P}_{\mathfrak{g}_s}(\rho)$ cannot vanish faster than 2^{-n}. Secondly, as $\mathcal{P}_{\mathfrak{g}}(\rho) \leq 1$, the number of qubits used in the radial register is bounded by $n_{\text{rad}} \leq n/2 + o(n)$. These restrictions correspond respectively to requirements on (a generalized notion of) the initial state entanglment [33] and the circuit expressibility [4], and are linked through the unifying Lie-algebraic theory of Ref. [6]. In Fig. 14.2 (reproduced with permission from Ref. [19]) we plot empirical results for the cost function gradient scaling under various choices of $n_{\text{rad}}/n_{\text{qubits}}$, the conclusions of which are consistent with the analytical findings. Under the constraints of Corollary 1, then, the rotationally equivariant models join the (short) list of QML architectures provably free from barren plateaus [11, 21]. This is an extremely encouraging finding, but nonetheless the true test of machine learning models is to benchmark their performance in practice on a real dataset. The results of such a test are reported in Fig. 14.3 (reproduced with permission from Ref. [19]), where it is found that, for a specific task in which rotation symmetry is present, the rotationally-equivariant models indeed drastically outperform their generic counterparts, despite being in principle strictly less expressive. One can interpret this as a significant validation of the above arguments.

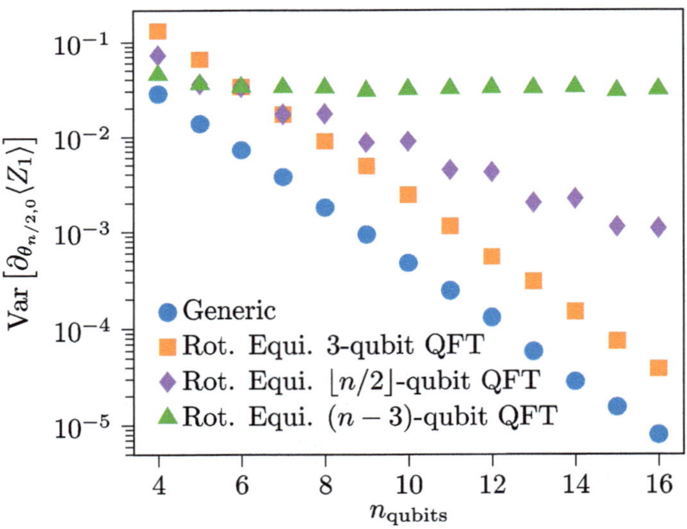

Fig. 14.2 Gradient Scaling. The variance of the derivative of a measured observable with respect to a parameter in the middle of the models of Fig. 14.1b under different assignments of the qubits to encoding radial and rotational information. In all cases the variances are estimated by averaging over 1000 randomly initialized models. When the number n_{rad} of qubits not involved in the quantum Fourier transform grows faster than logarithmically with n_{qubits} the model suffers from a barren plateau, as exhibited by the exponential vanishing of the cost function gradients. Reproduced with permission from Ref. [19]

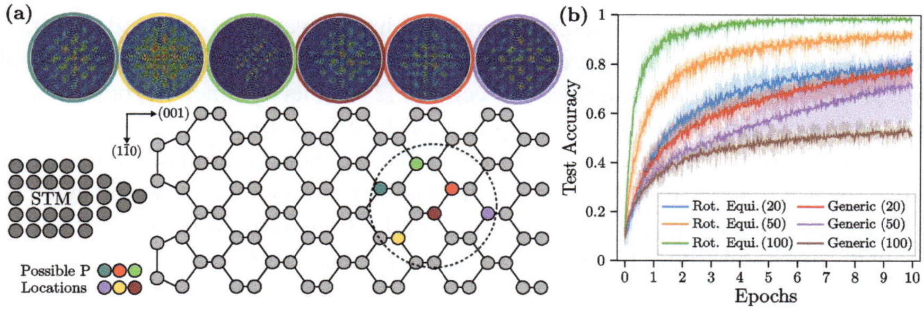

Fig. 14.3 Pinpointing Atomic Impurities with Symmetry-Informed QML. In order to numerically test the performance of the models of this section in practice, Ref. [19] considered a classification task which involves determining the location of phosphorus donor atoms implanted into silicon by analyzing their STM images. **a** P donors in Si create remarkably rich STM images which depend strongly on their precise location in the lattice. The classification problem is to determine the correct lattice site from ten possibilities corresponding to positions between $3a_0$ and $4.5a_0$ beneath the surface (with $a_0 \approx 0.54$nm the silicon lattice constant) given a (simulated, see Appendix B) STM image. **b** Each model configuration is randomly initialized and trained five times, with the mean accuracy achieved on a validation set of 500 examples throughout training plotted, and the area between the best and worst performing of the training runs shaded. The architecture and number of layers in each model is indicated; in the rotationally equivariant case the operators on the angular register are taken to be a chain of nearest-neighbor CZ gates. The rotationally equivariant models drastically outperform their generic counterparts on this task, despite possessing fewer trainable parameters and having access to only a fraction of the total Hilbert space. Moreover, we see that increasing the depth of the generic models actually adversely affects their performance in practice, due to the increasing issues with trainability and given only a finite set of training examples. The training set consists of 7500 examples, equally distributed among the classes. Reproduced with permission from Ref. [19]

14.3.3 Summary and Outlook

Much as early classical neural networks suffered from significant trainability issues [34] that were solved only after considerable experimentation and a sequence of increasingly optimized architectural decisions, early QML models based on generic circuits look likely to be replaced by carefully designed constructions. In the absence of large-scale fault-tolerant quantum computers on which to easily experiment with different types of QML architectures in the many qubit regime, symmetry principles represent a promising guide to current research [8–18] that has now lead to the discovery of multiple architectures which are provably trainable [11, 21, 22]. In particular, it seems likely to continue to be fruitful to combine these ideas with the increased understanding of the nature of QML models afforded by their rapidly developing Lie-algebraic theory [6, 7], through which one can study the complexity of models as they interpolate between the trainable and untrainable regimes, as characterized by the dimensionality of the dynamical Lie algebra.

Of course, there are many more symmetries that are yet to be explored. In future work, for example, it would be extremely interesting to explore generalizations of the QFT-based technique described in Sect. 14.3.2 to other symmetry groups, which in many cases admit Fourier transforms that can be efficiently evaluated on quantum computers [31]. With the encoding set up so that the induced representation on the Hilbert space is given as in this work by the regular representation, it would again be (block) diagonalised by the corresponding Fourier transform. In the case of a non-abelian group non-trivial blocks will appear, and the set of allowable equivariant operators in the variational ansatz will no longer be arbitrary diagonal operators, but instead (as a result of Schur's lemma) those operators which are multiples of the identity on each of the irreducible components in the decomposition of the regular representation. For any abelian symmetry group, arbitrary diagonal operators will continue to be equivariant. More generally the deep connections between Fourier analysis and representation theory may lead to some very interesting variational algorithms—QFTs play a central role in many fault tolerant quantum algorithms with provable speed ups; by connecting these to the representation theory (and therefore symmetries) of groups it may be possible to convert these algorithms into geometric QML algorithms which also enjoy provable speed ups.

Finally, it must be acknowledged that imposing restrictions on the expressibility of QML models introduces the potential for efficient classical simulations, which can become possible, for example, in the case of a small dynamical Lie algebra [35, 36], or in the presence of sufficiently constraining symmetry [37]. The compromises that can be explored through families of models such as those explored here, which can interpolate between easily trainable but efficiently classically simulable circuits through to classically intractable but barren-plateau-ridden circuits, then constitute an interesting research direction in the ongoing search for practical quantum advantage in machine learning.

References

1. Noether E (1918) Invariante variationsprobleme, math-phys. Klasse, pp 235–257
2. Kobayashi M, Maskawa T (1973) CP-violation in the renormalizable theory of weak interaction. Prog Theor Phys 49:652–657 (1973)
3. Bronstein MM, Bruna J, Cohen T, Veličković P (2021) Geometric deep learning: grids, groups, graphs, geodesics, and gauges. arXiv:2104.13478
4. Holmes Z, Sharma K, Cerezo M, Coles PJ (2022) Connecting ansatz expressibility to gradient magnitudes and barren plateaus. PRX Quantum 3(1):010313
5. Larocca M, Czarnik P, Sharma K, Muraleedharan G, Coles PJ, Cerezo M (2022) Diagnosing barren plateaus with tools from quantum optimal control. Quantum 6:824
6. Ragone M, Bakalov BN, Sauvage F, Kemper AF, Ortiz Marrero C, Larocca M, Cerezo M (2024) A lie algebraic theory of barren plateaus for deep parameterized quantum circuits. Nat Commun 15(1):7172
7. Fontana E, Herman D, Chakrabarti S, Kumar N, Yalovetzky R, Heredge J, Sureshbabu SH, Pistoia M (2023) The adjoint is all you need: Characterizing barren plateaus in quantum ansätze. arXiv:2309.07902

8. Meyer JJ, Mularski M, Gil-Fuster E, Mele AA, Arzani F, Wilms A, Eisert J (2023) Exploiting symmetry in variational quantum machine learning. PRX Quantum 4(1):010328
9. Ragone M, Nguyen QT, Schatzki L, Braccia P, Larocca M, Sauvage F, Coles PJ, Cerezo M (2022) Representation theory for geometric quantum machine learning. arXiv:2210.07980
10. Nguyen QT, Schatzki L, Braccia P, Ragone, Coles PJ, Sauvage F, Larocca M, Cerezo M (2022) Theory for equivariant quantum neural networks. arXiv:2210.08566
11. Schatzki L, Larocca M, Nguyen QT, Sauvage F, Cerezo M (2024) Theoretical guarantees for permutation-equivariant quantum neural networks. npj Quantum Information, vol 10, no 1, p 12
12. Sauvage F, Larocca M, Coles PJ, Cerezo M (2024) Building spatial symmetries into parameterized quantum circuits for faster training. Quantum Sci Technol 9(1):015029
13. Skolik A, Cattelan M, Yarkoni S, Bäck T, Dunjko V (2023) Equivariant quantum circuits for learning on weighted graphs. npj Quantum Information, vol 9, no 1, p 47
14. Larocca M, Sauvage F, Sbahi FM, Verdon G, Coles PJ, Cerezo M (2022) Group-invariant quantum machine learning. PRX Quantum 3:030341 Sep
15. Zheng H, Ravi GS, Wang H, Setia K, Chong FT, Liu J (2022) Benchmarking variational quantum circuits with permutation symmetry. arXiv:2211.12711
16. West MT, Sevior M, Usman M (2023) Reflection equivariant quantum neural networks for enhanced image classification. Mach Learn Sci Technol 4(3):035027
17. Chang SY, Grossi M, Le Saux B, Vallecorsa S (2023) Approximately equivariant quantum neural network for p4m group symmetries in images In: 2023 IEEE international conference on quantum computing and engineering (QCE), vol 01, pp 229–235
18. East RD, Alonso-Linaje G, Park C-Y (2023) All you need is spin: Su (2) equivariant variational quantum circuits based on spin networks. arXiv:2309.07250
19. West MT, Heredge J, Sevior M, Usman M (2024) Provably trainable rotationally equivariant quantum machine learning. PRX Quantum 5(3):030320
20. Heredge J, Hill C, Hollenberg L, Sevior M (2023) Permutation invariant encodings for quantum machine learning with point cloud data. arXiv:2304.03601
21. Pesah A, Cerezo M, Wang S, Volkoff T, Sornborger AT, Coles PJ (2021) Absence of barren plateaus in quantum convolutional neural networks. Phys Rev X 11(4):041011
22. West MT, Heredge J, Sevior M, Usman M (2023) Provably trainable rotationally equivariant quantum machine learning. arXiv:2311.05873
23. Shao S-H (2023) What's done cannot be undone: Tasi lectures on non-invertible symmetry. arXiv:2308.00747
24. Peruzzo A, McClean J, Shadbolt P, Yung M-H, Zhou X-Q, Love PJ, Aspuru-Guzik A, O'brien JL (2014) A variational eigenvalue solver on a photonic quantum processor. Nat Commun 5(1):1–7
25. Cerezo M, Arrasmith A, Babbush R, Benjamin SC, Endo S, Fujii K, McClean JR, Mitarai K, Yuan X, Cincio L, Coles PJ (2021) Variational quantum algorithms. Nat Rev Phys 3(1):625–644
26. Tang HL, Shkolnikov V, Barron GS, Grimsley HR, Mayhall NJ, Barnes E, Economou SE (2021) qubit-adapt-vqe: an adaptive algorithm for constructing hardware-efficient ansätze on a quantum processor. PRX Quantum 2(2):020310
27. M. Zaheer, S. Kottur, S. Ravanbakhsh, B. Poczos, R. R. Salakhutdinov, and A. J. Smola, "Deep sets," in *Advances in Neural Information Processing Systems* (I. Guyon, U. V. Luxburg, S. Bengio, H. Wallach, R. Fergus, S. Vishwanathan, and R. Garnett, eds.), vol. 30, Curran Associates, Inc., 2017
28. H. Maron, O. Litany, G. Chechik, and E. Fetaya, "On learning sets of symmetric elements," in *Proceedings of the 37th International Conference on Machine Learning* (H. D. III and A. Singh, eds.), vol. 119 of *Proceedings of Machine Learning Research*, pp. 6734–6744, PMLR, 13–18 Jul 2020
29. J. Heredge, M. West, L. Hollenberg, and M. Sevior, "Non-unitary quantum machine learning," *arXiv preprint* arXiv:2405.17388, 2024
30. W. Fulton and J. Harris, *Representation Theory: A First Course*. Springer, 1991

31. Childs AM, van Dam W (2010) Quantum algorithms for algebraic problems. Rev Mod Phys 82:1–52 Jan
32. M. T. West, A. C. Nakhl, J. Heredge, F. M. Creevey, L. C. Hollenberg, M. Sevior, and M. Usman (2024) Drastic circuit depth reductions and preserved adversarial robustness by approximate encoding for quantum machine learning. Intelligent Computing 3:100
33. Cerezo M, Sone A, Volkoff T, Cincio L, Coles PJ (2021) Cost function dependent barren plateaus in shallow parametrized quantum circuits. Nat Commun 12(1):1–12
34. Bengio Y, Simard P, Frasconi P (1994) Learning long-term dependencies with gradient descent is difficult. IEEE Trans Neural Networks 5(2):157–166
35. M. L. Goh, M. Larocca, L. Cincio, M. Cerezo, and F. Sauvage, "Lie-algebraic classical simulations for variational quantum computing," *arXiv preprint* arXiv:2308.01432, 2023
36. M. Cerezo, M. Larocca, D. García-Martín, N. L. Diaz, P. Braccia, E. Fontana, M. S. Rudolph, P. Bermejo, A. Ijaz, S. Thanasilp, *et al.*, "Does provable absence of barren plateaus imply classical simulability? Or, why we need to rethink variational quantum computing," *arXiv preprint* arXiv:2312.09121, 2023
37. Anschuetz ER, Bauer A, Kiani BT, Lloyd S (2023) Efficient classical algorithms for simulating symmetric quantum systems. Quantum 7:1189

Chapter 15
Characterizing Privacy of Quantum Variational Circuit Models

Jamie Heredge

Abstract In this chapter, we focus on data privacy in machine learning and explore the robustness of quantum machine learning models against data privacy breaches. The primary motivation is to identify a novel form of quantum advantage by discovering quantum machine learning models that offer enhanced robustness against privacy breaches compared to classical machine learning models. While this line of research is important in its own right, it also serves as an example for demonstrating how Lie-algebraic techniques can be employed in practice to characterize the privacy of quantum machine learning models.

15.1 Introduction

The importance of data privacy has grown significantly over time, particularly in the context of machine learning models that work with sensitive datasets. Privacy concerns are prevalent across various applications, such as financial records [1, 2], healthcare information [3–5], and location data [6]. Each of these domains presents its own unique challenges and considerations regarding privacy. Additionally, the widespread adoption of stringent legal frameworks on a global scale [7] has heightened the urgency to enhance data privacy measures.

The concept of privacy in quantum machine learning is not only intriguing but also holds the promise of a novel form of quantum advantage. By leveraging the unique properties of quantum systems, researchers aim to develop models that are inherently robust against privacy breaches. This chapter serves as a comprehensive exploration of the privacy characteristics of VQC models, employing Lie-algebraic techniques to provide insights into their privacy guarantees. Through this investigation, we aim to establish a framework for understanding the conditions under which quantum models can offer superior privacy protections, thereby contributing to the broader discourse on data privacy in the age of quantum computing.

J. Heredge (✉)
Global Technology Applied Research, JP Morgan Chase, New York, NY, USA
e-mail: jamie.heredge@jpmorgan.com

M. Usman (ed.), *Quantum Robustness in Artificial Intelligence*, Quantum Science and Technology, https://doi.org/10.1007/978-3-032-11153-1_15

Distributed learning frameworks, such as federated learning [8–10] (which will be the focus of this chapter), offer enhanced computational efficiency and the promise of improved privacy in machine learning tasks. To illustrate the federated learning framework, consider a scenario where a large technology company aims to train machine learning models using data from millions of customers. The conventional approach would involve users sending their data to a central server, where a model could be trained on this aggregated data. However, this data may be sensitive, and customers might be unwilling to share it, or there may be strict legal requirements prohibiting the collection of certain types of data. Is it possible to train a machine learning model without directly accessing the data? With the federated learning framework, this is indeed possible.

In federated learning, each customer receives the model architecture and weights, allowing them to process their data locally on their own device. This enables them to compute the gradient of the model with respect to their data for the given weights. Instead of transmitting their confidential data, individual clients send their computed gradients to the central server. These gradients are then aggregated by the central server and used to update the model weights of the base model, which are subsequently distributed back to all users, updating all models globally. This iterative process continues until the training is complete. Not only does this approach distribute the computational load across numerous devices, but it also ensures that individual clients do not have to share their data, only the gradients derived from their data.

While the federated learning framework initially seems to address the issue of data privacy, there is a significant caveat. Since the central server is aware of the model architecture and weights, as well as the gradients sent by an individual user, it may be possible to reverse engineer the confidential data from the gradients. This process of recovering data from model gradients is known as gradient inversion. Research has demonstrated that neural networks are susceptible to gradient inversion-based attacks, which can reconstruct the original input data [11–15].

To mitigate this vulnerability, several classical techniques have been developed to enhance the privacy of distributed learning models. These include methods based on gradient encryption [16], the introduction of artificial noise in gradients to apply differential privacy techniques [10], and strategies that utilize batch training for gradient mixing [17]. Despite their mitigative effects, these techniques are not entirely foolproof. They may still leak some input information, impose significant computational overhead during model training in a distributed setting, or result in diminished model performance. To give an example, one method that can be applied is adding random noise to the gradients. For example, one method to enhance privacy is by adding random noise to the gradients. This approach increases privacy by making gradient inversion more challenging. However, it can also lead to a reduction in model performance, as the added noise may interfere with the effectiveness of the model's learning process.

In this chapter, we will summarize results investigating the inherent privacy of VQC models [18, 19] by exploring a fundamental question: *Given the gradients of a quantum machine learning model, how challenging is it to reconstruct the original classical data inputs?* We aim to discover if something about the quantum nature of the model provides some inherent protection. This chapter offers a comprehensive overview of strategies for extracting the key properties of VQCs and ensuring robust privacy guarantees. It presents the main theorems, algorithms, and findings from Heredge et al. [19], which established the mathematical foundation for examining the privacy of VQC models through a Lie-algebraic lens.

15.2 Case Study: High-Frequency Encodings

To give an introductory case study, we will consider the work of Kumar et al. [18]. In this work, it was suggested that highly expressive product encoding maps along with an overparameterized hardware efficient ansatz (HEA) as the variational layer would require an exponential amount of resources (in terms of the number of qubits n) for an attacker to learn the input from the gradients.

They used an encoding circuit where each feature in a d-dimensional data input would be encoded into m qubits through a single rotation gate on each qubit with a scaling factor of 5^{k-1} where $k \in [1, m]$. This results in a total of dm qubits to encode the information using only single-qubit rotation gates. This type of encoding circuit is called a Fourier tower encoding and can be written as

$$V(\mathbf{x}) = \bigotimes_{j=1}^{d} \bigotimes_{k=1}^{m} R_X(5^{k-1} x_j), \tag{15.1}$$

where R_X is a X rotation gate and $x_j \in \mathbf{x}$ is a single feature of the classical datapoint $\mathbf{x}$. The result of this encoding is that when the gradient of the loss function is written as a Fourier series, it results in a frequency spectrum with all possible integer values with a total spectrum size of $\frac{5^{md}-1}{2}$ [18]. This is then combined with an overparameterized hardware efficient ansatz for the variational layer of the VQC circuit, which is a circuit architecture that is widely believed to be difficult to simulate classically. The exponentially large Fourier spectrum leads to a loss landscape in which finding a solution numerically is difficult due to an exponential amount of local minima which end up exponentially close together.

They considered numerical attacks in which one may attempt to guess an input $\mathbf{x}'$ and pass it into the model to find a gradient guess C' which can then be used in a technique called numerical gradient inversion. The loss of the gradient inversion model is the difference between this guessed gradient C' and the known gradients C. If this loss goes to 0, then the guessed gradient matches the known gradient exactly, and it can be concluded that the guess $\mathbf{x}'$ is a good guess for the true confidential data $\mathbf{x}$. As the number of qubits per data feature m is increased in the Fourier tower encoding, this leads to exponentially larger frequencies in the gradient spectrum, which complicates the loss landscape, see Figs. 15.1 and 15.2.

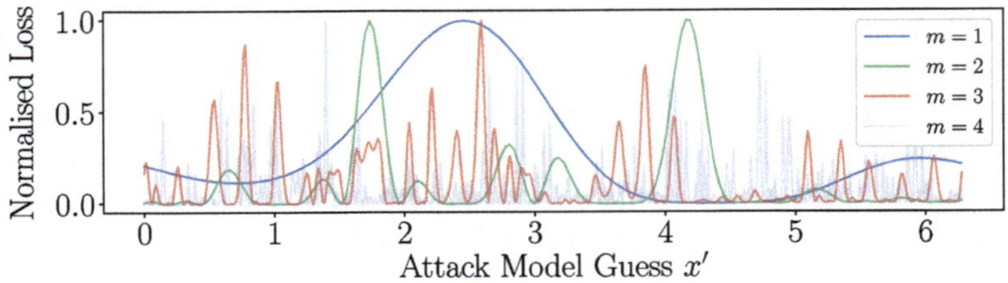

Fig. 15.1 Visualization of the loss landscape of the gradient inversion model attack as the attack model guess x' is varied in the univariate case $d = 1$. The loss is plotted for models with differing numbers of qubits per input m. The shape of the loss landscape is influenced by the highest frequency terms which scale exponentially with m. The plotted loss is normalized for readability. The landscape contains exponentially more local minima as m is increased, making a numerical attack more difficult [18]

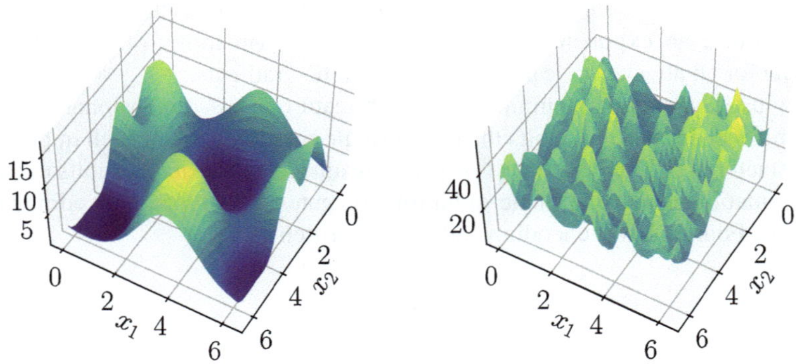

Fig. 15.2 Gradient inversion attack loss landscape for (Left) $m = 1$ and (Right) $m = 2$ for the multivariate case $d = 2$ with two inputs $x' = (x_1, x_2)$. One can see that increasing m significantly increases the difficulty of navigating the loss function required for gradient inversion [18]

The study by [18], while pioneering in its theoretical analysis of the privacy aspects of a specific Variational Quantum Circuit (VQC) model architecture, presents certain notable limitations. One major drawback is that overparameterization of a Hardware Efficient Ansatz (HEA) can result in an untrainable model. This occurs because the model rapidly mixes to a 2-design [20], leading to the barren plateau phenomenon [21], where gradients vanish and hinder effective training. Furthermore, the introduction of high-frequency terms in the encoding map may not be a uniquely quantum effect. Classical machine learning models can also benefit from initially loading data with high-frequency feature maps [22]. Despite these limitations, the case study serves as a catalyst for further exploration into the privacy of VQC models. In the remainder of this chapter, we will shift focus from specific architectures to characterizing privacy in generic VQC models. By establishing this framework, we will revisit the case study and provide more concrete critiques regarding the practicality of the encoding and variational circuit suggested in Kumar et al. [18].

15.3 Definitions and Framework

In this section, we will describe the mathematical framework that underpins Variational Quantum Circuit (VQC) models, with a particular focus on their privacy characteristics. This chapter aims to provide a comprehensive framework for characterizing privacy breaches in VQC models, utilizing Lie-algebraic techniques to explore the potential for both weak and strong privacy breaches. The notations used in this chapter are summarized in Sect. 15.9.

15.3.1 Variational Quantum Circuits for Machine Learning

A Variational Quantum Circuit (VQC) is structured in the following manner. We begin with a d-dimensional input vector $\mathbf{x} \in \mathcal{X} \subset \mathbb{R}^d$, which is loaded into the quantum encoding circuit $V(\mathbf{x})$ of n qubits resulting in a feature map with the input state mapping:

$$\rho(\mathbf{x}) = V(\mathbf{x})|0\rangle^{\otimes n}\langle 0|^{\otimes n} V(\mathbf{x})^{\dagger}. \tag{15.2}$$

This operation loads the input vector of dimension d into a Hilbert space $\mathcal{H} = (\mathbb{C}^2)^{\otimes n}$ of dimension $\dim(\rho(\mathbf{x})) = 2^n$. We consider the common scenario where the number of qubits in a given algorithm is proportional to the input vector dimension d. The quantum state $\rho(\mathbf{x})$ is then passed through a variational circuit ansatz $U(\boldsymbol{\theta})$ defined as

$$\mathbf{U}(\boldsymbol{\theta}) = \prod_{k=1}^{D} e^{-i\theta_k \mathbf{H}_{\nu(k)}}, \tag{15.3}$$

which is parameterized by a vector of variational parameters $\boldsymbol{\theta} = [\theta_1, \cdots, \theta_D]$, where D is the total number of variational parameters. Here $\{\mathbf{H}_1, \ldots, \mathbf{H}_N\}$ are the set of N Hermitian generators of the circuit U. Note that there may be more variational parameters than there are Hermitian generators, in which case multiple θ_k values can be linked to the same Hermitian generator. For this reason the generator assignment map $\nu : [D] \to [N]$ is used to assign each variational parameter θ_k to its corresponding generator $\mathbf{H}_{\nu(k)}$. Under this notation, multiple distinct variational parameters can use the same generator element, albeit the generator will be applied at different points in the circuit. This is the case for repeated layers of a variational ansatz, where for L repeated layers one would have $D = NL$ and $\nu(k) = ((k-1) \bmod N) + 1$. This structure is notably general, as it encompasses several common ansatz structures, including the hardware efficient ansatz, the quantum alternating operator ansatz, and the Hamiltonian variational ansatz, among others, as highlighted in [23].

The parameterized state $\rho(\mathbf{x})$ is processed through a variational circuit denoted by $U(\boldsymbol{\theta})$, followed by the measurement of an observable $\mathbf{O} \in \mathcal{H}$. For a given set of parameters $\boldsymbol{\theta}$, the output of the variational quantum circuit model is represented as the expectation value of $\mathbf{O}$ with the parameterized state:

$$y_{\boldsymbol{\theta}}(\mathbf{x}) = \mathrm{Tr}(\mathbf{U}^{\dagger}(\boldsymbol{\theta})\boldsymbol{O}\mathbf{U}(\boldsymbol{\theta})\rho(\mathbf{x})). \tag{15.4}$$

For the task of optimizing the variational quantum circuits, the model output is fed into the desired cost function $\texttt{Cost}(\boldsymbol{\theta}, \mathbf{x})$ which is subsequently minimized to obtain

$$\boldsymbol{\theta}^* = \arg\min_{\boldsymbol{\theta}} \texttt{Cost}(\boldsymbol{\theta}, \mathbf{x}), \tag{15.5}$$

where $\boldsymbol{\theta}^*$ are the optimized parameter values.

The typical optimization process involves calculating the gradient of the cost function with respect to the parameters $\boldsymbol{\theta}$, which requires computing the gradient with respect to the model output $y_{\boldsymbol{\theta}}(\mathbf{x})$:

$$C_j = \frac{\partial y_{\boldsymbol{\theta}}(\mathbf{x})}{\partial \theta_j}, \quad j \in [D]. \tag{15.6}$$

Going forward, we will directly deal with the recoverability of input $\mathbf{x}$ given C_j, instead of working with specific cost functions. If one wishes to work with specific cost functions then it is important to notice that using chain rule we can see that the

$$\tilde{C}_j = \frac{\partial \texttt{Cost}(y_\theta, y)}{\partial \theta_j} = \frac{\partial \texttt{Cost}(y_\theta, y)}{\partial y_\theta} \cdot \frac{\partial y_\theta}{\partial \theta_j}. \tag{15.7}$$

If the value of $\frac{\partial \texttt{Cost}(y_\theta, y)}{\partial y_\theta}$ can be calculated, which would be possible if we were given the output of the model y_θ, along with the data labels y, then it would be possible to directly convert between $\tilde{C}_j$ and C_j. When this is not the case, it is still possible to attempt a solution, although it requires one additional gradient compared to usual. We show this by considering that the $\frac{\partial \texttt{Cost}(y_\theta, y)}{\partial y_\theta}$ term, although unknown, will be the same for all θ_j. Hence we can eliminate this term considering ratios of the known gradients defined as

$$R_j \equiv \frac{\tilde{C}_j}{\tilde{C}_1} = \frac{\frac{\partial y_\theta}{\partial \theta_j}}{\frac{\partial y_\theta}{\partial \theta_1}}. \tag{15.8}$$

This allows us to write new equations for all $j > 1$ without this unknown term as

$$\frac{\partial y_\theta}{\partial \theta_j} = R_j \cdot \frac{\partial y_\theta}{\partial \theta_1}, \tag{15.9}$$

hence in the notation used throughout this chapter

$$C_j = R_j \cdot C_1. \tag{15.10}$$

We included the above argument for completeness, but going forward we shall instead ignore any specific cost function and focus solely on the recoverability of input $\mathbf{x}$ given C_j in order to simplify the notation.

15.3.2 Lie Theoretic Framework

Lie-algebraic techniques are ubiquitous in quantum computing, and are especially relevant for the study of VQC models. This chapter provides an important example of how these techniques can be used in practice to provide privacy characterizations of a wide range of VQC models. We begin by reviewing both introductory and recent findings on the Lie theoretic framework for variational quantum circuits, which are important to this chapter. For a comprehensive review of this subject, readers are directed to [23, 24].

Definition 1 (Dynamical Lie Algebra [19]) The dynamical Lie algebra (DLA) $\mathfrak{g}$ for an ansatz $\mathbf{U}(\boldsymbol{\theta})$ of the form of Eq. 15.3 is defined as the real span of the Lie closure of the generators of U

$$\mathfrak{g} = \text{span}_{\mathbb{R}}\langle i\mathbf{H}_1, \cdots, i\mathbf{H}_N\rangle_{\text{Lie}}, \tag{15.11}$$

where the closure is defined by taking all possible nested commutators of $S = \{i\mathbf{H}_1, \cdots, i\mathbf{H}_N\}$. In other words, it is the set of elements obtained by taking the commutation between elements of S until no further linearly independent elements are obtained.

Consider a simple quantum system with two generators, $i\mathbf{H}_1$ and $i\mathbf{H}_2$. The DLA would include all possible combinations of these generators through commutation, such as $[i\mathbf{H}_1, i\mathbf{H}_2]$, $[i\mathbf{H}_1, [i\mathbf{H}_1, i\mathbf{H}_2]]$,$[i\mathbf{H}_2, [i\mathbf{H}_1, i\mathbf{H}_2]]$, and so on. Note that eventually these nested commutators may equal 0 or give elements already in the DLA set, in which case the DLA has a finite dimension. If we take a single qubit with two rotation gates X and Y, we see that the Pauli generators of the gates will be $\{i\mathsf{X}, i\mathsf{Y}\}$, taking the commutator will give us $i\mathsf{Z}$ and any further nested commutator calculations will equal 0 or an element already in the set, meaning the DLA of this circuit will be $\{i\mathsf{X}, i\mathsf{Y}, i\mathsf{Z}\}$.

Definition 2 (Dynamical Lie Group [19]) The dynamical Lie group $\mathcal{G}$ for an ansatz $\mathbf{U}(\boldsymbol{\theta})$ of the form of Eq. 15.3 is determined by the DLA $\mathfrak{g}$ such that

$$\mathcal{G} = e^{\mathfrak{g}}, \tag{15.12}$$

where $e^{\mathfrak{g}} := \{e^{i\mathbf{H}}, \quad i\mathbf{H} \in \mathfrak{g}\}$ and is a subgroup of $\text{SU}(2^n)$. For generators in $\mathfrak{g}$, the set of all $\mathbf{U}(\boldsymbol{\theta})$ of the form Eq. 15.3 generates a dense subgroup of $\mathcal{G}$.

Relating this back to a quantum computing example we know that an X rotation gate for a single qubit will be of the form $U_X(\theta) = e^{i\theta X} \in \text{SU}(2)$, and this is generated by the Pauli X matrix, which is the Hermitian operator for this particular gate. We see in the Dynamical Lie Group that the generators of the circuit determine the type of unitary operators that can be implemented in the circuit.

Definition 3 (Adjoint Representation [19]) The Lie algebra adjoint representation is the following linear action: $\forall \mathbf{K}, \mathbf{H} \in \mathfrak{g}$,

$$\mathrm{ad}_{\mathbf{H}}\mathbf{K} := [\mathbf{H}, \mathbf{K}] \in \mathfrak{g}, \tag{15.13}$$

and the Lie group adjoint representation is the following linear action: $\forall \mathbf{U} \in \mathcal{G}, \forall \mathbf{H} \in \mathfrak{g}$,

$$\mathrm{Ad}_{\mathbf{U}}\mathbf{H} := \mathbf{U}^{\dagger}\mathbf{H}\mathbf{U} \in \mathfrak{g}. \tag{15.14}$$

The adjoint representation describes how elements of the Lie algebra transform under the action of the Lie group. It is a crucial concept for understanding how quantum states evolve.

Definition 4 (DLA basis [19]) The basis of the DLA is denoted as $\{i\mathbf{B}_\alpha\}_\alpha$, $\alpha \in \{1, \cdots, \dim(\mathfrak{g})\}$, where $\mathbf{B}_\alpha$ are Hermitian operators and form an orthonormal basis of $\mathfrak{g}$ with respect to the Frobenius inner product.

Any observable $\mathbf{O}$ is said to be entirely supported by the DLA whenever $i\mathbf{O} \in \mathfrak{g}$ or, in other words,

$$\mathbf{O} = \sum_\alpha \mu_\alpha \mathbf{B}_\alpha, \tag{15.15}$$

where μ_α is coefficient of support of $\mathbf{O}$ in the basis $\mathbf{B}_\alpha$.

Definition 5 (Lie Algebra Supported Ansatz [25]) A Lie Algebra Supported Ansatz (LASA) is a periodic ansatz of the form of Eq. 15.3 of a VQC where the measurement operator $\boldsymbol{O}$ is completely supported by the DLA $\mathfrak{g}$ associated with the generators of $U(\boldsymbol{\theta})$, that is:

$$i\boldsymbol{O} \in \mathfrak{g}. \tag{15.16}$$

In addition to its connections to the trainability of a VQC, this condition also implies that $\forall\boldsymbol{\theta}$, $U^{\dagger}(\boldsymbol{\theta})i\boldsymbol{O}U(\boldsymbol{\theta}) \in \mathfrak{g}$, which enables us to express the evolution of the observable $\boldsymbol{O}$ in terms of elements of $\mathfrak{g}$. This is key to some simulation algorithms that are possible for polynomial-sized DLAs [26, 27], which will become important concepts in discussions found later in this chapter.

15.3.3 Input Recoverability Definitions

In this section, we explore the concept of recovering classical input data by analyzing the gradients $\{C_j\}_{j=1}^{D}$ of a Variational Quantum Circuit (VQC). This exploration is structured to separately consider the encoding and variational components of a quantum variational model.

A useful concept in machine learning is the creation of data *snapshots*. These snapshots are compact and efficient representations of the input data's feature map encoding. Essentially, a snapshot retains sufficient information to substitute for the full feature map encoded data, enabling the training of a machine learning model for a distinct task with the same data, but without the need to explicitly know that the input data was passed through the feature map.

Consider a machine learning model that processes images. Instead of using the entire image, a snapshot might capture essential features such as edges or color distributions, allowing the model to perform tasks like classification or recognition without needing the full-image data.

In quantum computing, snapshots of expectation values are particularly useful for simulating quantum circuits. For example, in methods such as $\mathfrak{g}$-sim [27], these snapshots are used as input vectors for classical simulators. The simulator can then process these vectors efficiently under certain conditions, recreating the operation of a variational quantum circuit.

For our discussion of privacy it will become useful to classify the process of input data $\mathbf{x}$ recovery into two stages: the first concerns recovering snapshots of the quantum state $\rho(\mathbf{x})$ (Eq. 15.2) from the gradients, which involves only considering the variational part of the circuit (Fig. 15.3).

Definition 6 (Snapshot Recovery [19]) Given the gradients $C_j, \quad j \in [D]$ as defined in Eq. 15.6 as well as the parameters $\boldsymbol{\theta} = [\theta_1, \cdots, \theta_D]$, we consider a VQC to be snapshot recoverable if there exists an efficient $O(poly(d, \frac{1}{\epsilon}))$ classical polynomial-time algorithm to recover the vector $\mathbf{e}_{\text{snap}}$ such that

$$|[\mathbf{e}_{\text{snap}}]_\alpha - \text{Tr}(\mathbf{B}_\alpha \rho(\mathbf{x}))| \leq \epsilon, \ \forall \alpha \in [\dim(\mathfrak{g})], \tag{15.17}$$

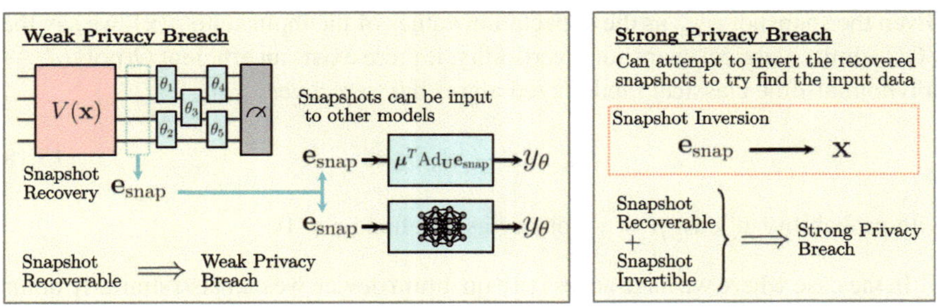

Fig. 15.3 Overview of the general framework and definitions. Weak privacy breach corresponds to attacks where expectation value snapshots of the data are retrieved. These can be used as inputs to other models, without explicitly needing the exact data, allowing one to potentially learn characteristics of the data. If these snapshots can then be further inverted to retrieve the input data $\mathbf{x}$ explicitly, we say the attack has succeeded in a strong privacy breach [19]

for some $\{\mathbf{B}_\alpha\}$ forming an Frobenius-orthonormal basis of the DLA $\mathfrak{g}$ corresponding to $U(\boldsymbol{\theta})$ in Eq. 15.3, and the above holds for any $\epsilon > 0$. We call $\mathbf{e}_{\text{snap}}$ the snapshot of $\mathbf{x}$.

In essence, $\mathbf{e}_{\text{snap}}$ represents the orthogonal projection of the input quantum state $\rho(\mathbf{x})$ onto the Dynamical Lie Algebra (DLA) of the ansatz. Consequently, the elements within $\mathbf{e}_{\text{snap}}$ are the sole contributors from the input state that influence the generation of the model output $y_{\boldsymbol{\theta}}(\mathbf{x})$, as defined in Eq. 15.4.

We regard the retrieval of the snapshot $\mathbf{e}_{\text{snap}}$ from a quantum state $\rho(\mathbf{x})$ as a *weak privacy* breach. This is because the snapshot can be utilized to train the Variational Quantum Circuit (VQC) model for other learning tasks involving the same dataset $\mathbf{x}$, without requiring access to the actual data.

For example, consider a VQC that classifies images of cats and dogs. If an image of your pet dog has been encoded into a quantum state and the expectation snapshot $\mathbf{e}_{\text{snap}}$ of this state is discovered, then this snapshot could be passed through a VQC model. The model could then identify information about your pet (i.e., it can determine that it is a dog, rather than a cat) even though it may not be possible to fully reconstruct the image itself. This would therefore be a form of weak privacy breach.

Next, we explore a more severe form of privacy breach, where the input data $\mathbf{x}$ can be completely reconstructed. Once the snapshot $\mathbf{e}_{\text{snap}}$ has been obtained, the subsequent step involves inverting this snapshot to retrieve the original data $\mathbf{x}$. This process is largely dependent on the encoding component of the circuit. In our previous example, this would correspond to being able to fully recreate the pixels that form the image of your pet dog, which would constitute a strong privacy breach. In our definition of snapshot inversion, we identify two scenarios that allow for distinct solution approaches: one involves snapshot inversion using purely classical methods, while the other employs snapshot inversion techniques that can leverage quantum samples.

Definition 7 (Classical Snapshot Invertible Model [19])
Given the snapshot $\mathbf{e}_{\text{snap}}$ as the expectation values of the input state $\rho(\mathbf{x})$, we say that VQC admits classical snapshot invertibility, if there exists an efficient $\mathcal{O}(\text{poly}(d, \frac{1}{\epsilon}))$ polynomial-time classical randomized algorithm to recover

$$\mathbf{x}' : \|\mathbf{x}' - \mathbf{x}\|_2 \leq \epsilon, \tag{15.18}$$

with probability at least $p = \frac{2}{3}$, for any user defined $\epsilon > 0$.

In the case where we have access to a quantum device we can also similarly define

Definition 8 (Quantum-Assisted Snapshot Inversion [19]) Given the snapshot $\mathbf{e}_{\text{snap}}$ as the expectation values of the input state $\rho(\mathbf{x})$, and the ability to query $\text{poly}(d, \frac{1}{\epsilon})$ number of samples from the encoding circuit V to generate snapshots $\mathbf{e}'_{\text{snap}}$ for any given input $\mathbf{x}'$, we say that VQC admits quantum-assisted snapshot invertibility, if there exists an efficient $O(\text{poly}(d, \frac{1}{\epsilon}))$ polynomial-time classical randomized algorithm to recover

$$\mathbf{x}' : \|\mathbf{x}' - \mathbf{x}\|_2 \leq \epsilon, \tag{15.19}$$

with probability at least $p = \frac{2}{3}$, for any user defined $\epsilon > 0$.

In this chapter, we focus on input recoverability by considering the conditions under which VQC would admit snapshot recovery and the conditions it would admit snapshot invertibility. Considering these two steps individually allows us to delineate the exact mechanisms that contribute to the overall recovery of the input.

It is also important to note that recovering the inputs of a Variational Quantum Circuit (VQC) may be limited by periodicity. Specifically, there might only exist a classical polynomial-time algorithm capable of recovering $\tilde{\mathbf{x}} = \mathbf{x} + \mathbf{k}\pi$ with ϵ-closeness, where $\mathbf{k} \in \mathbb{Z}$. This limitation arises because the encodings generated by quantum feature maps inherently include trigonometric terms, which introduce periodicity. In the most general scenario, this means that recovering $\mathbf{x}$ might only be feasible up to some periodicity. However, this constraint can be relaxed if the quantum feature map is assumed to be injective, allowing for more precise recovery of the original input data.

Figure 15.4 shows a diagram that highlights the Lie-algebraic simulation method [27] along with specifications of the input recovery framework as defined in this work.

15.4 Snapshot Recovery

In this section, we explore the concept of snapshot recovery, a critical aspect of understanding privacy breaches in Variational Quantum Circuit (VQC) models. Snapshot recovery, as defined in this chapter, refers to the process of reconstructing the expectation values of a quantum state from the gradients of a VQC model. This process is crucial because it allows an attacker to potentially use these recovered snapshots to train other models on the same data, without direct access to the original input data. The ability to recover these snapshots from the gradients represents a weak privacy breach, as it enables the extraction of meaningful information from the encoded quantum states. This concept of *weak privacy breaches* specifically focuses on the recovery of input snapshots as introduced in Definition 6. As the term suggests, the objective is to recover the vector $\mathbf{e}_{\text{snap}}$ for a given Schmidt orthonormal basis $\{\mathbf{B}_\alpha\}_{\alpha \in \dim(\mathfrak{g})}$ of the DLA associated with the VQC ansatz $\mathbf{U}(\boldsymbol{\theta})$. This is under the assumption that the attacker has access to the following information:

1. D gradient information updates $C_j = \frac{\partial y_{\boldsymbol{\theta}}(\mathbf{x})}{\partial \theta_j}$, $j \in [D]$ as defined in Eq. 15.6.
2. Ansatz architecture $\mathbf{U}(\boldsymbol{\theta})$ presented as an ordered sequence of Hermitian generators $\{\theta_k, \mathbf{H}_{\nu(k)}\}_{k=1}^{D}$, where $\mathbf{H}_{\nu(k)}$ is expressed as a polynomial (in the number of qubits) linear combination of Pauli strings.

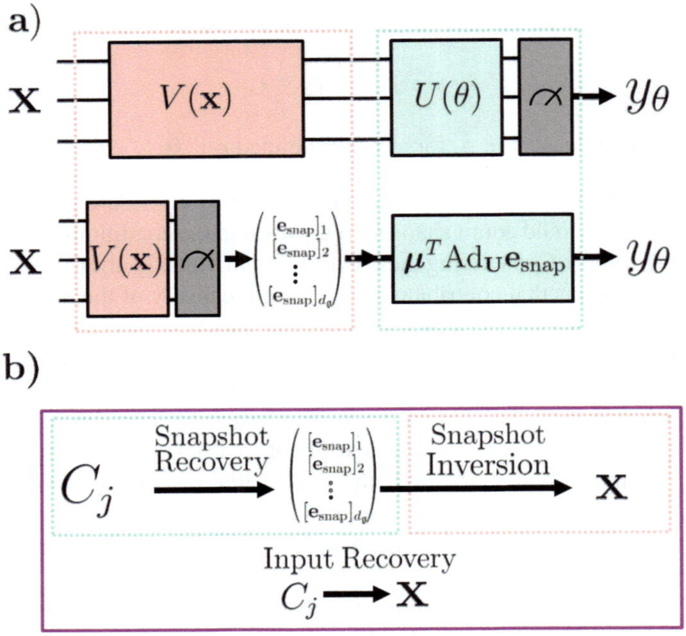

Fig. 15.4 **a** Visualization of the difference between the circuit implementation of a variational quantum model and a Lie-algebraic simulation procedure of the same model [27]. In the Lie-algebraic simulation framework [27], input data $\mathbf{x}$ is encoded into a quantum circuit using $V(\mathbf{x})$; however, the measurements are then performed on this encoded state and used to form a vector of snapshot expectation values. This vector of snapshot expectation values can then be passed as inputs to a classical simulator that uses the adjoint form of $U(\theta)$, which can be performed with resources scaling with the dimension of the DLA formed by the generators of $U(\theta)$. **b** In this work, we assess the ability to recover an input $\mathbf{x}$ from gradients C_j. This can be broken into two parts: firstly, the snapshot $\mathbf{e}_{\text{snap}}$ must be recovered from the gradients C_j, which corresponds to reversing the Lie-algebraic simulation step. Secondly, the recovered snapshot $\mathbf{e}_{\text{snap}}$ must be inverted to find the original data $\mathbf{x}$, which requires finding the values of $\mathbf{x}$ that when input into $V(\mathbf{x})$ will give the same snapshot values $\mathbf{e}_{\text{snap}}$. If both snapshot recovery and snapshot inversion can be performed, then it admits efficient input recovery [19]

3. Measurement operator $\mathbf{O}$ which satisfies the LASA condition according to Definition 5 and expressed as a polynomial (in the number of qubits) linear combination of Pauli strings.

By recovering these snapshots, an attacker can train the VQC model for other learning tasks that effectively extract the same information from the input states $\rho(\mathbf{x})$, all without needing access to the actual data. The core element of the snapshot recoverability algorithm leverages the $\mathfrak{g}$-sim framework [26, 27], which we briefly review in the following subsection. In doing so, we also clarify some previously implicit assumptions to construct a system of linear equations that can be solved to recover $\mathbf{e}_{\text{snap}}$, as detailed in Algorithm 2.

15.4.1 Review of Lie-Algebraic Simulation Framework

We begin by examining the $\mathfrak{g}$-sim framework [26, 27], which is used for the classical computation of the cost function and gradients of VQCs, particularly when the observable is within the DLA of the selected ansatz. This framework specifically evolves the expectation values of observables through the adjoint representation. Nonetheless, for this method to be efficient, it is essential that the dimension of the DLA ($\dim(\mathfrak{g})$) increases only polynomially with the number of qubits.

The first step in the $\mathfrak{g}$-sim framework is to construct an orthonormal basis for the DLA $\mathfrak{g}$, using the parameters $\{\theta_k, \mathbf{H}_{\nu(k)}\}_{k=1}^{D}$. Algorithm 1 provides a well-known method for this task. The process involves calculating pairwise commutators until no further linearly independent elements can be found. Since all operators are represented in the Pauli basis, the orthogonal projectors and norm calculations required by Algorithm 1 can be performed efficiently. If the dimension of the DLA is $\mathcal{O}(\mathrm{poly}(n))$, then the iteration complexity, which is the number of sets of commutators computed, is polynomial in n. However, it is crucial to recognize that the elements forming our approximation for the DLA basis might have exponential support on the Pauli basis, due to the computation of new pairwise commutators at each step. Thus, for the entire procedure to remain efficient, it is vital that the nested commutators of the generators $\mathbf{H}_{\nu(k)}$ do not have exponential support on the Pauli basis and therefore must satisfy the following condition:

Definition 9 (Slow Pauli Expansion [19]) A set of Hermitian generators $\{\mathbf{H}_1, \ldots, \mathbf{H}_N\}$ on n-qubits expressed as linear combinations of $\mathcal{O}(\mathrm{poly}(\dim(\mathfrak{g})))$ Pauli strings satisfies the slow Pauli expansion condition if $\forall r \in [N]$, $[\mathbf{H}_r, [\cdots, [\mathbf{H}_2, \mathbf{H}_1]]]$ can be expressed as a linear combination of $\mathcal{O}(\mathrm{poly}(\dim(\mathfrak{g})))$ Pauli strings.

This condition appears to be satisfied in many examples of polynomial-sized DLA VQC models, however, it is uncertain how robust this assumption is in general. This implies that the attacks we describe might not be feasible for all VQCs that meet the polynomial DLA condition, and therefore privacy preservation might still be achievable.

Algorithm 1 : Finding DLA basis [19]

Require: Hermitian circuit generators $\{H_1, \ldots, H_N\}$, all elements are linear combinations of polynomially-many Pauli strings

Ensure: $\mathcal{A}''' = \{\mathbf{B}_1, \ldots, \mathbf{B}_{\dim(\mathfrak{g})}\}$ as the basis for the DLA $\mathfrak{g}$

1. Let $\mathcal{A} = \{H_1, \ldots, H_N\}$, with all elements represented in the Pauli basis.
2. Repeat until breaks
   a. Compute pairwise commutators of elements of $\mathcal{A}$ into $\mathcal{A}'$
   b. Orthogonally project $\mathcal{A}'$ onto orthogonal complement of $\mathcal{A}$ in $\mathfrak{g}$
   c. Set new $\mathcal{A}''$ to be $\mathcal{A}$ plus new orthogonal elements. If no new elements, break.
3. Perform Gram–Schmidt on $\mathcal{A}$ forming $\mathcal{A}'''$.
4. Return $\mathcal{A}'''$.

Given the orthonormal basis $\mathbf{B}_\alpha$ for $\mathfrak{g}$, under the LASA condition, we can express $\mathbf{O} = \sum_{\alpha \in [\dim(\mathfrak{g})]} \mu_\alpha \mathbf{B}_\alpha$, and hence we can write the output as

$$y_{\boldsymbol{\theta}}(\mathbf{x}) = \mathrm{Tr}(\mathbf{U}^\dagger(\boldsymbol{\theta})\mathbf{O}\mathbf{U}(\boldsymbol{\theta})\rho(\mathbf{x})) = \sum_\alpha \mathrm{Tr}(\mu_\alpha \mathbf{U}^\dagger \mathbf{B}_\alpha \mathbf{U}\rho(\mathbf{x}))$$
$$= \sum_\alpha \mathrm{Tr}(\mu_\alpha \mathrm{Ad}_{\mathbf{U}}(\mathbf{B}_\alpha)\rho(\mathbf{x})). \tag{15.20}$$

In addition, given the form of $\mathbf{U}$, we can express $\mathrm{Ad}_{\mathbf{U}}$ as

$$\mathrm{Ad}_{\mathbf{U}} = \prod_{k=1}^{D} e^{-\theta_k \mathrm{ad}_{i\mathbf{H}_{\nu(k)}}}. \tag{15.21}$$

We can also compute the structure constants for our basis $\mathbf{B}_\alpha$, which is the collection of $\dim(\mathfrak{g}) \times \dim(\mathfrak{g})$ matrices for the operators $\mathrm{ad}_{i\mathbf{B}_\alpha}$. As a result of linearity, we also have the matrix for each $\mathrm{ad}_{i\mathbf{H}}$ for $\mathbf{H} \in \mathfrak{g}$ in the basis $\mathbf{B}_\alpha$. Then by performing matrix exponentiation and multiplying $\dim(\mathfrak{g}) \times \dim(\mathfrak{g})$ we can compute the matrix for $\mathrm{Ad}_{\mathbf{U}}$.

Using the above, the model output may be written as

$$y_\theta = \sum_{\alpha,\beta} \mu_\alpha [\mathrm{Ad}_{\mathbf{U}}]_{\alpha\beta} \mathrm{Tr}(\mathbf{B}_\beta \rho(\mathbf{x})) = \boldsymbol{\mu}^T \mathrm{Ad}_{\mathbf{U}} \mathbf{e}_{\mathrm{snap}}, \tag{15.22}$$

where $\mathbf{e}_{\mathrm{snap}}$ is a vector of expectation values of the initial state, i.e., $[\mathbf{e}_{\mathrm{snap}}]_\beta = \mathrm{Tr}[\mathbf{B}_\beta \rho(\mathbf{x})]$.

Similar to the cost function, the circuit gradient can also be computed via $\mathfrak{g}$-sim. Let

$$C_j = \frac{\partial y_\theta}{\partial \theta_j} = \boldsymbol{\mu}^T \frac{\partial \mathrm{Ad}_{\mathbf{U}}}{\partial \theta_j} \mathbf{e}_{\mathrm{snap}} =: \chi^{(j)} \cdot \mathbf{e}_{\mathrm{snap}}, \tag{15.23}$$

where the adjoint term differentiated with respect to θ_j can be written as

$$\frac{\partial \mathrm{Ad}_{\mathbf{U}}}{\partial \theta_j} = \left[\prod_{k=j}^{D} e^{\theta_k \mathrm{ad}_{i\mathbf{H}_{\nu(k)}}}\right] \mathrm{ad}_{i\mathbf{H}_{\nu(j)}} \left[\prod_{k=1}^{j} e^{\theta_k \mathrm{ad}_{i\mathbf{H}_{\nu(k)}}}\right]. \tag{15.24}$$

The components of $\chi^{(j)}$ can be expressed as

$$\chi_\beta^{(j)} = \sum_\alpha \mu_\alpha \left[\frac{\partial \mathrm{Ad}_{\mathbf{U}}}{\partial \theta_j}\right]_{\alpha,\beta}, \tag{15.25}$$

allowing C_j terms to be represented in a simplified manner as

$$C_j = \sum_{\beta=1}^{\dim(\mathfrak{g})} \chi_\beta^{(j)} [\mathbf{e}_{\text{snap}}]_\beta. \tag{15.26}$$

The key feature of this setup is that the matrices and vectors involved have dimension $\dim(\mathfrak{g})$. Consequently, for a polynomial-sized Dynamical Lie Algebra (DLA), the simulation time will scale polynomially, allowing model outputs to be calculated in polynomial time [27]. Specifically, the matrices for each $\text{ad}_{i\mathbf{H}_k}$ in the basis $\{\mathbf{B}_l\}$ and $\text{Ad}_\mathbf{U}$ are polynomial in this context.

This Lie-algebraic simulation technique was introduced to demonstrate efficient methods for simulating Lie-Algebraic Supported Ansatz (LASA) circuits with polynomially sized DLA. In this work, we utilize the framework to investigate the snapshot recovery of variational quantum algorithms. Based on the above discussion, the proof of the following theorem is self-evident (due to the fact that all matrices involved in the calculation would become polynomial in size).

Theorem 1 *(Complexity of* $\mathfrak{g}$*-sim [19]) If ansatz family* $\mathbf{U}(\boldsymbol{\theta})$ *with an observable* $\mathbf{O}$ *satisfies both the LASA condition and Slow Pauli Expansion, then the cost function and its gradients can be simulated with complexity* $O(poly(\dim(\mathfrak{g})))$ *using a procedure that at most queries a quantum device a polynomial number of times to compute the* $\dim(\mathfrak{g})$*-dimensional snapshot vector* $\mathbf{e}_{snap}$.

This theorem means that in the right configuration, the challenge of simulating the variational component of a VQC scales as $O(\text{poly}(\dim(\mathfrak{g})))$. Consequently, if a variational circuit possesses a DLA dimension that scales polynomially, the circuit can be efficiently simulated classically in polynomial time, possibly eliminating the need for a quantum device. However, it's worth noting that a quantum device might still be required initially to determine the expectation value snapshots [28].

15.4.2 Snapshot Recovery Algorithm

With the framework for the $\mathfrak{g}$-sim [27] established, we focus on how snapshots $\mathbf{e}_{\text{snap}}$ of the input data can be recovered using the VQC model gradients C_j, with the process detailed in Algorithm 2. In particular, the form of Eq. 15.26 allows a setup leading to the recovery the snapshot vector $\mathbf{e}_{\text{snap}}$ from the gradients $\{C_j\}_{j=1}^D$, but requires the ability to solve the system of D linear equations given by $\{C_j\}$ with $\dim(\mathfrak{g})$ unknowns $[\mathbf{e}_{\text{snap}}]_{\beta \in \dim(\mathfrak{g})}$. The following theorem formalizes the complexity of recovering the snapshots from the gradients.

Algorithm 2 : Snapshot Recovery (Source Heredge et al [19])

Require: Observable $\mathbf{O}$ such that $i\mathbf{O} \in \mathfrak{g}$, generators $\{\mathbf{H}_{\nu(k)}\}_{k=1}^{D}$, ordered sequence $(\{\theta_k, \mathbf{H}_{\nu(k)}\})_{k=1}^{D}$, and gradients $C_j = \frac{\partial y_{\theta}(\mathbf{x})}{\partial \theta_j}$, $j \in [D]$ for some unknown classical input $\mathbf{x}$.

Ensure: Snapshot $\mathbf{e}_{\text{snap}}$ for $\mathbf{x}$

1. Run Algorithm 1 to obtain an orthonormal basis for the DLA $\{\mathbf{B}_\beta\}_{\beta\in[\dim(\mathfrak{g})]}$
2. For $\beta \in [\dim(\mathfrak{g})]$, compute the $\dim(\mathfrak{g}) \times \dim(\mathfrak{g})$ matrix $\text{ad}_{i\mathbf{B}_\beta}$
3. For $k \in [D]$, compute the coefficients of $\mathbf{H}_{\nu(k)}$ in the basis $\{\mathbf{B}_\beta\}_{\beta\in[\dim(\mathfrak{g})]}$, which gives us $\text{ad}_{i\mathbf{H}_{\nu(k)}}$
4. For $k \in [D]$, compute the $\dim(\mathfrak{g}) \times \dim(\mathfrak{g})$ matrix exponential $e^{\theta_k \text{ad}_{i\mathbf{H}_{\nu(k)}}}$
5. For $j \in [D]$ compute the $\dim(\mathfrak{g}) \times \dim(\mathfrak{g})$ matrix
$$\frac{\partial \text{Ad}_{\mathbf{U}}}{\partial \theta_j} = \left[\prod_{k=j}^{D} e^{\theta_k \text{ad}_{i\mathbf{H}_{\nu(k)}}}\right] \text{ad}_{i\mathbf{H}_{\nu(j)}} \left[\prod_{k=1}^{j} e^{\theta_k \text{ad}_{i\mathbf{H}_{\nu(k)}}}\right]. \tag{15.27}$$
6. For $\beta \in [\dim(\mathfrak{g})]$, compute the coefficients μ_β of $\mathbf{O}$ in the basis $\{\mathbf{B}_\beta\}_{\beta\in[\dim(\mathfrak{g})]}$
7. For $j \in [D]$, $\beta \in [\dim(\mathfrak{g})]$, compute
$$\chi_\beta^{(j)} = \sum_\alpha \mu_\alpha \left[\frac{\partial \text{Ad}_{\mathbf{U}}}{\partial \theta_j}\right]_{\alpha,\beta}, \tag{15.28}$$
and construct $D \times \dim(\mathfrak{g})$ matrix $\mathbf{A}$ with $\mathbf{A}_{rs} = \chi_s^{(r)}$.
8. Solve the following linear system,
$$[C_1, \ldots, C_D]^{\mathsf{T}} = \mathbf{A}\mathbf{y}, \tag{15.29}$$
and return $\mathbf{y}$ as the snapshot $\mathbf{e}_{\text{snap}}$.

Theorem 2 *(Snapshot Recovery [19]) Given the requirements specified in Algorithm 2, along with the assumption that the number of variational parameters $D \geq \dim(\mathfrak{g})$, where $\dim(\mathfrak{g})$ is the dimension of the DLA $\mathfrak{g}$, the VQC model admits snapshot $\mathbf{e}_{snap}$ recovery with complexity scaling as $O(poly(\dim(\mathfrak{g})))$.*

Proof Firstly, we note that given the gradients C_j and parameters $\theta_{j\in[D]}$, the only unknowns are the components of the vector $\mathbf{e}_{\text{snap}}$ of length $\dim(\mathfrak{g})$. Therefore, it is necessary to have $\dim(\mathfrak{g})$ equations in total; otherwise, the system of equations would be underdetermined and it would be impossible to find a unique solution. The number of equations depends on the number of gradients and, therefore, the number of variational parameters in the model, hence the requirement that $D \geq \dim(\mathfrak{g})$.

Assuming now that we deal with the case that there are $D \geq \dim(\mathfrak{g})$ variational parameters of the VQC model, we can therefore arrive at a determined system of equations. The resulting system of simultaneous equations can be written in a matrix form as

$$\begin{pmatrix} C_1 \\ C_2 \\ \vdots \\ C_D \end{pmatrix} = \begin{pmatrix} \chi_1^{(1)} & \chi_2^{(1)} & \cdots & \chi_{\dim(\mathfrak{g})}^{(1)} \\ \chi_1^{(2)} & \chi_2^{(2)} & \cdots & \chi_{\dim(\mathfrak{g})}^{(2)} \\ \vdots & \vdots & \ddots & \vdots \\ \chi_{\dim(\mathfrak{g})}^{(D)} & \chi_{\dim(\mathfrak{g})}^{(D)} & \cdots & \chi_{\dim(\mathfrak{g})}^{(D)} \end{pmatrix} \begin{pmatrix} [\mathbf{e}_{\text{snap}}]_1 \\ [\mathbf{e}_{\text{snap}}]_2 \\ \vdots \\ [\mathbf{e}_{\text{snap}}]_{\dim(\mathfrak{g})} \end{pmatrix} \tag{15.30}$$

In order to solve the system of equations highlighted in Eq. 15.30 to obtain $\mathbf{e}_{\text{snap}}$, we first need to compute the coefficients $\{\chi_\beta^{(j)}\}_{j\in[D],\beta\in[\dim(\mathfrak{g})]}$. This can done by the $\mathfrak{g}$-sim procedure highlighted in the previous section and in steps 1–7 in Algorithm 2 with complexity $\mathcal{O}(\text{poly}(\dim(\mathfrak{g})))$. The next step is to solve the system of equations, i.e., step 8 of Algorithm 2, which can be solved using Gaussian elimination procedure incurring a complexity $\mathcal{O}(\dim(\mathfrak{g})^3)$ [29]. Thus, the overall complexity of recovering the snapshots from the gradients is $\mathcal{O}(\text{poly}(\dim(\mathfrak{g})))$. This completes the proof. □

What does this tell us about privacy? If the dimension of the Dynamical Lie Algebra (DLA) is exponentially large in the number of qubits, i.e., $\dim(\mathfrak{g}) = \mathcal{O}(\exp(n))$, then the system of equations required would necessitate exponentially many gradients and, consequently, exponentially many trainable parameters to be fully determined. While this does not necessarily guarantee privacy for the model (as numerical attacks or other approaches may still succeed), it would prevent the closed-form solution snapshot recovery approach via simultaneous equations. For the system of simultaneous equations to be solvable, it is necessary that $\dim(\mathfrak{g}) = \mathcal{O}(\text{poly}(n))$. With this requirement met, it becomes feasible to solve the system of equations in Eq. 15.30 in polynomial time and retrieve the snapshot vector $\mathbf{e}_{\text{snap}}$. Thus, a model is considered snapshot recoverable if the DLA dimension scales polynomially with d.

This result is crucial and far-reaching in the study of Quantum Machine Learning (QML) privacy, as it precisely outlines the conditions in the variational part of the circuit that allows a weak privacy breach. If the DLA dimension scales polynomially, a weak privacy breach is possible, meaning the expectation values can be recovered. Later, we will explore the implications of this on circuit design, particularly in relation to the connection between DLA dimension and barren plateaus.

15.5 Snapshot Invertibility

In this section, we explore the concept of snapshot invertibility, a crucial aspect of understanding strong privacy breaches in Variational Quantum Circuit (VQC) models. Snapshot invertibility refers to the ability to reconstruct the original input data from the recovered snapshot expectation values of a quantum state. This process represents a strong privacy breach, as it allows an attacker to fully recover the confidential input data, thereby compromising the privacy of the quantum machine learning model.

The challenge of snapshot inversion lies in the encoding architecture of the VQC model. While snapshot recovery focuses on extracting expectation values from model gradients, snapshot invertibility dives deeper into the encoding mechanisms that could potentially allow the reconstruction of the original data. In this section, we examine various encoding strategies and their susceptibility to inversion attacks, highlighting the conditions under which strong privacy breaches can occur.

A crucial prerequisite for implementing our snapshot inversion techniques is the ability to compute the expectation values $\mathrm{Tr}(\rho(\mathbf{x}')\mathbf{B}_k), \ \ \forall k \in [\dim(\mathfrak{g})]$ for some hypothesized input $\mathbf{x}'$. This capability serves as a distinguishing factor between classical snapshot inversion and quantum-assisted snapshot inversion. It is widely acknowledged that determining expectation values for specific observables is generally less computationally intensive than classically simulating the state $\rho(\mathbf{x})$ [30]. Consequently, even if the state $\rho(\mathbf{x})$ is challenging to simulate classically, it may still be feasible to carry out snapshot inversion using classical methods. In scenarios where quantum assistance is available, the expectation values $\mathrm{Tr}(\rho(\mathbf{x}')\mathbf{B}_k)$ can be determined through appropriate measurements of the encoding circuit $V(\mathbf{x}')$, making quantum-assisted snapshot inversion a viable approach.

In the first subsection, we explore inversion attacks that are pertinent to widely used feature maps, specifically utilizing insights into the locality of the encoding circuit. These feature maps share a common trait: by focusing on a subset of the inputs, it becomes possible to express $\rho(\mathbf{x})$ or its expectations in a simplified manner. The following subsection explores more arbitrary encoding schemes, approaching the issue as a black-box optimization problem. In general, snapshot inversion can be challenging or even impossible, despite the snapshots being efficiently retrievable or the feature map being classically simulatable. Our emphasis will be on identifying sufficient conditions for executing snapshot inversion, which will inform strategies for enhancing privacy.

15.5.1 Snapshot Inversion for Local Encodings

For efficiency purposes, a common encoding strategy in QML is to encode elements of the input vector $\mathbf{x}$ using local quantum gates, often employing single-qubit rotations. Typically, the majority of the circuit's complexity is concentrated either in the variational section or through non-parameterized entangling operations within the feature map. In this section, we examine methods for recovering components of the vector $\mathbf{x}$ using snapshot vectors, up to periodicity, when the feature map encodes each x_j using local quantum interactions. Specifically, we establish constraints on the permissible level of interaction between qubits used to encode each x_j. Additionally, we demonstrate that the frequency with which the feature map can encode a single x_j must be sufficiently limited. Although these conditions may seem stringent, it is important to note that they are met by some commonly used encodings, such as the *Pauli product feature map* and the *Fourier tower map* [18] (which was seen in the initial case study).

For the Pauli product encoding, we will show that a completely classical snapshot inversion attack is possible. An example of a Pauli product encoding is given below:

$$\bigotimes_{j_1}^{n} \rho_j(x_j) = \bigotimes_{j_1}^{n} R_{\mathsf{X}}(x_j)|0\rangle\langle 0|R_{\mathsf{X}}(-x_j), \tag{15.31}$$

where R_{X} is the parameterized Pauli X rotation gate. The Fourier tower map is similar to Eq. (15.31) but utilizes a parallel data re-uploading scheme, i.e.:

$$\bigotimes_{j=1}^{d} \Big(\bigotimes_{l=1}^{m} R_{\mathsf{X}}(5^{l-1} x_j) \Big), \tag{15.32}$$

where $n = dm$, with m being the number of qubits used to encode a single dimension of the input.

1. Pauli Product Encoding: The first attack that we present will specifically target Eq. (15.31). However, the attack does apply to the Fourier tower map as well. More generally, the procedure applies to any parallel data re-uploading schemes of the form:

$$\bigotimes_{j=1}^{d} \Big(\bigotimes_{l=1}^{m} R_{\mathsf{X}}(\alpha_l x_j) \Big). \tag{15.33}$$

We explicitly utilize Pauli X rotations, but a similar result holds for Y or Z. For Pauli operator P, let $\mathsf{P}_j := i\mathbb{I}^{\otimes(j-1)} \otimes \mathsf{P} \otimes \mathbb{I}^{\otimes(n-j)}$.

The significant part of Algorithm 3 is that it returns a closed-form solution for the input data, hence it can in theory be calculated exactly. Hence given it was possible to recover the snapshot from the gradients, it will then be possible to recover the original data inputs too.

Theorem 3 *(Pauli Encoding Snapshot Inversion [19]) Suppose that the polynomial DLA and slow Pauli expansion (Definition 9) conditions are satisfied. Also, suppose that we are given a snapshot vector* $\mathbf{e}_{snap}(\mathbf{x})$ *for a VQC with trainable portion* $\mathbf{U}(\theta)$ *with DLA* $\mathfrak{g}$ *and Pauli product feature encoding (Eq. (15.31)) and the corresponding DLA basis elements* $(\mathbf{B}_k)_{k=1}^{\dim(\mathfrak{g})}$*. The classical Algorithm 3 outputs an* ϵ *estimate of* x_j*, up to periodicity, or outputs FAILURE, with time* $O(poly(n)\log(1/\epsilon))$*.*

Algorithm 3 : Classical Snapshot Inversion for Pauli Product Encoding (Source Heredge et al [19])

Require: Snapshot vector $\mathbf{e}_{\text{snap}}(\mathbf{x})$ of dimension $\dim(\mathfrak{g}) = \mathcal{O}(\text{poly}(n))$ corresponding to a basis $(\mathbf{B}_k)_{k=1}^{\dim(\mathfrak{g})}$ of DLA $\mathfrak{g}$. Each $\mathbf{B}_k$ is expressed as a linear combination of $\mathcal{O}(\text{poly}(n))$ Pauli strings. Snapshot inversion is being performed for a VQC model that utilized a trainable portion of $\mathbf{U}(\theta)$ with DLA $\mathfrak{g}$ and Pauli product encoding Equation (15.31). Index $j \in [d]$, $\epsilon < 1$

Ensure: An ϵ estimate of the j-th component x_j of the data input $\mathbf{x} \in \mathbb{R}^d$ up to periodicity, or output FAILURE.

if $i\mathsf{Z}_j \in \mathfrak{g}$ **then**
  $\alpha \leftarrow 1, \beta \leftarrow 0$
  $\mathbf{W} \leftarrow \mathsf{Y}_j$
else if $i\mathsf{Y}_j \in \mathfrak{g}$ **then**
  $\alpha \leftarrow 1, \beta \leftarrow 0$
  $\mathbf{W} \leftarrow \mathsf{Y}_j$
else

1. Determine set of Pauli strings required to span elements $(i\mathbf{B}_k)_{k=1}^{\dim(\mathfrak{g})}$ and denote the set $\mathcal{P}_{\mathfrak{g}}$.
2. $\mathcal{P}_{\mathfrak{g}} \leftarrow \mathcal{P}_{\mathfrak{g}} \cup \{\mathsf{Z}_j, \mathsf{Y}_j\}$, $|\mathcal{P}_{\mathfrak{g}}| = \mathcal{O}(\text{poly}(n))$ by assumption. Reduce $\mathcal{P}_{\mathfrak{g}}$ to a basis.
3. Let $\mathbf{C}$ be a $|\mathcal{P}_{\mathfrak{g}}| \times \dim(\mathfrak{g})$ matrix whose k-th column corresponds to the components of $i\mathbf{B}_k$ in the basis $\mathcal{P}_{\mathfrak{g}}$.
4. Let $\mathbf{A}$ be a $|\mathcal{P}_{\mathfrak{g}}| \times 2$ whose first column contains a 1 in the row corresponding to Z_j and whose second column contains a 1 in the row corresponding to Y_j.
5. Perform a singular value decomposition on $\mathbf{A}^{\mathsf{T}}\mathbf{C}$, and there are at most two nonzero singular values r_1, r_2.

  if $r_1 \neq 1$ and $r_2 \neq 1$ **then**
    return FAILURE
  else
    1. $\mathbf{W} \leftarrow$ singular vector with singular value 1.
    2. Expand $i\mathbf{W}$ in basis $(i\mathsf{Z}_j, i\mathsf{Y}_j)$ record components as α and β respectively.
  end if
end if

1. Expand $i\mathbf{W}$ in basis $(\mathbf{B}_k)_{k=1}^{\dim(\mathfrak{g})}$, and record components as γ_k.
2. Compute

$$\tilde{x}_j = \cos^{-1}\left[\frac{2}{\text{sign}(\alpha)\sqrt{\alpha^2+\beta^2}}\sum_{k=1}^{\dim(\mathfrak{g})}\gamma_k[\mathbf{e}_{\text{snap}}]_k\right] - \tan^{-1}(\beta/\alpha). \tag{15.34}$$

3. **return** $\tilde{x}_j$.

Proof Steps 1-5 in Algorithm 3 can be performed in $O(\text{poly}(n))$ classical time due to the polynomial DLA and slow Pauli expansion conditions. The purpose of step 5 is to compute the angles between the linear subspaces $\mathfrak{g}$ and $\text{span}_{\mathbb{R}}\{i\mathsf{Z}_j, i\mathsf{Y}_j\}$. This is to identify if there is any intersection, i.e., if $\exists\alpha, \beta$ such that $\alpha i\mathsf{Z}_j + \beta i\mathsf{Y}_j \in \mathfrak{g}$, which is identified by singular values equal to 1. The algorithm cannot proceed if the intersection is trivially empty as the snapshot vector does not provide the required measurement to obtain x_j efficiently with this scheme. From now on, we suppose that such an element was found.

We can without loss of generality just focus on the one-qubit reduced density matrix for x_j. In this case, using Bloch sphere representation:

$$\rho_j(x_j) = R_{\mathsf{X}}(x_j)|0\rangle\langle 0|R_{\mathsf{X}}(-x_j) = \frac{\mathbb{I} - \sin(x_j)\mathsf{Y} + \cos(x_j)\mathsf{Z}}{2}, \tag{15.35}$$

such that

$$\begin{aligned} \text{Tr}([\alpha\mathsf{Z}_j + \beta\mathsf{Y}_j]\rho_j(x_j)) &= \frac{\alpha}{2}\cos(x_j) - \frac{\beta}{2}\sin(x_j) \\ &= \frac{\text{sign}(\alpha)}{2}\sqrt{\alpha^2+\beta^2}\cos(x_j + \tan^{-1}(\beta/\alpha)). \end{aligned} \tag{15.36}$$

However, by assumption, $\gamma_k \in \mathbb{R}$ such that

$$\begin{aligned} \alpha i\mathsf{Z}_j + \beta i\mathsf{Y}_j &= \sum_{k=1}^{\dim(\mathfrak{g})} \gamma_i \mathbf{B}_k \\ \implies \text{Tr}([\alpha\mathsf{Z}_j + \beta\mathsf{Y}_j]\rho_j(x_j)) &= \sum_{k=1}^{\dim(\mathfrak{g})} \gamma_k [\mathbf{e}_{\text{snap}}]_k. \end{aligned} \tag{15.37}$$

So to recover x_j, we only need to solve

$$\sum_{k=1}^{\dim(\mathfrak{g})} \gamma_k [\mathbf{e}_{\text{snap}}]_k = d\frac{\text{sign}(\alpha)}{2}\sqrt{\alpha^2+\beta^2}\cos(x_j + \tan^{-1}(\beta/\alpha)), \tag{15.38}$$

which after rearranging allows the recovery of

$$x_j = \cos^{-1}\left[\frac{2}{\text{sign}(\alpha)\sqrt{\alpha^2+\beta^2}}\sum_{k=1}^{\dim(\mathfrak{g})} \gamma_k [\mathbf{e}_{\text{snap}}]_k\right] - \tan^{-1}(\beta/\alpha), \tag{15.39}$$

up to periodicity. By the polynomial DLA and slow Pauli expansion conditions (i.e., all DLA basis elements are expressed as linear combinations of Paulis), we can compute γ_k in $O(\text{poly}(n)\log(1/\epsilon))$ time. □

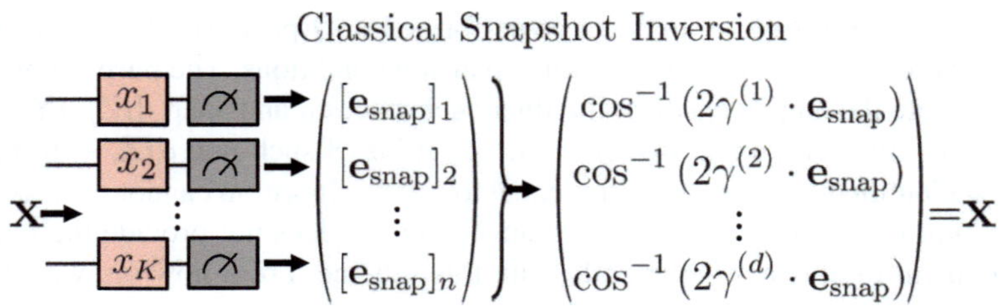

Fig. 15.5 A product map encoding, whereby each input variable x_j is encoded into an individual qubit, and the snapshot used by the model corresponds to single-qubit measurements of the DLA basis elements. In this setting, the snapshot is trivial to invert and find the original data using the relation $x_j = \cos^{-1}\left(2\gamma^{(j)} \cdot \mathbf{e}_{\text{snap}}\right)$ [19]

For illustrative purposes, we show in Fig. 15.5 the snapshot inversion process for the special case where $i\mathsf{Z}_j \in \mathfrak{g}$, i.e.:

$$x_j = \cos^{-1}\left(2\gamma^{(j)} \cdot \mathbf{e}_{\text{snap}}\right), \tag{15.40}$$

for $i\mathsf{Z}_j = \sum_{k=1}^{\dim(\mathfrak{g})} \gamma_k^{(j)} \mathbf{B}_k$. The general parallel data re-uploading case can be handled by applying the procedure to only one of the rotations that encodes at x_j at a time, checking to find one that does not cause the algorithm to return FAILURE.

2. General Pauli Encoding: We now present a more general procedure that applies to feature maps that use serial data re-uploading and multi-qubit Paulis. However, we introduce a condition that ensures that each x_j is locally encoded. More generally, we focus our discussion on encoding states that may be written as a tensor product of Ω subsystems, i.e., multipartite states.

$$\rho(\mathbf{x}) = \bigotimes_{J \in \mathcal{P}} \rho_J(\mathbf{x}), \tag{15.41}$$

where $\dim(\mathbf{x}_J)$ is constant. The procedure is highlighted in Algorithm 4 and requires solving a system of polynomial equations.

In addition, the procedure may not be completely classical as quantum assistance may be required to compute certain expectation values of $\rho_J(\mathbf{x})$, specifically with respect to the DLA basis elements. For simplicity, the algorithm and the theorem characterizing the runtime ignore potential errors in estimating these expectations. If classical estimation is possible, then we can potentially achieve a $\mathcal{O}(\text{poly}(\log(1/\epsilon)))$ scaling. However, if we must use quantum, then we will incur a $\mathcal{O}(1/\epsilon)$ (due to amplitude estimation) dependence, which can be significant. Theorem 4 presents the attack complexity ignoring these errors.

Algorithm 4: Snapshot Inversion for General Pauli Encodings (Source Heredge et al [19])

Require: Snapshot vector $\mathbf{e}_{\text{snap}}(\mathbf{x})$ of dimension $\dim(\mathfrak{g}) = O(\text{poly}(n))$ corresponding to a basis $(\mathbf{B}_k)_{k=1}^{\dim(\mathfrak{g}))}$ of DLA $\mathfrak{g}$. Each $\mathbf{B}_k$ is expressed as a linear combination of $O(\text{poly}(n))$ Pauli strings. Snapshot inversion is being performed for a VQC model that utilized a trainable portion of $\mathbf{U}(\theta)$ with DLA $\mathfrak{g}$ and separable encoding Equation (15.41) with qubit partition $\mathcal{P}$. Index $j \in [d]$, $\epsilon < 1$
Ensure: An ϵ estimate of the j-th component x_j of the data input $\mathbf{x} \in \mathbb{R}^d$ up to periodicity

1. Find a ρ_J for $J \in \mathcal{P}$ that depends on x_j. Let R denote the number of Pauli rotations in the circuit for preparing ρ_J that involve x_j.
2. For each $k \in [\dim(g)]$, compute $\text{Tr}(\mathbf{B}_k\rho_J(\mathbf{x}))$ and $\text{Tr}(\mathbf{B}_k\rho_{J^c}(\mathbf{x}))$.
3. Determine the set $\mathcal{S}_J = \{k : \text{Tr}(\mathbf{B}_k\rho_J(\mathbf{x})) \neq 0 \;\&\; \text{Tr}(\mathbf{B}_k\rho_{J^c}(\mathbf{x})) = 0\}$, $J^c := [n] - J$.

if $\mathcal{S}_J < \dim(\mathbf{x}_J)$ **then**
  return FAILURE
else
  1. For each $k \in \mathcal{S}_J$ evaluate $\text{Tr}(\mathbf{B}_k\rho_J(\mathbf{x}))$ at $M = 2R^{\dim(\mathbf{x}_J)} + 1$ points, $\mathbf{x}_r \in \{\frac{2\pi r}{2R+1} : r = -R, \ldots R\}^{\dim(\mathbf{x}_J)}$
  2. For each k, solve linear system

$$\text{Tr}(\mathbf{B}_k\rho_J(\mathbf{x}_r)) = \alpha_0 + \sum_{r\in[R]^{\dim(\mathbf{x}_J)}} \alpha_r e^{i\mathbf{r}\cdot\mathbf{x}_r}$$

  for α's.
  3. Consider the polynomial system:

$$[\mathbf{e}_{\text{snap}}]_k = \Re\left[\alpha_0 + \sum_{r\in[R]^{\dim(\mathbf{x}_J)}} \alpha_{\mathbf{r}} \prod_{j=1}^{\dim(\mathbf{x}_J)} (T_{r_j}(u_j) + iv_j U_{r_j-1}(u_j))\right],$$

$$\text{with } k \in \mathcal{S}_J, \tag{15.42}$$

$$u_j^2 + v_j^2 = 1, j \in J, \tag{15.43}$$

  where $u_j = \cos(x_j)$, $v_k = \sin(x_j)$ and T_r, U_r relate to Chebyshev polynomials.
  4. Apply Buchberger's to obtain a Gröbner basis for the system.
  5. Back substitution and univariable root-finding algorithm [31] (e.g., Jenkins-Traub [32]) to obtain $\tilde{\mathbf{x}}_J$.
  6. **return** $\tilde{\mathbf{x}}_j$
end if

Theorem 4 *(General Pauli Encoding Snapshot Inversion [19]) Suppose that the feature encoding state $\rho(\mathbf{x})$ is a multipartite state, specifically there exists a partition $\mathcal{P}$ of qubits $[n]$ such that*

$$\rho(\mathbf{x}) = \bigotimes_{J\in\mathcal{P}} \rho_J(\mathbf{x}),$$

where we define $\mathsf{x}_J \subseteq \mathbf{x}$ to be components of $\mathbf{x}$ on which ρ_J depends. In addition, we have as input an $O(poly(n))$-dimensional snapshot vector $\mathbf{e}_{snap}$ with respect to a known basis $\mathbf{B}_k$ for the DLA of the VQC.

Suppose that for $\rho_J(\mathbf{x})$ the following conditions are satisfied:

- *dim*$(\mathbf{x}_J) = \mathcal{O}(1)$*;*
- *each x_k is encoded at most $R = \mathcal{O}(poly(n))$ times in, potentially multi-qubit, Pauli rotations;*
- *and the set $\mathcal{S}_J = \{k : Tr(\mathbf{B}_k\rho_J(\mathbf{x})) \neq 0 \;\&\; Tr(\mathbf{B}_k\rho_{J^c}(\mathbf{x})) = 0\}$ has cardinality at least dim(x_J), where $J^c := [n] - J$,*

then the model admits quantum-assisted snapshot inversion for recovering x_J. Furthermore, a classical snapshot inversion can be performed if $\forall k$, $Tr(\mathbf{B}_k\rho_J(\mathbf{x}))$ can be evaluated classically for all $\mathbf{x}$. Overall, ignoring error in estimating $Tr(\mathbf{B}_k\rho_J(\mathbf{x}))$, with the chosen parameters, this leads to a $\mathcal{O}(poly(n, \log(1/\epsilon)))$ algorithm.

Proof Given that each x_k is encoded with multi-qubit Pauli rotations, i.e., possible eigenvalues are 1 and -1, it is well known [33] that the following holds:

$$f_k(\mathsf{x}_J) = \mathrm{Tr}(\mathbf{B}_k\rho_J(\mathbf{x})) = \alpha_0 + \sum_{r\in[R]^{\dim(\mathsf{x}_J)}} \alpha_r e^{i\mathbf{r}\cdot\mathbf{x}_r}, \forall k \in \mathcal{S}, \tag{15.44}$$

and $\mathrm{Tr}(\mathbf{B}_k\rho_J(\mathbf{x}))$ is real. The set $\mathcal{S}_J$ is to ensure that we can isolate a subsystem where $\dim(\mathsf{x}_J)$ is constant.

To ensure that the number of terms is $\mathcal{O}(\mathrm{poly}(n))$ it suffices to restrict to $\dim(\mathsf{x}_J) = \mathcal{O}(\log(n))$, $R = \mathcal{O}(\log(n))$. The α coefficients can be computed by evaluating $\mathrm{Tr}(\mathbf{B}_k\rho_J(\mathbf{x}))$ at $2R^{\dim(\mathsf{x}_J)} + 1 = \mathcal{O}(\mathrm{poly}(n))$ different points $\mathbf{x}'$. Depending on whether $\mathrm{Tr}(\mathbf{B}_k\rho_J(\mathbf{x}))$ can be evaluated classically or quantumly implies whether this falls under classical or quantum-assisted snapshot inversion. This leads to a system of $\dim(\mathsf{x}_J)$ equations in x_J:

$$[\mathbf{e}_{\text{snap}}]_k = f_k(\mathsf{x}_J), k = 1, \ldots, \dim(\mathfrak{g}). \tag{15.45}$$

Using the Chebyshev polynomials T_n, U_n of the first and second kind, respectively, we can express the system as a system of polynomial equations with additional constraints:

$$[\mathbf{e}_{\text{snap}}]_k = \Re\left[\alpha_0 + \sum_{r\in[R]^{\dim(\mathsf{x}_J)}} \alpha_{\mathbf{r}} \prod_{j=1}^{\dim(\mathsf{x}_J)} (T_{r_j}(u_j) + i v_j U_{r_j-1}(u_j))\right],$$

$$\text{with } k \in \mathcal{S}_J, \tag{15.46}$$

$$u_j^2 + v_j^2 = 1, j \in J, \tag{15.47}$$

where $u_j = \cos(x_j)$, $v_k = \sin(x_j)$. In addition, we use the Chebyshev polynomials defined as $\cos(n\theta) = T_n(\cos(\theta))$ and $\sin(\theta)U_{n-1}(\cos(\theta)) = \sin(n\theta)$. By our assumption that the DLA is polynomial, we have $\mathcal{O}(\mathrm{poly}(n))$ equations in $2\dim(\mathsf{x}_J) = \mathcal{O}(\log\log(n))$ unknowns.

If all conditions until now are satisfied, we will have successfully written down a system of determined simultaneous equations. The precise details regarding results from computational geometry are outside the scope of this chapter, but we shall briefly mention the core arguments. Considering bounds from computational geometry we note that in the worst case of Buchberger's algorithm [34] the degrees of a reduced Gröbner basis are bounded by

$$M = 2\Big(\frac{\Delta^2}{2} + \Delta\Big)^{2^{Q-2}}, \tag{15.48}$$

where Δ is the maximum degree of any polynomial and Q is the number of unknown variables [35]. For a system of linear equations, it can be shown that a worst-case degree bound grows double exponentially in the number of variables [36]. The maximum degree of any equation in Eq. 15.46 is $\Delta = R^{\dim(\mathbf{x}_J)}$, and $Q = 2\dim(\mathbf{x}_J)$ such that

$$M = O(R^{2\dim(\mathbf{x}_J)2^{\dim(\mathbf{x}_J)}}), \tag{15.49}$$

so for our chosen conditions the maximum degree is bounded by $M = O(\text{poly}(n))$.

Buchberger's algorithm provides a Gröbner basis in which backsubstitution could be used to solve equations in one variable. Numerical methods for solving polynomials in one variable generally scale polynomially in the degree. For solving each univariate polynomial at each step of the backsubstitution, we can apply a polynomial root-finding method, such that Jenkins-Traub [32], which can achieve at least quadratic global convergence (converge from any initial point and at a rate that is at least $\log\log(1/\epsilon)$). This leads to an overall $O(\text{poly}(n, \log(1/\epsilon))$ algorithm, ignoring error in estimating $\text{Tr}(\mathbf{B}_j\rho_J)$. □

In essence, this theorem suggests that if a circuit has an encoding structure resulting in a separable state, then snapshot inversion is feasible, allowing for the recovery of input data from the snapshots. Furthermore, if the model also permits a weak privacy breach, where snapshots can be derived from the model's gradients, then it becomes possible to retrieve the snapshots from the gradients and, subsequently, the confidential data from these snapshots. This scenario indicates that a strong privacy breach is possible.

15.5.2 Snapshot Inversion for Generic Encodings

In the general scenario where $\dim(\mathfrak{g}) = O(\text{poly}(n))$, and it is not clear how to efficiently leverage our understanding of the circuit, one can employ black-box optimization methods to identify an $\mathbf{x}$ that yields the desired snapshot signature. In this setting we effectively guess inputs, find their snapshot values, and keep adjusting our guess until the snapshots match their target known values. To simplify, let's assume we limit our search to the range $[-1, 1]^d$. We begin with an initial estimate for the input parameters, represented as $\mathbf{x}'$, and use these to compute the expected snapshot values $\mathrm{Tr}[\mathbf{B}_k\rho(\mathbf{x}')]$. A cost function is then formulated to compare these values to the actual snapshot, denoted as $\mathbf{e}_{\text{snap}}$. For instance, the mean squared error can serve as the cost function:

$$f(\mathbf{x}') = \|\mathbf{e}_{\text{snap}} - (\mathrm{Tr}[\mathbf{B}_k\rho(\mathbf{x}')])_{k=1}^{\dim(\mathfrak{g})}\|_2^2 = \sum_{k\in[\dim(\mathfrak{g})]} \left([\mathbf{e}_{\text{snap}}]_k - \mathrm{Tr}[\mathbf{B}_k\rho(\mathbf{x}')]\right)^2. \tag{15.50}$$

The objective is to solve the optimization problem $\min_{\mathbf{x}'\in[-1,1]^d} f(\mathbf{x}')$. For general encoding maps, this problem appears to require a black-box optimization approach, where complexity is assessed based on the evaluations of f or possibly its gradient. However, in our context, the importance of identifying approximate local minima is unclear. Therefore, to ensure privacy breakage, it seems necessary to conduct an exhaustive grid search. Nonetheless, for the sake of completeness, we discuss results on first-order methods that can yield approximate local minima.

We start by reviewing some of the well-known results for black-box optimization. We recall Lipschitz continuity by

Definition 10 (L-Lipschitz Continuous Function) A function $f : \mathbb{R}^d \to \mathbb{R}$ is said be L-Lipschitz continuous if there exists a real positive constant $L > 0$ for which

$$|f(\mathbf{x}) - f(\mathbf{y})| \leq L\|\mathbf{x} - \mathbf{y}\|_2.$$

When we view the quantum circuit as a black-box L-Lipschitz function and consider $\mathbf{x}'$ within a convex, compact set with diameter P (such as $[-1, 1]^d$, which has a diameter of $2\sqrt{d}$), we can roughly estimate an upper bound for L by the highest frequency component of the multidimensional trigonometric series for f. This component can be exponentially large in terms of n. In this scenario, the number of function evaluations needed to find $\mathbf{x}'$ such that $|\mathbf{x} - \mathbf{x}'|_2 \leq \epsilon$ would scale as

$$O\left(P\left(\frac{L}{\epsilon}\right)^d\right), \tag{15.51}$$

this is complexity associated with grid search [37]. Consequently, even with a constant L, this task becomes computationally formidable, as it is exponential in $d = \Theta(n)$.

As previously discussed, it is feasible to employ first-order methods to develop an algorithm that is effectively independent of dimension for locating an approximate local minimum. Let's revisit the concept of β-smoothness, defined as follows:

Definition 11 (β-Smooth Function) A differentiable function $f : \mathbb{R}^d \to \mathbb{R}$ is said be β-smooth if there exists a real positive constant $\beta > 0$ for which

$$\|\nabla f(x) - \nabla f(y)\|_2 \leq \beta\|x - y\|_2.$$

If gradients of the cost function with respect to each parameter are available, then employing perturbed gradient descent [38] would approximately require

$$\tilde{O}\left(\frac{PL\beta}{\epsilon^2}\right), \tag{15.52}$$

function and gradient evaluations for an L-Lipschitz function that is β-smooth to find an approximate local minimum. The process involves calculating gradients by evaluating expectation values of ρ, then using methods such as finite-difference approximation or the parameter-shift rule for certain gate sets [33] to calculate the gradients.

Even if finding an approximate local minimum does not directly reveal significant information about $\mathbf{x}$, up to periodicity, it is still feasible to make this task difficult for an adversary. Typically, the encoding circuit will yield expectation values that include trigonometric terms. For example, consider a univariate scenario with a single trigonometric function $f(x) = \sin(\omega x)$, where ω represents the frequency. This function is ω-Lipschitz continuous, and its gradient is ω^2-Lipschitz continuous. Thus, when analyzing the scaling of gradient-based methods as described in Eq. 15.52, we find that the frequency of the trigonometric terms significantly impacts the ability to find a solution. As a result, if a frequency that scales exponentially, such as $\omega = O(\exp(n))$, is selected, snapshot inversion becomes exponentially more challenging using this approach.

Notably, if we refer back to the case study [18] discussed previously, we see that this indeed relied on exponentially scaling frequencies in the Fourier series decompositions of the model to achieve privacy. This is because if the feature map includes high-frequency components, such as the Fourier Tower map discussed in [18], then both β and L can scale as $O(\exp(n))$.

15.5.3 Expectation Value Landscape Numerical Results

While it is difficult to make definitive claims about privacy in the black-box context, we will show the numerical results of [19] here in order to provide insight into how encoding circuit structure affects the complexity of the loss landscape. One can consider two types of encoding that may be considered difficult to classically simulate. One corresponds to a single-qubit R_X rotation gate (but sandwiched between some fixed non-parameterized gates which are difficult to classically simulate) as shown in Fig. 15.6. This scenario would be challenging to fully simulate classically due to the presence of arbitrary unitary matrices. However, the outcome essentially involves taking measurements on an unknown basis. With just a few samples of x, it is possible to reconstruct the graph as a single frequency sinusoidal relationship. Consequently, the distance between the stationary points is $r = \pi$ for any value of n, which corresponds to a frequency $\omega = \frac{r}{\pi} = 1$, irrespective of n. As a result, this circuit demonstrates constant frequency scaling that is independent of n, making it

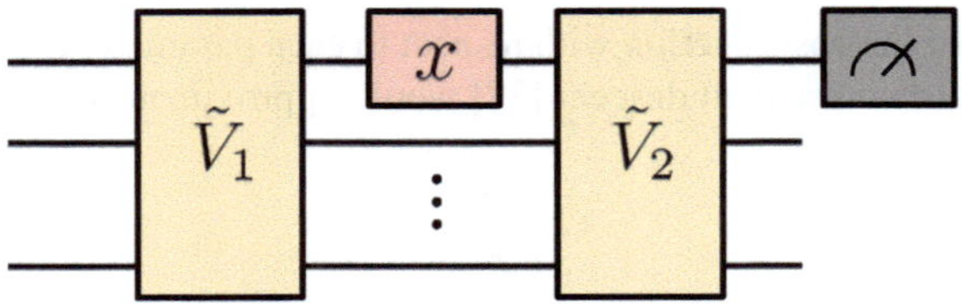

Fig. 15.6 Encoding circuit diagram showing a single-qubit R_X rotation gate parameterized by the univariate parameter x, but with arbitrary 2^n-dimensional unitaries applied before and after the x parameterized gate. Despite being hard to simulate analytically, the expectation value e_{in} varies as a simple sinusoidal function in x, regardless of the total number of qubits n. [19]

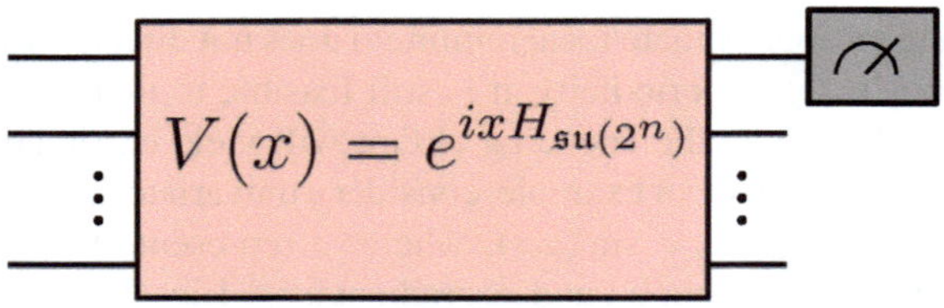

Fig. 15.7 Encoding circuit diagram showing a SU(2^n) gate parameterized by a univariate parameter x [19]

potentially straightforward for gradient-based methods to recover approximate local minima. The motivation behind choosing this example is to demonstrate that even in cases where the encoding circuit is difficult to classically simulate, it may still be possible to perform snapshot inversion.

In contrast, Fig. 15.7 demonstrates a circuit where x parameterizes an SU(2^n) gate. This results in a highest frequency term that is exponentially increasing. We will compare the expectation landscapes of these two encoding circuits, which are both difficult to simulate classically, in general, when measuring the same expectation value.

We now show visually the expectration value landscapes, as the input parameters are varied, in both the one-dimensional (Fig. 15.8) and two-dimensional case (Fig. 15.9). While these numerical results do not provide a definitive conclusion, they are insightful in illustrating that two encoding circuits, both challenging to simulate classically, can yield vastly different landscapes. One circuit may initially seem much more susceptible to numerical attacks than the other.

It's important to highlight that this field is rapidly evolving, with recent progress being made toward developing even more efficient attack strategies. For instance, a recent study has successfully integrated gradient estimation with the finite-difference method and adaptive low-pass filtering, further optimized using a Kalman filter, to enhance numerical attacks on VQC models [39]. The use of these filters may help perform attacks even when the loss landscape contains high-frequency terms. Therefore the discussion in this chapter does not rule out the potential for more advanced methods to compromise privacy. Instead, it emphasizes the conditions under which the methods proposed in [40] are guaranteed to lead to privacy breaches.

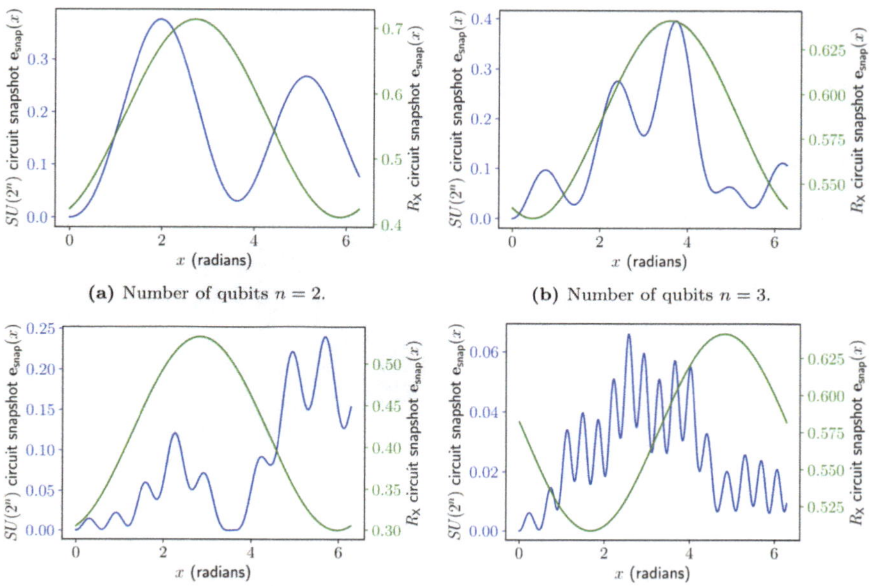

(a) Number of qubits $n = 2$.

(b) Number of qubits $n = 3$.

(c) Number of qubits $n = 4$.

(d) Number of qubits $n = 5$.

Fig. 15.8 Comparison of how expectation value of the measurement of Z_1 varies with univariate x for both the model parameterized using a single R_X rotation gate as detailed in Fig. 15.6 and the model parameterized using an $SU(2^n)$ gate as detailed in Fig. 15.7, for varying amounts of qubits [19]

15.6 Connections Between Trainability and Privacy in VQC

The main results of this chapter can be summarized as

- Weak Privacy Breach—If the DLA has polynomial scaling dimension, then snapshots can be retrieved from the gradients.
- Strong Privacy Breach—If, in addition, the encoding circuit adopts certain specific architectures (such as separable states), then classical input data can be retrieved from the snapshots.

While the results shown for the strong privacy breach case apply only to a more limited selection of architectures, it is important to note that the condition for a weak privacy breach is very general, as the DLA dimension can be calculated for many different VQC architectures. It may then seem a natural but naïve approach toward privacy would be to construct VQC circuits in which the variational circuit Hermitian generators form a DLA with an exponentially scaling dimensionality. While this doesn't guarantee privacy outright, it would prevent the weak privacy breach snapshot recovery that we discussed previously. However, although this would provide protection against this type of weak privacy breach, it would most likely result in a model that requires exponential time to train, thereby limiting its applicability

Fig. 15.9 Comparison of how expectation value of the measurement of Z_1 varies with a two-dimensional input (x_1, x_2) for both the model parameterized using a single R_X rotation gate as detailed in Fig. 15.6 and the model parameterized using an $SU(2^n)$ gate as detailed in Fig. 15.7, for varying amounts of qubits [19]

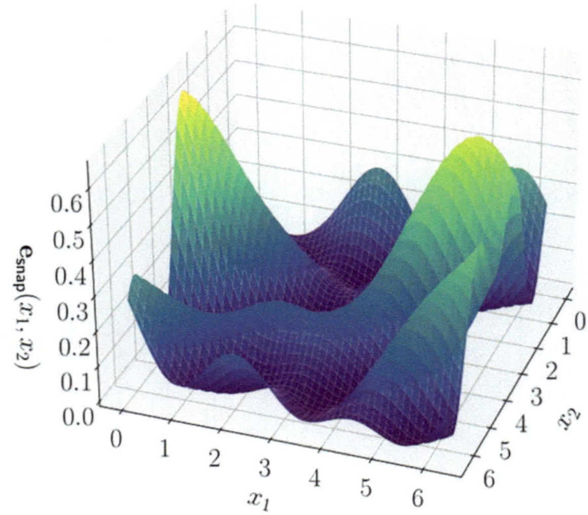

(a) Number of qubits $n = 2$.

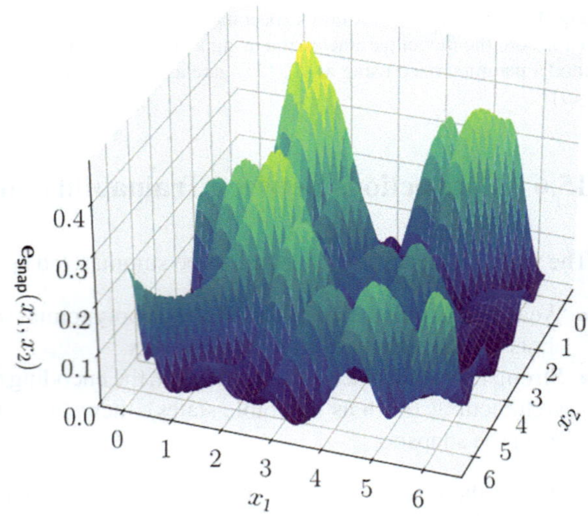

(b) Number of qubits $n = 3$.

in real-life situations. This problem arises due to the phenomenon known as barren plateaus, where the training landscape of the model becomes very flat, requiring an exponentially scaling number of evaluations to determine the correct gradient directions. In fact, by examining VQC models through a Lie-algebraic framework, the presence of barren plateaus has been directly linked to the circuit having an exponentially large DLA [25, 41]. The result of combining these findings with the results covered in this chapter, initially reported in [19], is that the cases in which a weak privacy breach is averted correspond to models that have barren plateaus and are therefore inefficient to train in practice, leading to the following corollary:

Corollary 1 (Trainability and Snapshot Recoverability [19]) *Any trainable VQC on n qubits that satisfies the LASA condition in Definition 5 fulfills the slow Pauli expansion condition as highlighted in Definition 9, and has a DLA $\mathfrak{g}$ whose dimension scales as $O(poly(n))$ would admit snapshot recoverability with complexity $O(poly(n))$.*

As an exponentially scaling DLA dimension generally leads to barren plateaus [25, 41], this implies that, in the majority of cases, any model that is trainable will also admit weak privacy breaches. This is a very fundamental result in the privacy of VQC models. While further research may be able to highlight exceptions to this rule by examining the assumptions used in more detail, it remains that the majority of VQC circuits currently studied will either admit weak privacy breaches or exhibit barren plateaus, both of which are undesirable qualities in privacy-enhanced VQC model design. The privacy of variational models beyond the LASA case is tied to a broader question in the field, specifically whether there exist quantum variational models that cannot be classically simulated and do not exhibit barren plateaus [28]. The generalization of quantum neural network loss landscapes beyond the LASA case has been explored with the introduction of the Jordan algebra supported ansatz (JASA) [42]. However, as of the time of writing, considerations of privacy within this broader framework remain an area for further research.

15.7 Revisiting Case Study

We will now revisit the case study presented at the start of this chapter based on the work [18]. We note that this work indeed examined direct input recovery and concluded that the gradients generated by $C_j(\mathbf{x})$ would form a loss landscape dependent on the highest frequency ω generated by the encoding circuit, indicating that exponentially scaling frequencies led to models that take exponential time to recover the input using quantum-assisted direct input recovery. This is because the Fourier tower map encoding circuit used in [18] was designed such that ω scales exponentially to provide privacy, this was done by using m qubits in a sub-register per data input x_j, with the single-qubit rotation gates parameterized by an exponentially scaling amount. The encoding can be defined as

$$\bigotimes_{j=1}^{d}\Big(\bigotimes_{l=1}^{m} R_X(5^{l-1}x_j)\Big). \tag{15.53}$$

Therefore, the gradient contained terms with the highest frequency scaling exponentially, resulting in a model where gradient descent methods required exponential time. These high-frequency terms offer protection against direct numerical inversion attacks. Now, let's explore the issue in terms of both weak and strong privacy breaches.

The case study utilized an overparameterized HEA as the variational circuit, which corresponds to an exponentially large DLA dimension. Consequently, a weak privacy breach is not feasible according to Theorem 2. However, this suggests that this circuit configuration would exhibit barren plateaus, making it inefficient to train in reality. While the exponentially large DLA dimension shields this model from weak privacy breaches, it also renders the model untrainable, undermining the utility of the approach proposed in [18].

Next, let's consider the privacy of the method used in [18] if the variational circuit was replaced with a polynomial-sized DLA dimension variational circuit. In this scenario, the model would be trainable but would also be susceptible to weak privacy breaches. However, how does the encoding procedure fare in this context? The encoding effectively involves single gate rotations with exponentially growing scaling factors, placing it within the Pauli product encoding regime. According to Theorem 3, we can derive a closed-form solution for the inputs, given the snapshots, up to a periodicity. Although qubits with high-frequency factors may introduce some obscurity due to high periodicity, it's important to note that every input feature is passed into at least one qubit where the prefactor is 1. This means a closed-form solution can be obtained without periodicity and a strong privacy breach would be possible. This implies that the encoding method using in [18] does not provide inherent protection against strong privacy breaches and the model only managed to maintain privacy due to its HEA component, which simultaneously rendered the mode untrainable. By analyzing this case study using our fresh perspective and tools, we can effectively identify errors and swiftly assess whether future modifications will result in a successful solution (Table 15.1).

15.8 Conclusion

In this chapter, we have explored the intricate relationship between privacy and trainability in Variational Quantum Circuit (VQC) models, focusing on the potential for privacy breaches through snapshot recovery and inversion techniques. Our exploration has revealed that the algebraic structure of VQC models plays a pivotal role in determining their susceptibility to privacy attacks. Specifically, we have demonstrated that a polynomial scaling dimension of the Dynamical Lie Algebra (DLA) is a key factor in enabling snapshot recovery from model gradients, which constitutes a weak privacy breach. This finding highlights a fundamental trade-off between

Table 15.1 Summary of results on the privacy guarantees and complexity provided by attack models on various VQC models studied in this chapter. We consider two privacy breach scenarios involving VQCs: *weak privacy* breach and *strong privacy* breach for classical or quantum-assisted polynomial-time methods. Weak privacy breach concerns the recovery of the meaningful snapshots of the input encoded state, allowing training VQC models for distinct learning tasks without requiring access to the input. Strong privacy breach concerns subsequently inverting the snapshots to recover the original input. We consider the snapshot invertibility for local Pauli encoding map which admits an efficient (polynomial in the number of qubit n) algorithm if the requirements stated in the table are met. For the case of generic encoding maps where the VQC is considered as a black-box L-Lipschitz function, snapshot invertibility requires performing the grid search which scales exponentially in the input dimension d, and thus it rules out efficient privacy breaches. Source in this table Heredge et al. [19]

Privacy breach	Description	Complexity	Requirements
Weak	Snapshot recovery (Sect. 15.4)	Algorithm 2: $O(\text{poly}(\dim(\mathfrak{g})))$	$O(\text{poly}(n))$ sized DLA + LASA condition (Definition 5)+ Slow Pauli Expansion (Def 9)
Strong	Snapshot inversion for local Pauli encoding (Sect. 15.5.1)	Algorithm 4: $O(\text{poly}(n, 1/\epsilon)$	Snapshot recovery requirement + Separable state with $\rho_J(\mathbf{x})$ parameterized by subset $\mathbf{x}_J \subseteq \mathbf{x}$ • $\dim(\mathbf{x}_J) = O(1)$ • each x_k is encoded at most $R = O(\text{poly}(n))$ times • Snapshot components with non-zero overlap w.r.t. $\rho_J(\mathbf{x}_J)$ has cardinality at least $\dim(\mathbf{x}_J)$.
Strong	Snapshot inversion for generic encoding (Sect. 15.5.2)	Grid Search : $O\left(\left(\frac{L}{\epsilon}\right)^d\right)$	Recovery cost function is L-Lipschitz leading to efficient privacy breach being not possible

privacy and trainability, as models with polynomial DLA dimensions are generally more trainable but also more vulnerable to privacy breaches.

Furthermore, we have examined the conditions under which snapshot inversion, a strong privacy breach, is possible. Our analysis of local multi-qubit Pauli encoding and generic encoding schemes has shown that certain architectures are more susceptible to inversion attacks, questioning their suitability for privacy-focused applications.

Our re-evaluation of previous studies, such as [18], through the lens of this new Lie-algebraic framework, has provided valuable insights into the limitations and potential improvements in privacy mechanisms for VQC models. The use of exponentially large DLA dimensions, while effective in preventing input data recovery, can lead to untrainable models, highlighting the need for a balanced approach in model design. While high-frequency terms seem to be useful in the context of numerical or black-box attacks and should likely be considered in future privacy-enhanced VQC designs.

In conclusion, this chapter offers a comprehensive framework for assessing and enhancing the privacy features of VQC models, providing strategic guidelines for the design of quantum circuits that prioritize security while maintaining trainability. As the field of quantum machine learning continues to evolve, these findings, which were initially reported in [19], pave the way for more robust model designs that is hoped may one day achieve a useful privacy advantage over classical methods, ultimately contributing to the development of effective and secure quantum machine learning applications.

15.9 Notations

See Table 15.2.

Table 15.2 Table of notation

Symbol	Meaning
$\mathbf{x}$	Data input
θ	Variational parameters
$y_\theta(\mathbf{x})$	Model output
C_j	Gradient of Model output wrt θ_j
n	Total number of qubits
d	Dimension of input vector
D	Number of variational parameters
N	Number of Hermitian generators
$V(\mathbf{x})$	Encoding circuit
$\mathbf{U}(\theta)$	Variational circuit
O	Measurement operator in VQC model
μ_α	Coefficients of O when written in terms of supported DLA basis terms
$\mathfrak{g}$	Dynamical Lie algebra
Ω	Maximum number of sub-register partitions
$\mathbf{B}_k$	Basis of DLA
B_k	Portion of basis DLA basis element that only acts on a given sub-register
H_k	Hermitian gate generators of the encoding map circuit
m	Number of qubits in sub-register
$\rho_J(\mathbf{x}_J)$	Constituent state in a separable state
Ω	Maximum number of constituent states in a separable state
A_k	Element of constituent separable state density matrix sum expansion
λ	Terms when expanding constituent density matrix $\rho_J(\mathbf{x}_J)$
P	Diameter of sphere of possible $\mathbf{x}$ values in $\mathbb{R}^d$
ϵ	User-defined acceptable error

Acknowledgements This chapter is adapted from the following publications:
Heredge, J., Kumar, N., Herman, D. et al. Characterizing privacy in quantum machine learning. npj Quantum Inf **11**, 80 (2025). https://doi.org/10.1038/s41534-025-01022-z [19].
Kumar, N., Heredge J., Li, C. et al. Expressive variational quantum circuits provide inherent privacy in federated learning arXiv:2309.13002 (2023). https://arxiv.org/abs/2309.13002 [18]
This chapter was prepared for informational purposes by the Global Technology Applied Research Center of JPMorgan Chase. This chapter is not a product of the Research Department of JPMorganChase or its affiliates. Neither JPMorgan Chase nor any of its affiliates makes any explicit or implied representation or warranty and none of them accept any liability in connection with this paper, including, without limitation, with respect to the completeness, accuracy, or reliability of the information contained herein and the potential legal, compliance, tax, or accounting effects thereof. This document is not intended as investment research or investment advice, or as a recommendation, offer, or solicitation for the purchase or sale of any security, financial instrument, financial product or service, or to be used in any way for evaluating the merits of participating in any transaction.

References

1. T. Liu, Z. Wang, H. He, W. Shi, L. Lin, W. Shi, R. An, C. Li, Efficient and secure federated learning for financial applications (2023)
2. T. Awosika, R.M. Shukla, B. Pranggono, Transparency and privacy: the role of explainable ai and federated learning in financial fraud detection (2023)
3. Kaissis G, Makowski MR, Rückert D, Braren RF (2020) Secure, privacy-preserving and federated machine learning in medical imaging. Nat Mach Intell 2:305–311
4. S. Ahamed, N. Nishant, A. Selvaraj, N. Gandhewar, S.A, B.K, Investigating privacy-preserving machine learning for healthcare data sharing through federated learning. Sci. Temp. **14**, 1308–1315 (2023)
5. E. Aguiar, C. Jr, A. Traina, Security and privacy in machine learning for health systems: Strategies and challenges. Yearbook Med. Inf. **32**, 269–281 (2023)
6. C. Park, T. Choi, T. Kim, M. Cho, J. Hong, M. Choi, J. Choo, Fedgeo: privacy-preserving user next location prediction with federated learning (2023)
7. Albrecht JP (2016) How the gdpr will change the world. European Data Protect Law Rev 2:287–289
8. A. Brauneck, L. Schmalhorst, M. Majdabadi, M. Bakhtiari, U. Völker, C. Saak, J. Baumbach, L. Baumbach, G. Buchholtz, Federated machine learning in data-protection-compliant research. Nat. Mach. Intell. **5**, 2–4 (2023)
9. J. Tong, C. Luo, M. Islam, N. Sheils, J. Buresh, M. Edmondson, P. Merkel, E. Lautenbach, R. Duan, Y. Chen, Distributed learning for heterogeneous clinical data with application to integrating covid-19 data across 230 sites. NPJ Digit. Med. **5**, 76 (2022)
10. H.B. McMahan, E. Moore, D. Ramage, S. Hampson, B.A. Y. Arcas, Communication-efficient learning of deep networks from decentralized data (2023)
11. L. Zhu, Z. Liu, S. Han, Deep leakage from gradients, in *Advances in Neural Information Processing Systems*, vol. 32 (Curran Associates, Inc., 2019)
12. Y. Huang, S. Gupta, Z. Song, K. Li, S. Arora, Evaluating gradient inversion attacks and defenses in federated learning, in *Advances in Neural Information Processing Systems*, ed. by A. Beygelzimer, Y. Dauphin, P. Liang, J.W. Vaughan (2021)
13. B. Zhao, K.R. Mopuri, H. Bilen, idlg: improved deep leakage from gradients (2020). arXiv:2001.02610
14. J. Geiping, H. Bauermeister, H. Dröge, M. Moeller, Inverting Gradients—How easy is it to break privacy in federated learning?, in *Advances in Neural Information Processing Systems 33: Annual Conference on Neural Information Processing Systems 2020, NeurIPS 2020, 6–12 Dec 2020, virtual* (2020)

15. H. Yin, A. Mallya, A. Vahdat, J. Alvarez, J. Kautz, P. Molchanov, See through gradients: image batch recovery via gradInversion, in *2021 IEEE/CVF Conference on Computer Vision and Pattern Recognition (CVPR)*, pp. 16332–16341 (2021)
16. Phong LT, Aono Y, Hayashi T, Wang L, Moriai S (2018) Privacy-preserving deep learning via additively homomorphic encryption. IEEE Trans Inf Forensics Secur 13(5):1333–1345
17. S. Eloul, F. Silavong, S. Kamthe, A. Georgiadis, S.J. Moran, Enhancing privacy against inversion attacks in federated learning by using mixing gradients strategies (2022). arXiv:2204.12495
18. N. Kumar, J. Heredge, C. Li, S. Eloul, S.H. Sureshbabu, M. Pistoia, Expressive variational quantum circuits provide inherent privacy in federated learning (2023). arXiv:2309.13002
19. J. Heredge, N. Kumar, D. Herman, S. Chakrabarti, R. Yalovetzky, S.H. Sureshbabu, C. Li, M. Pistoia, Characterizing privacy in quantum machine learning. NPJ Quantum Inf **11**, 80 (2025)
20. J. Haah, Y. Liu, X. Tan, Efficient approximate unitary designs from random pauli rotations (2024). arXiv:2402.05239
21. McClean JR, Boixo S, Smelyanskiy VN, Babbush R, Neven H (2018) Barren plateaus in quantum neural network training landscapes. Nat Commun 9(1):4812
22. M. Tancik, P. P. Srinivasan, B. Mildenhall, S. Fridovich-Keil, N. Raghavan, U. Singhal, R. Ramamoorthi, J. T. Barron, and R. Ng, "Fourier features let networks learn high frequency functions in low dimensional domains," 2020
23. Larocca M, Czarnik P, Sharma K, Muraleedharan G, Coles PJ, Cerezo M (2022) Diagnosing barren plateaus with tools from quantum optimal control. Quantum 6:824
24. M. Ragone, P. Braccia, Q. T. Nguyen, L. Schatzki, P. J. Coles, F. Sauvage, M. Larocca, and M. Cerezo, "Representation theory for geometric quantum machine learning," *arXiv preprint* arXiv:2210.07980, 2022
25. E. Fontana, D. Herman, S. Chakrabarti, N. Kumar, R. Yalovetzky, J. Heredge, S. H. Sureshbabu, and M. Pistoia, "The adjoint is all you need: Characterizing barren plateaus in quantum ans\" atze," *arXiv preprint* arXiv:2309.07902, 2023
26. Somma R, Ortiz G, Barnum H, Knill E, Viola L (2004) Nature and measure of entanglement in quantum phase transitions. Phys Rev A 70(4):042311
27. M. L. Goh, M. Larocca, L. Cincio, M. Cerezo, and F. Sauvage, "Lie-algebraic classical simulations for variational quantum computing," *arXiv preprint* arXiv:2308.01432, 2023
28. M. Cerezo, M. Larocca, D. García-Martín, N. Diaz, P. Braccia, E. Fontana, M. S. Rudolph, P. Bermejo, A. Ijaz, S. Thanasilp, *et al.*, "Does provable absence of barren plateaus imply classical simulability? or, why we need to rethink variational quantum computing," *arXiv preprint* arXiv:2312.09121, 2023
29. Grcar JF (2011) Mathematicians of gaussian elimination. Notices of the AMS 58(6):782–792
30. Suzuki R, Mitarai K, Fujii K (2022) Computational power of one- and two-dimensional dual-unitary quantum circuits. Quantum 6:631 Jan
31. J. Nocedal and S. J. Wright, *Numerical optimization*. Springer, 1999
32. Jenkins MA, Traub JF (1970) A three-stage variable-shift iteration for polynomial zeros and its relation to generalized rayleigh iteration. Numer Math 14(3):252–263
33. Wierichs D, Izaac J, Wang C, Lin CY-Y (2022) General parameter-shift rules for quantum gradients. Quantum 6:677 Mar
34. B. Buchberger, *Gröbner Bases: An Algorithmic Method in Polynomial Ideal Theory*, pp. 184–232. Reidel Publishing Company, 01 1985
35. Dubé TW (1990) The structure of polynomial ideals and gröbner bases. SIAM J Comput 19(4):750–773
36. Mayr EW, Meyer AR (1982) The complexity of the word problems for commutative semigroups and polynomial ideals. Adv Math 46(3):305–329
37. Y. Nesterov *et al.*, *Lectures on convex optimization*, vol. 137. Springer, 2018
38. C. Jin, R. Ge, P. Netrapalli, S. M. Kakade, and M. I. Jordan, "How to escape saddle points efficiently," 2017

39. G. Papadopoulos, S. Eloul, Y. Satsangi, J. Heredge, N. Kumar, C.-F. Chen, and M. Pistoia, "A numerical gradient inversion attack in variational quantum neural-networks," 2025
40. J. Heredge, N. Kumar, D. Herman, S. Chakrabarti, R. Yalovetzky, S. H. Sureshbabu, C. Li, and M. Pistoia, "Prospects of privacy advantage in quantum machine learning," 2024. https://doi.org/10.48550/arXiv.2405.08801
41. M. Ragone, B. N. Bakalov, F. Sauvage, A. F. Kemper, C. O. Marrero, M. Larocca, and M. Cerezo, "A Unified Theory of Barren Plateaus for Deep Parametrized Quantum Circuits," *arXiv*, 9 2023
42. E. R. Anschuetz, "A unified theory of quantum neural network loss landscapes," 2025

Chapter 16
Robust Verification of Quantum States in the Adversarial Scenario

Zihao Li, Huangjun Zhu, and Masahito Hayashi

Abstract Efficient and robust verification of quantum states in the adversarial scenario is essential for various quantum information processing tasks, including blind quantum computation and quantum networks. This chapter reviews the basic framework and protocols for addressing this problem. Thanks to these protocols, many important quantum states can be verified in the adversarial scenario using local projective measurements with unexpected high efficiency and robustness.

16.1 Introduction

Reliable and trustworthy quantum state preparation is a cornerstone for developing quantum technologies, such as quantum computers, communication networks, sensors, and measurement devices. In practice, it is usually difficult to prepare entangled quantum states perfectly, so it is crucial to efficiently verify whether a quantum system is prepared in the desired target state. This task, known as quantum state verification (QSV), lies at the heart of the important research field of quantum characterization, verification, and validation (QCVV) [1–5].

Z. Li · H. Zhu
State Key Laboratory of Surface Physics and Department of Physics, Fudan University, Shanghai, China
e-mail: zihaoli20@fudan.edu.cn

H. Zhu
e-mail: zhuhuangjun@fudan.edu.cn

M. Hayashi (✉)
School of Data Science, The Chinese University of Hong Kong, Longgang District, Shenzhen, China
e-mail: hmasahito@cuhk.edu.cn

International Quantum Academy, Futian District, Shenzhen, China

Graduate School of Mathematics, Nagoya University, Nagoya, Japan

M. Usman (ed.), *Quantum Robustness in Artificial Intelligence*, Quantum Science and Technology, https://doi.org/10.1007/978-3-032-11153-1_16

Although the goal appears straightforward, QSV is highly challenging due to the fragility of quantum states and limitations of measurement tools. Traditional tomographic approaches [47–49] for this task are resource-consuming for systems with more than ten qubits. Even with popular approaches, such as direct fidelity estimation [45, 46] or classical shadow estimation [51–53], their scaling behaviors are still not satisfactory when restricted to local measurements. In this context, this review provides a detailed overview of theoretical advances in QSV, discusses significant results, and identifies open questions.

The history of QSV is closely tied to the development of quantum mechanics. Early studies on the basics of quantum theory highlighted the importance of preparing and measuring states. Nevertheless, the formulation of QSV as a specific research problem progressed slowly. In the early days of quantum mechanics, the focus was mostly on understanding the average values of observables of quantum systems, often using statistical ensembles. The concept of verifying a specific quantum state, especially a maximally entangled state, received less attention.

The rise of quantum information science at the end of the last century brought a big change. As researchers saw the potential of quantum systems for information processing tasks, the need for reliable state preparation and verification became very important. The growth of quantum computing, in particular, led to a lot of research on QSV, because the correct operation of a quantum computer depends on accurately setting up and controlling qubits. Early works in this area focused on finding efficient ways for describing quantum states, such as quantum state estimation [6, 7]. While quantum state estimation is powerful, it requires a huge number of measurements that scales exponentially with the system size, making it impractical for large quantum systems. This problem motivates the search for more efficient QSV methods.

In early works [8, 9], Hayashi and collaborators studied QSV for the maximally entangled state under the framework of hypothesis testing. The proposed method can be realized using only local measurements and has been demonstrated in experiments [10, 11]. The paper [8] introduced a test based on group symmetry with discrete measurements on a single copy of the state. The paper [9] used hypothesis testing with a binomial distribution under the condition that the states are independent and identically distributed (i.i.d.). It also examines this testing method from the perspective of group symmetry. Later, the paper [19] explored how to simplify this method using discrete measurements.

The QSV problem becomes more complicated in the adversarial scenario, where the states to be verified are prepared by a potentially malicious adversary and do not satisfy the i.i.d. assumption. In this case, the above method cannot be used to verify general stabilizer states, including graph states. To address this problem, the paper [12] proposed a stabilizer test for verifying graph states in the adversarial scenario. This test is limited to two-colorable graph states, but can be extended to multi-colorable graph states. Furthermore, the paper [12] pointed out that verification of graph states in the adversarial scenario can lead to a protocol for verifying blind measurement-based quantum computation.

However, the paper [12] has the following issues: (i) It does not handle the case in which the state is noisy, i.e., it is not robust. (ii) It does not cover general states beyond stabilizer states. (iii) The choices of several parameters of the strategy are restricted. To address issue (i), the later study [13] combined the idea of fault-tolerant computation with state verification. The method proposed in [13] allows checking if the error that occurs is part of a set that can be corrected. To tackle issue (ii), the papers [15, 17, 18] proposed several methods for verifying more general states, such as general hypergraph states. Due to issue (iii), the test in [8, 12] cannot be written as a simple measurement operator, which blocks the application of useful randomized tests. In contrast, the later paper [14] introduced a useful randomized test based on randomly choosing a stabilizer. The test in [14] can be written as a simple measurement operator and can be seen as a direct extension of the test in [8]. Later, the studies [20–22] fully resolved issues (ii) and (iii), creating a general QSV recipe for the adversarial scenario.

The best way to verify a state depends on the specific form of the target state. The verification of many important classes of quantum states in the adversarial scenario has been studied, including bipartite pure states [23], GHZ states [26, 27], graph states [12, 18], hypergraph states [15, 17, 21], weighted graph states [35], continuous-variable states [33], ground states of local Hamiltonians [17, 30], and antisymmetric basis states [28]. Recently, to address issues (i) and (iii), the paper [36] extended the method by [20–22] to cases where the target state has a certain symmetry. When the measurement device is not trusted in addition to the state, a more complex theory, called self-testing, is needed. This topic has been actively studied in various papers [39–43].

This chapter reviews the basic framework and protocols for QSV in the adversarial scenario, primarily based on papers [21, 22, 36]. The rest of this chapter is structured as follows: In Sect. 16.2, we review the basic framework of QSV in the nonadversarial scenario. In Sect. 16.3, we review the framework of QSV in the adversarial scenario, and introduce the general QSV recipe developed in [21, 22]. In Sect. 16.4, we clarify the importance of robustness in QSV, and introduce the robust QSV protocol proposed in [36] for the adversarial scenario. In Sect. 16.5, we summarize and discuss potential future directions.

16.2 Quantum State Verification in the Nonadversarial Scenario

In this section, we review the basic framework of QSV in the nonadversarial scenario. The results presented here provide a reference baseline for comparing QSV in the adversarial scenario discussed in later sections.

In the nonadversarial scenario, we consider a quantum device that is designed to produce the target state $|\Psi\rangle$ in a Hilbert space $\mathcal{H}$. In practice, however, the device actually produces i.i.d. states σ over N runs, which may deviate from $|\Psi\rangle$ due to noise. The task is to verify whether σ is sufficiently close to the target state, where the amount of closeness is quantified by the infidelity of σ with respect to $|\Psi\rangle$, that is, $\epsilon_\sigma := 1 - \langle\Psi|\sigma|\Psi\rangle$. To achieve this task, the verifier Alice performs a series of tests on the produced states. Each test is a two-outcome measurement $\{E_l, \mathbb{I} - E_l\}$ on a single state σ, which is randomly chosen classically with probability μ_l from a set of measurements. If the first measurement outcome E_l is obtained, we say that the state passes the test. After N runs, the device is accepted by Alice if all tests are passed; otherwise, it is rejected.

With the above procedure, Alice aims to achieve two basic goals of *completeness* and *soundness*. The completeness means that Alice does not mistakenly reject if the device produces copies of the ideal state $|\Psi\rangle$. To guarantee this condition, the test operator E_l is required to satisfy the condition $E_l|\Psi\rangle = |\Psi\rangle$ for all l, ensuring that the target state can always pass each test. Intuitively, the soundness means that Alice does not mistakenly accept if σ is far from the target state. Concretely, it means the following: if the infidelity ϵ_σ is larger than a certain threshold ϵ, then the probability that Alice accepts should be less than some small value δ. Here the threshold $0 < \epsilon < 1$ is called the target infidelity and $0 < \delta < 1$ is called the significance level. These two parameters specify the target verification precision.

The efficiency of verification is characterized by the number N of tests required to achieve a given precision. To determine this number, first note that the average probability that σ passes the test reads $\mathrm{tr}(\Omega\sigma)$, where $\Omega := \sum_l \mu_l E_l$ is called the verification operator or a strategy. In addition, given ϵ_σ as the infidelity of σ, the maximum passing probability of σ is [21, 22]

$$\max_{\tau:\langle\Psi|\tau|\Psi\rangle=1-\epsilon_\sigma} \mathrm{tr}(\Omega\tau) = 1 - [1 - \lambda(\Omega)]\epsilon_\sigma = 1 - \nu(\Omega)\epsilon_\sigma, \tag{16.1}$$

where $\lambda(\Omega)$ is the second largest eigenvalue of Ω, and $\nu(\Omega) := 1 - \lambda(\Omega)$ denotes the spectral gap between the largest and second largest eigenvalues of Ω.[1] Accordingly, the probability that the produced states pass all N tests is upper bounded by $[1 - \nu(\Omega)\epsilon_\sigma]^N$. Therefore, to achieve the soundness condition, i.e., to ensure that the infidelity ϵ_σ is at most ϵ with significance level δ, it suffices to guarantee $[1 - \nu(\Omega)\epsilon]^N \leq \delta$. The minimum number of tests required to satisfy this condition is

$$N_{\mathrm{NA}}(\epsilon, \delta, \Omega) = \left\lceil \frac{\ln \delta}{\ln[1 - \nu(\Omega)\epsilon]} \right\rceil \approx \frac{\ln \delta^{-1}}{\nu(\Omega)\epsilon}, \tag{16.2}$$

where the approximation holds when $\nu(\Omega)\epsilon \ll 1$.

Equation (16.2) highlights that the efficiency of QSV in the nonadversarial scenario is determined by the corresponding spectral gap $\nu(\Omega)$. To minimize the resource cost, it is essential to maximize the spectral gap by cleverly constructing the

[1] The largest eigenvalue of Ω is 1 since $E_l|\Psi\rangle = |\Psi\rangle\forall l$. The maximization in Eq. (16.1) is achieved iff the state τ is supported in the subspace of $\mathcal{H}$ associated with the largest two eigenvalues of Ω.

strategy Ω. If there is no restriction on the measurements, the optimal strategy consists of one single test $\{|\Psi\rangle\langle\Psi|, \mathbb{I} - |\Psi\rangle\langle\Psi|\}$, which yields $\Omega = |\Psi\rangle\langle\Psi|$ and $\nu(\Omega) = 1$. However, this test is difficult to realize especially when $|\Psi\rangle$ is a multipartite entangled quantum state. In practical scenarios, the verifier is often constrained to specific types of measurements, such as measurements based on local operations and classical communication (LOCC), which limits the achievable spectral gap.

Up to now, efficient verification strategies based on LOCC have been constructed for various classes of pure states, including maximally entangled states [8, 9, 19], bipartite pure states [14, 23–25], stabilizer states (including graph states) [12, 14, 16, 20, 21, 26], hypergraph states [21], weighted graph states [35], Dicke states [28, 29], ground states of local Hamiltonians [30, 31], and certain continuous-variable states [32], see [4, 5] for overviews. Moreover, a recent work [44] showed that for almost all n-qubit pure target states $|\Psi\rangle \in \mathcal{H}$, one can construct an efficient strategy Ω by using LOCC, whose inverse spectral gap $\nu(\Omega)^{-1}$ grows at most quadratically in n. This fact, together with Eq. (16.2), implies that the target state can be verified with only $O(n^2\epsilon^{-1} \ln \delta^{-1})$ tests in the nonadversarial scenario. Such a resource cost is much lower than traditional approaches based on tomography [48, 49] and direct fidelity estimation [45, 46].

16.3 Quantum State Verification in the Adversarial Scenario

Next, we turn to QSV in the adversarial scenario. Recall that in the nonadversarial scenario, the quantum states prepared by the device are i.i.d. However, this assumption no longer holds in the adversarial scenario, where the device is controlled by a potentially malicious adversary Bob and cannot be trusted. Efficient and reliable QSV in this scenario is challenging but crucial to various applications that entail high-security requirements, such as blind quantum computation and quantum networks. In this section, we introduce a general recipe for QSV in the adversarial scenario. With this recipe, any pure state can be verified efficiently in the adversarial scenario as long as it can be verified efficiently in the nonadversarial scenario, with only a constant-factor increase in resource consumption. The main results presented here are based on [21, 22].

16.3.1 Formulation

In the adversarial scenario, the device controlled by Bob is supposed to produce $N + 1$ copies of the target state $|\Psi\rangle \in \mathcal{H}$, but actually generates a quantum state ρ on the entire system $\mathcal{H}^{\otimes(N+1)}$. If Bob is malicious, then he can mess up the verifier Alice by generating an arbitrary correlated or even entangled state ρ. Our goal is to verify whether the reduced state on one system (for application) is close enough to the target state by performing tests on other N systems. It should be noted that Alice cannot select a fixed system, such as the last one, as the target for application. If so, then a clever Bob could deceive Alice by preparing good states on the first N systems while producing a bad state on the last one. To eliminate this security loophole, our verification recipe runs as follows: After receiving the state ρ, Alice randomly selects N systems and apply a verification strategy Ω to test each of them. If all N tests are passed, Alice accepts the reduced state σ_{N+1} on the remaining system; otherwise, she rejects.

Since the tests are randomly performed on N systems of ρ, in the following we may assume that ρ is a permutation-invariant state without loss of generality. Then the probability that Alice accepts is given by

$$p(\rho) = \mathrm{tr}[(\Omega^{\otimes N} \otimes \mathbb{I})\rho]. \tag{16.3}$$

Once accepting, the conditional reduced state on the remaining system can be written as [assuming $p(\rho) > 0$]

$$\sigma_{N+1} = \frac{1}{p(\rho)} \mathrm{tr}_{1,2,\ldots,N}[(\Omega^{\otimes N} \otimes \mathbb{I})\rho], \tag{16.4}$$

where $\mathrm{tr}_{1,2,\ldots,N}$ is the partial trace over the first N systems. Then the fidelity of σ_{N+1} with respect to the target state reads

$$F(\rho) = \langle\Psi|\sigma_{N+1}|\Psi\rangle = \frac{f(\rho)}{p(\rho)}, \tag{16.5}$$

where

$$f(\rho) = \mathrm{tr}[(\Omega^{\otimes N} \otimes |\Psi\rangle\langle\Psi|)\rho]. \tag{16.6}$$

Similar to the nonadversarial scenario, here Alice also hopes to achieve the basic goals of *completeness* and *soundness*. The completeness means that Alice always accepts if Bob produces $N+1$ ideal states $|\Psi\rangle$. Since $|\Psi\rangle$ can always pass each test, this condition is automatically guaranteed. The soundness is still characterized by the target infidelity ϵ and significance level δ and is stated as follows: if the infidelity $1 - F(\rho)$ of the reduced state is greater than ϵ, then the acceptance probability $p(\rho)$ should be less than δ. To characterize the soundness condition in a mathematical way, we define the following figure of merit for $0 < \delta < 1$:

$$\bar{\epsilon}(\Omega, N, \delta) := 1 - \min_{\rho}\{F(\rho) \mid p(\rho) \geq \delta\}, \tag{16.7}$$

where the minimization is taken over all permutation-invariant states ρ on $\mathcal{H}^{\otimes(N+1)}$. This quantity is called the *guaranteed infidelity*, because Alice can make sure (with significance level δ) that the reduced state has infidelity at most $\bar{\epsilon}(\Omega, N, \delta)$ with the target state once accepting; it is nondecreasing in δ by definition. Using this quantity, the condition of soundness can be expressed as an inequality $\bar{\epsilon}(\Omega, N, \delta) \leq \epsilon$.

To specify the overhead of QSV in the adversarial scenario, we define $N(\epsilon, \delta, \Omega)$ as the minimum number of tests required to achieve the soundness condition, namely:

$$N(\epsilon, \delta, \Omega) := \min\{N \geq 1 \,|\, \bar{\epsilon}(\Omega, N, \delta) \leq \epsilon\}, \quad 0 < \epsilon, \delta < 1. \tag{16.8}$$

This number is the main figure of merit of concern in our setting; it cannot be smaller than $N_{\mathrm{NA}}(\epsilon, \delta, \Omega)$ defined in Eq. (16.2) [21, 22], since it is more difficult to verify a quantum state in the adversarial scenario than nonadversarial scenario.

16.3.2 Computation of the Guaranteed Infidelity

According to the above discussions, the computation of the guaranteed infidelity $\bar{\epsilon}(\Omega, N, \delta)$ given in Eq. (16.7) plays a pivotal role in analyzing the soundness of our protocol and determining the minimum number of tests required. In this section, we present a general method for addressing this task. Specifically, we show that the quantum optimization problem in Eq. (16.7) can be effectively reduced to a classical optimization problem, significantly simplifying the calculation.

Suppose the verification operator Ω for the target state $|\Psi\rangle$ has spectral decomposition $\Omega = \sum_{j=1}^{D} \lambda_j \Pi_j$, where D is the dimension of the Hilbert space, λ_j are the eigenvalues of Ω arranged in decreasing order $1 = \lambda_1 > \lambda_2 \geq \cdots \geq \lambda_D \geq 0$, and Π_j are mutually orthogonal rank-1 projectors with $\Pi_1 = |\Psi\rangle\langle\Psi|$. Here the second largest eigenvalue $\lambda := \lambda_2$ and the smallest eigenvalue $\omega := \lambda_D$ deserve special attention because they determine the performance of Ω to a large extent, as we shall see later. Note that $p(\rho)$ defined in Eq. (16.3) and $f(\rho)$ defined in Eq. (16.5) only depend on the diagonal elements of ρ in the product basis constructed from the eigenbasis of Ω (as determined by Π_j). Hence, we may further assume that ρ is diagonal in this basis without loss of generality. In other words, ρ can be expressed as a mixture of tensor products of Π_j.

Let $\mathbf{k} = (k_1, k_2, \ldots, k_D)$ be a sequence of D nonnegative integers that sum up to $N+1$, and $\mathscr{S}_N$ be the set of all such sequences. For each $\mathbf{k} \in \mathscr{S}_N$, we can define a permutation-invariant diagonal density matrix $\rho_{\mathbf{k}}$ on $\mathcal{H}^{\otimes(N+1)}$ as follows:

$$\rho_{\mathbf{k}} := \mathscr{P}\left\{\Pi_1^{\otimes k_1} \otimes \Pi_2^{\otimes k_2} \otimes \cdots \otimes \Pi_D^{k_D}\right\}, \tag{16.9}$$

where $\mathscr{P}\{\cdot\}$ denotes the uniform mixture of all permutations in $\{\cdot\}$. Accordingly, any permutation-invariant diagonal density matrix ρ on $\mathcal{H}^{\otimes(N+1)}$ can be expressed as $\rho = \sum_{\mathbf{k}\in\mathscr{S}_N} c_{\mathbf{k}}\rho_{\mathbf{k}}$, where $c_{\mathbf{k}}$ form a probability distribution. Let $\boldsymbol{\lambda} := (\lambda_1, \ldots, \lambda_D)$, then $p(\rho)$, $f(\rho)$, and $F(\rho)$ can be written as follows:

$$p(\rho) = \sum_{\mathbf{k}\in\mathscr{S}_N} c_{\mathbf{k}}\eta_{\mathbf{k}}(\boldsymbol{\lambda}), \qquad f(\rho) = \sum_{\mathbf{k}\in\mathscr{S}_N} c_{\mathbf{k}}\zeta_{\mathbf{k}}(\boldsymbol{\lambda}), \tag{16.10}$$

where $\eta_{\mathbf{k}}(\boldsymbol{\lambda}) := p(\rho_{\mathbf{k}})$ and $\zeta_{\mathbf{k}}(\boldsymbol{\lambda}) := f(\rho_{\mathbf{k}})$ only depend on $\boldsymbol{\lambda}$ and $\mathbf{k}$. Consequently, the guaranteed infidelity given in Eq. (16.7) reduces to the following form:

$$\bar{\epsilon}(\Omega, N, \delta) = 1 - \min_{\{c_{\mathbf{k}}\}} \left\{ \frac{\sum_{\mathbf{k}\in\mathscr{S}_N} c_{\mathbf{k}}\zeta_{\mathbf{k}}(\boldsymbol{\lambda})}{\sum_{\mathbf{k}\in\mathscr{S}_N} c_{\mathbf{k}}\eta_{\mathbf{k}}(\boldsymbol{\lambda})} \,\middle|\, \sum_{\mathbf{k}\in\mathscr{S}_N} c_{\mathbf{k}}\eta_{\mathbf{k}}(\boldsymbol{\lambda}) \geq \delta \right\}, \tag{16.11}$$

where we have used the relation $F(\rho) = f(\rho)/p(\rho)$. Therefore, the original complicated quantum optimization problem in Eq. (16.7) has been transformed into a classical optimization problem, which is much easier to solve.

Let us take the simple example with $\Omega = |\Psi\rangle\langle\Psi|$ to demonstrate the effectiveness of the above method. In this case we have $\boldsymbol{\lambda} := (1, 0, \ldots, 0)$. Then a straightforward calculation reveals that $\eta_{\mathbf{k}}(\boldsymbol{\lambda})$ and $\zeta_{\mathbf{k}}(\boldsymbol{\lambda})$ only depend on the value of k_1:

$$\eta_{\mathbf{k}} = p(\rho_{\mathbf{k}}) = \begin{cases} 1 & k_1 = N+1, \\ \frac{1}{N+1} & k_1 = N, \\ 0 & k_1 \leq N-1, \end{cases} \qquad \zeta_{\mathbf{k}} = f(\rho_{\mathbf{k}}) = \begin{cases} 1 & k_1 = N+1, \\ 0 & k_1 \leq N. \end{cases} \tag{16.12}$$

Using this relation, Eq. (16.11) can be further simplified to the following form:

$$\bar{\epsilon}(\Omega, N, \delta) = 1 - \min_{\substack{c_1, c_2 \geq 0 \\ c_1 + c_2 \leq 1}} \left\{ \frac{c_1}{c_1 + \frac{c_2}{N+1}} \,\middle|\, c_1 + \frac{c_2}{N+1} \geq \delta \right\} = \begin{cases} 1, & \delta \leq \frac{1}{N+1}, \\ \frac{\delta - 1}{N\delta}, & \delta > \frac{1}{N+1}. \end{cases} \tag{16.13}$$

Therefore, to achieve the soundness condition $\bar{\epsilon}(\Omega, N, \delta) \leq \epsilon$, it suffices to choose a sufficiently large N. Concretely, given $0 < \epsilon, \delta < 1$, the minimum number of tests required to verify the pure state $|\Psi\rangle$ within infidelity ϵ and significance level δ reads

$$N(\epsilon, \delta, \Omega) = \min\{N \geq 1 \,|\, \bar{\epsilon}(\Omega, N, \delta) \leq \epsilon\} = \left\lceil \frac{1-\delta}{\delta\epsilon} \right\rceil \approx \frac{1}{\delta\epsilon}. \tag{16.14}$$

Here the scaling behavior with δ is $O(\delta^{-1})$, which is not satisfactory although the strategy $\Omega = |\Psi\rangle\langle\Psi|$ is optimal in the nonadversarial scenario. Fortunately, as we shall see shortly, the scaling with δ can be improved to $O(\ln \delta^{-1})$ by slightly adjusting the strategy.

16.3.3 Verification with Homogeneous Strategies

In this subsection we consider QSV in the adversarial scenario with homogeneous strategies. A strategy Ω for $|\Psi\rangle$ is called homogeneous [21, 22] if it has the form:

$$\Omega = |\Psi\rangle\langle\Psi| + \lambda(\mathbb{I} - |\Psi\rangle\langle\Psi|), \qquad 0 \leq \lambda < 1. \tag{16.15}$$

Among all verification strategies with a given spectral gap, the homogeneous strategy stands out as the most efficient one, making it an ideal choice in QSV. So far, homogeneous strategies based on LOCC have been constructed for many important quantum states [8, 9, 14, 19, 21, 23, 24, 26, 28, 34]. If there are no restrictions on the measurements, the homogeneous strategy Ω can always be realized by performing the test $P = |\Psi\rangle\langle\Psi|$ with probability $1 - \lambda$ and the trivial test with probability λ. By "trivial test" we mean the test operator is equal to the identity operator.

In the previous subsection, the results for the case of $\lambda = 0$ have been presented [see Eqs. (16.13) and (16.14)]. In the following, we focus on homogeneous strategies Ω with $0 < \lambda < 1$, in which case Ω is called nonsingular. Now it is natural to replace Ω with λ in the notation of the guaranteed infidelity $\bar{\epsilon}(\Omega, N, \delta)$ and the minimum number of tests $N(\epsilon, \delta, \Omega)$, since the homogeneous strategy in Eq. (16.15) is determined by λ. Using the technique introduced in Sect. 16.3.2, we can deduce many useful properties and analytical formulas for $\bar{\epsilon}(\lambda, N, \delta)$ and $N(\epsilon, \delta, \lambda)$ [21, 22]. Here we only present their monotonicity properties and informative bounds for the sake of intuition.

Proposition 1 *$\bar{\epsilon}(\lambda, N, \delta)$ is nonincreasing in δ for $0 < \delta < 1$ and in N for $N \geq 1$. $N(\epsilon, \delta, \lambda)$ is nonincreasing in δ for $0 < \delta < 1$ and in ϵ for $0 < \epsilon < 1$.*

Let $k_+ := \lceil \log_\lambda \delta \rceil$ and $k_- := \lfloor \log_\lambda \delta \rfloor$. The following two theorems provide tight bounds for the guaranteed infidelity and the minimum number of tests:

Theorem 1 *Suppose $0 < \lambda < 1$ and $\lambda^N \leq \delta < 1$. Then we have*

$$\frac{k_+}{k_+ + (N - k_+)\lambda} \leq \bar{\epsilon}(\lambda, N, \delta) \leq \frac{k_-}{k_- + (N - k_-)\lambda}. \tag{16.16}$$

Theorem 2 *Suppose $0 < \epsilon, \delta, \lambda < 1$. Then we have*

$$k_- + \left\lceil \frac{k_- F}{\lambda\epsilon} \right\rceil \leq N(\epsilon, \delta, \lambda) \leq \left\lceil \frac{\ln \delta}{\lambda\epsilon \ln \lambda} - \frac{\nu k_-}{\lambda} \right\rceil, \tag{16.17}$$

where $F = 1 - \epsilon$, $\nu = 1 - \lambda$, and the bounds are saturated when $\log_\lambda \delta$ is an integer.

According to Theorem 2, given $0 < \lambda < 1$, the minimum number of tests required by a homogeneous strategy in the adversarial scenario only scales as $O(\epsilon^{-1} \ln \delta^{-1})$. The scaling behaviors with respect to δ and ϵ are the same as the counterpart in the nonadversarial scenario (cf. Eq. (16.2)), and are thus optimal.

Next, we investigate the influence of the parameter $0 < \lambda < 1$ on the number of tests. In the nonadversarial scenario, a smaller λ means a smaller test number among homogeneous strategies. However, the situation is very different in the adversarial scenario. Here, we are mostly interested in the high-precision limit, which means $\delta, \epsilon \to 0$. In this case, we have $k_- \approx \log_\lambda \delta$, so Theorem 2 implies that

$$N(\epsilon, \delta, \lambda) \approx \frac{\ln \delta}{\lambda\epsilon \ln \lambda} = \frac{1}{\lambda \ln \lambda^{-1}} \cdot \frac{\ln \delta^{-1}}{\epsilon}. \tag{16.18}$$

This implies that the verification efficiency is characterized by the factor $(\lambda \ln \lambda^{-1})^{-1}$, as reflected in Fig. 16.1. Let e be the base of the natural logarithm. Then the factor is strictly decreasing in λ when $0 < \lambda < 1/\mathrm{e}$, strictly increasing in λ when $1/\mathrm{e} < \lambda < 1$, and attains its minimum value e when $\lambda = 1/\mathrm{e}$. Therefore, the homogeneous strategy with $\lambda = 1/\mathrm{e}$ is optimal in the high-precision limit, in which case we have

$$N(\epsilon, \delta, \lambda = \mathrm{e}) \approx \mathrm{e}\,\epsilon^{-1} \ln \delta^{-1}, \tag{16.19}$$

which is comparable to the counterpart $\epsilon^{-1} \ln \delta^{-1}$ for the nonadversarial scenario.

The above discussions demonstrate that the optimal strategy in the adversarial scenario significantly differs from that in the nonadversarial scenario. As a consequence, entangled measurements, which are essential in the latter scenario, may no longer be needed in the former scenario. For instance, in the case of bipartite pure states and GHZ states, the optimal homogeneous strategy with $\lambda = 1/\mathrm{e}$ can be constructed using solely local projective measurements [8, 9, 19, 26, 28, 34].

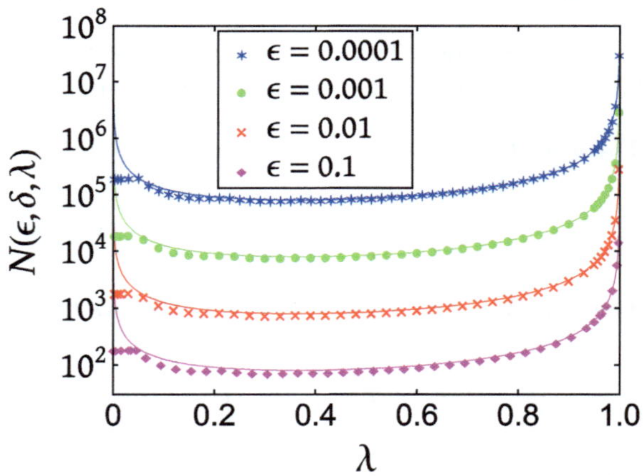

Fig. 16.1 Number of tests required to verify a pure state in the adversarial scenario within infidelity ϵ and significance level $\delta = 0.05$ using a homogeneous strategy characterized by λ. The curves correspond to the approximate formula in Eq. (16.18)

16.3.4 Verification with General Strategies

Many verification strategies proposed in the literature are not homogeneous. The performance of such a general strategy Ω in the adversarial scenario is mainly determined by its second largest eigenvalue λ (or, equivalently, $\nu = 1 - \lambda$) and the smallest eigenvalue ω [21, 22].

When $\omega = 0$, Ω is singular. In this case, the guaranteed infidelity can be bounded by $\bar{\epsilon}(\Omega, N, \delta) \le (1-\delta)/(N\nu\delta)$ [21, 22], which implies that

$$N(\epsilon, \delta, \Omega) = \min\{N \ge 1 \,|\, \bar{\epsilon}(\Omega, N, \delta) \le \epsilon\} \le \left\lceil \frac{1-\delta}{\nu\delta\epsilon} \right\rceil \approx \frac{1}{\nu\delta\epsilon}. \quad (16.20)$$

This is a generalization of Eq. (16.14). Here the scaling in δ is not good enough.

When $\omega > 0$, the strategy Ω is nonsingular. In this case, the following theorem provides informative bounds for the guaranteed infidelity and the number of tests [21, 22]:

Theorem 3 *Suppose* $0 < \delta, \epsilon < 1$ *and* Ω *is nonsingular, then we have*

$$\bar{\epsilon}(\Omega, N, \delta) \le \frac{h \ln(\omega\delta)^{-1}}{N + 1 + (\ln \lambda)^{-1} \ln(\omega\delta)^{-1} + h \ln(\omega\delta)^{-1}}, \quad (16.21)$$

$$\frac{hF \ln \delta^{-1}}{\epsilon} - \frac{1}{\lambda\epsilon} \le N(\epsilon, \delta, \Omega) \le \frac{h \ln(F\delta)^{-1}}{\epsilon}, \quad (16.22)$$

where $F = 1 - \epsilon$ *and* $h(\lambda, \omega) := (\min\{\lambda \ln \lambda^{-1}, \omega \ln \omega^{-1}\})^{-1}$.

According to Theorem 3, the test number required by a nonsingular strategy scales as $O(\epsilon^{-1} \ln \delta^{-1})$, which is better than that of a singular strategy, and achieves the same scaling behaviors with δ and ϵ as the counterpart for the nonadversarial scenario presented in Eq. (16.2). In the high-precision limit $\delta, \epsilon \to 0$, the upper and lower bounds in Eq. (16.22) are tight with respect to the relative deviation. So the test number can be approximated as follows:

$$N(\epsilon, \delta, \Omega) \approx \frac{h(\lambda, \omega) \ln \delta^{-1}}{\epsilon}. \quad (16.23)$$

Therefore, the efficiency of a nonsingular strategy in the high-precision limit is characterized by $h(\lambda, \omega)$. Notice that $h(\lambda, \omega)$ is not smaller than the factor $(\lambda \ln \lambda^{-1})^{-1}$, which characterizes the efficiency of a homogeneous strategy in the high-precision limit (cf. Eq. (16.18)). Hence, among all strategies with a given spectral gap, the homogeneous strategy is the most efficient one.

16.3.5 Recipe for Constructing Efficient Protocols

We have seen that the performance of a verification strategy Ω in the adversarial scenario is not good if the smallest eigenvalue ω of Ω is 0 or close to 0. Here we introduce a general recipe for solving this problem by using the trivial test. Based on a given strategy Ω, we can construct a new *hedged strategy* by performing the trivial test and Ω with probabilities p and $1 - p$, respectively. The resulting hedged verification operator has the following form:

$$\Omega_p = p\mathbb{I} + (1 - p)\Omega, \quad 0 \le p < 1. \tag{16.24}$$

The second largest and smallest eigenvalues of Ω_p read

$$\lambda_p = p + (1 - p)\lambda, \quad \omega_p = p + (1 - p)\omega, \tag{16.25}$$

where λ and ω are the second largest and smallest eigenvalues of Ω, respectively.

According to Eq. (16.22), to achieve target infidelity ϵ and significance level δ, the number of tests required by the strategy Ω_p is bounded by

$$N(\epsilon, \delta, \Omega_p) < \frac{h(p, \nu, \omega)\ln(F\delta)^{-1}}{\epsilon}, \tag{16.26}$$

where $F = 1 - \epsilon$, $\nu = 1 - \lambda$, $h(p, \nu, \omega) = \left(\min\{\lambda_p \ln \lambda_p^{-1}, \omega_p \ln \omega_p^{-1}\}\right)^{-1}$. To achieve a high efficiency, we need to cleverly choose the probability p such that the test number $N(\epsilon, \delta, \Omega_p)$ is as small as possible. The following theorem implies that $p = \nu/\mathrm{e}$ is a good choice to achieve this goal. This choice can be made without knowing the value of ω, while achieving a nearly optimal performance.

Theorem 4 *If $p = \nu/\mathrm{e}$, then*

$$N(\epsilon, \delta, \Omega_p) < \frac{h(\nu/\mathrm{e}, \nu, 0)\ln(F\delta)^{-1}}{\epsilon} \le \frac{\ln(F\delta)^{-1}}{(1 - \nu + \mathrm{e}^{-1}\nu^2)\nu\epsilon}. \tag{16.27}$$

Compared with the required number of tests $N_{\mathrm{NA}}(\epsilon, \delta, \Omega)$ in the nonadversarial scenario (cf. Eq. (16.2)), the overhead in the test number satisfies

$$\frac{N(\epsilon, \delta, \Omega_p)}{N_{\mathrm{NA}}(\epsilon, \delta, \Omega)} < \nu h(\nu/\mathrm{e}, \nu, 0)\frac{[\ln(1 - \nu\epsilon)^{-1}]\ln(F\delta)^{-1}}{\nu\epsilon \ln \delta^{-1}}. \tag{16.28}$$

Analysis shows that this upper bound increases monotonically with ϵ and ν, as illustrated in Fig. 16.2. In addition, when $\epsilon, \delta \le 0.1$, the bound is at most 3 [21, 22]. Therefore, any pure state can be verified efficiently in the adversarial scenario as long as it can be verified efficiently in the nonadversarial scenario. Moreover, since only a

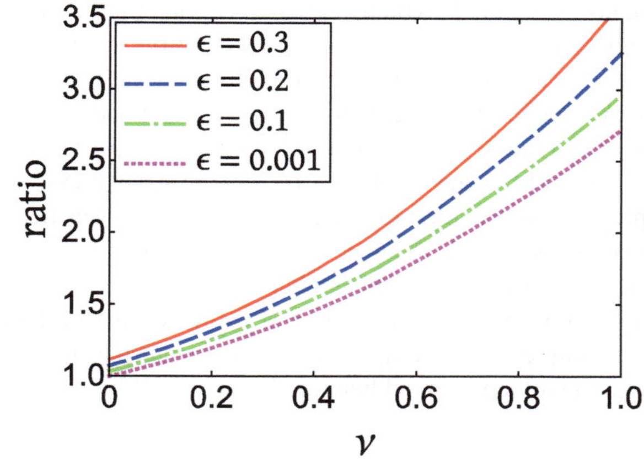

Fig. 16.2 Overhead of QSV in the adversarial scenario compared with the nonadversarial scenario. Each curve represents the upper bound in Eq. (16.28) for the ratio $N(\epsilon, \delta, \Omega_p)/N_{\mathrm{NA}}(\epsilon, \delta, \Omega)$, where significance level $\delta = 0.05$, $p = \nu/\mathrm{e}$, and ν is the spectral gap of Ω

trivial test is added, our recipe for constructing efficient protocols for the adversarial scenario requires the same measurement settings as required for the nonadversarial scenario.

16.4 Robust Quantum State Verification in the Adversarial Scenario

As mentioned in Sect. 16.3.1, a valid verification protocol in the adversarial scenario has to meet the basic requirements of completeness and soundness. The latter one is characterized by the target infidelity ϵ and significance level δ. Following [21, 22], we have introduced a general recipe for QSV in the adversarial scenario, whose sample cost $N = O(\epsilon^{-1} \ln \delta^{-1})$ achieves the optimal scaling behaviors in ϵ and δ.

However, this recipe does not consider the problem of *robustness*. To characterize the robustness of a protocol, we need to consider the situation in which the state preparation is controlled by honest Bob, who produces an i.i.d. quantum state ρ over the entire system $\mathcal{H}^{\otimes(N+1)}$, that is, ρ is a tensor power of the form $\rho = \tau^{\otimes(N+1)}$. Due to inevitable noise in state preparation, τ may not equal the target state $|\Psi\rangle\langle\Psi|$. Nevertheless, if the infidelity $\epsilon_\tau := 1 - \langle\Psi|\tau|\Psi\rangle$ is smaller than the target infidelity, i.e., $\epsilon_\tau < \epsilon$, then τ is still useful. To take robustness into account, such a state should be accepted with a high probability.

Recall that in the verification recipe introduced in the previous section, Alice accepts Bob's state only if all N tests are passed. However, this requirement would lead to poor robustness. Consider the homogeneous strategy Ω in Eq. (16.15) as an example. When Bob generates i.i.d. states τ with infidelity ϵ_τ, the probability that τ passes each test (associated with Ω) is $\mathrm{tr}(\Omega\tau) = 1 - \nu\epsilon_\tau$, where $\nu = 1 - \lambda$ denotes the spectral gap of the strategy. Consequently, the probability that Alice accepts τ

reads $(1-\nu\epsilon_\tau)^N$, which decreases exponentially with the test number N, and is not satisfactory. For high-precision verification, we have $\delta, \epsilon \ll 1$ and $N \gg 1$, so it is difficult to get τ accepted even if Bob is honest. Since the acceptance probability is very small, if Alice wants to accept τ at least once, she needs to repeat the verification protocol many times, which substantially increases the actual sample cost [36].

To address the limitation of lacking robustness, in this section, we extend the previous verification recipe to construct a new verification protocol based on homogeneous strategies. This new protocol allows state acceptance even when some tests fail, thereby significantly enhancing robustness. Moreover, our protocol can still achieve optimal scaling behaviors with respect to the target precision δ and ϵ, and the sample cost is comparable to the counterpart in the nonadversarial scenario. The main results presented here are based on [36].

16.4.1 Robust Verification Protocol and Guaranteed Infidelity

Our new protocol for verifying $|\Psi\rangle \in \mathcal{H}$ in the adversarial scenario runs as follows; see Fig. 16.3 for an illustration. The verification recipe introduced in the previous section can be viewed as a special case of this protocol with parameter $k = 0$.

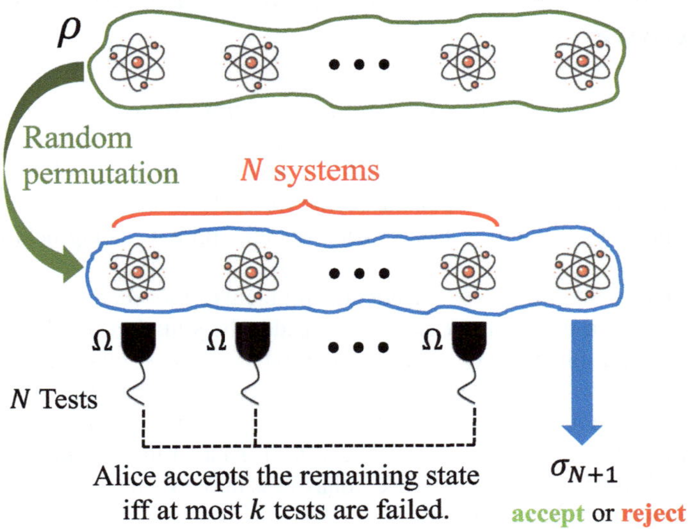

Fig. 16.3 Schematic view of our verification protocol. Here the state ρ generated by Bob might be arbitrarily correlated or entangled. To verify the target state, Alice first randomly permutes all $N+1$ systems, and then uses a strategy Ω to test each of the first N systems. Finally, she accepts the reduced state on the remaining system iff at least $N-k$ tests are passed

1. Bob produces a state ρ on the whole space $\mathcal{H}^{\otimes(N+1)}$ and sends it to Alice.
2. After receiving the state, Alice randomly permutes the $N+1$ systems of ρ, and applies the homogeneous strategy Ω given in Eq. (16.15) to test the first N systems.
3. Alice chooses an integer $0 \le k \le N-1$, called the number of allowed failures. If at most k failures are observed among the N tests, Alice accepts the reduced state σ_{N+1} on the remaining system; otherwise, she rejects it.

Since the state ρ has been randomly permuted before performing the tests, in the following discussions we can assume that ρ is permutation invariant without loss of generality. Similar to Eq. (16.3), the probability that Alice accepts ρ reads

$$p_k(\rho) = \sum_{i=0}^{k} \binom{N}{i} \operatorname{tr}\left(\left[\Omega^{\otimes(N-i)} \otimes \overline{\Omega}^{\otimes i} \otimes \mathbb{I}\right]\rho\right), \tag{16.29}$$

where $\overline{\Omega} := \mathbb{I} - \Omega$. Denote by σ_{N+1} the reduced state on the remaining system when at most k failures are observed. Then the fidelity between σ_{N+1} and the target state $|\Psi\rangle$ reads [assuming $p_k(\rho) > 0$]:

$$F_k(\rho) = \frac{1}{p_k(\rho)} \sum_{i=0}^{k} \binom{N}{i} \operatorname{tr}\left(\left[\Omega^{\otimes(N-i)} \otimes \overline{\Omega}^{\otimes i} \otimes |\Psi\rangle\langle\Psi|\right]\rho\right). \tag{16.30}$$

Similar to Eq. (16.7), the actual verification precision is characterized by the following guaranteed infidelity with $0 < \delta < 1$:

$$\bar{\epsilon}_\lambda(k, N, \delta) := 1 - \min_{\rho} \left\{ F_k(\rho) \mid p_k(\rho) \ge \delta \right\}, \tag{16.31}$$

where the minimization is taken over permutation-invariant states ρ on $\mathcal{H}^{\otimes(N+1)}$, and λ is determined by Eq. (16.15). By definition, if Alice accepts the state prepared by Bob, then she can guarantee (with significance level δ) that the reduced state σ_{N+1} has infidelity at most $\bar{\epsilon}_\lambda(k, N, \delta)$ with $|\Psi\rangle$. Consequently, she can ensure that

$$|\operatorname{tr}(E\sigma_{N+1}) - \langle\Psi|E|\Psi\rangle| \le \sqrt{\bar{\epsilon}_\lambda(k, N, \delta)} \tag{16.32}$$

for any POVM element $0 \le E \le \mathbb{I}$, according to the relation between the fidelity and trace norm. That is, the deviation of any measurement outcome probability from the ideal value is not larger than $\sqrt{\bar{\epsilon}_\lambda(k, N, \delta)}$.

According to the above discussions, the guaranteed infidelity $\bar{\epsilon}_\lambda(k, N, \delta)$ is of central importance to analyzing the soundness of our protocol. Its analytical formula is complicated, and is presented in [36, Supplementary Note 1]. Let $\mathbb{Z}^{\ge j}$ be the set of integers larger than or equal to j. The monotonicity properties of $\bar{\epsilon}_\lambda(k, N, \delta)$ are clarified by the following proposition:

Proposition 2 *Suppose* $0 \le \lambda < 1$, $0 < \delta < 1$, $k \in \mathbb{Z}^{\ge 0}$, *and* $N \in \mathbb{Z}^{\ge k+1}$. *Then* $\bar{\epsilon}_\lambda(k, N, \delta)$ *is nonincreasing in* δ *and* N, *but nondecreasing in* k.

16.4.2 Verification with a Fixed Error Rate

With the new verification protocol, when honest Bob generates i.i.d. quantum states τ on $\mathcal{H}$ with infidelity ϵ_τ, the probability that Alice accepts τ reads

$$p^{\mathrm{iid}}_{N,k}(\tau) = B_{N,k}\big(1 - \mathrm{tr}(\Omega\tau)\big) = B_{N,k}(\nu\epsilon_\tau), \tag{16.33}$$

where N is the number of tests, k is the number of allowed failures, and $B_{N,k}(p) := \sum_{j=0}^{k}\binom{N}{j}p^j(1-p)^{N-j}$ represents the binomial cumulative distribution function. If k is chosen to be sublinear in N, that is, $k = o(N)$, then $p^{\mathrm{iid}}_{N,k}(\tau)$ decays to zero as N increases, which is not satisfactory. To achieve robust verification, it is preferable to choose a large value of k, so that the acceptance probability is sufficiently high.

Here, we set the number k to be proportional to the test number, that is, $k = \lfloor s\nu N\rfloor$, where $0 \le s < 1$ is the error rate, and $\nu = 1 - \lambda$ is the spectral gap of the strategy Ω. In this case, if the state τ has infidelity $\epsilon_\tau < s$, then $p^{\mathrm{iid}}_{N,k}(\tau)$ converges to one as N increases, which is desired. In addition, we can deduce the following theorem [36].

Theorem 5 *Suppose* $0 < s, \lambda < 1$, $0 < \delta \le 1/4$, *and* $N \in \mathbb{Z}^{\ge 1}$. *Then*

$$s - \frac{1}{\nu N} < \bar{\epsilon}_\lambda(\lfloor \nu s N\rfloor, N, \delta) \le s + \frac{1}{\nu\lambda}\sqrt{\frac{s\ln\delta^{-1}}{N}} + \frac{\ln\delta^{-1}}{2\nu^2\lambda N} + \frac{2}{\lambda N}. \tag{16.34}$$

Theorem 5 implies that $\bar{\epsilon}_\lambda(\lfloor \nu s N\rfloor, N, \delta)$ converges to the error rate s as the number N of tests grows large, see Fig. 16.4 for an illustration. Therefore, to achieve the soundness condition with target infidelity ϵ and significance level δ, which means $\bar{\epsilon}_\lambda(\lfloor \nu s N\rfloor, N, \delta) \le \epsilon$, it suffices to set s to be less than ϵ, and choose a sufficiently large N. Within a fixed error rate $0 \le s < \epsilon$, the following theorem provides an informative bound for the number of tests required to achieve this goal [36].

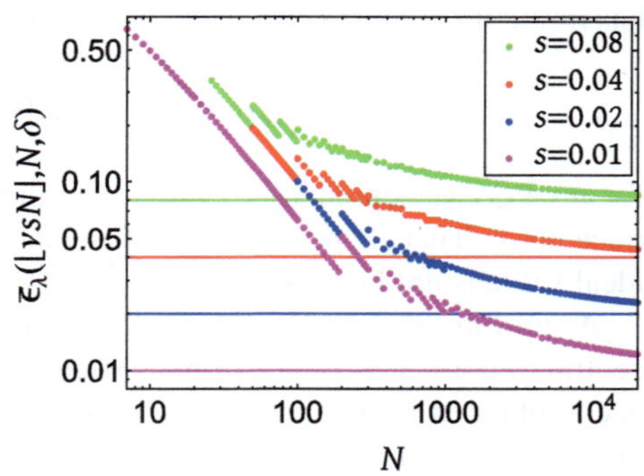

Fig. 16.4 Variations of the guaranteed infidelity with the test number N and error rate s. Here we set $\lambda = 1/2$ and significance level $\delta = 0.1$. The discrete points represent the exact values, and the horizontal lines represent several values of the error rate s

Theorem 6 *Suppose* $0 < \lambda < 1$, $0 \le s < \epsilon < 1$, *and* $0 < \delta \le 1/2$. *Then the inequality* $\bar{\epsilon}_\lambda(\lfloor \nu s N \rfloor, N, \delta) \le \epsilon$ *holds if the number* N *of tests satisfies*

$$N \ge \frac{\epsilon}{[\lambda\nu(\epsilon - s)]^2}\left(\ln \delta^{-1} + 4\lambda\nu^2\right). \tag{16.35}$$

In particular, if the ratio s/ϵ is a constant, then the resource overhead required by our protocol is only $O(\epsilon^{-1}\ln\delta^{-1})$. The scaling behaviors are the best we can expect.

16.4.3 The Number of Allowed Failures

Next, we consider the case in which the number N of tests is given. To construct a concrete verification protocol, we need to specify the number k of allowed failures so that the conditions of soundness and robustness are satisfied simultaneously. On the one hand, a small value of k is preferred to guarantee the soundness condition $\bar{\epsilon}_\lambda(k, N, \delta) \le \epsilon$ according to Proposition 2. On the other hand, a larger k is preferred to guarantee robustness, since the acceptance probability Eq. (16.33) in the i.i.d. case is increasing in k. Hence, we need to find a good balance between the two conflicting requirements to construct a robust and efficient verification protocol. The following proposition clarifies how large k can we choose to guarantee the soundness [36]:

Proposition 3 *Suppose* $0 < \lambda, \epsilon < 1$, $0 < \delta \le 1/4$, *and* $N, k \in \mathbb{Z}^{\ge 0}$. *If* $\nu\epsilon N \le k \le N - 1$, *then* $\bar{\epsilon}_\lambda(k, N, \delta) > \epsilon$. *If* $k \le l(\lambda, N, \epsilon, \delta)$, *then* $\bar{\epsilon}_\lambda(k, N, \delta) \le \epsilon$. *Here*

$$l(\lambda, N, \epsilon, \delta) := \left\lfloor \nu\epsilon N - \frac{\sqrt{N\epsilon \ln \delta^{-1}}}{\lambda} - \frac{\ln \delta^{-1}}{2\lambda\nu} - \frac{2\nu}{\lambda} \right\rfloor. \tag{16.36}$$

According to this proposition, the soundness condition can be achieved by setting the number of allowed failures $k = l(\lambda, N, \epsilon, \delta)$. Now we consider how much robustness can we achieve with this choice. Suppose honest Bob prepares i.i.d. quantum states τ on $\mathcal{H}$ with infidelity $0 < \epsilon_\tau < \epsilon$. Then in the large N limit, the probability that Alice accepts τ has the following asymptotic behavior [36]:

$$p^{\text{iid}}_{N,l}(\tau) = B_{N,l}(\nu\epsilon_\tau) = 1 - \exp\left[-D(\nu\epsilon \| \nu\epsilon_\tau)N + O(\sqrt{N})\right]. \tag{16.37}$$

Here $D(x\|y) := x\ln(\frac{x}{y}) + (1-x)\ln(\frac{1-x}{1-y})$ denotes the relative entropy, and l is a shorthand for $l(\lambda, N, \epsilon, \delta)$. As long as the number of tests N is suitably chosen, the probability of acceptance can be arbitrarily close to one, as illustrated in Fig. 16.5. Hence, our verification protocol is able to reach any degree of robustness.

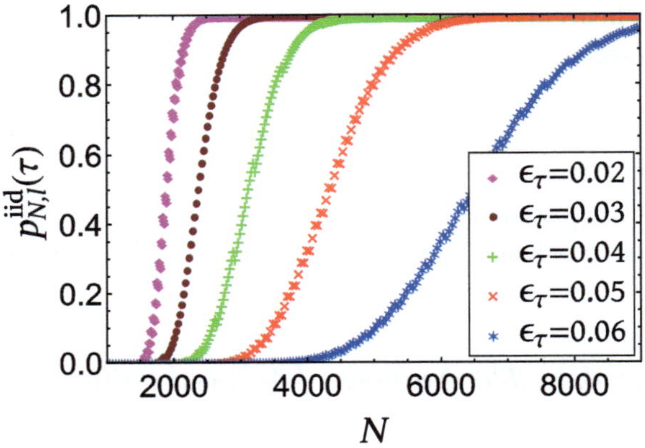

Fig. 16.5 The probability that Alice accepts i.i.d. quantum states τ on $\mathcal{H}$. Here we set $\lambda = 1/2$, infidelity $\epsilon = 0.1$, and significance level $\delta = 0.005$; ϵ_τ is the infidelity between τ and the target state $|\Psi\rangle$; and $l(\lambda, N, \epsilon, \delta)$ is the number of allowed failures defined in Eq. (16.36)

16.4.4 Sample Complexity of Robust Verification

Next, we study the sample cost required by our protocol to achieve a specified verification soundness and robustness. Let ρ be the state on $\mathcal{H}^{\otimes(N+1)}$ prepared by Bob, and let σ_{N+1} be the reduced state on the remaining system conditioned on the event that Alice accepts. To verify the target state within infidelity ϵ, significance level δ, and robustness $0 \le r < 1$, the following conditions should be satisfied:

1. **Soundness**: The probability that Alice accepts should be less than δ if the infidelity of σ_{N+1} with the target state is larger than ϵ.
2. **Robustness**: The probability that Alice accepts should be at least $1 - \delta$ if ρ is a tensor power of some quantum state τ on $\mathcal{H}$, which has infidelity $\epsilon_\tau \le r\epsilon$.

The tensor power structure in Condition 2 can be generalized: Instead of requiring identical copies of τ, we may consider independent states $\tau_1, \tau_2, \ldots, \tau_{N+1}$, each with infidelity at most $r\epsilon$. All our conclusions remain valid under this extension.

To satisfy the two conditions above, we must appropriately choose the number of tests N and the number k of allowed failures. To quantify the resource cost, we denote by $N_{\min}(\epsilon, \delta, \lambda, r)$ the minimum number of tests required for robust verification. It is defined as the smallest positive integer N for which there exists another integer $0 \le k \le N - 1$ such that the pair (N, k) fulfills the two conditions stated above. Note that the soundness and robustness conditions can be formally expressed as follows:

$$\bar{\epsilon}_\lambda(k, N, \delta) \le \epsilon, \qquad B_{N,k}(\nu r\epsilon) \ge 1 - \delta. \tag{16.38}$$

So the minimum test number can be written as follows:

$$N_{\min}(\epsilon, \delta, \lambda, r) := \min_{N,k} \left\{ N \,\middle|\, k, N \in \mathbb{Z}^{\ge 0}, k \le N - 1, \text{and Eq. (16.38) holds} \right\}. \tag{16.39}$$

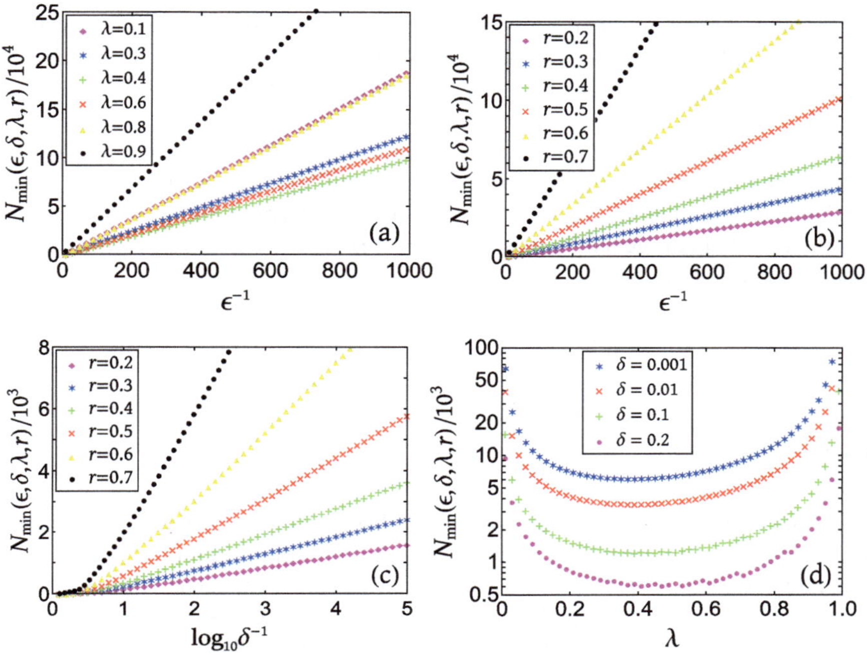

Fig. 16.6 Minimum number of tests $N_{\min}(\epsilon, \delta, \lambda, r)$ required for robust verification. **a** Variations of $N_{\min}$ with ϵ^{-1} and λ, where robustness $r = 1/2$ and significance level $\delta = 0.05$. **b** Variations of $N_{\min}$ with ϵ^{-1} and r, where $\delta = 0.05$ and $\lambda = 1/2$. **c** Variations of $N_{\min}$ with $\log_{10} \delta^{-1}$ and r, where $\lambda = 1/2$ and $\epsilon = 0.05$. **d** Variations of $N_{\min}$ with λ and δ, where $\epsilon = 0.05$ and $r = 1/2$

The value of $N_{\min}(\epsilon, \delta, \lambda, r)$ can be calculated numerically through an algorithm established in [36]. As shown in Fig. 16.6, when δ and r are fixed, $N_{\min}(\epsilon, \delta, \lambda, r)$ is inversely proportional to ϵ; when ϵ, r are fixed and δ approaches 0, $N_{\min}(\epsilon, \delta, \lambda, r)$ is proportional to $\ln \delta^{-1}$. In addition, Fig. 16.6d indicates that a homogeneous strategy Ω with small or large λ is not very efficient for robust verification, while any choice in the range $0.3 \leq \lambda \leq 0.5$ can achieve near-optimal performance.

The following theorem provides suitable choices for the number k of allowed failures and the test number N that can guarantee the conditions of soundness and robustness simultaneously [36].

Theorem 7 *Suppose $0 < \lambda, \epsilon < 1$, $0 < \delta \leq 1/2$, and $0 \leq r < 1$. Then the conditions of soundness and robustness in Eq.* (16.38) *hold as long as*

$$k = \left\lfloor \left(\frac{\lambda\sqrt{2\nu} + r}{\lambda\sqrt{2\nu} + 1} \right) \nu \epsilon N \right\rfloor, \qquad N \geq \left\lceil \left[\frac{\lambda\sqrt{2\nu} + 1}{\lambda\nu(1 - r)} \right]^2 \frac{\ln \delta^{-1} + 4\lambda\nu^2}{\epsilon} \right\rceil. \quad (16.40)$$

Notice that the lower bound for N in Eq. (16.40) also serves as an informative upper bound for $N_{\min}(\epsilon, \delta, \lambda, r)$ and thus clarifies the sample complexity of robust verification. As a consequence of Theorem 7, for any given degree of robustness $0 \leq r < 1$, only $O(\epsilon^{-1} \ln \delta^{-1})$ tests are required by our protocol, which still achieves the optimal scaling behaviors with respect to the target infidelity ϵ and significance level δ.

Numerical calculation shows that the bound for $N_{\min}(\epsilon, \delta, \lambda, r)$ in Eq. (16.40) is still a bit conservative, especially when r is small. That is, the actual sample cost is smaller than what can be theoretically guaranteed. Nevertheless, the bound is quite informative about the general trends. If we choose $r = 1/3$ and $\lambda = 1/2$, for example, then Theorem 7 implies that

$$N_{\min}(\epsilon, \delta, \lambda, r) \leq \lceil 81\, \epsilon^{-1}(\ln \delta^{-1} + 0.5) \rceil. \tag{16.41}$$

16.5 Discussion

Efficient verification of quantum states in the adversarial scenario is crucial to various applications that require high-security requirements. In this chapter, we reviewed our general QSV recipe [21, 22] for the adversarial scenario. The resource cost required by this recipe is comparable to the counterpart for the nonadversarial scenario. When homogeneous strategies are accessible, this recipe can be further extended to a robust and efficient QSV protocol [36]. For any given degree of robustness, the latter protocol achieves the optimal sample complexity with respect to the target infidelity and significance level. Compared with other QSV approaches, it can reduce the sample cost dramatically in a realistic scenario. Our robust protocol can readily be applied to verify many important quantum states in the adversarial scenario, including stabilizer states, graph states, and bipartite pure states, even if only local projective measurements are available.

There are some intriguing problems that merit further investigation. First, so far we have only considered the setting in which the measurement devices are controlled by the verifier and are trustworthy. It is also desirable to construct robust and efficient QSV protocols for the adversarial scenario with untrusted measurement devices. To this end, some device-independent (DI) verification protocols have been proposed [39–43]. To further reduce the overhead and improve the robustness, it might be helpful to combine our method with the *DI quantum state certification* approach developed in [50], which combines the idea of QSV and self-testing.

Second, although our QSV protocol is highly robust, it is not perfectly fault tolerant. For example, consider the case in which the target state is an n-qubit graph state $|G\rangle$, and the state preparation device produces i.i.d. state $[Z \otimes I^{\otimes(n-1)}]|G\rangle$, where only the first qubit of $|G\rangle$ undergoes a phase flip. Since this tiny local error can be easily corrected through quantum error correction, this state is expected to be accepted with high probability. However, our protocol cannot achieve this goal, since the produced state is orthogonal to the target state $|G\rangle$. In the future, by combining our robust QSV protocol with the verification approach in [13], we may achieve both high robustness and fault tolerance simultaneously in the adversarial scenario.

Acknowledgements Z.L. and H.Z. are supported by Shanghai Science and Technology Innovation Action Plan (Grant No. 24LZ1400200), Shanghai Municipal Science and Technology Major Project (Grant No. 2019SHZDZX01), and the National Key Research and Development Program of China (Grant No. 2022YFA1404204). M.H. is supported in part by the National Natural Science Foundation of China (Grant No. 62171212) and the General R&D Projects of 1+1+1 CUHK-CUHK(SZ)-GDST Joint Collaboration Fund (Grant No. GRDP2025-022).

References

1. Šupić I, Bowles J (2020) Self-testing of quantum systems: a review. Quantum 4:337
2. Eisert J et al (2020) Quantum certification and benchmarking. Nat Rev Phys 2:382–390
3. Carrasco J, Elben A, Kokail C, Kraus B, Zoller P (2021) Theoretical and experimental perspectives of quantum verification. PRX Quantum 2:010102
4. Kliesch M, Roth I (2021) Theory of quantum system certification. PRX Quantum 2:010201
5. Yu X-D, Shang J, Gühne O (2022) Statistical methods for quantum state verification and fidelity estimation. Adv Quantum Technol 5:2100126
6. Holevo AS (1982) Probabilistic and statistical aspects of quantum theory North-Holland, Amsterdam (Originally published in Russian, 1980)
7. Helstrom CW (1976) Quantum detection and estimation theory. Academic, New York
8. Hayashi M, Matsumoto K, Tsuda Y (2006) A study of LOCC-detection of a maximally entangled state using hypothesis testing. J Phys A: Math Gen 39:14427
9. Hayashi M (2009) Group theoretical study of LOCC-detection of maximally entangled state using hypothesis testing. New J Phys 11:043028
10. Hayashi M, Shi B-S, Tomita A, Matsumoto K, Tsuda Y, Jiang Y-K (2006) Hypothesis testing for an entangled state produced by spontaneous parametric down conversion. Phys Rev A 74:062321
11. Hayashi M, Tomita A, Matsumoto K (2008) Statistical analysis of testing of an entangled state based on the Poisson distribution framework. New J Phys 10:043029
12. Hayashi M, Morimae T (2015) Verifiable measurement-only blind quantum computing with stabilizer testing. Phys Rev Lett 115:220502
13. Fujii K, Hayashi M (2017) Verifiable fault tolerance in measurement-based quantum computation. Phys Rev A 96:030301(R)
14. Pallister S, Linden N, Montanaro A (2018) Optimal verification of entangled states with local measurements. Phys Rev Lett 120:170502
15. Morimae T, Takeuchi Y, Hayashi M (2017) Verification of hypergraph states. Phys Rev A 96:062321
16. Dangniam N, Han Y-G, Zhu H (2020) Optimal verification of stabilizer states. Phys Rev Res 2:043323
17. Takeuchi Y, Morimae T (2018) Verification of many-qubit states. Phys Rev X 8:021060
18. Takeuchi Y, Mantri A, Morimae T, Mizutani A, Fitzsimons JF (2019) Resource-efficient verification of quantum computing using Serfling's bound. npj Quantum Inf 5:27
19. Zhu H, Hayashi M (2019) Optimal verification and fidelity estimation of maximally entangled states. Phys Rev A 99:052346
20. Zhu H, Hayashi M (2019) Efficient verification of hypergraph states. Phys Rev Appl 12:054047
21. Zhu H, Hayashi M (2019) General framework for verifying pure quantum states in the adversarial scenario. Phys Rev A 100:062335
22. Zhu H, Hayashi M (2019) Efficient verification of pure quantum states in the adversarial scenario. Phys Rev Lett 123:260504
23. Li Z, Han Y-G, Zhu H (2019) Efficient verification of bipartite pure states. Phys Rev A 100:032316

24. Wang K, Hayashi M (2019) Optimal verification of two-qubit pure states. Phys Rev A 100:032315
25. Yu X-D, Shang J, Gühne O (2019) Optimal verification of general bipartite pure states. npj Quantum Inf 5:112
26. Li Z, Han Y-G, Zhu H (2020) Optimal verification of Greenberger-Horne-Zeilinger states. Phys Rev Appl 13:054002
27. Hayashi M, Koshiba T (2018) Secure modulo zero-sum randomness as cryptographic resource. Cryptology ePrint Archive, Paper 2018/802. https://eprint.iacr.org/2018/802
28. Li Z, Han Y-G, Sun H-F, Shang J, Zhu H (2021) Verification of phased Dicke states. Phys Rev A 103:022601
29. Liu Y-C, Yu X-D, Shang J, Zhu H, Zhang X (2019) Efficient verification of Dicke states. Phys Rev Appl 12:044020
30. Zhu H, Li Y, Chen T (2024) Efficient verification of ground states of frustration-free hamiltonians. Quantum 8:1221
31. Chen T, Li Y, Zhu H (2023) Efficient verification of Affleck-Kennedy-Lieb-Tasaki states. Phys Rev A 107:022616
32. Liu Y-C, Shang J, Zhang X (2021) Efficient verification of entangled continuous-variable quantum states with local measurements. Phys Rev Res 3:L042004
33. Wu Y-D, Bai G, Chiribella G, Liu N (2021) Efficient verification of continuous-variable quantum states and devices without assuming identical and independent operations. Phys Rev Lett 126:240503
34. Liu Y-C, Li Y, Shang J, Zhang X (2023) Verification of arbitrary entangled states with homogeneous local measurements. Adv Quantum Technol 6:2300083
35. Hayashi M, Takeuchi Y (2019) Verifying commuting quantum computations via fidelity estimation of weighted graph states. New J Phys 21:093060
36. Li Z, Zhu H, Hayashi M (2023) Robust and efficient verification of graph states in blind measurement-based quantum computation. npj Quantum Inf 9:115
37. Markham D, Krause A (2020) A simple protocol for certifying graph states and applications in quantum networks. Cryptography 4(1):3
38. Unnikrishnan A, Markham D (2022) Verification of graph states in an untrusted network. Phys Rev A 105:052420
39. Hayashi M, Koshiba T (2022) Quantum verifiable protocol for secure modulo zero-sum randomness. Quantum Inf Process 21:291
40. Hayashi M, Hajdušek M (2018) Self-guaranteed measurement-based blind quantum computation. Phys Rev A 97:052308
41. Bharti K, Ray M, Xu Z-P, Hayashi M, Kwek L-C, Cabello A (2022) Graph-theoretic approach for self-testing in Bell scenarios. PRX Quantum 3:030344
42. Mayers D, Yao A (2004) Self testing quantum apparatus. Quantum Inf Comput 4:273
43. McKague M, Yang TH, Scarani V (2012) Robust self-testing of the singlet. J Phys A 45:455304
44. Huang H-Y, Preskill J, Soleimanifar M (2024) Certifying almost all quantum states with few single-qubit measurements. In: 2024 IEEE 65th annual symposium on foundations of computer science (FOCS). arXiv:2404.07281
45. Flammia ST, Liu Y-K (2011) Direct fidelity estimation from few Pauli measurements. Phys Rev Lett 106:230501
46. da Silva MP, Landon-Cardinal O, Poulin D (2011) Practical characterization of quantum devices without tomography. Phys Rev Lett 107:210404
47. O'Donnell R, Wright J (2016) Efficient quantum tomography. In: Proceedings of the forty-eighth annual ACM symposium on theory of computing. ACM, New York
48. Gross D, Liu Y-K, Flammia ST, Becker S, Eisert J (2010) Quantum state tomography via compressed sensing. Phys Rev Lett 105:150401
49. Häffner H et al (2005) Scalable multiparticle entanglement of trapped ions. Nature 438:643
50. Gočanin A, Šupić I, Dakić B (2022) Sample-Efficient device-independent quantum state verification and certification. PRX Quantum 3:010317

51. Huang H-Y, Kueng R, Preskill J (2020) Predicting many properties of a quantum system from very few measurements. Nat Phys 16:1050
52. Mao C, Yi C, Zhu H (2025) Qudit shadow estimation based on the Clifford group and the power of a single magic gate. Phys Rev Lett 134:160801
53. Bertoni C, Haferkamp J, Hinsche M, Ioannou M, Eisert J, Pashayan H (2024) Shallow shadows: expectation estimation using low-depth random clifford circuits. Phys Rev Lett 133:020602

Chapter 17
Quantum Error Correction and Detection for Quantum Machine Learning

Eromanga Adermann, Haiyue Kang, Martin Sevior, and Muhammad Usman

Abstract At the intersection of quantum computing and machine learning, quantum machine learning (QML) is poised to revolutionize artificial intelligence. However, the vulnerability of the current generation of quantum computers to noise and computational error poses a significant barrier to this vision. While quantum error correction (QEC) offers a promising solution for almost any type of hardware noise, its application requires millions of qubits to encode even a simple logical algorithm, rendering it impractical in the near term. In this chapter, we examine strategies for integrating QEC and quantum error detection (QED) into QML under realistic resource constraints. We first quantify the resource demands of fully error-corrected QML and propose a partial QEC approach that reduces overhead while enabling error correction. We then demonstrate the application of a simple QED method, evaluating its impact on QML performance and highlighting challenges we have yet to overcome before we achieve fully fault-tolerant QML.

17.1 Introduction

Quantum machine learning (QML) [1–13] is an emerging field that promises performance improvements over classical counterparts, including training speed-ups, increased reliability, and superior feature extraction capabilities. In recent years, a variety of QML models [1–7, 14–36] have been developed and characterized for various datasets, with the most widely adopted model being the quantum variational classifier (QVC) [20]. The basic working principle of QVC-based QML is quite similar to classical machine learning in the sense that a loss function is iteratively

E. Adermann
Quantum Systems, CSIRO, Marsfield, NSW, Australia

H. Kang (✉) · M. Sevior · M. Usman
School of Physics, University of Melbourne, VIC, Parkville, Australia
e-mail: haiyuek@student.unimelb.edu.au

M. Usman
Quantum Systems, CSIRO, Clayton, VIC, Australia

M. Usman (ed.), *Quantum Robustness in Artificial Intelligence*, Quantum Science and Technology, https://doi.org/10.1007/978-3-032-11153-1_17

minimized, driving toward its global minimum, which is usually achieved in the form of fitting some highly non-trivial parametrized function.

However, the prevalence of noise in the current generation of quantum processors poses a significant challenge for practical-scale implementations of QML, as they lead to trainability issues arising from noise-induced barren plateaus [37] and performance degradations from noise accumulation in deep circuits. Quantum error correction (QEC) protocols are under development as a solution to hardware noise, but their extremely high spacetime overheads [38, 39], due primarily to magic state distillation, make them infeasible for near-term practical implementation.

As such, we are presented with the challenge of mitigating, detecting, and correcting errors that are bound to arise with implementations of QML on quantum hardware with physical resource limitations. To this end, we investigate practical strategies for integrating QEC into QML. We begin by detailing the resources required to realize full quantum error corrected quantum machine learning, to illustrate the extent of the challenge before us. We then explore a means of circumventing this challenge with the partial application of QEC, which allows for a significant reduction in spacetime overhead. We conclude this chapter by demonstrating the application of a simple error detection code to a QML algorithm and evaluating its effectiveness.

17.2 Resource Overhead for Quantum Error Correction

The implementation of quantum machine learning (QML) with quantum error correction (QEC) can address challenges associated with trainability and noise accumulation in deep quantum circuits. However, existing QEC protocols, such as surface codes, are expensive in terms of spacetime cost for QML architectures. Variational QML circuits rely on parametrized quantum gates, which are non-Clifford operations and typically require highly expensive logical T-gate implementation through magic state injection, which is based on the preparation of a logical magic state from a single physical qubit or short-distance qubit first [38]. Without the redundancy of multiple physical qubits, the logical T-gate error rate is close to the physical qubit error rate. Thus, one must mitigate the magic state through state distillation [38, 40, 41], yet this leads to an exponentially large spacetime overhead. To quantitatively address how large the cost is, we present an analysis of the spacetime overhead of executing a full 10-qubit quantum variational circuit with QEC.

To evaluate this overhead in a quantum neural network, we numerically estimate the scaling of the spacetime complexities of the full variational circuit with 10 qubits, estimated through the Azure quantum resource estimator [42, 43], which is based on Ref. [38], as shown in Table 17.1.

Paramatrised variational circuit

A parametrized variational circuit is one critical component of the workflow of quantum neural networks or variational algorithms. The circuit structure used for the spacetime overhead benchmark is described as follows:

First, the feature of the input data $\boldsymbol{x}_i$ is encoded into some quantum state $\rho_{\boldsymbol{x}_i} = C(\boldsymbol{x}_i)\,|0\rangle^{\otimes n} := |\psi(\boldsymbol{x}_i)\rangle$. For simplicity, here we ignore the cost of the embedding stage by assuming the circuit is insignificant compared to the inference stage.

Second, the state is evolved through a non-trivial unitary $U(\boldsymbol{\theta})$ with parameters $\boldsymbol{\theta} \in [-\pi, \pi]^D$, which is decomposed into a sequence of D unitaries with parameter θ_{lm} at layer l on qubit m

$$U(\boldsymbol{\theta}) = \prod_l W_l \prod_m e^{-i\theta_{lm}H_{lm}}, \tag{17.1}$$

where θ_{lm} and $H_{lm} = \hat{\boldsymbol{n}}_{lm} \cdot \boldsymbol{\sigma}$, $\boldsymbol{\sigma} = (X, Y, Z)^T$ are the training parameters for the single-qubit rotation gates $R_{lm} := e^{-i\theta_{lm}H_{lm}}$, and W_l is the entangling layer without parameters. For every R_{lm}, there are three degrees of freedom, which can be encoded into the Euler angles (α, β, γ),

$$R_{lm} = R_Z(\alpha_{lm})R_Y(\beta_{lm})R_Z(\gamma_{lm}). \tag{17.2}$$

The circuit design is illustrated in Fig. 17.1.

At the end of the circuit, to reduce the impact of ordinary barren plateaus [44], the system is measured with respect to the single-qubit Pauli-Z observables.

With the assistance of Microsoft Azure quantum resource estimator [42, 43] based on defect-based logical qubits introduced in Ref. [38], we simulate the total number of physical qubits and the runtime needed to implement a full variational circuit with

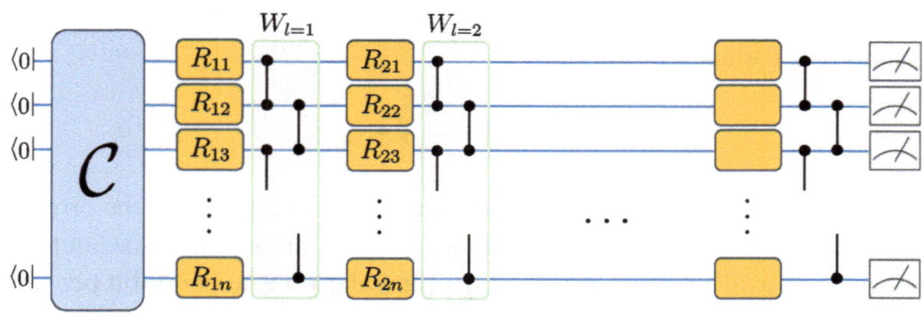

Fig. 17.1 The detailed circuit design of the variational circuit presented in this paper. The data is encoded into the circuit through amplitude encoding to minimize the requirement on the number of qubits and conserve memory for simulation. The parameterized unitary consists of multiple layers of unitaries, with each layer containing a sequence of single-qubit rotations with parameterized rotation axes and angles, denoted as R_{lm} as shorthand for $R_{mn_{lm}}(\theta_{lm})$, followed by a sequence of entangling Controlled-Z gates without trainable parameters. An error channel $\mathcal{E}$ is added after every ideal gate R_{lm}, constitutes $\tilde{R}_{lm}$. The inferred probabilities for each potential label are evaluated by measuring the Pauli-Z expected values on each of the qubits. This figure is adapted from Ref. [45]

Table 17.1 Summary on the cost of the full variational circuit with 10 algebraic qubits ($Q_{\text{alg}} = 10$) with different numbers of layers and logical qubits with or without distillation, evaluated by the Azure quantum resource estimator [42, 43]. For the situation that involves distillations, the figures are calculated based on a physical gate error rate of 10^{-3} for both the Clifford gates and T gates before distillation, while achieving a specific logical error rate for the whole variational circuit. For the situation without distillation, the cost of the T factory qubits is simply a number at the same level to the number of data qubits, since they share the same code distance, and there are a maximum of Q_{alg} logical $|T\rangle$ states to be injected in parallel. It is easy to see that the total overhead differs by almost 2 orders of magnitude between with and without distillation. To a large extent, the time cost is linear with respect to the number of QVC layers. This should still be the case without distillation, yet with a significantly smaller time cost per logical patch-cycle. This table is reproduced from Ref. [45] with permission

Error budget	QVC layers	logical qubit error rate (per cycle)	distilled T error rate
1×10^{-3}	50	3.00×10^{-10}	2.47×10^{-9}
1×10^{-3}	100	3.00×10^{-10}	2.47×10^{-9}
1×10^{-4}	50	3.00×10^{-11}	5.51×10^{-10}
1×10^{-4}	100	3.00×10^{-11}	5.51×10^{-10}

Spatial cost				Temporal cost		
Code distance	Data qubits	T factory qubits	T factory qubits (no distillation)	One logical cycle (μs)	No. logical cycles	Total runtime (ms)
15	13 500	1.746×10^6	$\sim 1.35 \times 10^4$	6.0	4 060	24
15	13 500	1.782×10^6	$\sim 1.35 \times 10^4$	6.0	8 410	50
17	17 340	1.746×10^6	$\sim 1.73 \times 10^4$	6.8	4 360	30
17	17 340	1.764×10^6	$\sim 1.73 \times 10^4$	6.8	8 710	59

or without distillations. Although this only demonstrates the behavior for defect-type surface codes, the logic still applies to other surface codes, such as the patch-type ones. Specifically, given a physical gate error rate of 10^{-3}, which includes both the physical Clifford gates and the raw magic state, the costs to achieve a full-circuit error budget of 10^{-3} and 10^{-4} are summarized as in Table 17.1. We note that the full-circuit error budget ϵ is equal to

$$\epsilon = \epsilon_{\text{log}} + \epsilon_{\text{dis}} + \epsilon_{\text{syn}}, \tag{17.3}$$

for contributions from logical, distillation, and synthesizing errors (the error of synthesizing rotation gates with arbitrary angles), which are equally distributed in our simulation. As a result, the required average logical error rate per qubit per stabilizer cycle ϵ_L or distilled T gate error rate ϵ_T must be such that it does not exceed its contribution to the error budget,

$$1 - (1 - \epsilon_L)^{N_L} \leq \epsilon_{\text{log}} \text{ or } 1 - (1 - \epsilon_T)^{N_T} \leq \epsilon_{\text{dis}}, \tag{17.4}$$

where N_T is the corresponding number of logical T gates, N_L total number of patch-cycles. In the limit where $\epsilon_L \ll 1$, Equation (17.4) approximates to

$$N_L \epsilon_L \leq \epsilon_{\log} \text{ or } N_T \epsilon_T \leq \epsilon_{\text{dis}}. \tag{17.5}$$

As a result, the maximum allowed logical qubit error rate per logical qubit per cycle is

$$\max_d \left(\epsilon_L = 0.03 \left(\frac{p}{p^*} \right)^{\frac{d+1}{2}} \right) \tag{17.6}$$

that also satisfies Eq. (17.4). We omit the next lower integer on $\lfloor \frac{d+1}{2} \rfloor$ as we always assume d is odd. p is the maximum physical error rate per syndrome round, $p^* = 0.01$ is the physical error rate threshold below which the error rate of the logical qubit is less than the error rate of the physical qubit, and 0.03 is the coefficient extracted numerically from simulations when fitting an exponential curve to model the relationship between the logical and physical error rate. We remark that, due to the discretized nature of the allowed logical error rate per qubit per cycle, the actual full-circuit failure probability is always less than the benchmarked error budget.

Through a careful analysis, we find that to simply execute a 10-logical qubit QVC circuit, millions of physical qubits are required in the T-state factory with code distance greater than 15 to achieve a reasonable full-circuit logical error rate, which is two orders of magnitude larger than without distillation. Specifically, for a 100-layer QVC, the cost skyrockets to 1.76×10^6 physical qubits ($6090d^2$ qubits with code distance 17) in order to achieve a reasonable whole-circuit error budget of 10^{-4} with a physical gate error rate of 10^{-3}. However, without distillation, it is expected that only $\sim 60d^2$ qubits are needed, which is comparable to the size of data qubits, for the T factory; therefore, maximally $\sim 120d^2$ qubits in total are necessary. However, the time cost should still scale linearly with respect to the number of layers in the circuit, but with a significantly smaller time cost per patch-cycle, which is the number of logical patches times the logical cycles. On the other hand, the time cost is largely a linear relationship with respect to the number of QVC layers. This should still be the case without distillation, yet with a significantly smaller time cost per logical patch-cycle. Overall, it is easy to see that the total overhead differs by almost two orders of magnitude with and without distillation.

17.3 Partial Quantum Error Correction

Since depolarizing noise maps to the image of the maximally mixed state, which loses all the quantum information in the long term, its mitigation cannot be accomplished through any self-adaptive mid-circuit manipulations such as dynamical decoupling [46], Pauli twirling [47], whereas other strategies, including zero-noise extrapolation [48], probabilistic error cancellation [49], usually require noise parameter characterizations of substantial overhead. Therefore, the necessity of quantum error correction for the quantum variational algorithm becomes apparent when dealing with the noise-induced barren plateaus of quantum neural network training.

However, we also want to avoid such an overhead of a full-scale QEC that restricts its implementation on near-term quantum processors as discussed in the previous subsection, but still preserving most of its robustness against the hardware noise and its induced barren plateaus. By taking a step back from a fully fault-tolerant surface code-based QEC, we propose a novel approach that relies on partial QEC. In our scheme, Clifford gates which includes Controlled-Zs are still kept to be error-corrected, but the T gates are left without any magic state distillation before their injection. By doing this, we can treat the logical variational circuit having its two-qubit gates, as long as they were induced from the Clifford group, with a negligible error rate. However, significant error rates remain for a general single-qubit unitary $U_{\hat{\boldsymbol{n}}}(\theta)$ after decomposing it into Clifford plus T gates with error rate $\sim 1-(1-\epsilon_T)^{\log_2(1/\epsilon)}$, where ϵ_T is the raw T gate error rate and the power $\log_2(1/\epsilon)$ is the number of T gates involved in achieving precision ϵ [50] of the rotation angle θ. An illustration of the physical implementation of partial-QEC is shown in Fig. 17.2.

Partial-QEC eliminates the major cost of QEC associated with non-Clifford operations, thereby reducing resource requirements from millions of qubits to merely a few

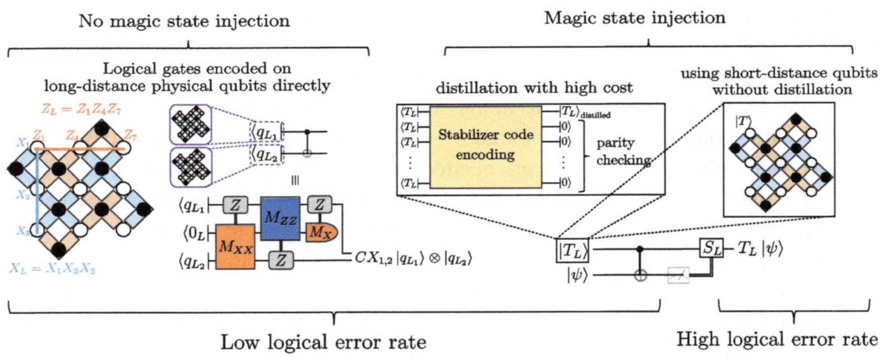

Fig. 17.2 Demonstration of how logical operators are implemented on patch-based surface codes introduced in Ref. [38]. White dots indicate data qubits, black dots indicate syndrome extraction qubits with orange and blue strips representing Pauli-X and Z stabilizers, respectively. For operators induced from the Clifford group, including X, Z, and CX gates, their logical operators can be encoded from many physical operators directly, without the need for ancilla qubits [38, 39]. For operators outside the Clifford group, such as the T gate, its logical operator T_L cannot be implemented directly, but must be teleported from an ancilla logical qubit in the state $|T_L\rangle = \frac{1}{\sqrt{2}}(|0_L\rangle + e^{i\pi/4}|1_L\rangle)$ via magic state injection. It turns out that $|T_L\rangle$ must be prepared from a single, physical qubit $|T\rangle$ state first, and then perform stabilizer measurements [39]. Or, one could choose to carry out magic state distillation with very high spacetime cost. **b** Without the redundancy of encoding one logical operator from multiple physical operators, the logical error rate for T gates is comparable to the physical T error rate. However, if the state is distilled properly, the logical gate error rate can be suppressed to the same level as other Clifford gates. This figure is adapted from Ref. [45]

thousand qubits. Our scheme is motivated by the fact that single-qubit parametrized gates are being trained, and therefore, any noise-induced disturbances will be overcome by the training process itself. This is partially inspired by classical machine learning studies, where it has recently been shown that neural networks can be trained regardless of noisy neurons [51, 52]. In later paragraphs, it demonstrates that QVC models can be trained with error-corrected Controlled-Zs and noisy single-qubit gates without encountering any trainability issues and without compromising final classification accuracy.

17.3.1 Trainability of QVC with Partial-QEC

To demonstrate the working of our scheme, we selected the 10-class MNIST dataset [53], which is a standard benchmark in the literature for classical and quantum machine learning models. We experimentally demonstrate the trainability of the QVC75 model (a model with 75 variational layers) for a single-qubit gate error rate comparable to that of state-of-the-art quantum computers, which is reinforced by gradient values above the shot noise. Our work can be generalized to deeper QVC models to further improve classification accuracies, as well as to novel architectures with provable trainability in the presence of hardware noise.

Assuming errors in a general parameterized single-qubit gate persist, and that two-qubit gates are error corrected with a negligible logical error rate, we introduce a depolarizing noise channel after every single-qubit unitary channel $\Lambda_i = R_{\hat{\boldsymbol{n}}_{lm}}(\theta_{lm}) \cdot R^{\dagger}_{\hat{\boldsymbol{n}}_{lm}}(\theta_{lm})$ on qubit i with strength p. The combined channel is expressed as $\tilde{\Lambda}_i := \mathcal{E}_i \circ \Lambda_i$, where

$$
\begin{aligned}
\mathcal{E}_i(\rho) &= (1-p)\rho + \frac{p}{3}\left(X_i\rho X_i + Y_i\rho Y_i + Z_i\rho Z_i\right), \\
\Lambda_i(\rho) &= U_i\rho U_i^{\dagger}
\end{aligned}
\tag{17.7}
$$

are the noise channel and unitary channel in Kraus representation, respectively. Since the noise channel acts on the logical space, p is interpreted as the combined logical X, Y, and Z-flip error rate. In practice, the channel would map the density operator after every decomposed gate of $R_{\hat{\boldsymbol{n}}_{lm}}$. However, since any unitary commutes with the depolarizing channel, it is convenient to treat it as in Eq. (17.7).

To evaluate the trainability of the variational circuit with only single-qubit noise, we used a learning rate of 0.005 and a total of 15000 images split into 50 images per batch in our training set, and 250 images in our test set, which are drawn and ordered randomly. The expectation values are evaluated on the Pauli-Z operators, and every circuit is executed with 10000 shots to simulate real hardware with non-zero shot noise and avoid potential saddle points [54].

As shown in Fig. 17.3a, the classification success rate attains a lower saturation more quickly when the depolarization strength increases. There is also a clear trend that both the gradients and the cost function landscape quickly flatten as the depo-

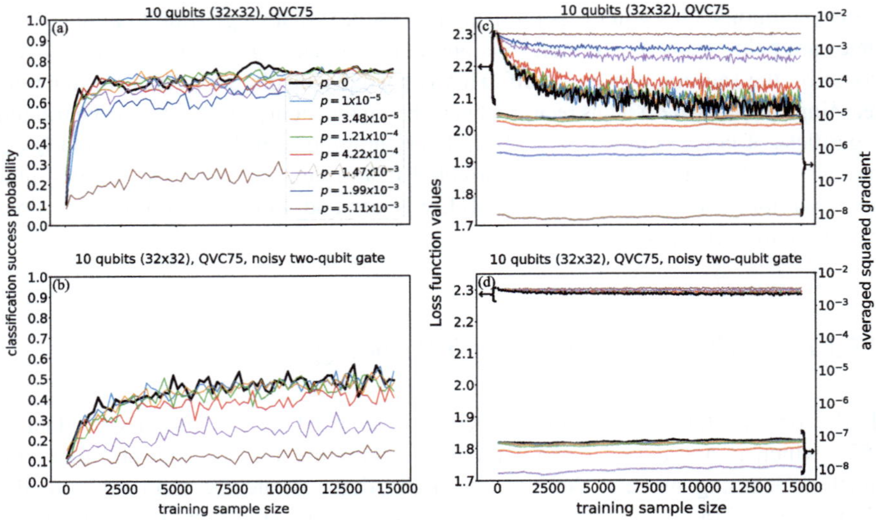

Fig. 17.3 The performance of the QVC in MNIST number classification problems with **a** 75 quantum layers but without a classical layer, and **b** 75 layers with a fully-connected classical layer and noisy two-qubit gate. The classification success rates inferred from the test datasets are plotted against the total number of images trained. The noise-free simulation is highlighted with black, thickened lines to contrast with noisy ones. Since we employ amplitude encoding, the number of pixels per image is exactly 2^n, which corresponds to the dimension of $\sqrt{2^n} \times \sqrt{2^n}$, where n is the number of qubits for the variational circuit. The legend denotes the strengths of the depolarizing channel from 0 to 5.11×10^{-3}. **c** The corresponding cost function values (left axis) and their average gradients squared (right axis), $\mathbb{E}_k\left((\nabla_{\theta_k}\mathcal{L})^2\right)$, of the trainable parameters (classical and quantum), where the arrows indicate the axis correspondence of the plots. A clear trend of flattening loss landscapes and vanishing gradients can be observed as the depolarizing strength increases. When the gradients drop to a scale around 10^{-8}, i.e., $p_{\text{depol}} = 5.11 \times 10^{-3}$, due to shot noise, the model becomes distinctively not trainable. **d** For the noisy two-qubit gate plot where the classical layer is included, the averaged cost function gradients with respect to all parameters, including both the quantum and classical layers, actually increase after being trained with more images. In contrast, the classical layer gradients by themselves are behaving as expected, as shown in the zoomed figure. This figure is reproduced from Ref. [60] with permission

larizing noise becomes stronger, as is evident in Figs. 17.3c and 17.4. These results agree with the prediction of [37], reinforcing the argument that depolarizing noise induces barren plateaus, which have vanishing gradients that scale exponentially with their strength.

Most importantly, as shown by the blue curves in Fig. 17.3a, c, despite shallower gradients and a smaller asymptotic success rate, the model is still trainable and above the shot noise at least for depolarizing strength of $p = 1.99 \times 10^{-3}$ with only a slight reduction in the classification success rate. This corresponds to a logical parametrized single-qubit gate $U_{\hat{\boldsymbol{n}}_{lm}}(\theta_{lm})$ with net error rate of 1.33×10^{-3}, a value can be achieved with underlying raw logical T gate error rate no more than $\epsilon_T = 10^{-4}$ by decomposing $U_{\hat{\boldsymbol{n}}_{lm}}(\theta_{lm})$ into Clifford + T.

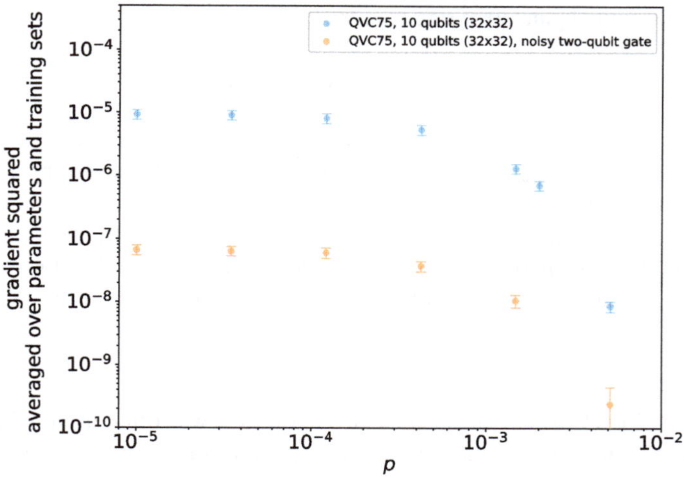

Fig. 17.4 The gradients squared averaged over all trainable parameters and all iterations in the training versus the depolarizing strength p in logarithmic scale. Error bar takes the standard deviation of the mean average among all iterations. This figure is reproduced from Ref. [60] with permission

Depolarizing noise strength and logical T gate error rate analysis
Given a depolarizing channel with a depolarizing noise strength p_{depol} added after an arbitrary single-qubit unitary U, the actual decay rate p is determined from the standard randomized benchmarking [55]. Specifically,

$$p := \int d\psi \operatorname{Tr}(|\psi\rangle\langle\psi| \tilde{U}^\dagger \circ \tilde{U}(|\psi\rangle\langle\psi|)) \tag{17.8}$$

where $\tilde{U}$ and $\tilde{U}^\dagger$ are the channels of the noisy U and $U^\dagger$ gates. Since the depolarizing channel commutes with any single-qubit unitaries, let $|\psi\rangle = V_\psi |0\rangle$, and absorb V_ψ into U as $W_\psi = UV_\psi$, we have

$$\begin{aligned} p &= \int d\psi \operatorname{Tr}\{V_\psi |0\rangle\langle 0| V_\psi^\dagger U^\dagger \mathcal{E}_{\text{depol}}^2 (UV_\psi |0\rangle\langle 0| V_\psi^\dagger U^\dagger) U\} \\ &= \int d\psi \operatorname{Tr}\{W_\psi |0\rangle\langle 0| W_\psi^\dagger \mathcal{E}_{\text{depol}}^2 (W_\psi |0\rangle\langle 0| W_\psi^\dagger)\} \\ &= \operatorname{Tr}\{|0\rangle\langle 0| \mathcal{E}_{\text{depol}}^2 (|0\rangle\langle 0|)\}. \end{aligned} \tag{17.9}$$

Substituting $\mathcal{E}_{\text{depol}}\left(\frac{I+Z}{2}\right) = \frac{I+\left(1-\frac{4p_{\text{depol}}}{3}\right)Z}{2}$ into Eq. (17.9), it is apparent that

$$p = \frac{1+\left(1-\frac{4p_{\text{depol}}}{3}\right)^2}{2}. \tag{17.10}$$

Therefore, the actual gate error rate is given by

$$r := 1 - p - (1 - p)/d, \tag{17.11}$$

where d is the dimension of the Hilbert space, for a single-qubit state $d = 2$. However, such an error rate for an arbitrary single-qubit unitary must be shared by a sequence of Clifford + T gates in the practice of error correction code, and the number of T gates required scales as $\sim 1.5 \log_2(1/\epsilon)$ for a precision of ϵ [56]. Therefore, based on the assumption of partial QEC that only considers the significant error rate for T gates, the relationship between the logical T gate error rate ϵ_T to the arbitrary rotation is given as

$$r \sim 1 - (1 - \epsilon_T)^{1.5 \log_2(1/\epsilon)}. \tag{17.12}$$

In literature [40, 41, 57], a physical error rate of 10^{-4} was often used, which arises from the expectation that it has a reasonably good error suppression factor, which scales as $O\left(\frac{p_{\text{threshold}}}{p}\right)$, and practical reachability. Since the partial QEC protocol would still require the physical gate error rate to be sufficiently below the threshold around 10^{-2} [38], at the time when full-scale QEC becomes feasible, a physical two-qubit error rate of 10^{-4} is fundamental and achievable, which is an order of magnitude lower than what has been achieved nowadays. According to the estimation in [58], which uses the SI1000 error model [59] by assuming that $p_2 = 10p_1$, we have $\frac{2}{3}p_2 \leq \epsilon_T \leq p_2$ for p_2 sufficiently small, where p_1, p_2 are the physical single and two-qubit gate error rates, respectively. Therefore, we use $\epsilon_T = 10^{-4}$ as the upper bound for $p_2 = 10^{-4}$. For a logical T gate with an raw error rate of $\epsilon_T = 10^{-4}$, the final U gate error rate can be derived as $r = 1.99 \times 10^{-3}$ with a small tolerance of $\epsilon = 10^{-4}$ in its rotation angle. This implies that the corresponding depolarizing channel strength is $p_{\text{depol}} = 2.99 \times 10^{-3}$.
In one of the latest literature [50], the authors improve the scaling of the number of T gates required to be $\sim \log_2(1/\epsilon)$ by introducing T_X gates that are $\pi/4$ rotations in X direction, but have the same cost as normal T gates. In this case, the actual gate error can be further relaxed to 1.33×10^{-3}, which corresponds to $p_{\text{depol}} = 1.99 \times 10^{-3}$. At this rate, we show that the model is still trainable in Fig. 17.3, where the classification rate is only slightly decreased.

Furthermore, as shown in Fig. 17.3b, d, we test the QVC model performance when the two-qubit gate noise is also turned on, which is aimed at simulating when the circuit is purely executed on physical qubits without error-correction, thereby demonstrating the necessity of partial QEC in balancing the model's trainability and overhead compared to no QEC. Similarly to the single-qubit gate noise channel, the two-qubit gate noise is added after the noise-free unitary: $\tilde{\Lambda}_i := \mathcal{E}_{(q_1,q_2)} \circ \Lambda_{(q_a,q_b)}$, where q_1, q_2 are the two qubits the gate is acting on. The overall noise channel $\mathcal{E}_{(q_1,q_2)}$ contains two subchannels: depolarizing channel $\mathcal{E}_{\text{depol}_i}$ on qubit i with the same structure as in Eq. (17.7), and crosstalk nearest-neighbor ZZ couplings $\mathcal{E}_{\text{crosstalk}_{(q_1,q_2)}}$ on qubits q_1 and q_2, where

$$\mathcal{E}_{\text{crosstalk}_{(q_1,q_2)}}[\bullet] = U_{Z_{q_1}Z_{q_2}} \bullet U^{\dagger}_{Z_{q_1}Z_{q_2}}. \tag{17.13}$$

The unitary $U_{Z_{q_1}Z_{q_2}} = \exp\left(-i\alpha Z_{q_1} Z_{q_2}\right)$ is simply a two-qubit unitary with generating Hamiltonian ZZ and angle of rotation 2α. Since multiple channels are involved, we describe each channel $\mathcal{E}_i := e^{\mathcal{L}_i}$ with Lindbladian generator $\mathcal{L}_i$ in the superoperator representation, where

$$\mathcal{E}_i = \sum_j K_{ij}^T \otimes K_{ij}, \tag{17.14}$$

K_{ij} are the Kraus operators of channel i. Therefore, the net Lindbladian

$$\begin{aligned} \mathcal{L} = {} & \mathcal{L}_{\text{depol}_{q_1}} + \mathcal{L}_{\text{depol}_{q_2}} \\ & + \mathcal{L}_{\text{crosstalk}_{(q_1-1,q_1)}} + \mathcal{L}_{\text{crosstalk}_{(q_2,q_2+1)}} \end{aligned} \tag{17.15}$$

is the sum of the Lindbladian generator of all noise channels for the master equation,

$$\frac{d\rho}{dt} = \mathcal{L}[\rho], \tag{17.16}$$

and have the solution $|\rho(t)\rangle\rangle = e^{\mathcal{L}t}|\rho(0)\rangle\rangle$, where $|\rho\rangle\rangle$ denotes the density operator in vectorized form. For a more realistic simulation, the depolarizing and crosstalk noise strengths are characterized by the IBM superconducting quantum computer *ibm_torino*.

When two-qubit gate noises are added back, the classification success rate is much worse. In all situations, the classification success rate never cross 0.5 and the average squared gradient $\mathbb{E}_k\left((\nabla_{\theta_k}\mathcal{L})^2\right)$ is below 10^{-7}. At the single-qubit depolarizing channel strength of $p = 1.47 \times 10^{-3}$, classification success rate barely increases and the model is almost non-trainable, whereas the model behaves properly for the case without the two-qubit gate noises. The comparison between noisy and noise-free two-qubit gates demonstrates the utility of partial QEC, which significantly improves the trainability of the machine learning model while still achieving a drastic overhead reduction by forgoing distillations.

17.4 Implementation of Quantum Error Detection with Quantum Machine Learning

In addition to the challenges associated with implementing QEC with QML on physical hardware, there are challenges associated with implementing QEC with QML in classical simulations due to the high computational overhead. For this reason, QML research has largely ignored the impact of noise, or focused on error mitigation rather than full quantum error correction. To address this research gap, we investigated the performance of a simple, parity-classifying QVC implemented with the [[4,2,2]] error-detecting stabilizer code within simulated noisy environments. To our knowledge, this work constitutes the first study on the implementation of a QML algorithm with a QEC code. Although our study was limited to error detection only, its results are generalisable to fully error-correcting codes and more complex QML algorithms.

17.4.1 Logical Encoding

We chose a very simple 2-qubit QVC for our experiments (displayed in Fig. 17.5), to minimize computational overhead. The QVC takes as input two qubits encoded in the basis encoding, and classifies their parity through measurement of the first qubit in the Z basis. The quantity measured is the expectation value over 1000 shots. We use only one rotational parameter, θ, to train the classifier, as any more than one leads to overfitting. The classifier is able to reach an accuracy of 1.0 within 100 training iterations.

We used the [4,2,2] 4-qubit Calderbank-Shor-Steane (CSS) stabilizer code for our error detection, as it is the simplest stabilizer code that protects against X and Z single-qubit errors [62]. It encodes two logical qubits using four physical qubits, and can only facilitate detection (not correction) of single-qubit errors. Errors are detected by taking measurements of ancilla qubits after applying stabilizers to the physical qubits, known as syndrome extraction.

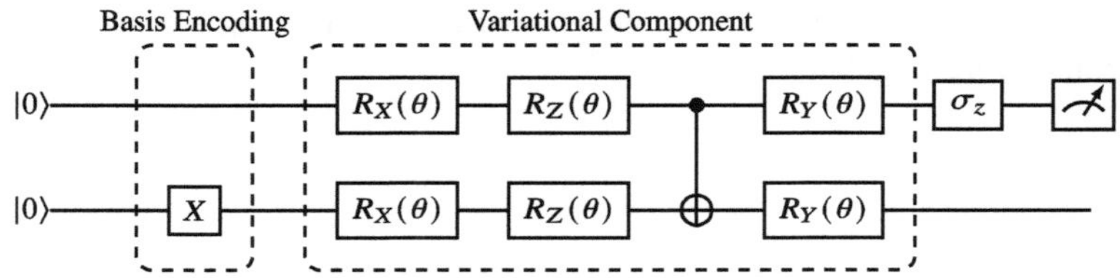

Fig. 17.5 The Variational Quantum Classifier (VQC) with an example input state of $|01\rangle$ and rotational parameter θ. This figure is adapted from Ref. [61]

We used the following mapping to encode two logical qubits with four physical qubits:

$$|00\rangle_L = \frac{1}{\sqrt{2}}(|0000\rangle + |1111\rangle) \quad (17.17)$$

$$|01\rangle_L = \frac{1}{\sqrt{2}}(|0011\rangle + |1100\rangle) \quad (17.18)$$

$$|10\rangle_L = \frac{1}{\sqrt{2}}(|0101\rangle + |1010\rangle) \quad (17.19)$$

$$|11\rangle_L = \frac{1}{\sqrt{2}}(|0110\rangle + |1001\rangle), \quad (17.20)$$

where the left-hand side represents the four different 2-logical qubit states (indicated by the subscript L), and the right-hand side represents the physical qubit states.

With these definitions, the CNOT gate in the QVC is logically encoded by a SWAP gate between the first two physical qubits. The logical encoding for the rotation gates require additional ancilla qubits (one ancilla per rotation gate to be encoded), where the ancilla qubits undergo the rotation instead of the physical qubits. This process generally requires a series of CNOT gates before and after the application of each rotation gate, which entangle the physical (and logical) qubits with the ancilla qubits.

In our logical encoding, the quantum state of the full qubit system (including ancillas), is always of the form:

$$|\Psi\rangle = \sum_{i=0}^{3} c_i \, |\psi_i\rangle_L \bigotimes_{j=0}^{n-1} |\phi_i\rangle_{a_j}, \quad (17.21)$$

where $|\psi_i\rangle_L$ represents each of the four possible logical basis states, c_i is the complex coefficient associated with the i-th logical basis state, $|\phi_i\rangle_{a_j}$ represents the state of the j-th ancilla qubit, a_j, associated with the i-th logical basis state, and n is the total number of ancillas that have been introduced into the system.

To illustrate, the full quantum state after the first two R_X rotations is given by:

$$\begin{aligned}|\Psi\rangle = &-i\mathrm{cs}\,|00\rangle_L\,|0\rangle_{a_1}\,|0\rangle_{a_2} + \mathrm{c}^2\,|01\rangle_L\,|0\rangle_{a_1}\,|1\rangle_{a_2} \\ &- \mathrm{s}^2\,|10\rangle_L\,|1\rangle_{a_1}\,|0\rangle_{a_2} - i\,\mathrm{sc}\,|11\rangle_L\,|1\rangle_{a_1}\,|1\rangle_{a_2}, \end{aligned} \quad (17.22)$$

where c and s are short-hand notations for $\cos(\theta)$ and $\sin(\theta)$, the logical states are subscripted with L and the ancilla states are subscripted with a_j. There are two ancilla states (a_1, a_2) invoked because two rotations have occurred. By the end of a logical operation, the ancilla qubits in each term reflect the logical state within the same term.

The steps we used to logically perform the double R_Y and R_X gates are outlined below:

1. **New ancilla initiation**: If no rotations have been performed in previous steps, initiate two new ancilla qubits to match the initial input logical state to the circuit. For example, if the initial input logical state is $|01\rangle_L$, then the two new ancillas, a_j and a_{j+1}, should respectively be in the states $|0\rangle$ and $|1\rangle$. If double rotations have been performed in previous steps, invoke the new ancilla qubits in the $|0\rangle$ state, then match the new ancilla states to the previous two ancilla states that were invoked to perform the last double rotation, by applying CNOT gates to the new ancilla qubits controlled by the two previous ancilla qubits.
2. **Change previous ancilla states**: If double rotations have been performed in previous steps, apply CNOT gates to each of the previously invoked ancilla qubits, controlled by the new ancilla qubits. For example, for newly invoked ancilla qubits a_j and a_{j+1}, we apply CNOT gates to any previous ancillas $a_{j-2}, a_{j-4}, \ldots, a_0$ and $a_{j-1}, a_{j-3}, \ldots, a_1$ controlled by a_j and a_{j+1} respectively.
3. **Change logical state**: Apply CNOT gates to physical qubits q_1 and q_3 controlled by newly initiated ancilla a_j, and CNOT gates to q_2 and q_3 controlled by newly initiated ancilla a_{j+1}.
4. **Apply rotation gate**: Apply the relevant rotation gates to each of the two ancillas.
5. **Undo logical state change:** Apply the same set of CNOT operations targeting the physical qubits and controlled by the two ancillas as was performed in Step 3.
6. **Undo previous ancilla state change**: Apply the same set of CNOT operations as in Step 2, targeting the ancillas invoked for previous rotations, and controlled by the newest ancilla states.

The steps for implementing the R_Z gates are much simpler and do not require as many CNOT gates between physical and ancilla qubits. We only match the newly introduced ancilla qubits to the previous two ancilla qubits, then apply the R_Z rotation gate to each new ancilla qubit. For the logical CNOT gate, we applied additional CNOT gates after the SWAP gate to ensure matching between the ancilla and logical states in each term of the full quantum state of the system.

The above steps can be used to logically implement single-qubit rotations, in which case we only need to introduce one ancilla qubit each time. Our approach can also be generalized to circuits with any number of rotation gates.

The logical circuit does not allow the direct Z basis measurement of the first logical qubit, so we conducted the logical equivalent by measuring the probability distribution across the 16 states spanned by the four physical qubits. Using these probabilities, we calculated the equivalent probability distribution for the four logical 2-qubit states, from which we determined the expectation value of the Z basis measurement of the first logical qubit.

17.4.2 Experimental Simulations

As stabilizer codes can only detect and correct combinations of X and Z errors, we only consider X, Y and Z errors for our noise models, in the form of probabilistic gate noise and depolarising environmental noise. This means that our noise models are inherently unable to capture the full range of noise and errors that might arise in physical NISQ systems. However, since our aim is to evaluate the effectiveness of the [[4,2,2]] stabilizer code in improving training outcomes, we only need to simulate noise that the code is theoretically capable of detecting.

We implement the gate noise model with single-qubit "error" gates applied after each single-qubit gate, where there is a probability (or Pauli Error Rate), given by p (with $0 < p < 1$), of either an X, Y or Z error occurring. Additionally, we apply single-qubit error gates after each 2-qubit gate, on the same qubits targeted by the 2-qubit gates, doubling the Pauli Error Rate compared to single-qubit gates, to better model the increased error rate for multi-qubit gates compared to single-qubit gates.

Our environmental noise model is a highly simplified model consisting of X,Y and Z errors. We apply Pauli errors at regular intervals to each physical and ancilla qubit at the same time. Applying noise at regular intervals mimics the cumulative build-up of errors in quantum circuits that occur as a result of environmental noise, where the specific regularity of the noise injections reflects the typical relaxation time and dephasing time of the system. Although this model does not include amplitude damping noise, dephasing noise, or the entire span of complex errors that could arise from realistic noise models, it is able to capture a range of alterations that may occur to the qubits as a result of energy loss to the system and decoherence, and is compatible with our choice of QEC code.

We ran simulations of the logically-encoded circuit under the gate noise and environmental noise models in a classical high-performance computing environment. The application of each stabilizer and syndrome extraction adds one extra qubit to the system, hence the resource overhead for the simulations ranged from 12 qubits to 20 qubits depending on the number of syndrome extraction rounds performed.

Since there are only four unique data points that can be used to train the VQC (namely, [0,0,0], [0,1,1], [1,0,1], and [1,1,0]), we duplicated the data to produce 24 samples for training the VQC. We used a batch size of 8 and ran the training for 100 iterations each, which was more than sufficient for convergence in a zero noise environment. Since the [4,2,2] stabilizer code can only detect errors, we discarded shots where at least one X or Z error was detected.

We applied the noise models to both ancilla and physical qubits in the system, but kept the syndrome extraction qubits noise-free. We chose a Pauli Error Rate ranging from 0.001 to 0.01 for both models, consistent with current NISQ device capabilities [63]. For the environmental noise model, we chose a Pauli error injection regularity of once every four gates, with the same Pauli Error Rates as used for the gate noise model.

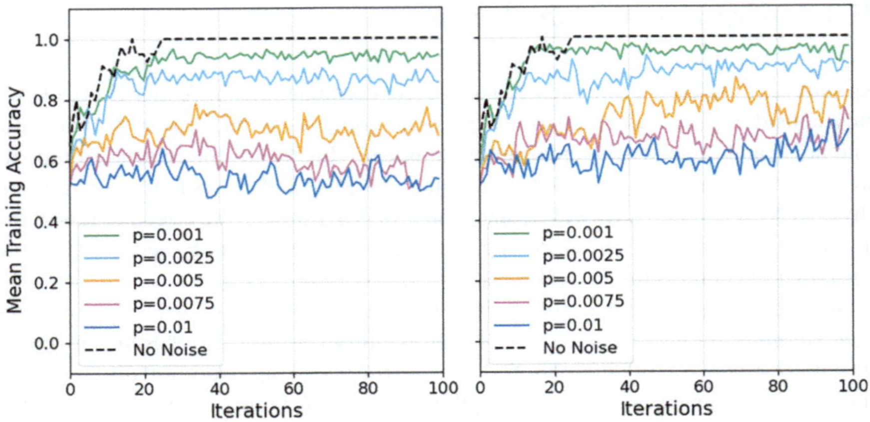

Fig. 17.6 Mean training accuracy of the logically-encoded QVC when training under different levels of gate (left) and environmental (right) noise, ranging from $p = 0.001 - 0.01$. The black dashed line indicates the accuracy obtained from training without noise. This figure is adapted from Ref. [61]

17.4.3 Impact of Noise on VQC Training and Performance

In Fig. 17.6, we show the evolution of the mean accuracy achieved during training under both noise models, with varying noise levels expressed as Pauli Error Rates and without error detection. The mean was calculated from 10 simulations with different starting seeds. As expected, higher noise levels produce lower mean final training accuracies. The QVC struggles to learn when $p \geq 0.005$, which is evident in how the training accuracy stays roughly constant throughout training. When noise levels satisfy $p \leq 0.0025$, learning appears hampered but not impossible.

Our results indicate that the simple QVC model is fairly susceptible to noise, which is likely due to its reliance on only two qubits to learn and make predictions. While more complex QML algorithms generally have greater learning capacities and may be more robust to noise and errors, they also require more qubits and gates, increasing the potential for errors. Consequently, a combination of error correction and mitigation is most likely needed to train and implement both simple and complex QML algorithms on NISQ devices (where noise levels can exceed $p = 0.005$).

17.4.4 Effectiveness of Error Detection on QVC Training and Performance

In Fig. 17.7, we display the effect of implementing the [4,2,2] stabilizer code with varying numbers of syndrome extraction rounds on the final training accuracy achieved after model convergence. In this and subsequent figures, we are interested

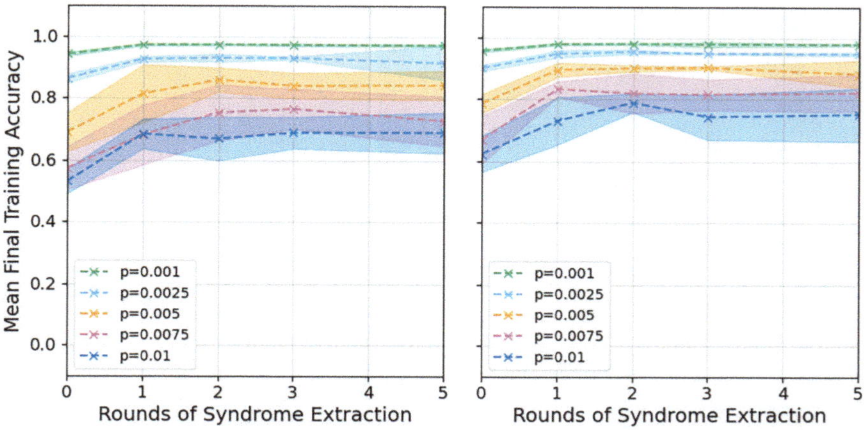

Fig. 17.7 Mean final training accuracy of the logically-encoded QVC under different levels of gate (left) and environmental (right) noise ranging from $p = 0.001 - 0.01$, and with different degrees of error detection implemented from 0 to 5 rounds of syndrome extractions. The mean final training accuracy is calculated from the training accuracies attained by the QVC over the final 40 iterations of training, after convergence. The standard deviation of the mean final training accuracy from 10 training runs is represented by the shaded regions. This figure is adapted from Ref. [61]

in the mean final training accuracy, which we calculated by taking the mean of the accuracies recorded over the last 40 iterations of training (as we can assume the training has stabilized by this point), and averaging this mean over 10 simulations of training. We also report the first standard deviation associated with this mean.

Under both noise models, we observe that for low noise (which we define as $p \leq 0.0025$), the training accuracy always improves with increasing number of syndrome extraction rounds and there is high consistency in the final training accuracy recorded across the 10 simulations. However, for higher levels of noise (specifically, $0.005 \leq p \leq 0.01$), the final training accuracy is quite variable (large spread in values), and occasionally worsens with more syndrome extractions.

There is also evidence of limitations in the effectiveness of error detection at low noise levels. We find that even for $p \leq 0.0025$, the increase in syndrome extraction rounds produces no significant increase in final training accuracy beyond two extraction rounds. The final accuracy does not reach 1.00 even at low noise levels and with five rounds of syndrome extractions.

These results suggest that there is noise in the logically-encoded circuit that syndrome measurements cannot detect. The only possible source of this noise is the ancilla qubits, which we do not apply any syndrome extractions to, but are entangled with the physical qubits via multiple CNOT gates.

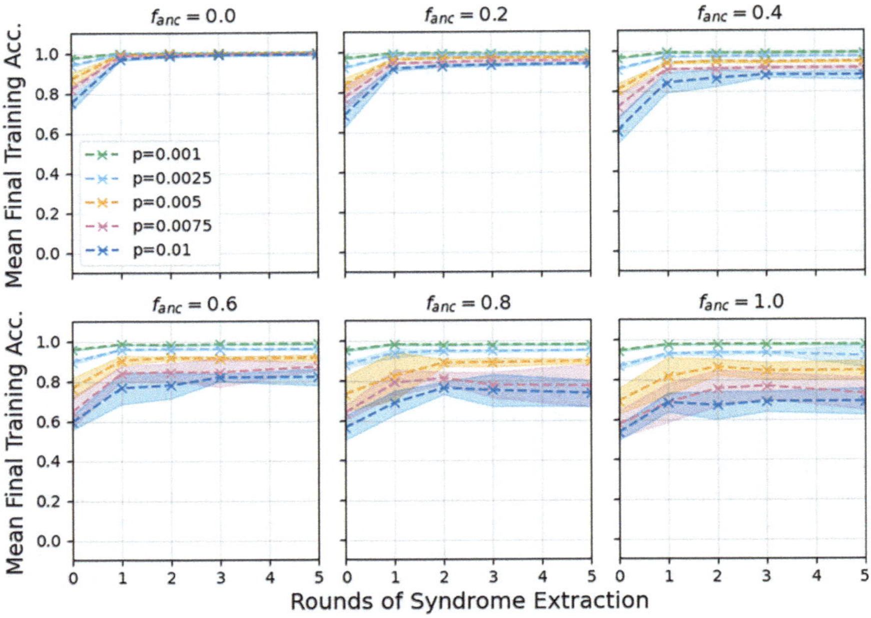

Fig. 17.8 Impact of ancilla qubit error rate on the mean final training accuracy, under the gate noise model with physical qubit error probability ranging from $p = 0.001 - 0.01$. Each subplot displays the variation in mean final training accuracy with the number of rounds of syndrome extraction, for a specific combination of ancilla and physical qubit error probabilities. The ancilla error rates are expressed as a fraction of the physical error rates, denoted by the fraction f_{anc}. The standard deviation in the meaning final training accuracies, each calculated from 10 training runs, is shown by the shaded regions. This figure is adapted from Ref. [61]

17.4.4.1 Ancilla Qubit Noise

To determine if the ancilla qubit errors are responsible for the limited effectiveness of the [4,2,2] stabilizer code in detecting errors, we trained the QVC under different levels of ancilla qubit noise. We show in Fig. 17.8 the evolution in the mean final training accuracy under the gate noise model, as the number of syndrome extractions increase, and with different levels of ancilla noise. We use f_{anc} to denote the fraction of the physical Pauli Error Rate that we apply to the ancilla qubits. For example, $f_{anc} = 0.5$ means that the ancilla Pauli Error Rate is half the physical Pauli Error Rate. As the results under the environmental noise model are very similar, we limit our analysis to the results obtained under the gate noise model only.

It is clear from the plot that as the ancilla error rate increases, the syndrome extractions become less effective and less reliable, confirming our earlier hypothesis. There is a greater spread in the final training accuracy, as well as an increase in non-monotonicity (which we suggest is a consequence of the greater variance), when the ancilla error rate is higher. Lower noise levels produce better training outcomes than higher noise levels for $f_{anc} \neq 0.0$; specifically there is a far smaller chance of exhibiting non-monotonicity in the final training accuracy as the number of syndrome extractions increase, and a far smaller variance in the final training accuracy. Notably, when there is no noise on the ancilla qubits, the error detection works as expected and even at the highest noise levels, the training accuracy reaches 1.0 within five rounds of syndrome extractions. As long as there is any noise on the ancilla qubits, the error detection loses effectiveness, and becomes unreliable at higher ancilla noise levels (i.e., $p_{anc} \geq 0.005$).

Thus, the high variability in final training accuracy at high ancilla error rates is very likely due to poor detection of ancilla errors. Since the ancilla qubits are entangled with the physical qubits, their errors spread easily into the physical qubits through the CNOT gates. While some of these errors will be detectable by syndrome measurements on the physical qubits, most of these errors will be undetectable. Hence, as the noise level increases, it becomes more difficult to protect the training through syndrome extractions, leading to the lower final accuracies and greater variance in its value.

The plateauing in the mean final training accuracies observed at lower levels of noise exists as long as there is any ancilla noise present. At higher levels of ancilla noise, the plateauing is hidden by the greater variance and non-monotonicity in the evolution of the final training accuracy. The plateau indicates that there is a limit for how many ancilla-caused errors can be detected and removed from the physical qubits by the stabilizer code for a given Pauli Error Rate, leaving only errors that syndrome measurements cannot detect. When that limit is reached, adding more syndrome extractions will not result in the detection of more errors, leading to the plateau.

17.4.4.2 Threshold Ancilla Error Rate

Our results for both noise models motivate the definition of a threshold Pauli Error Rate for ancilla qubits, such that when the error rate is larger than this threshold, error correction may not be effective. In particular, the addition of more syndrome extractions may not improve training, there is high variability in final training accuracy, and the maximum mean final training accuracy is considerably lower than 1.0 for the system.

In determining the threshold Pauli Error Rate for our system, we exclude ancilla error rates that produce a plateau at a final training accuracy of less than 0.90, even if high variability is not an issue. We deduce from our results a threshold Pauli Error Rate of $p = 0.003$ for the gate noise model, and $p = 0.004$ for the environmental depolarizing noise model. For comparison, the current lowest single-qubit gate error rate exhibited by a NISQ device is 0.15% [63], meaning that we may be able to run our simple QVC with the [4,2,2] code on the least noisy NISQ devices under special circumstances (for example, if the dominant noise is gate noise and environmental noise is very low). However, with the addition of real-world environmental noise to the gate noise, it is possible that the error threshold would be too low to run on currently available NISQ devices.

There is a parallel between the threshold we have defined and the Threshold Theorem for quantum error correction, which asserts that there is a critical error rate below which sufficiently good quantum error correction codes can successfully correct errors. For error rates above this threshold, errors accumulate too quickly for effective error correction. However, despite the parallels, the Threshold Theorem does not explicitly cover the phenomenon of ancilla errors spreading to the physical qubits and reducing the effectiveness of the error correcting code.

Though the threshold values we found are specific to our system and cannot be generalized to other systems, both the limit on the maximum training accuracy achievable and the existence of a threshold error rate for ancilla qubits should generalize to other combinations of QML algorithms and QEC codes. All QML algorithms contain rotation gates, and when implemented with QEC codes where ancilla qubits are needed for logically encoding rotation gates, thereby entangling the ancilla and physical qubits, we can expect error propagation between the ancilla and physical qubit registers. If the QEC code cannot correct the full range of errors that arise from such propagation, its effectiveness will be limited in noisy environments, especially above a threshold error rate. These results could also be relevant for implementing certain non-QML algorithms with QEC codes. This is because many QEC codes require ancilla qubits to logically encode rotation gates (and other non-Clifford gates), and since no known code can fully address the full spectrum of possible errors in practice, these implementations could also theoretically suffer from errors propagating from ancilla qubits to physical qubits.

Our findings highlight the need for additional considerations when applying techniques to achieve fault-tolerant implementations of QML algorithms. Fault-tolerant quantum machine learning may only be achievable with QEC if additional error mitigation techniques are used with it. For example, performing rotations with error mitigation instead of as logical operations (such as with zero-noise extrapolation [64] or dynamical gate error correction [65]), and using flag qubits normally applied to syndrome qubits [66, 67] in the ancilla register to minimize errors before they propagate, may help minimize uncorrectable errors.

17.5 Conclusion

In this chapter, we discussed the application of quantum machine learning, specifically the variational circuit, in the context of a noisy environment, where extensive simulation of a whole training process is lacking in most literature. Established on the theory of quantum error correction, we proposed a partial QEC protocol by foregoing magic state distillation. In this way, we balanced the conflicting needs of minimal mitigation overhead and effectiveness against depolarizing noise. The simulation reveals that to achieve a reasonable logical error rate, the spacetime overhead to execute a simple 10-qubit variational circuit is at least two orders of magnitude more expensive compared to using the raw magic states directly. Remarkably, we simulated the QVC model performance, particularly its trainability, under the setting of partial-QEC with single-qubit gate noise only. In the experiments for classifying the handwritten numbers of MNIST datasets, we presented clear evidence of the model's trainability with a depolarizing noise strength of $p = 1.47 \times 10^{-3}$, which corresponds to a gate error rate of 1.96×10^{-3}, a value higher than that of state-of-the-art quantum computers. Moreover, the poor performance when both single- and two-qubit noise was turned on further reinforces the necessity of partial quantum error correction instead of directly running on physical qubits.

Extending our investigation of QVC performance under noisy conditions, we directly incorporated quantum error detection into the training process and evaluated the resulting QVC accuracy. Specifically, we examined the performance of a simple parity-classifying 2-qubit QVC implemented with the [[4,2,2]] error-detecting code under Pauli noise. We demonstrated through classical simulations that even simple error detection codes can improve model performance. However, our findings suggest that the effectiveness of QEC is constrained by ancilla-induced error propagation, which sets a threshold on achievable QVC accuracy.

Given these limitations and the resource demands of implementing QEC with QML, it is clear that a purely QEC-based approach to fault-tolerant QML is not feasible. To minimize resource usage and maximize QEC effectiveness, we require a hybrid approach combining QEC, algorithmic design, and error mitigation strategies for the large-scale, practical implementation of QML on noisy quantum hardware.

Acknowledgements The authors greatly acknowledge the computational resources provided by the National Computing Infrastructure (NCI) through the National Computational Merit Allocation Scheme (NCMAS), the University of Melbourne's Research Computing Services, the Petascale Campus Initiative, and the Commonwealth Scientific and Industrial Research Organisation (CSIRO) High Performance Computing cluster. H.K. was supported by the Australian Government Research Training Program Scholarship, and E.A. was supported by the CSIRO Impossible Without You program.

References

1. Biamonte J, Wittek P, Pancotti N, Rebentrost P, Wiebe N, Lloyd S (2017) Quantum machine learning. Nature 549(7671):195–202
2. Schuld M, Sinayskiy I, Petruccione F (2015) An introduction to quantum machine learning. Contemp Phys 56(2):172–185
3. McClean JR, Romero J, Babbush R, Aspuru-Guzik A (2016) The theory of variational hybrid quantum-classical algorithms. New J Phys 18(2):023023
4. Peruzzo A, McClean J, Shadbolt P, Yung M-H, Zhou X-Q, Love PJ, Aspuru-Guzik A, O'brien JL (2014) A variational eigenvalue solver on a photonic quantum processor. Nat Commun 5(1):4213
5. Farhi E, Goldstone J, Gutmann S (2014) A quantum approximate optimization algorithm. arXiv:1411.4028
6. Lloyd S, Mohseni M, Rebentrost P (2014) Quantum principal component analysis. Nat Phys 10(9):631–633

7. Schuld M, Sinayskiy I, Petruccione F (2014) The quest for a quantum neural network. Quantum Inf Process 13:2567–2586
8. Wang Z, Wang F, Li L, Wang Z, van der Laan T, Leon RCC, Huang J-K, Usman M, Quantum kernel learning for small dataset modeling in semiconductor fabrication: application to ohmic contact. Adv Sci e06213
9. Tsang SL, West MT, Erfani SM, Usman M (2023) Hybrid quantum–classical generative adversarial network for high-resolution image generation. IEEE Trans Quantum Eng 4:1–19
10. Wang Z, van der Laan T, Usman M (2025) Self-adaptive quantum kernel principal component analysis for compact readout of chemiresistive sensor arrays. Adv Sci 12(15):2411573
11. Gill SS, Kumar A, Singh H, Singh M, Kaur K, Usman M, Buyya R (2022) Quantum computing: a taxonomy, systematic review and future directions. Softw: Pract Exp 52(1):66–114
12. Khatun A, Usman M (2025) Classical autoencoder distillation of quantum adversarial manipulations. Phy Rev Res 7:L042054
13. Khatun A, Usman M (2024) Quantum transfer learning with adversarial robustness for classification of high-resolution image datasets. Advanced Quantum Technologies 8:2400268
14. Gujju Y, Matsuo A, Raymond R (2024) Quantum machine learning on near-term quantum devices: current state of supervised and unsupervised techniques for real-world applications. Phys Rev Appl 21(6):067001
15. Wang D, Higgott O, Brierley S (2019) Accelerated variational quantum eigensolver. Phys Rev Lett 122(14):140504
16. Tilly J, Chen H, Cao S, Picozzi D, Setia K, Li Y, Grant E, Wossnig L, Rungger I, Booth GH et al (2022) The variational quantum eigensolver: a review of methods and best practices. Phys Rep 986:1–128, 2022
17. Cerezo M, Sharma K, Arrasmith A, Coles PJ (2022) Variational quantum state eigensolver. npj Quantum Inf 8(1):113
18. Beer K, Bondarenko D, Farrelly T, Osborne TJ, Salzmann R, Scheiermann D, Wolf R (2020) Training deep quantum neural networks. Nat Commun 11(1):808
19. Havlíček V, Córcoles AD, Temme K, Harrow AW, Kandala A, Chow JM, Gambetta JM (2019) Supervised learning with quantum-enhanced feature spaces. Nature 567(7747):209–212
20. Mitarai K, Negoro M, Kitagawa M, Fujii K (2018) Quantum circuit learning. Phys Rev A 98(3):032309
21. Farhi E, Neven H (2018) Classification with quantum neural networks on near term processors. arXiv:1802.06002
22. Lloyd S, Weedbrook C (2018) Quantum generative adversarial learning. Phys Rev Lett 121(4):040502
23. Dallaire-Demers P-L, Killoran N (2018) Quantum generative adversarial networks. Phys Rev A 98(1):012324
24. Huggins W, Patil P, Mitchell B, Whaley KB, Stoudenmire EM (2019) Towards quantum machine learning with tensor networks. Quantum Sci Technol 4(2):024001
25. McArdle S, Endo S, Aspuru-Guzik A, Benjamin SC, Yuan X (2020) Quantum computational chemistry. Rev Mod Phys 92(1):015003
26. Bermejo P, Braccia P, Rudolph MS, Holmes Z, Cincio L, Cerezo M (2024) Quantum convolutional neural networks are (effectively) classically simulable. arXiv:2408.12739
27. West MT, Sevior M, Usman M (2023) Reflection equivariant quantum neural networks for enhanced image classification. Mach Learn: Sci Technol 4(3):035027
28. West MT, Heredge J, Sevior M, Usman M (2024) Provably trainable rotationally equivariant quantum machine learning. PRX Quantum 5(3):030320
29. Heredge J, West M, Hollenberg L, Sevior M (2025) Nonunitary quantum machine learning. Phys Rev Appl 23(4):044046
30. Wang Z, van der Laan T, Usman M (2024) Quantum kernel principal components analysis for compact readout of chemiresistive sensor arrays. arXiv:2409.00115
31. McClean JR, Boixo S, Smelyanskiy VN, Babbush R, Neven H (2018) Barren plateaus in quantum neural network training landscapes. Nat Commun 9(1):4812

32. West MT, Tsang S-L, Low JS, Hill CD, Leckie C, Hollenberg LCL, Erfani SM, Usman M (2023) Towards quantum enhanced adversarial robustness in machine learning. Nat Mach Intell 5(6):581–589
33. West MT, Erfani SM, Leckie C, Sevior M, Hollenberg LCL, Usman M (2023) Benchmarking adversarially robust quantum machine learning at scale. Phys Rev Res 5(2):023186
34. Parrish RM, Hohenstein EG, McMahon PL, Martínez TJ (2019) Quantum computation of electronic transitions using a variational quantum eigensolver. Phys Rev Lett 122(23):230401
35. Olivera-Atencio ML, Lamata L, Morillo M, Casado-Pascual J (2023) Quantum reinforcement learning in the presence of thermal dissipation. Phys Rev E 108(1):014128
36. Olivera-Atencio ML, Lamata L, Casado-Pascual J (2023) Benefits of open quantum systems for quantum machine learning. Adv Quantum Technol, p 2300247
37. Wang S, Fontana E, Cerezo M, Sharma K, Sone A, Cincio L, Coles PJ (2021) Noise-induced barren plateaus in variational quantum algorithms. Nat commun 12(1):6961
38. Fowler AG, Mariantoni M, Martinis JM, Cleland AN (2012) Surface codes: towards practical large-scale quantum computation. Phys Rev A—Atom Mol Opt Phys 86(3):032324
39. Fowler AG, Gidney C (2018) Low overhead quantum computation using lattice surgery. arXiv:1808.06709
40. Litinski D (2019) A game of surface codes: large-scale quantum computing with lattice surgery. Quantum 3:128
41. Litinski D (2019) Magic state distillation: not as costly as you think. Quantum 3:205
42. https://github.com/microsoft/qsharp/tree/main/resource_estimator
43. Beverland ME, Murali P, Troyer M, Svore KM, Hoefler T, Kliuchnikov V, Low GH, Soeken M, Sundaram A, Vaschillo A (2022) Assessing requirements to scale to practical quantum advantage. arXiv:2211.07629
44. Cerezo M, Sone A, Volkoff T, Cincio L, Coles PJ (2021) Cost function dependent barren plateaus in shallow parametrized quantum circuits. Nat Commun 12(1):1791
45. Kang H, Kim Y, Adermann E, Sevior M, Usman M (2025) Almost fault–tolerant quantum machine learning with drastic overhead reduction. Quantum Sci Technol. arXiv:2507.18954
46. Uhrig GS (2007) Keeping a quantum bit alive by optimized π-pulse sequences. Phys Rev Lett 98(10):100504
47. Wallman JJ, Emerson J (2016) Noise tailoring for scalable quantum computation via randomized compiling. Phys Rev A 94(5):052325
48. Giurgica-Tiron T, Hindy Y, LaRose R, Mari A, Zeng WJ (2020) Digital zero noise extrapolation for quantum error mitigation. In: 2020 IEEE international conference on quantum computing and engineering (QCE), pp 306–316. IEEE
49. Van Den Berg E, Minev ZK, Kandala A, Temme K (2023) Probabilistic error cancellation with sparse pauli–lindblad models on noisy quantum processors. Nat Phys 19(8):1116–1121
50. Gidney C (2025) How to factor 2048 bit RSA integers with less than a million noisy qubits. arXiv:2505.15917
51. Audhkhasi K, Osoba O, Kosko B (2016) Noise-enhanced convolutional neural networks. Neural Netw 78:15–23
52. Jin C, Netrapalli P, Ge R, Kakade SM, Jordan MI (2021) On nonconvex optimization for machine learning: gradients, stochasticity, and saddle points. J ACM (JACM) 68(2):1–29
53. Deng L (2012) The mnist database of handwritten digit images for machine learning research [best of the web]. IEEE Signal Process Mag 29(6):141–142
54. Liu J, Wilde F, Mele AA, Jin X, Jiang L, Eisert J (2025) Stochastic noise can be helpful for variational quantum algorithms. Phys Rev A 111(5):052441
55. Magesan E, Gambetta JM, Emerson J (2011) Scalable and robust randomized benchmarking of quantum processes. Phys Rev Lett 106(18):180504
56. Kliuchnikov V, Lauter K, Minko R, Paetznick A, Petit C (2023) Shorter quantum circuits via single-qubit gate approximation. Quantum 7:1208
57. Chamberland C, Noh K (2020) Very low overhead fault-tolerant magic state preparation using redundant ancilla encoding and flag qubits. npj Quantum Inf 6(1):91

58. Li Y (2015) A magic state's fidelity can be superior to the operations that created it. New J Phys 17(2):023037
59. Gidney C, Newman M, Fowler A, Broughton M (2021) A fault-tolerant honeycomb memory. Quantum 5:605
60. Kang H, Harper B, Usman M, Sevior M (2025) Time-adaptive single-shot crosstalk detector on superconducting quantum computer. arXiv:2502.14225
61. Adermann E, Suzuki H, Usman M (2025) Variational quantum machine learning with quantum error detection, Apr 2025. arXiv:2504.06775
62. Vaidman L, Goldenberg L, Wiesner S (1996) Error prevention scheme with four particles. Phys Rev A 54(3):R1745
63. Arute F, Arya K, Babbush R, Bacon D, Bardin JC, Barends R, Biswas R, Boixo S, Brandao FGSL, Buell DA et al (2019) Quantum supremacy using a programmable superconducting processor. Nature 574(7779):505–510
64. Pascuzzi VR, He A, Bauer CW, De Jong WA, Nachman B (2022) Computationally efficient zero-noise extrapolation for quantum-gate-error mitigation. Phys Rev A 105(4):042406
65. Khodjasteh K, Viola L (2009) Dynamical quantum error correction of unitary operations with bounded controls. Phys Rev A-Atom Mol Opt Phys 80(3):032314
66. Chamberland C, Zhu G, Yoder TJ, Hertzberg JB, Cross AW (2020) Topological and subsystem codes on low-degree graphs with flag qubits. Phys Rev X 10(1):011022
67. Chao R, Reichardt BW (2020) Flag fault-tolerant error correction for any stabilizer code. PRX Quantum 1(1):010302

Chapter 18
Noise, Errors, and Adversarial Robustness in Quantum Machine Learning

Alex Jin and Manas Mukherjee

Abstract Quantum machine learning (QML) systems face unique challenges from both intrinsic quantum noise and adversarial threats. In the context of QML systems, bit-flip errors can occur due to imperfections in quantum gates, stray electromagnetic fields, or thermal fluctuations. These errors are particularly damaging in noisy intermediate-scale quantum (NISQ) devices, where error correction is either absent or limited. In this paper, we provide a structured overview of error types, theoretical frameworks for quantifying errors and robustness of quantum machine learning models, experimental results, and strategies to model and leverage these noises.

18.1 Introduction

Quantum devices are inherently noisy. Unlike classical digital systems, where bits can be preserved and manipulated with high fidelity, qubits are susceptible to a wide array of physical imperfections. These include decoherence, gate errors, measurement errors, cross-talk between qubits, and stochastic fluctuations in the environment. Such noise not only degrades the performance of quantum algorithms but also fundamentally alters their security landscape. In particular, when adversarial attacks are deployed against quantum models running on noisy hardware, the nature of these attacks can differ dramatically from the idealized, noiseless case.

The vulnerability of classical neural networks to adversarial attacks has been extensively studied over the past decade. Techniques such as Fast Gradient Sign Method (FGSM), Projected Gradient Descent (PGD), and Carlini-Wagner (CW) attacks have revealed that even imperceptible perturbations to inputs can cause dramatic misclassifications. Translating these ideas to quantum systems is nontrivial due to several key differences: quantum data may exist in Hilbert space representations,

A. Jin (✉) · M. Mukherjee
Centre for Quantum Technologies, National University of Singapore, Singapore, Singapore
e-mail: e1051010@u.nus.edu

M. Mukherjee
e-mail: cqtmukhe@nus.edu.sg

M. Usman (ed.), *Quantum Robustness in Artificial Intelligence*, Quantum Science and Technology, https://doi.org/10.1007/978-3-032-11153-1_18

the models themselves may be implemented via parameterized quantum circuits (PQCs), and the measurements yield probabilistic outputs. Moreover, the effects of noise are not merely additive or linear: they can introduce non-unitary dynamics that complicate the attacker's ability to predict or control model outputs.

Despite these complications, researchers have begun developing quantum analogs of adversarial attacks. For example, perturbations to quantum input states, small shifts in gate parameters, or alterations to circuit topology can significantly influence output distributions. Notably, studies have demonstrated that even small adversarial perturbations to quantum states can degrade classification performance in variational quantum classifiers (VQCs). However, these studies typically assume idealized quantum models—i.e., noiseless circuits executed on perfect hardware. This assumption limits their relevance in the noisy intermediate-scale quantum (NISQ) era, where gate fidelity and coherence times remain fundamental bottlenecks.

In reality, quantum circuits are executed under a significant noise budget. Every quantum gate applied introduces some error; every qubit held in superposition experiences some decoherence; and every measurement collapses with inherent stochasticity. Consequently, even in the absence of an adversary, predictions made by QML models may already be prone to error. Introducing adversarial perturbations into this fragile environment may either exacerbate vulnerabilities or, counterintuitively, obscure the effectiveness of the attack due to random noise masking its effects. A nuanced understanding of this interaction is crucial. On the one hand, noise may aid adversarial robustness by introducing randomness that blunts targeted attacks. On the other hand, noise can amplify the sensitivity of the model to small input perturbations, effectively lowering the barrier for successful adversarial manipulation. This ambiguous role of noise poses important questions: Does noise inherently make QML models more robust, or more vulnerable, to adversarial inputs? Are there specific noise that contribute more to this dynamic? How can quantum adversarial robustness be meaningfully evaluated in the presence of noise?

In this chapter, we aim to clarify these questions through an examination of adversarial robustness under realistic noise assumptions. We begin by introducing a taxonomy of noise channels relevant to quantum computation, ranging from gate-level stochastic errors to system-level decoherence. Building on this, we consider several case studies involving variational quantum classifiers trained on quantum-encoded classical datasets. We explore how adversarial attacks interact in the presence of realistic hardware noise, using simulations calibrated to NISQ processors. These examples reveal nontrivial patterns: in some instances, noise dampens the adversarial effect, while in others, it destabilizes the model enough that even benign inputs lead to misclassification.

A key insight emerging from this work is that quantum adversarial robustness is an inherently noise-sensitive phenomenon. Unlike classical systems, where deterministic computation provides a relatively stable backdrop for adversarial analysis, quantum systems exist at the intersection of coherence and randomness. This interplay mandates a fundamentally different approach to security and reliability. Rather than merely translating classical methods into the quantum setting, robust quantum AI must be built with noise in mind.

18.2 Types of Errors in QML Systems

18.2.1 Bit-Flip Errors

Bit-flip errors cause a qubit to change state from $|0\rangle$ to $|1\rangle$, or vice versa. These errors are modeled by the Pauli-X operator, X, which acts as a NOT gate in the computational basis:

$$X|0\rangle = |1\rangle, \quad X|1\rangle = |0\rangle.$$

In the context of a quantum machine learning model, bit-flip errors can occur due to imperfections in quantum gates, stray electromagnetic fields, or thermal fluctuations. Such errors are especially damaging in noisy intermediate-scale quantum (NISQ) devices, where error correction is either absent or limited.

Bit-flip noise can be modeled as a probabilistic channel:

$$\mathcal{E}(\rho) = (1 - p)\rho + pX\rho X,$$

where ρ is the qubit's density matrix and p is the probability of a bit-flip error occurring.

From a robustness perspective, bit-flip errors are adversarial in nature if they systematically push a QML model toward incorrect outputs. For example, a bit-flip in the measurement qubit of a quantum classifier could invert the predicted label. Assessing and mitigating this behavior is crucial in adversarial robustness studies.

18.2.2 Phase-Flip Errors

Phase-flip errors alter the phase of qubit superpositions without changing the probability of measurement outcomes in the computational basis. These errors are modeled by the Pauli-Z operator, Z:

$$Z|0\rangle = |0\rangle, \quad Z|1\rangle = -|1\rangle.$$

While the amplitudes remain unchanged, the relative phase shift can impact quantum interference and thus the output of a QML model. These errors typically arise due to fluctuations in control parameters or decoherence in isolated systems.

The phase-flip noise channel can be described as

$$\mathcal{E}(\rho) = (1 - p)\rho + pZ\rho Z.$$

In QML, phase-flip errors may distort the learned decision boundary or affect entangled states used in variational circuits, ultimately degrading model performance.

18.2.3 Depolarization Noise

Depolarization noise causes a qubit to be replaced by the maximally mixed state with some probability p, effectively erasing all information:

$$\rho \to (1-p)\rho + \frac{p}{3}(X\rho X + Y\rho Y + Z\rho Z).$$

This type of noise represents a uniform and isotropic error model and is often used as a worst-case benchmark in theoretical studies. It degrades both amplitude and phase information, making it particularly harmful in QML applications that rely on coherent information propagation.

Depolarization can arise due to random gate errors, measurement imprecision, or environmental coupling. From an adversarial robustness viewpoint, its stochastic nature challenges both detection and defense mechanisms.

18.2.4 Amplitude Damping

Amplitude damping represents energy dissipation, modeling the loss of energy from a quantum system to its environment. It captures phenomena such as spontaneous emission, where a qubit in the excited state $|1\rangle$ decays to the ground state $|0\rangle$.

The amplitude damping channel is expressed using Kraus operators:

$$E_0 = \begin{pmatrix} 1 & 0 \\ 0 & \sqrt{1-\gamma} \end{pmatrix}, \quad E_1 = \begin{pmatrix} 0 & \sqrt{\gamma} \\ 0 & 0 \end{pmatrix}, \quad \mathcal{E}(\rho) = E_0 \rho E_0^\dagger + E_1 \rho E_1^\dagger,$$

where γ is the probability of decay.

This error type causes asymmetry, disproportionately affecting the $|1\rangle$ component, making it especially problematic for algorithms relying on balanced superpositions. In QML, this can lead to bias in learned decision boundaries or impair encoding of input distributions.

18.2.5 Phase Damping

Phase damping describes the loss of quantum coherence without energy exchange. It models pure dephasing due to environmental entanglement with the system, which destroys phase relationships while leaving populations unchanged.

The Kraus operators for phase damping are:

$$E_0 = \sqrt{1-p}\, I, \quad E_1 = \sqrt{p}\begin{pmatrix}1 & 0\\ 0 & 0\end{pmatrix}, \quad E_2 = \sqrt{p}\begin{pmatrix}0 & 0\\ 0 & 1\end{pmatrix},$$

where p is the decoherence probability.

Though distinct from phase-flip errors (which apply discrete Z gates), phase damping produces a similar outcome: the decay of off-diagonal density matrix elements. This results in an effective reduction of superposition and quantum interference, crucial to QML models.

18.3 Quantifying Robustness and Resilience of QML Models

Quantum robustness can be formalized through two main lenses: noise resilience and adversarial robustness. The former concerns itself with a model's ability to maintain its performance when the input data are corrupted by *random unstructured perturbations*—commonly referred to as noise. The noise can stem from sensor errors, environmental factors, transmission corruption, or quantum decoherence. The latter describes a model's ability to resist *maliciously crafted perturbations* designed to mislead the model into making incorrect predictions. These intentional perturbations are usually small and imperceptible, but highly effective.

Let $\mathscr{H}$ be a finite-dimensional Hilbert Space and let $\mathscr{S}(\mathscr{H})$ be the set of density operators acting on $\mathscr{H}$. For the rest of the chapter, we assume a **Quantum Classifier** to be a map $f : \mathscr{S}(\mathscr{H}) \to \mathscr{K}$ that maps states $\rho \in \mathscr{S}(\mathscr{H})$ to a set of K labels, $\mathscr{K} = \{1, ..., K\}$. In the following, we will outline the protocols for quantifying adversarial robustness and noise resilience under this definition.

18.3.1 Quantum Hypothesis Testing

Quantum classifiers, and more generally machine learning classifiers, can be seen as performing *implicit hypothesis testing*—deciding between hypotheses like "the input image is a zero" versus "the input image is a one" in the case of the MNIST dataset. A recent study [1] has shown that adversarial robustness and noise resilience of QML models can be formulated under the asymmetric hypothesis testing setting.

Let ρ and ρ' be density operators (quantum states) associated to hypotheses H_0 and H_1, respectively. Following the quantum Neyman-Pearson lemma [6], the most powerful test between ρ and ρ' at significance α uses the projector $\Pi_\lambda = \{\rho' - \lambda\rho > 0\}$ where λ is fixed by

$$\mathrm{tr}(\Pi_\lambda \rho) = \alpha.$$

Consequently, we can set a maximum bound for type-I error (false positive rate) and optimize for a minimal type-II error (false negative rate). We will now outline how this setup can be used to define protocols for quantifying adversarial robustness and noise resilience.

18.3.2 Epsilon-Robustness

A QML model is said to be ε-robust (or have $1 - \varepsilon$ fidelity) for a state ρ, if small perturbations (e.g., a slightly perturbed quantum state ρ') lead to bounded changes in the output prediction. This mirrors the classical robustness notion against ε-bounded adversarial examples. The authors in [1] have derived the quantum ϵ-robustness bound using experimental measurements A and B—the two most measured output states during an experimental run. Mathematically, given a quantum classifier f and a state ρ, define:

$$r_F = \frac{1}{2}(1 + \sqrt{1 - p_B - p_A(1 - 2p_B) + 2\sqrt{p_A p_B (1 - p_A)(1 - p_B)}}), \quad (18.1)$$

where $1 \geq p_A \geq p_B \geq 0$ are the highest two values of the classifier's output probability vector. Then,

$$\forall \rho' \text{ such that } F(\rho', \rho) > r_F, \ f(\rho') = f(\rho),$$

where F is the Uhlmann Fidelity, which measures the closeness between two quantum states. The upshot of the above result is that for any quantum classifier, a single measurement of any state ρ allows us to bound the behavior of neighboring states as well.

Using the computed minimum robustness fidelity, r_F, we may compute the certified test set accuracy. Given a set of labeled test data $\mathscr{T} = \{(\rho_i, y_i)\}_{i=1}^{|\mathscr{T}|}$, the certified test set accuracy at fidelity $1 - \epsilon$ is defined as

$$\frac{1}{|\mathscr{T}|} \sum_{(\rho, y) \in \mathscr{T}} \mathbb{1}\{\mathbb{A}(\rho) = y \wedge r_F(\rho) \leq 1 - \epsilon\} \quad (18.2)$$

Equation (18.2) extends the notion to an entire set of states, allowing us to answer the question of what is the maximum perturbation ϵ that our quantum system is capable of tolerating before measured states are altered.

18.3.3 Resilience Against Depolarization Noise

Depolarization noise presents a uniform and isotropic threat to QML models, as it symmetrically affects all basis directions on the Bloch sphere. To quantify resilience under depolarization, we consider the classifier's output stability under the application of the depolarizing channel:

$$\mathcal{E}_{\text{DP}}(\rho) = (1-p)\rho + \frac{p}{3}(X\rho X + Y\rho Y + Z\rho Z)$$

where p represents the probability of depolarization. Now let f be a quantum classifier and let ρ be a benign state. Let $T(\rho, \mathcal{E}_{\text{DP}}(\rho))$ be the trace distance between the benign state and the depolarized state. We define

$$r_{\text{DP}}(p) = \frac{p}{2(1-p)}\left(\sqrt{\frac{p_A}{1-p_A}} - 1\right)$$

where p_A is the probability of most likely class the classifier predicts.

It has been shown [1] that robustness is guaranteed for any $T(\rho, \mathcal{E}_{\text{DP}}(\rho)) < r_{DP}(\rho)$. Intuitively, this means that for small depolarization noise or attack p, the classifier's output is guaranteed to remain stable. But as $p \to 1$, the state becomes maximally mixed and outputs tend to the model's prior bias. In experiments, $r_{\text{DP}}(p)$ is often approximated by sampling over a state many times.

Hence, resilience against depolarization noise captures a model's ability to make stable predictions under worst-case isotropic decoherence, which is critical in realistic quantum hardware settings.

18.4 Experimental Implementation

To illustrate the practical effects of noise and adversarial robustness, we evaluate QML classifiers on two public datasets: Iris Flower Dataset [2] and Pima Indians Diabetes Dataset [3].

The Iris Flower Dataset contains 150 samples of iris flowers from three species—setosa, versicolor, and virginica—with four numerical features: sepal length, sepal width, petal length, and petal width. The goal is typically to classify the species of an iris based on these measurements. We will be using the Iris Dataset to illustrate the protocols introduced in the previous section.

The Pima Indian Diabetes Dataset contains medical records for 768 female patients of Pima Indian heritage, all aged 21 or older, with the objective of predicting the onset of type 2 diabetes. The dataset includes eight numerical features such as glucose concentration, blood pressure, body mass index (BMI), insulin levels, and number of pregnancies, along with a binary target variable indicating the presence or absence of diabetes. Its real-world clinical relevance and moderate complexity make it a popular choice for testing classification algorithms and evaluating predictive health models. We note that this dataset is slightly more difficult than the Iris dataset, usually requiring a significant amount of pre-processing in order for models to achieve high classification accuracies. We will treat the dataset as is and avoid doing any pre-processing.

18.4.1 Data Re-Uploading Classifier

Dataset: throughout this subsection we evaluate on the *Iris Flowers* dataset [2]. This subsection empirically tests the robustness bound of Eq. (18.1) under bit-flip, phase-flip and depolarization noise on the Iris dataset. An efficient QML classifier can be constructed using a single qubit under the data re-uploading architecture [4, 5], where data are encoded through angle embedding and measurement results determine binary classification.

Single Qubit Data Re-uploading Classifier Example:
We will train a binary model using the Adam Optimizer with cross-entropy loss. We exclude the iris virginica and split the remaining data into 5:2:3 ratio for training, validation, and testing. Since this is a relatively simple dataset, training the model for 20 epochs allows the best model to achieve 100% accuracyon the training, validation and test sets (Fig. 18.1).

```
Epoch:  0 | Loss: 0.901 | Train acc: 0.580 | Valid acc: 0.400
Epoch:  1 | Loss: 0.331 | Train acc: 0.940 | Valid acc: 0.850
...
Epoch: 20 | Loss: 0.055 | Train acc: 1.000 | Valid acc: 1.000

# Evaluation with best model
Train acc 1.00 | Valid acc 1.00 | Test acc : 1.00
```

Listing 1: Single-Qubit Data Re-uploading classification with PennyLane

```python
import pennylane as qml
from pennylane import numpy as np

dev = qml.device("default.mixed", wires=1)

@qml.qnode(dev)
def qcircuit(x, y, weight, bias):
    for i in range(0, len(weight), 2):
        omega = 0
        theta = np.dot(x, weight[i]) + bias[i]
        phi = np.dot(x, weight[i+1]) + bias[i+1]
        qml.Rot(omega, theta, phi, wires=0)
    return qml.expval(qml.Hermitian(y, wires=[0]))

```

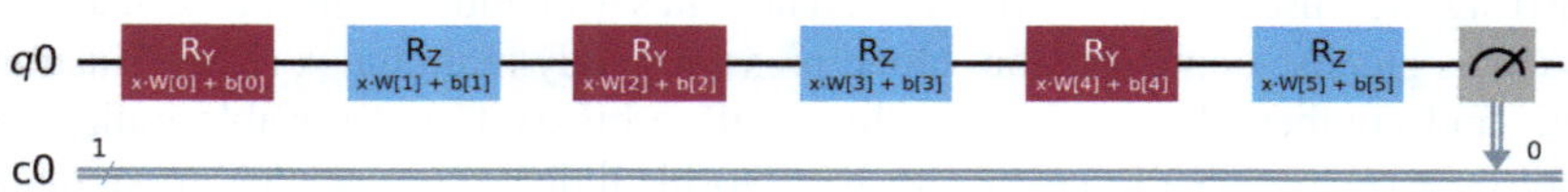

Fig. 18.1 Single Qubit Data Re-uploading circuit: example code with visualization

We will now proceed to quantify the model's adversarial robustness and noise resilience. To proceed we will fix an α value (maximum type-I error) of 0.1, an ε-value of 0.1, and a circuit sampling size of 100 for each state in the test data. As we can see from Fig. 18.2, the protocol was able to certify all points in the test set. Following Eq. 18.2, we may compute the certified test set accuracy as 1.0 at robustness fidelity 0.9.

Let us take a look at how the model performs against depolarization noise under the protocol defined in the previous section. We will assume a depolarization probability of 0.2. Following the definition of R_{DP}, we can see the results in Fig. 18.3. Although

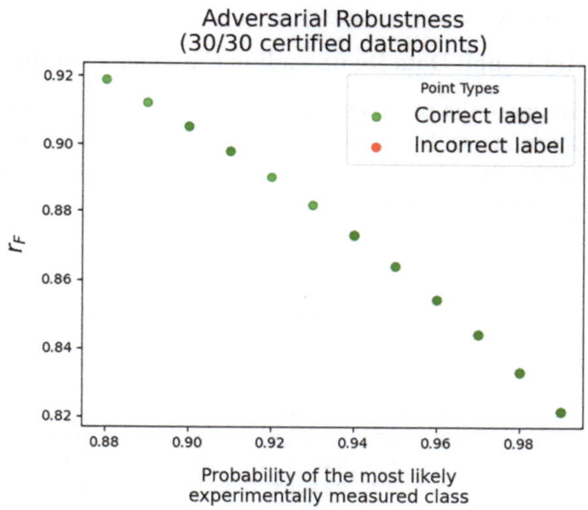

Fig. 18.2 Adversarial robustness of the test set certified at fidelity 0.9. All points are colored green as we achieved 100% test accuracy

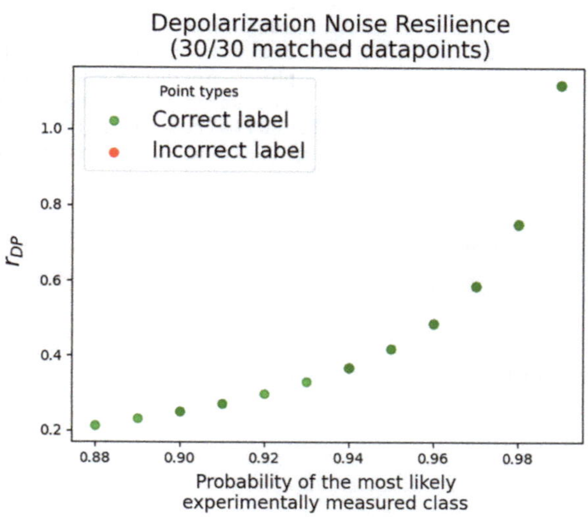

Fig. 18.3 Noise resilience analysis of all 30 test set data points. All points are colored green as we achieved 100% test accuracy

all 30 points from the test set are classified correctly, they show an increasing amount of noise tolerance as the predicted value's fidelity increases.

18.4.2 *Multi-qubit Entangled Classifier*

Dataset: throughout this subsection we evaluate on the *Pima Indians Diabetes* dataset [3].
We now evaluate the effect of entanglement in a 2-qubit variational circuit. An example architecture might include R_y encodings followed by CNOT gates and parameterized R_z rotations.

Multi-qubit Data Re-uploading Classifier Example:

Similar to the previous example, we will train a multi-qubit model using the Adam Optimizer with cross-entropy loss. We split the data into 5:2:3 ratio for training, validation and testing. We will train the model for 20 epochs and use the best model to evaluate on the test set (Fig. 18.4).

```
import pennylane as qml
from pennylane import numpy as np

dev = qml.device("default.mixed", wires=2)

@qml.qnode(dev)
def qcircuit(x, y, weights, biases, n_blocks=2):
    for block in range(n_blocks):
        weight = weights[block]
        bias = biases[block]
        for i in range(0, len(weight), 2):
            omega = 0
            theta = np.dot(x, weight[i]) + bias[i]
            phi = np.dot(x, weight[i+1]) + bias[i+1]
            qml.Rot(omega, theta, phi, wires=0)
        qml.CNOT(wires=[0,1])
    return qml.expval(qml.Hermitian(y, wires=[0]))

```

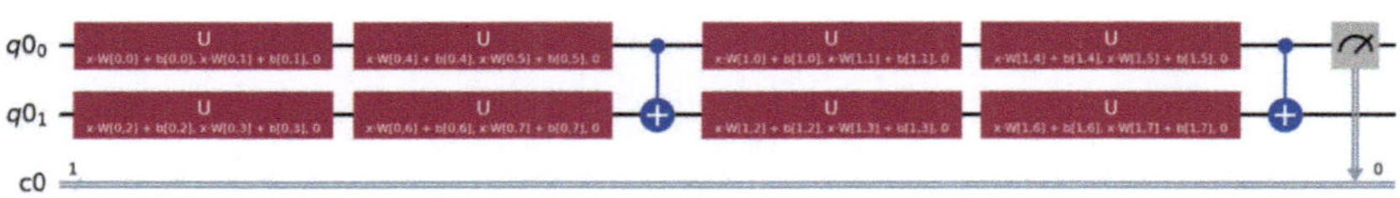

Fig. 18.4 Multi-qubit Data Re-uploading classification: example code with visualization

```
Epoch:  0 | Loss: 0.771 | Train acc: 0.359 | Valid acc: 0.307
Epoch:  1 | Loss: 0.577 | Train acc: 0.674 | Valid acc: 0.705
...
Epoch: 20 | Loss: 0.442 | Train acc: 0.768 | Valid acc: 0.797

# Evaluation with best model
Train acc 0.768 | Valid acc 0.797 | Test acc : 0.788
```

Listing 2: Multi-Qubit Data Re-uploading classification with PennyLane

We can see that the model struggles to achieve higher than 80% training or validation accuracy and the final test accuracy evaluated using the best model achieves 78.8% accuracy. This is expected as it is known to be a more challenging dataset. We will now proceed to examine the trained model's adversarial robustness and noise resilience.

In Fig. 18.5, we can see that we have fewer certified datapoints compared to the example in the previous subsection. Moreover, this observation agrees with the model's inability to learn a good separation curve. Moreover, in Fig. 18.6, we can see that there are a lot more points with r_{DP} values below 0.2, which is the minimum value in Fig. 18.3. This reflects that these are points near the model's decision boundary and can be easily flipped by any noise or attack.

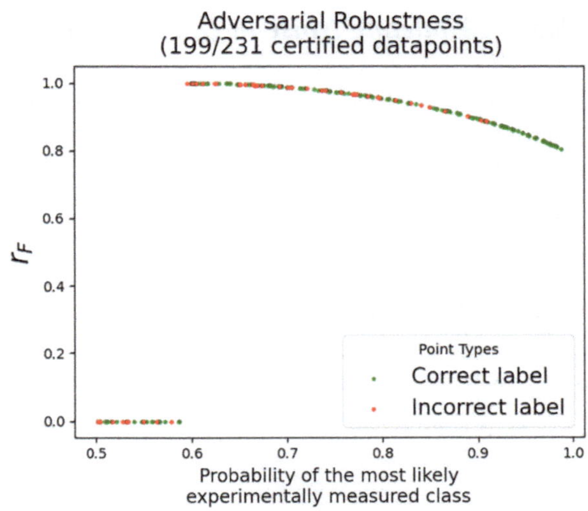

Fig. 18.5 Adversarial robustness of the test set certified at probability 0.9. r_F values of 0 represents uncertified points under this protocol

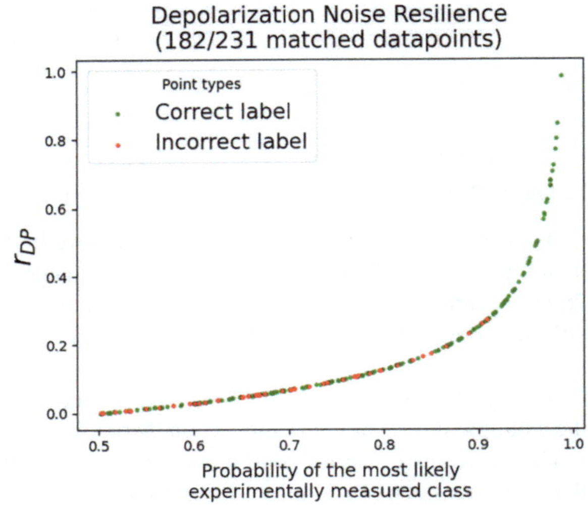

Fig. 18.6 Noise resilience analysis of all test set data points

18.5 Modeling and Leveraging Noise in QML Systems

18.5.1 Modeling Noise in QML Systems

Integrating noise into quantum simulation is essential for accurately modeling the behavior of real-world quantum systems, which are inherently prone to decoherence, gate imperfections, and measurement errors. Noise modeling can be integrated into circuit simulators or included during training. One approach is noise injection: simulate noisy gate operations. We provide an example noisy circuit below that simulates bit-flip noises. Other types of noises can be implemented in a similar fashion.

Example: Adding bit-flip noise models into a training loop:

```
import pennylane as qml
from pennylane import numpy as np

dev = qml.device("default.mixed", wires=1)

@qml.qnode(dev)
def qcircuit(x, y, weight, bias):
    for i in range(0, len(weight), 2):
        omega = 0
        theta = np.dot(x, weight[i]) + bias[i]
        phi = np.dot(x, weight[i+1]) + bias[i+1]
        qml.Rot(omega, theta, phi, wires=0)
        qml.BitFlip(p, wires=0)
    return qml.expval(qml.Hermitian(y, wires=[0]))
```

Listing 3: Single Qubit Data Re-uploading classification with PennyLane

18.5.2 *Leveraging Noise in QML Systems*

We are now ready to combine the results from the previous sections. We will use Iris Flower dataset to illustrate the effect of training using noisy circuits. And we will proceed to compare the performance of models, trained with and without noise, under adversarial attacks.

In Fig. 18.7, we plot the loss on the training set over epochs for models with and without noise injection. We can clearly see the effect of different noises on the convergence of our cross-entropy loss function, as well as how different noise types affect our training curve. However, looking at the training loss curve alone does not

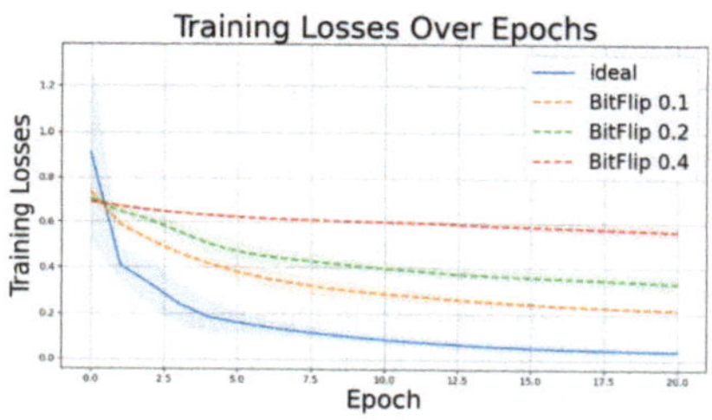

(a) Bit Flip

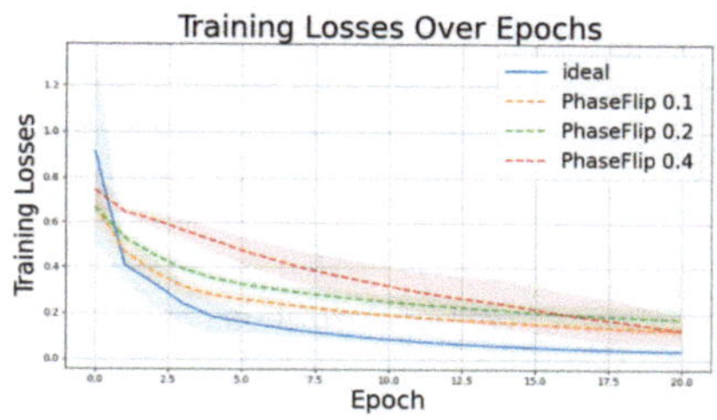

(b) Phase Flip

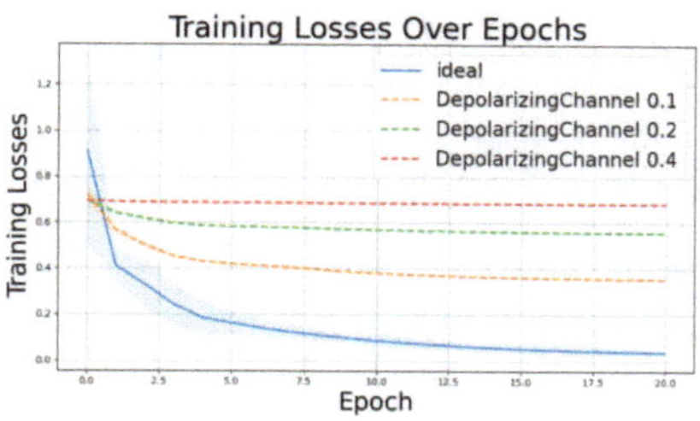

(c) Depolarization

(d) Amplitude Damping

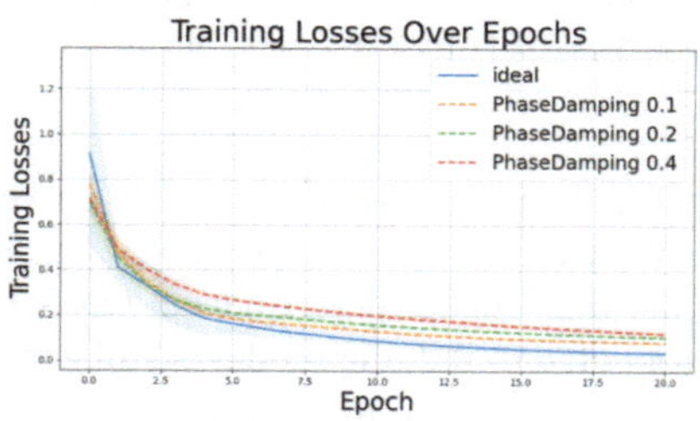

(e) Phase Damping

Fig. 18.7 Training losses (cross-entropy) of models being trained under different noise injections. Solid blue lines represent noiseless ideal models. Dashed lines represent models injected with different noise probabilities from 10 to 40 percent. Shaded regions around solid and dashed lines represent 1 standard deviation of error

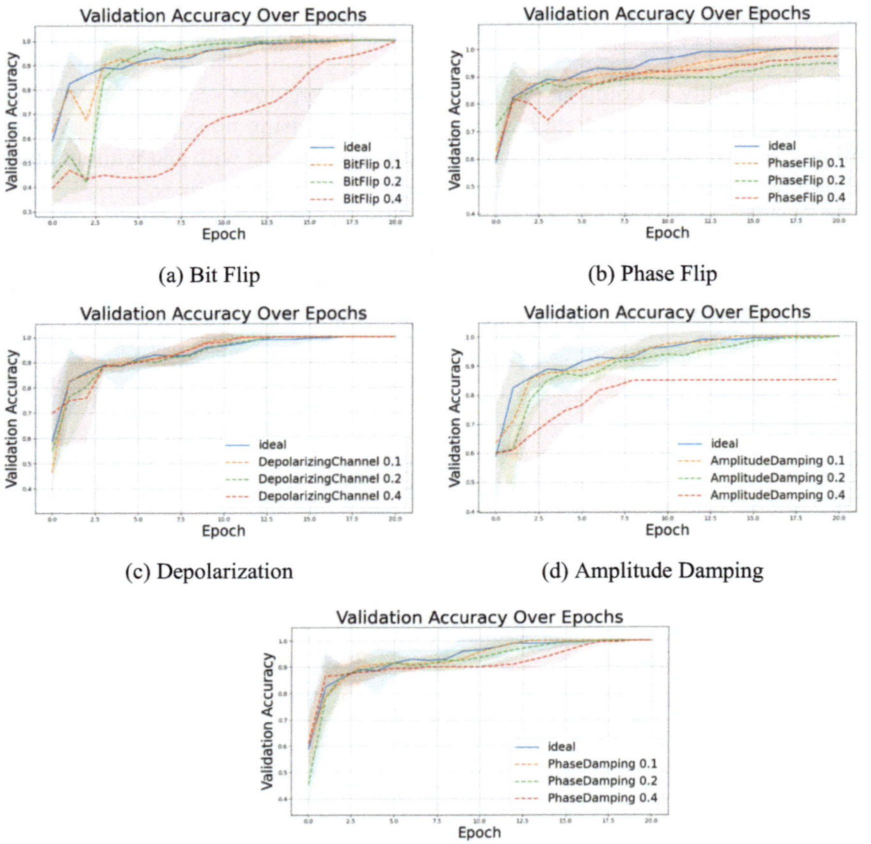

(a) Bit Flip

(b) Phase Flip

(c) Depolarization

(d) Amplitude Damping

(e) Phase Damping

Fig. 18.8 Validation accuracy of models being trained under different noise injections. Solid blue lines represent noiseless ideal models. Dashed lines represent models injected with different noise probabilities from 10 to 40 percent. Shaded regions around solid and dashed lines represent 1 standard deviation of error, measured over thirty runs with different random seeds

tell the full story. In Fig. 18.8, we plot the training curve via validation accuracy as a different proxy. We can see that certain noises, such as depolarization noise, have little effect on the final model performance. On the other hand, we also see that under bit-flip errors of $p \leq 0.2$, the model is still capable of achieving close to 100% validation accuracy, and seem to converge better than the noiseless version. More interestingly, we note that the effect of different noise types on a model is both architecture and dataset dependent.

18.5.3 Adversarial Training

Adversarial attacks in classical machine learning are broadly categorized based on the attacker's knowledge, goals, and manipulation techniques. Many of these attack types transfer conceptually in the context of QML. Unlike classical models, however, QML can be enhanced by training against perturbed inputs, some of which are naturally integrated into the quantum devices themselves. A few examples include circuit parameter perturbations, noisy input states, and gate-level noises.

Example Black-Box Attack:

```
def zoo_attack(f, x, y, epsilon=0.075, steps=100, h=1e-2, alpha
    =0.1):
    x_adv = x.copy()
    d = len(x)

    for step in range(steps):
        grad = np.zeros_like(x)

        for i in range(d):
            e_i = np.zeros_like(x)
            e_i[i] = 1.0

            f_plus = f(x_adv + h * e_i)
            f_minus = f(x_adv - h * e_i)

            grad[i] = (f_plus[y] - f_plus[1 - y] -
                       (f_minus[y] - f_minus[1 - y])) / (2 * h)

        x_adv -= alpha * grad
        x_adv = np.clip(x_adv, x - epsilon, x + epsilon)

        y_pred = np.argmax(f(x_adv))
        if y_pred != y:
            break

    return x_adv
```

Listing 4: zeroth order optimization based attack

Let us now examine the *zoo_attack* subroutine on trained models from Fig. 18.8a.

In Table 18.1, we go through such an example in detail against models trained on the Iris dataset. We begin by training a model, M_0 on the noiseless dataset. Next we train models, M_1 to M_F, under different flavors of noise, including bit-flip errors, phase-flip errors, depolarization noises, amplitude damping, and phase damping. From the table, we can see that robust training is especially effective when incorporating device noise. QML models trained with these techniques demonstrate improved performance in some noisy and most adversarial regimes.

Table 18.1 Model accuracy under zoo_attack with epsilon = 0.075, steps = 100, h = 0.01, and alpha = 0.1. These hyper-parameters were tuned for the noiseless model for reliable convergence. For the middle column, a single adversarial attack is first trained on M_0, and applied toward all 16 trained models. For the last column, 15 distinct attacks are trained on M_1 to M_F and applied on each model, respectively

Model	Accuracy	Adversarial example from M_0	Targeted adversarial example
M_0	Noiseless	0.55	–
M_1	BitFlip p = 0.1	0.70	0.65
M_2	BitFlip p = 0.2	0.65	0.30
M_3	BitFlip p = 0.4	0.75	0.60
M_4	PhaseFlip p = 0.1	0.75	0.60
M_5	PhaseFlip p = 0.2	0.75	0.55
M_6	PhaseFlip p = 0.4	0.70	0.80
M_7	Depolarization p = 0.1	0.70	0.60
M_8	Depolarization p = 0.2	0.70	0.60
M_9	Depolarization p = 0.4	0.70	0.60
M_A	AmplitudeDamping p = 0.1	0.70	0.60
M_B	AmplitudeDamping p = 0.2	0.75	0.60
M_C	AmplitudeDamping p = 0.4	0.60	0.55
M_D	PhaseDamping p = 0.1	0.70	0.70
M_E	PhaseDamping p = 0.2	0.70	0.55
M_F	PhaseDamping p = 0.4	0.70	0.35

18.6 Conclusion

Noise and errors in quantum systems pose serious threats to the robustness of QML models, yet also offer a rich domain for co-optimizing resilience. In this chapter, we have investigated the intricate interplay between quantum noise, circuit errors, and adversarial robustness in quantum machine learning (QML). Our analysis demonstrates that quantum noise, while traditionally viewed as a source of computational error, can act analogously to a natural defense mechanism against adversarial perturbations in certain settings. This phenomenon, akin to classical randomized smoothing, highlights the nontrivial role of decoherence and gate imperfections in shaping robustness behavior.

Our study reveals that the impact of quantum noise is not uniformly beneficial or detrimental; rather, its influence on adversarial robustness is highly dependent on the circuit architecture, the entanglement structure, and the adversary's capabilities. These insights have important implications for the design of secure quantum AI systems. As fault-tolerant quantum computing remains an ongoing challenge, understanding and exploiting the nuances of noise-induced effects can inform more resilient model construction in the NISQ era. Moreover, our findings suggest a future avenue in adversarial training and certification under realistic, noisy conditions, where robustness becomes not just a byproduct of regularization, but an engineered feature of the quantum learning process itself.

References

1. Weber M, Liu N, Li B, Zhang C, Zhao Z (2021) Optimal provable robustness of quantum classification via quantum hypothesis testing. npj Quantum Inf. 7(1):76. https://doi.org/10.1038/s41534-021-00431-0
2. Fisher RA (1936) Iris. UCI machine learning repository. https://doi.org/10.24432/C56C76
3. Smith JW, Everhart JE, Dickson WC, Knowler WC, Johannes RS (1988) Pima Indians diabetes dataset. UCI machine learning repository. https://archive.ics.uci.edu/ml/datasets/pima+indians+diabetes
4. Pérez-Salinas A, Cervera-Lierta A, Gil-Fuster E, Latorre JI (2020) Data re-uploading for a universal quantum classifier. Quantum 4:226. https://quantum-journal.org/papers/q-2020-02-06-226/
5. Dutta T, Pérez-Salinas A, Cheng JPS, Latorre JI, Mukherjee M (2022) Single-qubit universal classifier implemented on an ion-trap quantum device. Phys Rev A 106(1):012411. https://doi.org/10.1103/PhysRevA.106.012411
6. Audenaert K, Nussbaum M, Szkoła A, Verstraete F (2008) Asymptotic error rates in quantum hypothesis testing. Commun Math Phys 279:251–283

References

FSC
www.fsc.org
MIX
Papier aus verantwortungsvollen Quellen
Paper from responsible sources
FSC® C105338

If you have any concerns about our products,
you can contact us on
ProductSafety@springernature.com

In case Publisher is established outside the EU,
the EU authorized representative is:
Springer Nature Customer Service Center GmbH
Europaplatz 3, 69115 Heidelberg, Germany

Printed by Libri Plureos GmbH
in Hamburg, Germany